Edited by Philip Garrou, Mitsumasa Koyanagi, and Peter Ramm

Handbook of 3D Integration

3D Process Technology

Verlag GmbH & Co. KGaA

Editors

Dr. Philip Garrou
Microelectronic Consultants of North Carolina
3021 Cornwallis Road
Research Triangle Park, NC 27709
USA

Mitsumasa Koyanagi
Tohoku University
New Industry Creation Hatchery Center
6-6-10 Aza-Aoba, Aramaki
Sendai 980-8579
Japan

Dr. Peter Ramm
Fraunhofer EMFT
Device and 3D Integration
Hansastraße 27d
80686 München
Germany

All books published by **Wiley-VCH** are carefully produced. Nevertheless, authors, editors, and publisher do not warrant the information contained in these books, including this book, to be free of errors. Readers are advised to keep in mind that statements, data, illustrations, procedural details or other items may inadvertently be inaccurate.

Library of Congress Card No.: applied for

British Library Cataloguing-in-Publication Data
A catalogue record for this book is available from the British Library.

Bibliographic information published by the Deutsche Nationalbibliothek
The Deutsche Nationalbibliothek lists this publication in the Deutsche Nationalbibliografie; detailed bibliographic data are available on the Internet at <http://dnb.d-nb.de>.

© 2014 Wiley-VCH Verlag GmbH & Co. KGaA, Boschstr. 12, 69469 Weinheim, Germany

All rights reserved (including those of translation into other languages). No part of this book may be reproduced in any form – by photoprinting, microfilm, or any other means – nor transmitted or translated into a machine language without written permission from the publishers. Registered names, trademarks, etc. used in this book, even when not specifically marked as such, are not to be considered unprotected by law.

Print ISBN: 978-3-527-33466-7
ePDF ISBN: 978-3-527-67013-0
ePub ISBN: 978-3-527-67012-3
mobi ISBN: 978-3-527-67011-6
oBook ISBN: 978-3-527-67010-9

Cover Design Adam Design, Weinheim, Germany

Typesetting Thomson Digital, Noida, India

Printing and Binding Markono Print Media Pte Ltd, Singapore

Printed on acid-free paper

Edited by
*Philip Garrou, Mitsumasa Koyanagi,
and Peter Ramm*

Handbook of 3D Integration

Related Titles

Brand, O., Dufour, I., Heinrich, S.M., Josse, F. (eds.)

Resonant MEMS

Principles, Modeling, Implementation and Applications

2014

Print ISBN: 978-3-527-33545-9

Iannacci, J.

Practical Guide to RF-MEMS

2013

Print ISBN: 978-3-527-33564-0

Ramm, P., Lu, J.J., Taklo, M.M. (eds.)

Handbook of Wafer Bonding

2012

Print ISBN: 978-3-527-32646-4

Saile, V., Wallrabe, U., Tabata, O., Korvink, J.G. (eds.)

LIGA and its Applications

2009

Print ISBN: 978-3-527-31698-4

Garrou, P., Bower, C., Ramm, P. (eds.)

Handbook of 3D Integration

Technology and Applications of 3D Integrated Circuits

2008

Print ISBN: 978-3-527-33265-6

Contents

List of Contributors *XVII*

1 3D IC Integration Since 2008 *1*
Philip Garrou, Peter Ramm, and Mitsumasa Koyanagi
1.1 3D IC Nomenclature *1*
1.2 Process Standardization *2*
1.3 The Introduction of Interposers (2.5D) *4*
1.4 The Foundries *6*
1.4.1 TSMC *6*
1.4.2 UMC *7*
1.4.3 GlobalFoundries *7*
1.5 Memory *7*
1.5.1 Samsung *7*
1.5.2 Micron *8*
1.5.3 Hynix *9*
1.6 The Assembly and Test Houses *9*
1.7 3D IC Application Roadmaps *10*
References *11*

2 Key Applications and Market Trends for 3D Integration and Interposer Technologies *13*
Rozalia Beica, Jean-Christophe Eloy, and Peter Ramm
2.1 Introduction *13*
2.2 Advanced Packaging Importance in the Semiconductor Industry is Growing *16*
2.3 3D Integration-Focused Activities – The Global IP Landscape *18*
2.4 Applications, Technology, and Market Trends *22*
References *32*

3 Economic Drivers and Impediments for 2.5D/3D Integration *33*
Philip Garrou
3.1 3D Performance Advantages *33*
3.2 The Economics of Scaling *33*

3.3	The Cost of Future Scaling 34
3.4	Cost Remains the Impediment to 2.5D and 3D Product Introduction 37
3.4.1	Required Economics for Interposer Use in Mobile Products 38
3.4.2	Silicon Interposer Pricing 38
	References 40

4	**Interposer Technology** 41
	Venky Sundaram and Rao R. Tummala
4.1	Definition of 2.5D Interposers 41
4.2	Interposer Drivers and Need 42
4.3	Comparison of Interposer Materials 44
4.4	Silicon Interposers with TSV 45
4.5	Lower Cost Interposers 48
4.5.1	Glass Interposers 48
4.5.1.1	Challenges in Glass Interposers 49
4.5.1.2	Small-Pitch Through-Package Via Hole Formation and Ultrathin Glass Handling 49
4.5.1.3	Metallization of Glass TPV 51
4.5.1.4	Reliability of Copper TPVs in Glass Interposers 52
4.5.1.5	Thermal Dissipation of Glass 53
4.5.1.6	Glass Interposer Fabrication with TPV and RDL 53
4.5.2	Low-CTE Organic Interposers 53
4.5.3	Polycrystalline Silicon Interposer 55
4.5.3.1	Polycrystalline Silicon Interposer Fabrication Process 56
4.6	Interposer Technical and Manufacturing Challenges 57
4.7	Interposer Application Examples 58
4.8	Conclusions 60
	References 61

5	**TSV Formation Overview** 65
	Dean Malta
5.1	Introduction 65
5.2	TSV Process Approaches 67
5.2.1	TSV-Middle Approach 68
5.2.2	Backside TSV-Last Approach 68
5.2.3	Front-Side TSV-Last Approach 69
5.3	TSV Fabrication Steps 70
5.3.1	TSV Etching 70
5.3.2	TSV Insulation 71
5.3.3	TSV Metallization 71
5.3.4	Overburden Removal by CMP 72
5.3.5	TSV Anneal 73
5.3.6	Temporary Carrier Wafer Bonding and Debonding 74
5.3.7	Wafer Thinning and TSV Reveal 74

5.4	Yield and Reliability	75
	References	76
6	**TSV Unit Processes and Integration**	**79**
	Sesh Ramaswami	
6.1	Introduction	79
6.2	TSV Process Overview	80
6.3	TSV Unit Processes	82
6.3.1	Etching	82
6.3.2	Insulator Deposition with CVD	83
6.3.3	Metal Liner/Barrier Deposition with PVD	84
6.3.4	Via Filling by ECD of Copper	84
6.3.5	CMP of Copper	85
6.3.6	Temporary Bonding between Carrier and Device Wafer	86
6.3.7	Wafer Backside Thinning	86
6.3.8	Backside RDL	87
6.3.9	Metrology, Inspection, and Defect Review	87
6.4	Integration and Co-optimization of Unit Processes in Via Formation Sequence	88
6.5	Co-optimization of Unit Processes in Backside Processing and Via-Reveal Flow	89
6.6	Integration and Co-optimization of Unit Processes in Via-Last Flow	91
6.7	Integration with Packaging	92
6.8	Electrical Characterization of TSVs	92
6.9	Conclusions	96
	References	97
7	**TSV Formation at ASET**	**99**
	Hiroaki Ikeda	
7.1	Introduction	99
7.2	Via-Last TSV for Both D2D and W2W Processes in ASET	103
7.3	TSV Process for D2D	105
7.3.1	Front-Side Bump Forming	106
7.3.2	Attach WSS and Thinning	106
7.3.3	Deep Si Etching from the Backside	107
7.3.4	Liner Deposition	107
7.3.5	Removal of SiO_2 at the Bottom of Via	107
7.3.6	Barrier Metal and Seed Layer Deposition by PVD	110
7.3.7	Cu Electroplating	110
7.3.8	CMP	110
7.3.9	Backside Bump	111
7.3.10	Detach WSS	111
7.3.11	Dicing	112
7.4	TSV Process for W2W	113
7.4.1	Polymer Layer Coat and Development	114

7.4.2	Barrier Metal and Seed Layer Deposition *114*
7.4.3	Cu Plating *114*
7.4.4	CMP *115*
7.4.5	First W2W Stacking (Face to Face) *116*
7.4.6	Wafer Thinning and Deep Si Etching *116*
7.4.7	TSV Liner Deposition and SiO_2 Etching of Via Bottom *117*
7.4.8	Barrier Metal and Seed Layer Deposition and Cu Plating *117*
7.4.9	CMP *117*
7.4.10	Next W2W Stacking *118*
7.5	Conclusions *119*
	References *119*

8 Laser-Assisted Wafer Processing: New Perspectives in Through-Substrate Via Drilling and Redistribution Layer Deposition *121*
Marc B. Hoppenbrouwers, Gerrit Oosterhuis, Guido Knippels, and Fred Roozeboom

8.1	Introduction *121*
8.2	Laser Drilling of TSVs *121*
8.2.1	Cost of Ownership Comparison *121*
8.2.2	Requirements for an Industrial TSV Laser Driller *123*
8.2.3	Drilling Strategy *124*
8.2.3.1	Mechanical *124*
8.2.3.2	Optical *125*
8.2.4	Experimental Drilling Results *126*
8.3	Direct-Write Deposition of Redistribution Layers *126*
8.3.1	Introduction on Redistribution Layers *126*
8.3.2	Direct-Write Characteristics *127*
8.3.3	Direct-Write Laser-Induced Forward Transfer *128*
8.3.4	LIFT Results *130*
8.4	Conclusions and Outlook *131*
	References *132*

9 Temporary Bonding Material Requirements *135*
Rama Puligadda

9.1	Introduction *135*
9.2	Technology Options *136*
9.2.1	Tapes and Waxes *136*
9.2.2	Chemical Debonding *136*
9.2.3	Thermoplastic Bonding Material and Slide Debonding *136*
9.2.4	Debonding Using Release Layers *137*
9.3	Requirements of a Temporary Bonding Material *138*
9.4	Considerations for Successful Processing *139*
9.4.1	Application of the Temporary Bonding Adhesive to the Device Wafer and Bonding to Carrier *139*
9.4.2	Moisture and Contaminants on Surface *139*

9.4.3	Total Thickness Variation	140
9.4.4	Squeeze Out	140
9.5	Surviving the Backside Process	141
9.5.1	Edge Trimming	142
9.5.2	Edge Cleaning	142
9.5.3	Temperature Excursions in Plasma Processes	143
9.5.4	Wafer Warpage due to CTE Mismatch	143
9.6	Debonding	144
9.6.1	Debonding Parameters in Slide-Off Debonding	144
9.6.2	Mechanical Damage to Interconnects	144
	References	145

10 Temporary Bonding and Debonding – An Update on Materials and Methods 147
Wilfried Bair

10.1	Introduction	147
10.2	Carrier Selection for Temporary Bonding	148
10.3	Selection of Temporary Bonding Adhesives	151
10.4	Bonding and Debonding Processes	152
10.5	Equipment and Process Integration	155
	References	156

11 ZoneBOND®: Recent Developments in Temporary Bonding and Room-Temperature Debonding 159
Thorsten Matthias, Jürgen Burggraf, Daniel Burgstaller, Markus Wimplinger, and Paul Lindner

11.1	Introduction	159
11.2	Thin Wafer Processing	159
11.2.1	Thin Wafer Total Thickness Variation	161
11.2.2	Wafer Alignment	163
11.3	ZoneBOND Room-Temperature Debonding	163
11.4	Conclusions	165
	References	166

12 Temporary Bonding and Debonding at TOK 167
Shoji Otaka

12.1	Introduction	167
12.2	Zero Newton Technology	168
12.2.1	The Wafer Bonder	168
12.2.2	The Wafer Debonder	170
12.2.3	The Wafer Bonder and Debonder Equipment Lineups	170
12.2.4	Adhesives	170
12.2.5	Integration Process Performance	172
12.3	Conclusions	174
	References	174

13	**The 3M™ Wafer Support System (WSS)** *175*	
	Blake Dronen and Richard Webb	
13.1	Introduction *175*	
13.2	System Description *175*	
13.3	General Advantages *177*	
13.4	High-Temperature Material Solutions *178*	
13.5	Process Considerations *180*	
13.5.1	Wafer and Adhesive Delamination *180*	
13.5.2	LTHC Glass Delamination *181*	
13.6	Future Directions *181*	
13.6.1	Thermal Stability *181*	
13.6.2	Elimination of Adhesion Control Agents *182*	
13.6.3	Laser-Free Release Layer *183*	
13.7	Summary *183*	
	Reference *184*	
14	**Comparison of Temporary Bonding and Debonding Process Flows** *185*	
	Matthew Lueck	
14.1	Introduction *185*	
14.2	Studies of Wafer Bonding and Thinning *186*	
14.3	Backside Processing *186*	
14.4	Debonding and Cleaning *188*	
	References *189*	
15	**Thinning, Via Reveal, and Backside Processing – Overview** *191*	
	Eric Beyne, Anne Jourdain, and Alain Phommahaxay	
15.1	Introduction *191*	
15.2	Wafer Edge Trimming *192*	
15.3	Thin Wafer Support Systems *194*	
15.3.1	Glass Carrier Support System with Laser Debonding Approach *196*	
15.3.2	Thermoplastic Glue Thin Wafer Support System – Thermal Slide Debondable System *196*	
15.3.3	Room-Temperature, Peel-Debondable Thin Wafer Support Systems *197*	
15.4	Wafer Thinning *198*	
15.5	Thin Wafer Backside Processing *202*	
15.5.1	Via-Middle Thin Wafer Backside Processing: "Via-Reveal" Process *202*	
15.5.1.1	Mechanical Via Reveal *202*	
15.5.1.2	"Soft" Via Reveal *202*	
15.5.2	Via-Last Thin Wafer Backside Processing *203*	
	References *205*	

16	**Backside Thinning and Stress-Relief Techniques for Thin Silicon Wafers** *207*	
	Christof Landesberger, Christoph Paschke, Hans-Peter Spöhrle, and Karlheinz Bock	
16.1	Introduction *207*	
16.2	Thin Semiconductor Devices *207*	
16.3	Wafer Thinning Techniques *208*	
16.3.1	Wafer Grinding *209*	
16.3.2	Wet-Chemical Spin Etching *210*	
16.3.3	CMP Polishing *211*	
16.3.4	Plasma Dry Etching *212*	
16.3.5	Dry Polish *213*	
16.3.6	Chemical–Mechanical Grinding (CMG) *214*	
16.4	Fracture Tests for Thin Silicon Wafers *214*	
16.5	Comparison of Stress-Relief Techniques for Wafer Backside Thinning *216*	
16.6	Process Flow for Wafer Thinning and Dicing *220*	
16.7	Summary and Outlook on 3D Integration *222*	
	References *223*	
17	**Via Reveal and Backside Processing** *227*	
	Mitsumasa Koyanagi and Tetsu Tanaka	
17.1	Introduction *227*	
17.2	Via Reveal and Backside Processing in Via-Middle Process *227*	
17.3	Backside Processing in Back-Via Process *232*	
17.4	Backside Processing and Impurity Gettering *234*	
17.5	Backside Processing for RDL Formation *237*	
	References *239*	
18	**Dicing, Grinding, and Polishing (Kiru Kezuru and Migaku)** *241*	
	Akihito Kawai	
18.1	Introduction *241*	
18.2	Grinding and Polishing *241*	
18.2.1	Grinding General *241*	
18.2.1.1	Grinding Method *241*	
18.2.1.2	Rough Grinding and Fine Grinding *242*	
18.2.1.3	The Grinder Polisher *243*	
18.2.2	Thinning *243*	
18.2.2.1	Stress Relief *245*	
18.2.2.2	Die Attach Film *246*	
18.2.2.3	All-in-One System *246*	
18.2.2.4	Dicing Before Grinding *246*	
18.2.3	Grinding Topics for 3DIC Such as TSV Devices *246*	
18.2.3.1	Wafer Support System *246*	

18.2.3.2	Edge Trimming 247
18.2.3.3	Grinding to Improve Flatness 248
18.2.3.4	Higher Level of Cleanliness 248
18.2.3.5	Via Reveal 249
18.2.3.6	Planarization 249
18.3	Dicing 250
18.3.1	Blade Dicing General 250
18.3.1.1	Dicing Method 250
18.3.1.2	Blade Dicing Point 250
18.3.1.3	Blade 251
18.3.1.4	Optimization of Process Control 252
18.3.1.5	Dicer 252
18.3.1.6	Dual Dicing Applications 252
18.3.2	Thin Wafer Dicing 253
18.3.3	Low-k Dicing 254
18.3.4	Other Laser Dicing 254
18.3.4.1	Ablation 254
18.3.4.2	Laser Full Cut Application 255
18.3.4.3	Stealth Dicing (SD) 256
18.3.5	Dicing Topics for 3D-IC Such as TSV 257
18.3.5.1	Cutting of Chip on Chip (CoC) and Chip on Wafer (CoW) 258
18.3.5.2	Singulation of CoW and Wafer on Wafer (WoW) 259
18.4	Summary 260
	Further Reading 260
19	**Overview of Bonding and Assembly for 3D Integration** 261
	James J.-Q. Lu, Dingyou Zhang, and Peter Ramm
19.1	Introduction 261
19.2	Direct, Indirect, and Hybrid Bonding 262
19.3	Requirements for Bonding Process and Materials 263
19.4	Bonding Quality Characterization 267
19.5	Discussion of Specific Bonding and Assembly Technologies 269
19.6	Summary and Conclusions 273
	References 274
20	**Bonding and Assembly at TSMC** 279
	Douglas C.H. Yu
20.1	Introduction 279
20.2	Process Flow 280
20.3	Chip-on-Wafer Stacking 281
20.4	CoW-on-Substrate (CoWoS) Stacking 283
20.5	CoWoS Versus CoCoS 283
20.6	Testing and Known Good Stacks (KGS) 284
20.7	Future Perspectives 285
	References 285

21	**TSV Packaging Development at STATS ChipPAC** *287*	
	Rajendra D. Pendse	
21.1	Introduction *287*	
21.2	Development of the 3DTSV Solution for Mobile Platforms *289*	
21.3	Alternative Approaches and Future Developments *293*	
	References *294*	
22	**Cu–SiO$_2$ Hybrid Bonding** *295*	
	Léa Di Cioccio, S. Moreau, Loïc Sanchez, Floriane Baudin, Pierric Gueguen, Sebastien Mermoz, Yann Beilliard, and Rachid Taibi	
22.1	Introduction *295*	
22.2	Blanket Cu–SiO$_2$ Direct Bonding Principle *296*	
22.2.1	Chemical–Mechanical Polishing Parameters *296*	
22.3	Aligned Bonding *299*	
22.3.1	Wafer-to-Wafer Bonding *299*	
22.3.2	Die-to-Wafer Bonding in Pick-and-Place Equipment *299*	
22.3.3	Die-to-Wafer by the Self-Assembly Technique *300*	
22.4	Blanket Metal Direct Bonding Principle *302*	
22.5	Electrical Characterization *304*	
22.5.1	Wafer-to-Wafer and Die-to-Wafer Copper-Bonding Electrical Characterization *304*	
22.5.2	Reliability *307*	
22.5.3	Thermal Cycling *307*	
22.5.4	Stress Voiding (SIV) Test on 200 °C Postbonding Annealed Samples *308*	
22.5.5	Package-Level Electromigration Test *309*	
22.6	Conclusions *310*	
	References *311*	
23	**Bump Interconnect for 2.5D and 3D Integration** *313*	
	Alan Huffman	
23.1	History *313*	
23.2	C4 Solder Bumps *315*	
23.3	Copper Pillar Bumps *316*	
23.4	Cu Bumps *319*	
23.5	Electromigration *320*	
	References *322*	
24	**Self-Assembly Based 3D and Heterointegration** *325*	
	Takafumi Fukushima and Jicheol Bea	
24.1	Introduction *325*	
24.2	Self-Assembly Process *325*	
24.3	Key Parameters of Self-Assembly on Alignment Accuracies *327*	

24.4	How to Interconnect Self-Assembled Chips to Chips or Wafers *328*	
24.4.1	Flip-Chip-to-Wafer 3D Integration *329*	
24.4.2	Reconfigured-Wafer-to-Wafer 3D Integration *331*	
	References *332*	

25 High-Accuracy Self-Alignment of Thin Silicon Dies on Plasma-Programmed Surfaces *335*
Christof Landesberger, Mitsuru Hiroshima, Josef Weber, and Karlheinz Bock

25.1	Introduction *335*
25.2	Principle of Fluidic Self-Alignment Process for Thin Dies *335*
25.3	Plasma Programming of the Surface *336*
25.4	Preparation of Materials for Self-Alignment Experiments *337*
25.5	Self-Alignment Experiments *338*
25.6	Results of Self-Alignment Experiments *339*
25.7	Discussion *341*
25.8	Conclusions *342*
	References *343*

26 Challenges in 3D Fabrication *345*
Douglas C.H. Yu

26.1	Introduction *345*
26.2	High-Volume Manufacturing for 3D Integration *346*
26.3	Technology Challenges *346*
26.4	Front-Side and Backside Wafer Processes *346*
26.5	Bonding and Underfills *350*
26.6	Multitier Stacking *352*
26.7	Wafer Thinning and Thin Die and Wafer Handling *353*
26.8	Strata Packaging and Assembly *356*
26.9	Yield Management *359*
26.10	Reliability *360*
26.11	Cost Management *362*
26.12	Future Perspectives *362*
	References *364*

27 Cu TSV Stress: Avoiding Cu Protrusion and Impact on Devices *365*
Eric Beyne, Joke De Messemaeker, and Wei Guo

27.1	Introduction *365*
27.2	Cu Stress in TSV *365*
27.3	Mitigation of Cu Pumping *368*
27.4	Impact of TSVs on FEOL Devices *371*
	References *378*

28	**Implications of Stress/Strain and Metal Contamination on Thinned Die** *379*	
	Kangwook Lee and Mariappan Murugesan	
28.1	Introduction *379*	
28.2	Impacts of Cu Contamination on Device Reliabilities in Thinned 3DLSI *379*	
28.3	Impacts of Local Stress and Strain on Device Reliabilities in Thinned 3DLSI *386*	
28.3.1	Microbump-Induced Stresses in Stacked LSIs *387*	
28.3.2	Microbump-Induced TMS in LSI *388*	
28.3.3	Microbump-Induced LMS *389*	
	References *391*	
29	**Metrology Needs for 2.5D/3D Interconnects** *393*	
	Victor H. Vartanian, Richard A. Allen, Larry Smith, Klaus Hummler, Steve Olson, and Brian Sapp	
29.1	Introduction: 2.5D and 3D Reference Flows *393*	
29.2	TSV Formation *394*	
29.2.1	TSV Etch Metrology *395*	
29.2.2	Liner, Barrier, and Seed Metrology *397*	
29.2.3	Copper Fill Metrology (TSV Voids) *399*	
29.2.4	Cross-Sectional SEM (Focused Ion Beam Milling Sample Preparation) *400*	
29.2.5	X-Ray Microscopy and CT Inspection *400*	
29.2.6	Stress Metrology in Cu and Si *402*	
29.3	MEOL Metrology *404*	
29.3.1	Edge Trim Inspection *405*	
29.3.2	Bond Voids and Bond Strength Metrology *406*	
29.3.2.1	Acoustic Microscopy: Operation *407*	
29.3.2.2	Acoustic Microscopy for Defect Inspection and Review *407*	
29.3.2.3	Other Bond Void Detection Techniques *408*	
29.3.3	Bond Strength Metrology *409*	
29.3.4	Bonded Wafer Thickness, Bow, and Warp *410*	
29.3.4.1	Chromatic White Light *411*	
29.3.4.2	Infrared Interferometry *412*	
29.3.4.3	White Light Interferometry (or Coherence Scanning Interferometry) *414*	
29.3.4.4	Laser Profiling *415*	
29.3.4.5	Capacitance Probes *416*	
29.3.4.6	Differential Backpressure Metrology *417*	
29.3.4.7	Acoustic Microscopy for Measuring Bonded Wafer Thickness *417*	
29.3.5	TSV Reveal Metrology *418*	
29.4	Assembly and Packaging Metrology *420*	
29.4.1	Wafer-Level C4 Bump and Microbump Metrology and Inspection *421*	

	29.4.2	Package-Level Inspection: Scanning Acoustic Microscopy *422*
	29.4.3	Package-Level Inspection: X-Rays *424*
	29.5	Summary *426*
		References *427*

Index *431*

List of Contributors

Richard A. Allen
National Institute of Standards
and Technology (NIST)
100 Bureau Drive
Gaithersburg, MD 20899
USA

Wilfried Bair
SUSS MicroTec AG
Schleißheimer Str. 90
85748 Garching
Germany

Floriane Baudin
CEA/LETI
Department of Heterogeneous
Integration on Silicon
17, rue des Martyrs
38054 Grenoble CEDEX 9
France

Jicheol Bea
Tohoku University
New Industry Creation Hatchery
Center (NICHe)
6-6-10 Aza-Aoba, Aramaki
Sendai 980-8579
Japan

Rozalia Beica
Yole Developpement
Le Quartz
75, cours Emile Zola
69100 Lyon-Villeurbanne
France

Yann Beilliard
CEA/LETI
Department of Heterogeneous
Integration on Silicon
17, rue des Martyrs
38054 Grenoble CEDEX 9
France

Eric Beyne
IMEF
Kapeldreef 75
3001 Leuven
Belgium

Karlheinz Bock
Fraunhofer Research Institution for
Solid State Technologies EMFT
Hansastrasse 27d
80686 Munich
Germany

Jürgen Burggraf
E. Thallner GmbH
EV Group
DI-Erich-Thallner-Straße 1
4782 Sankt Florian am Inn
Austria

Daniel Burgstaller
E. Thallner GmbH
EV Group
DI-Erich-Thallner-Straße 1
4782 Sankt Florian am Inn
Austria

Joke De Messemaeker
IMEF
Kapeldreef 75
3001 Leuven
Belgium

Léa Di Cioccio
CEA/LETI
Department of Heterogeneous
Integration on Silicon
17, rue des Martyrs
38054 Grenoble CEDEX 9
France

Blake Dronen
3M Electronics Markets Materials
Building 225-3S-06
St. Paul, MN 55144-1000
USA

Jean-Christophe Eloy
Yole Developpement
Le Quartz
75, cours Emile Zola
69100 Lyon-Villeurbanne
France

Takafumi Fukushima
Tohoku University
New Industry Creation Hatchery
Center (NICHe)
6-6-10 Aza-Aoba, Aramaki
Sendai 980-8579
Japan

Philip Garrou
Microelectronic Consultants of
North Carolina
3021 Cornwallis Road
Research Triangle Park, NC 27709
USA

Pierric Gueguen
CEA/LETI
Department of Heterogeneous
Integration on Silicon
17, rue des Martyrs
38054 Grenoble CEDEX 9
France

Wei Guo
IMEF
Kapeldreef 75
3001 Leuven
Belgium

Mitsuru Hiroshima
Panasonic Factory Solutions Co., Ltd.
2-7 Matsuba-cho, Kadoma
Osaka 571-8502
Japan

Marc B. Hoppenbrouwers
Dutch Organization of Applied
Scientific Research (TNO)
5600 HE Eindhoven
The Netherlands

Alan Huffman
RTI International
3040 East Cornwallis Road
Post Office Box 12194
Research Triangle Park, NC 27709
USA

Klaus Hummler
SEMATECH
257 Fuller Road
Albany, NY 12203
USA

Hiroaki Ikeda
ASET
1-28-38 Shinkawa, Chuo-ku
Tokyo 104-0033
Japan

Anne Jourdain
IMEF
Kapeldreef 75
3001 Leuven
Belgium

Akihito Kawai
DISCO CORPORATION
13-11 Omori-Kita 2 Chome, Ota-ku
Tokyo 143-8580
Japan

Guido Knippels
Advanced Laser Separation
International (ALSI)
Platinawerf 20-G
6641 TL Beuningen
The Netherlands

Mitsumasa Koyanagi
Tohoku University
New Industry Creation Hatchery Center
6-6-10 Aza-Aoba, Aramaki
Sendai 980-8579
Japan

Christof Landesberger
Fraunhofer Research Institution for Solid State Technologies EMFT
Hansastrasse 27d
80686 Munich
Germany

Kangwook Lee
Tohoku University
New Industry Creation Hatchery Center
6-6-10 Aza-Aoba, Aramaki
Sendai 980-8579
Japan

Paul Lindner
E. Thallner GmbH
EV Group
DI-Erich-Thallner-Straße 1
4782 Sankt Florian am Inn
Austria

James J.-Q. Lu
Rensselaer Polytechnic Institute
Department of Electrical, Computer, and Systems Engineering
110 8th St.
Troy, NY 12160
USA

Matthew Lueck
RTI International
3040 East Cornwallis Road
Post Office Box 12194
Research Triangle Park, NC 27709
USA

Dean Malta
RTI International
3040 East Cornwallis Road
Post Office Box 12194
Research Triangle Park, NC 27709
USA

Thorsten Matthias
E. Thallner GmbH
EV Group
DI-Erich-Thallner-Straße 1
4782 Sankt Florian am Inn
Austria

S. Moreau
CEA/LETI
Department of Heterogeneous Integration on Silicon
17, rue des Martyrs
38054 Grenoble CEDEX 9
France

Sebastien Mermoz
CEA/LETI
Department of Heterogeneous Integration on Silicon
17, rue des Martyrs
38054 Grenoble CEDEX 9
France

Mariappan Murugesan
Tohoku University
New Industry Creation Hatchery Center
6-6-10 Aza-Aoba, Aramaki
Sendai 980-8579
Japan

Steve Olson
State University of NY (SUNY) at Albany
College of Nanoscience and Engineering (CNSE)
Albany, NY 12203
USA

Gerrit Oosterhuis
Dutch Organization of Applied Scientific Research (TNO)
5600 HE Eindhoven
The Netherlands

Shoji Otaka
TOK (Tokyo Ohka Kogya) Co. Ltd.
150 Nakamaruko, Nakahara-ku
Kawasaki 211-0012
Japan

Christoph Paschke
Fraunhofer Research Institution for Solid State Technologies EMFT
Hansastrasse 27d
80686 Munich
Germany

Rajendra D. Pendse
STATS ChipPAC
#05-17/20 Techpoint
10 Ang Mo Kio Street
Singapore 569059
Singapore

Alain Phommahaxay
IMEF
Kapeldreef 75
3001 Leuven
Belgium

Rama Puligadda
Brewer Science
2401 Brewer Drive
Rolla, MO 65401
USA

Sesh Ramaswami
Applied Materials, Inc.
Silcon Systems Group
974 East Arques Avenue
Sunnyvale, CA 94085
USA

Peter Ramm
Fraunhoffer EMFT
Device and 3D Integration
Hansastrasse 27d
80686 Munich
Germany

Fred Roozeboom
Dutch Organization of Applied
Scientific Research (TNO)
5600 HE Eindhoven
The Netherlands

Loïc Sanchez
CEA/LETI
Department of Heterogeneous
Integration on Silicon
17, rue des Martyrs
38054 Grenoble CEDEX 9
France

Brian Sapp
SEMATECH
257 Fuller Road
Albany, NY 12203
USA

Larry Smith
SEMATECH
257 Fuller Road
Albany, NY 12203
USA

Hans-Peter Spöhrle
Fraunhofer Research Institution for
Solid State Technologies EMFT
Hansastrasse 27d
80686 Munich
Germany

Venky Sundaram
Georgia Tech
School of Electrical and
Computational Engineering
777 Atlantic Drive NW
Atlanta, GA 3033-0250
USA

Rachid Taibi
CEA/LETI
Department of Heterogeneous
Integration on Silicon
17, rue des Martyrs
38054 Grenoble CEDEX 9
France

Tetsu Tanaka
Tohoku University
Graduate School of Biomedical
Engineering
6-6-01 Aza-Aoba, Aramaki
Sendai 980-8579
Japan

Rao R. Tummala
Georgia Tech
School of Electrical and
Computational Engineering
777 Atlantic Drive NW
Atlanta, GA 3033-0250
USA

Victor H. Vartanian
SEMATECH
257 Fuller Road
Albany, NY 12203
USA

Richard Webb
3M Electronics Markets Materials
Building 225-3S-06
St. Paul, MN 55144-1000
USA

Josef Weber
Fraunhofer Research Institution for
Modular Solid State Technology
EMFT
Hansastrasse 27d
80686 Munich
Germany

Markus Wimplinger
E. Thallner GmbH
EV Group
DI-Erich-Thallner-Straße 1
4782 Sankt Florian am Inn
Austria

Douglas C.H. Yu
TSMC
No.8, Li-Hsin Rd. Vl, Science Park
Hsinchu
Taiwan 300-78
P. R. China

Dingyou Zhang
Rensselaer Polytechnic Institute
Department of Electrical, Computer
and Systems Engineering
110 8th St.
Troy, NY 12160
USA

1
3D IC Integration Since 2008

Philip Garrou, Peter Ramm, and Mitsumasa Koyanagi

In Volume 1, we covered some of the history of the development of the 3D integrated circuit (3D IC) concept and we direct you to that chapter for such content [1].

Since the first two volumes of the *Handbook of 3D Integration* appeared in 2008, significant progress has been made to bring 3D IC technology to commercialization. This chapter will attempt to summarize some of the key developments during that period.

We previously described 3D IC integration as "an emerging, system level integration architecture wherein multiple strata (layers) of planar devices are stacked and interconnected using through-silicon (or other semiconductor material) vias (TSV) in the Z direction" as depicted schematically in Figure 1.1a and in cross section in Figure 1.1b [1].

With the continued pressure to miniaturize portable products and the near universal agreement that scaling as we have known it is soon coming to an end [2], a perfect storm has been created. The response to this dilemma at both the device and the package level has been to move into the third dimension.

It is commonly accepted that chip stacks wire-bonded down to a common laminate base and stacked packages such as package-on-package (PoP) are categorized as "3D packaging." Transistor design has also gone vertical [3] as Intel [4] and others move to "finfet" stacked transistor structures at the 22 nm generation. These are compared pictorially in Figure 1.2.

In Figure 1.3, we compare system-on-chip (SoC), 3D packaging, and 3D IC with through-silicon via (TSV) in various performance categories [5].

1.1
3D IC Nomenclature

Since 2008 there have been attempts to further refine the nomenclature for 3D IC integration, although it has not yet been universally adopted in publications. In 2009 the International Technology Roadmap for Semiconductors (ITRS)

Handbook of 3D Integration: 3D Process Technology, First Edition.
Edited by Philip Garrou, Mitsumasa Koyanagi, and Peter Ramm.
© 2014 Wiley-VCH Verlag GmbH & Co. KGaA. Published 2014 by Wiley-VCH Verlag GmbH & Co. KGaA.

1 3D IC Integration Since 2008

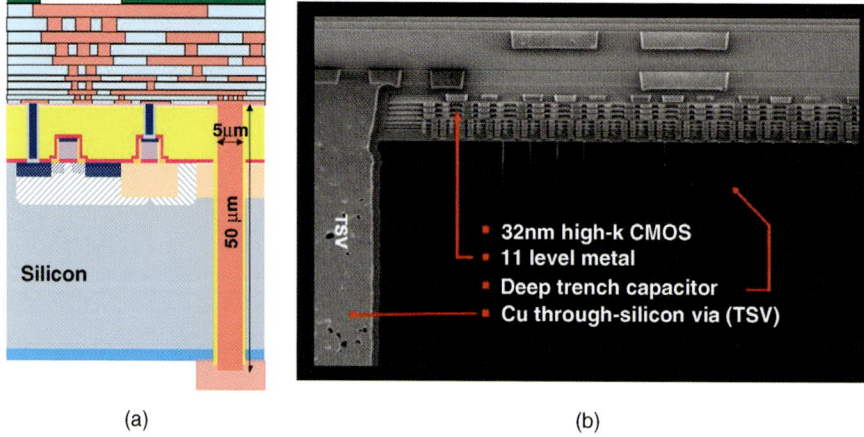

Figure 1.1 3D IC with TSV: (a) schematic (courtesy of IMEC) and (b) cross section (courtesy of IBM). Note that the IBM cross section is connected at a higher (fatter) on chip interconnect level.

proposed the following nomenclature in an attempt to define the possible different levels of connections possible as circuits are deconstructed onto separate strata (see Table 1.1) [6].

1.2
Process Standardization

3D IC requires three new pieces of technology: (1) insulated conductive vias through a thinned silicon substrate (i.e., TSV); (2) thinning and handling technology for wafers as thin as 50 μm or less; (3) technology to assemble and package such thinned chips.

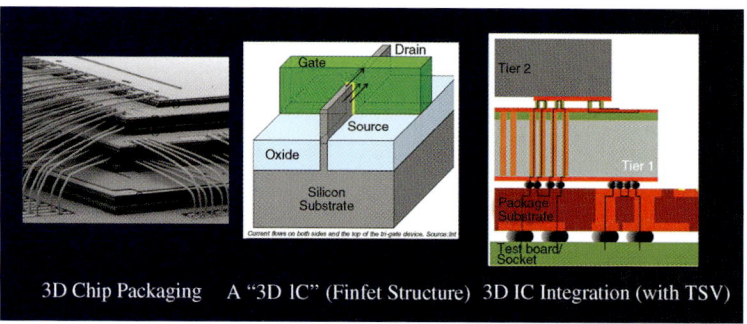

Figure 1.2 3D packaging, 3D finfet transistors, and 3D IC integration.

	On-chip SoC 	PoP 	WB stack 	TSV stack 
Memory capacity	−	++	++	++
Memory bandwidth	++	0	+	+++
Power consumption	++	0	+	+++
Form factor	0	+	++	+++
Cost	−−−	−	++	??

Figure 1.3 Comparison of SoC, 3D packaging, and 3D IC [5].

Table 1.1 2009 ITRS roadmap [6].

Level	Suggested name	Supply chain	Key characteristics
Package	3D packaging (3D-P)	OSAT assembly printed circuit board (PCB)	• Traditional packaging of interconnect technologies, for example, wire-bonded die stacks, package-on-package stacks • Also includes die in PCB integration • No through-Si vias
Bond-pad	3D wafer-level package (3D-WLP)	Wafer-level packaging	• WLP infrastructure, such as RDL and bumping • 3D interconnects are processed after the IC fabrication, "post IC passivation" (via-last process). Connections on bond-pad level • TSV density requirements follow bond-pad density roadmaps
Global	3D stacked integrated circuit/3D system-on-chip (3D-SIC/3D-SoC)	Wafer fab	• Stacking of large circuit blocks (tiles, IP blocks, memory banks), similar to an SoC approach but having circuits physically on different layers • Unbuffered I/O drivers (low C, little or no ISD protection on TSVs) • TSV density requirement significantly higher than 3D-WLP: Pitch requirement down to 4–16 μm

(continued)

Table 1.1 (Continued)

Level	Suggested name	Supply chain	Key characteristics
Intermediate	3D-SIC	Wafer fab	• Stacking of smaller circuit blocks, parts of IP blocks stacked in vertical dimensions • Mainly wafer-to-wafer stacking • TSV density requirements very high: Pitch requirement down to 1–4 µm
Local	3D IC	Wafer fab	• Sticking of transistor layers • Common back-end-of-line (BEOL) interconnect stack on multiple layers of front-end-of-line (FEOL) • Requires 3D connections at the density level of local interconnects

In the mid-2000s, practitioners were bewildered by the multitude of proposed technical routes to 3D IC. It has become clear, since then, that for most applications, the preferred process flow is what has been called a "via-middle" approach, where the TSVs are inserted after front-end transistor formation and early on during the on-chip interconnect process flow. This requires that TSVs are manufactured in back end of fab, not during or after the assembly process. This requires that TSV fabrication will be done by vertically integrated IDMs or foundries. TSV technology appears to be stabilized as depicted in Figure 1.4 and Table 1.2.

1.3
The Introduction of Interposers (2.5D)

Many believe the introduction of interposers (also known as 2.5D) was due to the failure of 3D IC, but this is not the case. Interposers were and are needed due to the lack of chip interface standardization and the need for a better thermal solution than is currently available for some 3D stacking situations.

The term "2.5D" is usually credited to Ho Ming Tong from Advanced Semiconductor Engineering (ASE), who in 2009 (or even earlier) declared that we might need an intermediate step toward 3D since the infrastructure and standards were not ready yet. The silicon interposer, Tong felt, would get us a major part of the way there, and could be ready sooner than 3D technology, thus the term "2.5D," which immediately caught on with other practitioners [7].

2.5D interposers resemble silicon multichip module technology of the 1990s, with the addition of TSV [8]. In today's applications, they provide high-density redistribution layers (RDLs), so the chips can be connected either through the interposer or next to each other on the top surface of the interposer as shown in

1.3 The Introduction of Interposers (2.5D)

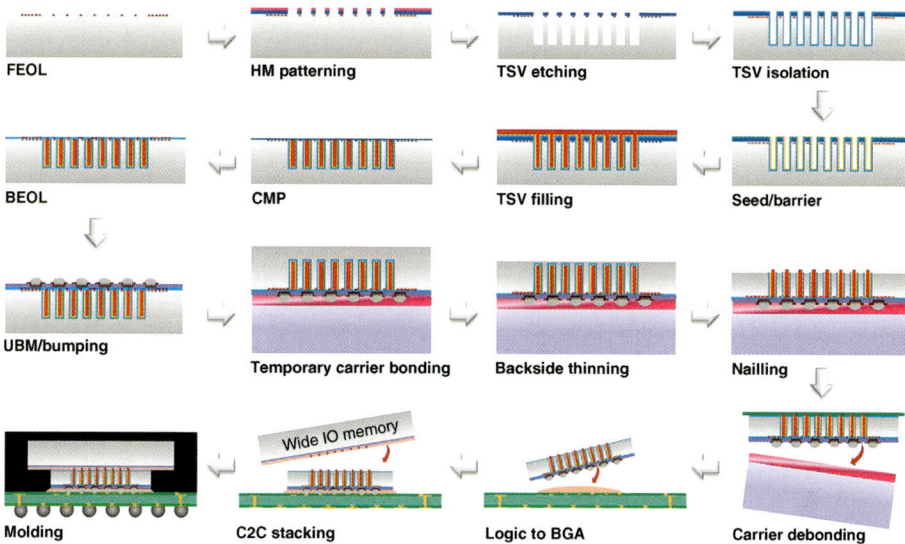

Figure 1.4 Standard 3D IC process flow. Courtesy of Yole Developpement.

Figure 1.5. The latter is the superior thermal solution since all chips can be attached to a heat sink for cooling.

Interposers will add cost and probably will not be a broadly accepted solution for low-cost mobile products, which would prefer straight 3D stacking [9].

Table 1.2 Standard 3D IC process flow options.

Process	Preferred option	Alternative options available			
TSV formation	Bosch deep reactive ion etching (DRIE)	Laser			
TSV Insulation	SiO$_2$	Polymer			
Conductor	Cu	W	pSi		
Process flow	Via-middle	Via-last (backside)[a]	Via-first (for pSi)	Via-last (front side)	
Stacking Bonding	IMC	Cu–Cu	Oxide bonding	Polymer bonding	Hybrid bonding (oxide–metal or polymer–metal)
Thin wafer handling	On carrier	On stack			

a) Preferred flow for CMOS image sensors.

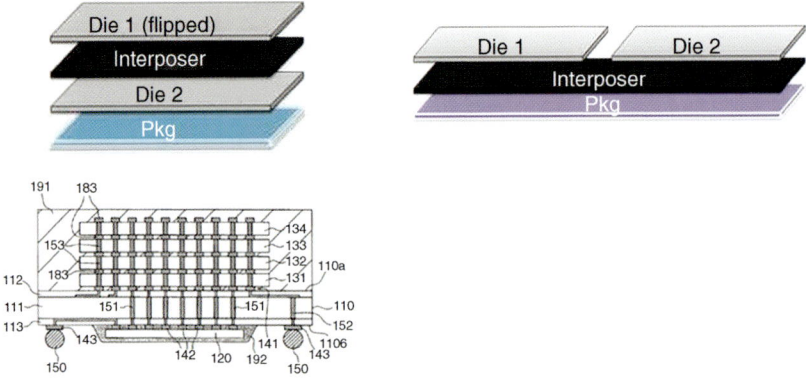

Figure 1.5 Interposer configurations.

1.4
The Foundries

1.4.1
TSMC

In October 2012, TSMC announced the readiness of their 2.5D CoWoS™ (chip-on-wafer-on-substrate) technology within their "Open Innovation Platform®" and made public their reference flows supporting CoWoS. Several EDA companies including Cadence, Mentor, Synopsys, and Ansys were announced as partners in the CoWoS reference flow [10]. Their first public CoWoS demonstrator vehicle (Figure 1.6) included logic and DRAM in a single module using the wide I/O interface [11].

Early TSMC customers reportedly included Xilinx, AMD, Nvidia, Qualcomm, Texas Instruments, Marvell, and Altera [12], with Xilinx being the first to production in late 2011.

Figure 1.6 2.5D TSMC demonstrator vehicle [11].

Reportedly due to "... the numerous technical challenges that make the conventional collaboration infrastructure more difficult" for 2.5 and 3D IC, TSMC has taken the position of being responsible for the full process (chip design and fabrication through module test).

1.4.2 UMC

UMC announced in the spring of 2011 that it had acquired production equipment for TSV and other 3D IC technologies. In 2013 UMC and STATS ChipPAC announced a jointly developed TSV-enabled 3D IC chip stack consisting of a Wide I/O memory test chip stacked upon a TSV-embedded 28 nm processor test chip [13].

1.4.3 GlobalFoundries

GlobalFoundries (GF) announced installation of TSV production tools for 20 nm technology wafers in their Fab 8 New York facility. The first full-flow silicon with TSVs was expected to start running in the third quarter of 2012 [14].

In contrast to TSMC, which announced a one-stop-shop turnkey line that included all of the assembly and test steps traditionally handled by outsourced semiconductor assembly and test (OSAT) facilities, UMC and GlobalFoundries indicated a preference to work under the open ecosystem model where they would handle TSV fabrication (Cu, via-middle) and other front-end steps while chips from various vendors would be back-end processed (i.e., temporary bonding/debonding, thinning, assembly, and test by their OSAT partners).

1.5 Memory

DRAM performance is constrained by the capacity of the data channel that sits between the memory and the processor. No matter how much faster the DRAM chip itself gets, the channel typically chokes due to the lack of transfer capacity; that is, they require more bandwidth. Wide I/O memory has been developed as the solution to this bandwidth problem [15].

Also, as more and more memory is required for a given application, power consumption also becomes important to both portable products and server farms, which need special cooling to keep them from overheating. Samsung reports that TSV-based RDIMM shows a 32% decrease in power consumption versus LRDIMM at 1333 Mbps [16].

1.5.1 Samsung

In late 2010, Samsung, who first revealed 3D TSV stacked memory prototypes in 2006, announced 40 nm, 8 GB RDIMM based on 4 Gb, 1.5 V, 40 nm DDR3

1 3D IC Integration Since 2008

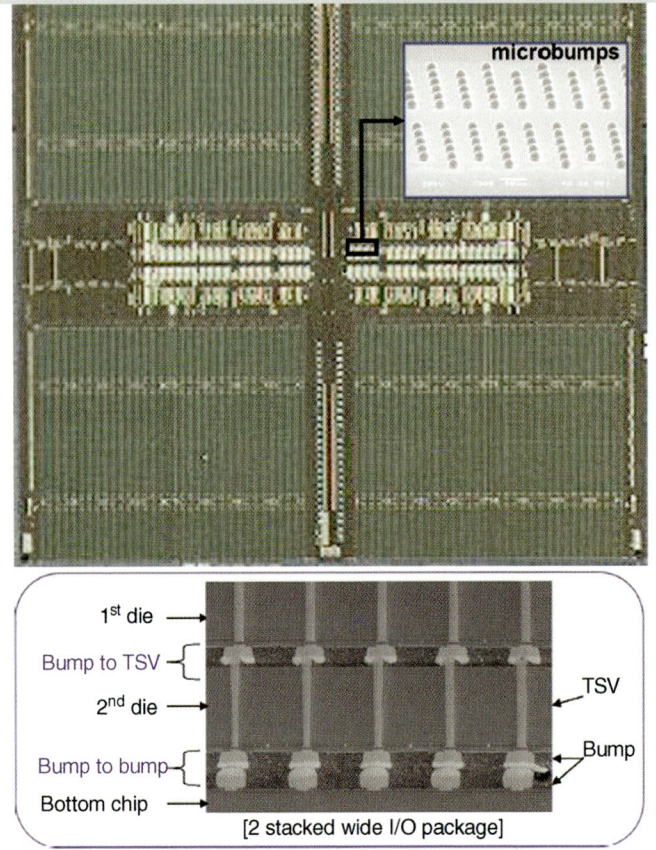

Figure 1.7 1.2 V 12.8 GB s^{-1} 2 Gb mobile wide I/O DRAM with 4 × 128 I/Os using TSV-based stacking [18].

memory chips operating at 1333 MHz and 3D TSV chip-stacking technology [17]. In 2011, they announced the development of wide I/O 1 Gb DRAM (Figure 1.7) [18]. Samsung has not announced any commercial memory products as of late 2013. Samsung is a member of the Micron hybrid memory cube consortium.

1.5.2
Micron

Micron developed a "hybrid memory cube" (HMC), which is a stack of multiple thinned memory dies sitting atop a logic chip bonded together using TSV (Figure 1.8). This greatly increases available DRAM bandwidth by leveraging the large number of I/O pins available through TSVs. The controller layer in the HMC allows a higher speed bus from the controller chip to the CPU and the thinned and TSV connected memory layers mean memory can be packed more densely in a given volume. The HMC requires about 10% of the volume of a DDR3 memory

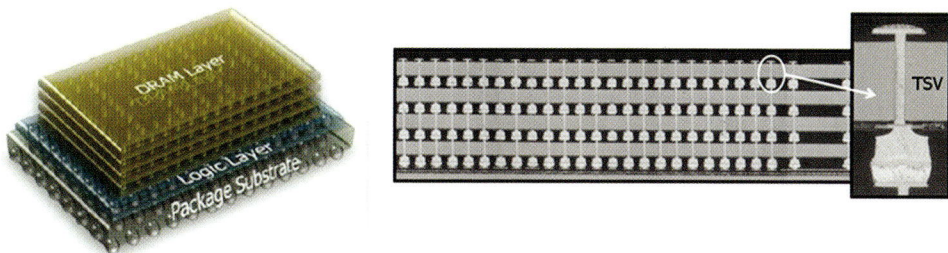

Figure 1.8 Micron hybrid memory cube (HMC) [19].

module. It is claimed that the technology provides 15× the performance of a DDR3 module, uses 70% less energy per bit than DDR3, and occupies 90% less space than today's RDIMMs. Micron has announced that they will be manufacturing the memory layers and have contracted IBM to manufacture the logic layer. Commercialization is scheduled for 2013–2014.

HMC electrical performance is compared to other DRAM modules in Table 1.3.

1.5.3
Hynix

Hynix reported that they expect "2 and 4 chip memory stacks with TSV to be in commercial production in 2014 and graphics solutions on interposers soon thereafter" [21].

1.6
The Assembly and Test Houses

Amkor was involved with commercial 3D IC assembly as part of their TSMC Xilinx program [22]. ASE, SPIL, and Powertech are all boosting 3D IC package and test capacity. SPIL (Siliconware) announced the instillation of dual damascene processing for high density interposers in 2013 [23]. Powertech, which has been in a 3D IC joint development program with Elpida (now Micron) and UMC for several years, announced volume production of 3D IC packaging and test capability in 2013.

Table 1.3 Comparison of Micron HMC to DDR memories [20].

Technology	VDD	BW (Gb s^{-1})	Power (W)
SDRAM PC133 1 Gb module	3.3	1.06	4.96
DDR-333 1 Gb module	2.5	2.66	5.48
DDR2-667 2 Gb module	1.8	5.34	5.18
DDR3-1333 2 Gb module	1.5	10.66	5.52
DDR4-2667 4 Gb module	1.2	21.34	6.60
HMC Gen 1 512 Mb cube	1.2	128.0	10.73

1 3D IC Integration Since 2008

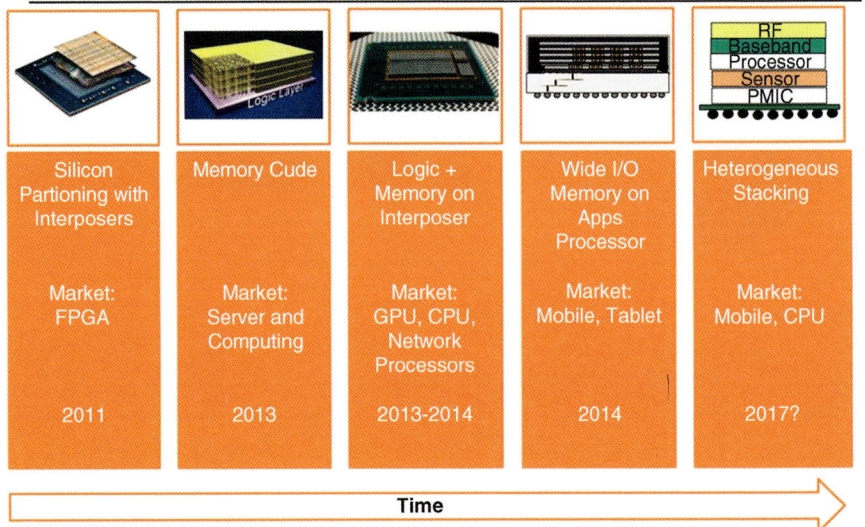

Figure 1.9 3D IC application timing. Courtesy of GlobalFoundries 2012.

1.7
3D IC Application Roadmaps

The first commercial application has been field-programmable gate arrays (FPGAs) with Xilinx (commercial) [22] and Altera (developing) [24] interposer-based solutions with TSMC.

Looking at the roadmap of GlobalFoundries in Figure 1.9, we see that 2.5D graphics processor modules and 3D application processors with baseband and/or memory should be coming soon.

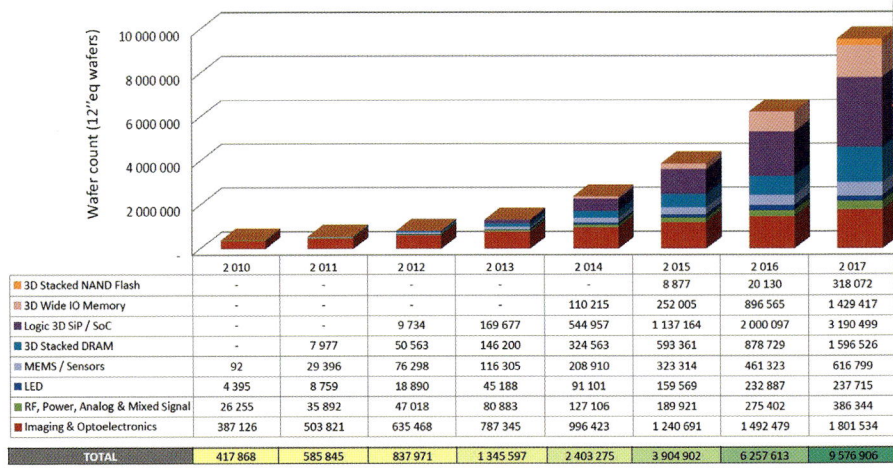

	2010	2011	2012	2013	2014	2015	2016	2017
3D Stacked NAND Flash	-	-	-	-	-	8 877	20 130	318 072
3D Wide IO Memory	-	-	-	-	110 215	252 005	896 565	1 429 417
Logic 3D SiP / SoC	-	-	9 734	169 677	544 957	1 137 164	2 000 097	3 190 499
3D Stacked DRAM	-	7 977	50 563	146 200	324 563	593 361	878 729	1 596 526
MEMS / Sensors	92	29 396	76 298	116 305	208 910	323 314	461 323	616 799
LED	4 395	8 759	18 890	45 188	91 101	159 569	232 887	237 715
RF, Power, Analog & Mixed Signal	26 255	35 892	47 018	80 883	127 106	189 921	275 402	386 344
Imaging & Optoelectronics	387 126	503 821	635 468	787 345	996 423	1 240 691	1 492 479	1 801 534
TOTAL	417 868	585 845	837 971	1 345 597	2 403 275	3 904 902	6 257 613	9 576 906

Figure 1.10 TSV chip wafer forecast 2010–2017. Courtesy of Yole Developpement.

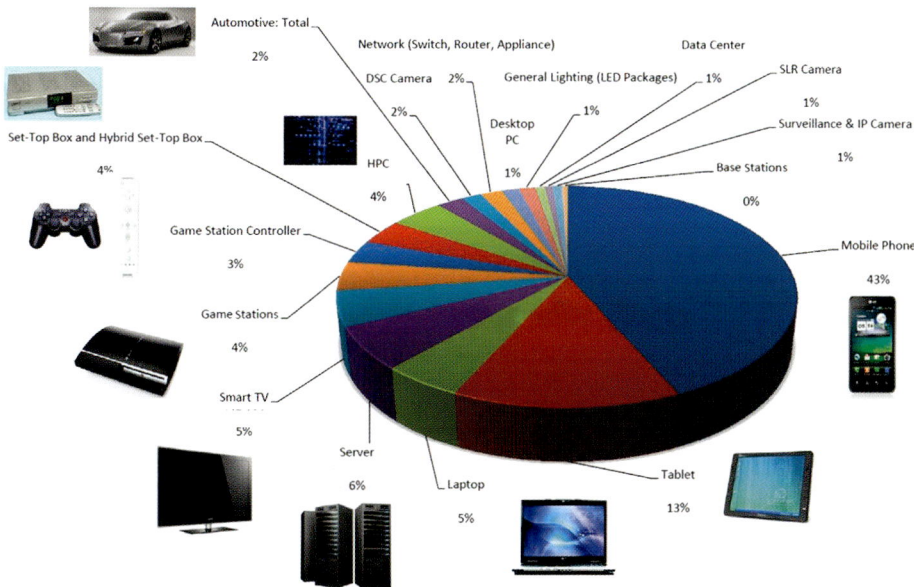

Figure 1.11 Global TSV chip end applications in 2017. Courtesy of Yole Developpement.

The latest projections by Yole Developpement are shown in Figures 1.10 and 1.11. 3D IC volume is expected to increase to nearly 10 MM wafers over the next 4 years with major increases in stacked memory, wide I/O DRAM, and logic plus memory system-in-packages (SiPs).

Examining 2.5/3D device product insertion, we see that most of these devices will eventually be incorporated in the smartphone and tablet markets.

References

1 Garrou, P. (2008) Chapter 1: Introduction to 3D integration, in *Handbook of 3D Integration*, vol. 1 (eds. P. Garrou, C. Bower, and P. Ramm), Wiley-VCH Verlag GmbH, Weinheim.
2 Heron, N.Z. (2008) Why is CMOS scaling coming to an end. 3rd International Design & Test Workshop, pp. 98–103.
3 Adee, S. (2007) Transistors go vertical. *IEEE Spectrum*, 44, 14.
4 James, D. (2012) Intel Ivy Bridge unveiled the first commercial tri-gate, high-k, metalgate CPU. IEEE Custom Integrated Circuit Conference, p. 1.
5 Garrou, P. (2011) 3D integration – are we there yet. Suss MicroTech Workshop, Semicon 2011.
6 Interconnect (2009) International Technology Roadmap for Semiconductors. Available at www.itrs.net/links/2009ITRS/2009Chapters_2009Tables/2009_Interconnect.pdf (accessed December 31, 2013).
7 Garrou, P. (2012) 3.5D interposers to someday replace PWBs – TSMC; GF engaging with 3D customers; Intel predicts Consolidation. Solid State Technology: Insights from the Leading Edge, 102, May 19, 2012.

8 Garrou, P. and Turlik, I. (1998) *Multichip Module Handbook*, McGraw-Hill.
9 Garrou, P. (2013) Interposer supply/ecosystem examined at IMAPS Device Packaging Conference. Available at www.electroiq.com/articles/ap/2012/03/interposer-supply-ecosystem-examined-atimaps-device-packaging.html (accessed December 31, 2013).
10 Garrou, P. (2012) TSMC officially ready for 2.5D, Apple order impact on TSMC. Solid State Technology: Insights from the Leading Edge, 122, November 11, 2012. Available at http://electroiq.com/insights-from-leading-edge/2012/11/iftle-122-tsmc-officially-ready-for-2-5d-apple-order-impact-on-tsmc/ (accessed December 31, 2013).
11 EDA 360 Insider (2012) 3D Thursday: want to see a closeup of the TSMC 3D IC test vehicle? June 6, 2012. Available at http://eda360insider.wordpress.com/2012/06/06/3d-thursday-want-to-see-a-closeup-of-the-tsmc-3d-ic-test-vehicle/ (accessed December 31, 2013).
12 Liu, K. (2012) TSC builds up 3DIC assembling capability to vie for lucrative business. *Taiwan Economic News*, August 14, 2012. Available at http://news.cens.com/cens/html/en/news/news_inner_40986.html (accessed December 31, 2013).
13 Garrou, P. (2013) UMC/SCP memory on logic; SEMI Europe 3D Summit Part 2. Solid State Technology: Insights from the Leading Edge, 135, February 12, 2013. Available at http://electroiq.com/insights-from-leading-edge/2013/02/iftle-135-umc-scp-memory-on-logic-semi-europe-3d-summit-part-2/ (accessed December 31, 2013).
14 Garrou, P. (2013) GlobalFoundries 2.5/3D at 20 nm; Intel Haswell GT3; UMC/SCP prototype details. Solid State Technology: Insights from the Leading Edge, 142, April 9, 2013. Available at http://electroiq.com/insights-from-leading-edge/2013/04/iftle-142-globalfoundries-2-5-3d-at-20nm-intel-haswell-gt3-umc-scp-prototype-details/ (accessed December 31, 2013).
15 JEDEC (2012) JEDEC publishes breakthrough standard for wide I/O mobile DRAM. Available at http://www.jedec.org/news/pressreleases/jedec-publishes-breakthrough-standard-wide-io-mobile-dram (accessed December 31, 2013).
16 Schauss, G. (2011) Samsung memory solution for HPC. Available at http://www.ena-hpc.org/2011/talks/schauss-slides.pdf (accessed December 31, 2013).
17 Kang, U. (2009) 8 Gb DDR3 DRAM using TSV technology. IEEE ISSCC.
18 Kim, J.S. (2011) A 1.2 V 12.8 GB/s 2 Gb mobile wide-I/O DRAM with 4_128 I/O's using TSV-based stacking. IEEE ISSCC.
19 Jeddeloh, J. and Keeth, B. (2012) New DRAM architecture increases density and performance. 2012 Symposium on VLSI Technology, p. 87.
20 Pawlowski, J.T. (2011) Hybrid memory cube: breakthrough DRAM performance with a fundamentally re-architected DRAM subsystem. Proceedings of the 23rd Hot Chips Symposium.
21 Suh, M. (2013) TSV application in memory devices. RTI 3D Architectures for Semiconductor Integration and Packaging Conference, Burlingame CA, December 2013.
22 Saban, K. (2010) Xilinx stacked silicon interconnect technology delivers breakthrough FPGA capacity, bandwidth, and power efficiency. Available at //www.xilinx.com/support/documentation/white_papers/wp380_Stacked_Silicon_Interconnect_Technology.pdf (accessed December 31, 2013).
23 Ma, M. (2013) The expanding role of OSATS in the era of system integration. ConFab 2013, Las Vegas, NV.
24 Rahman, A. *et al.* (2013) Application requirements of system-in-package FPGAs. RTI 3D Architectures for Semiconductor Integration and Packaging Conference, Burlingame, CA, December 2013.

2
Key Applications and Market Trends for 3D Integration and Interposer Technologies

Rozalia Beica, Jean-Christophe Eloy, and Peter Ramm

2.1
Introduction

Improvements in microprocessors and memory devices, through scaling of CMOS technologies have, for more than four decades, consistently followed Moore's law. While miniaturization is expected to continue, the benefits achieved through scaling, in terms of functionality, performance, and cost, are starting to weaken. Corresponding shrinking of transistors will probably only keep up until 2022 [1]. Fueled by research and developments of new materials and processes (various nanomaterials, carbon nanotubes, graphene, III–V compound semiconductors, and Ge channels), integration and architecture schemes, and so on (Figure 2.1), a growing trend in the industry is seen in moving from CMOS to package and system architecture value-added proposition (see Figure 2.1) [2,3].

For example, in the case of system-on-chip (SoC), which is a platform for horizontal integration of various devices (memory, logic, analog/RF), increased performance is achieved through CMOS technology and transistor miniaturization. All the devices integrated in SoC have to be manufactured using the same technology node that results in unnecessary increased costs. Horizontal integration also increases the total surface area of the system, increasing the form factor and added costs associated with the increase of the silicon surface (Figure 2.2).

As shown in Figure 2.2, system-in-package (SiP) architectures, in comparison to 2D SoC, bring several advantages, including reduced cost, increased performance, and smaller size. SiP and 3D ICs are achieving value from the various functions vertically integrated within the same package and their value

Handbook of 3D Integration: 3D Process Technology, First Edition.
Edited by Philip Garrou, Mitsumasa Koyanagi, and Peter Ramm.
© 2014 Wiley-VCH Verlag GmbH & Co. KGaA. Published 2014 by Wiley-VCH Verlag GmbH & Co. KGaA.

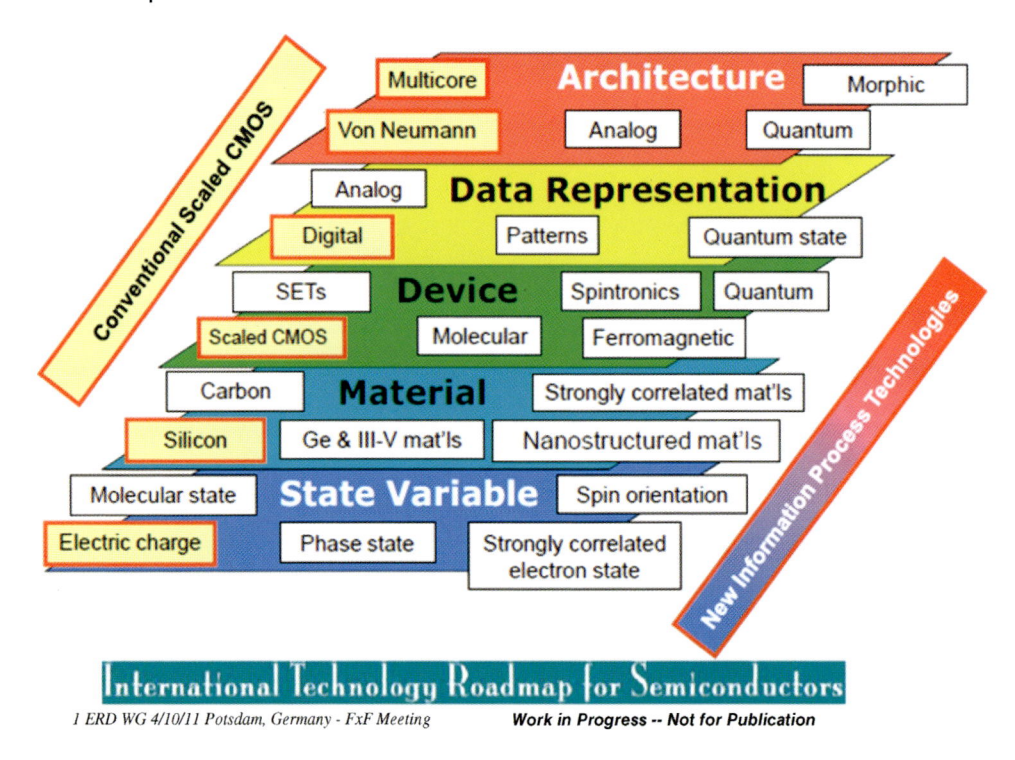

Figure 2.1 The taxonomy of emerging research information processing devices. The yellow boxes outlined with red show current CMOS technologies, based on electric charge and binary computational state variables; the remaining boxes are innovative technologies and ideas under development to continue to support further miniaturization and scaling [3]. Courtesy of ITRS.

proposition is not limited to CMOS scaling. There is no need for the devices package in a SiP to use the same CMOS technology nodes, rather each device (either being a memory, logic, analog, MEMS, passive component, etc.) is manufactured using optimum technology for the particular function it brings to the package. SiP and vertical integration enables heterogeneous integration of disparate technologies (digital and nondigital) reusing already proven and reliable technologies and designs from existing products, thus giving significant advantage over SoC-type integration. 3D IC integration, benefiting by the stacking of known good dies [5], can combine various functions on different lithography nodes, devices processed on different wafer sizes, and in different wafer fabs by different players in the industry.

This functional diversification, based on 3D integration, is known as "More than Moore" (MtM) (Figure 2.3).

2D SoC
"All-in-One chip system integration"

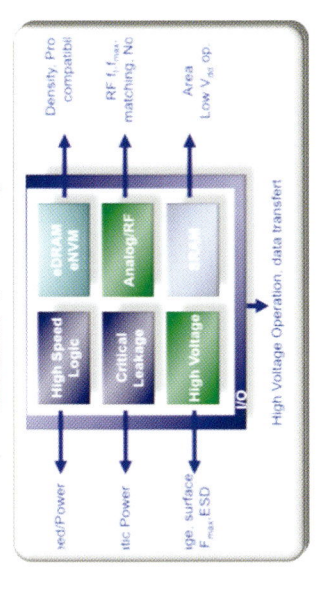

All functions on 28nm lithography
→ Chip area ↑, Cost ↑

3D IC
"Deintegrated and Reintegrated SoC"

MEMS	130 nm	200 mm
Memory	45 nm	300 mm
Logic	22 nm ?	450 mm ?
Analog	90 nm	300 mm

→ Cost ↓, Performance ↑, Size ↓

Evolution or Revolution?

Thanks to 3D, heterogeneous functions are integrated:
- On different lithography nodes
- On different wafer sizes
- In different wafer fabs
- By different players

Figure 2.2 2D SoC and 3D IC comparison [4]. Courtesy of Yole Developpement.

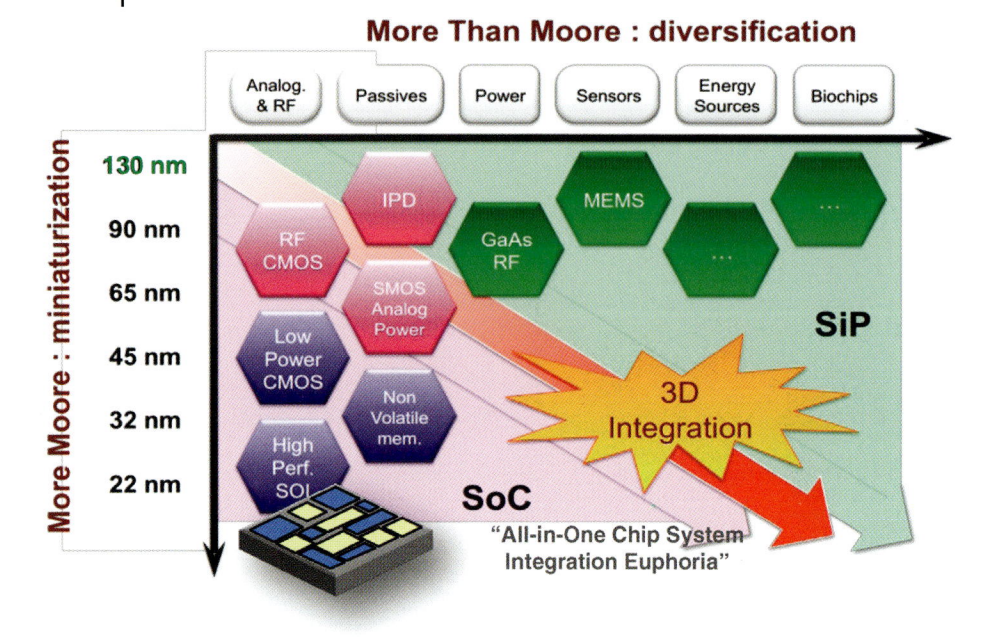

Figure 2.3 3D integration – enabling miniaturization and diversification [4]. Courtesy of Yole Developpement.

2.2
Advanced Packaging Importance in the Semiconductor Industry is Growing

As the semiconductor industry moves toward the systematic integration of stacked heterogeneous chips, 3D integration is expanding its reach in advanced packaging. 3D stacking includes chip-level, device-level, and wafer-level approaches.

One may ask, "Why package?" Packaging is done for speed, power, cost, and size advantages. Advanced packaging is transitioning to high-performance, high-density, low-cost collective wafer-level packaging techniques.

Several platforms have been developed in advanced packaging. For the past 40 years, the semiconductor packaging industry has continuously worked on developing new technologies and platforms to bridge the increasing I/O interconnect gap between the quickly decreasing silicon geometries (driven by Moore's law) and the slower speed at which the printed circuit geometries are shrinking (Figure 2.4).

This created opportunities for several packaging technologies to be developed over the years. Looking back at the major developments every 10 years, we can see a large variety of different technologies developed, from through-hole technology in the 1970s; surface mount devices (SMDs) in the 1980s; chip-scale packaging (CSP),

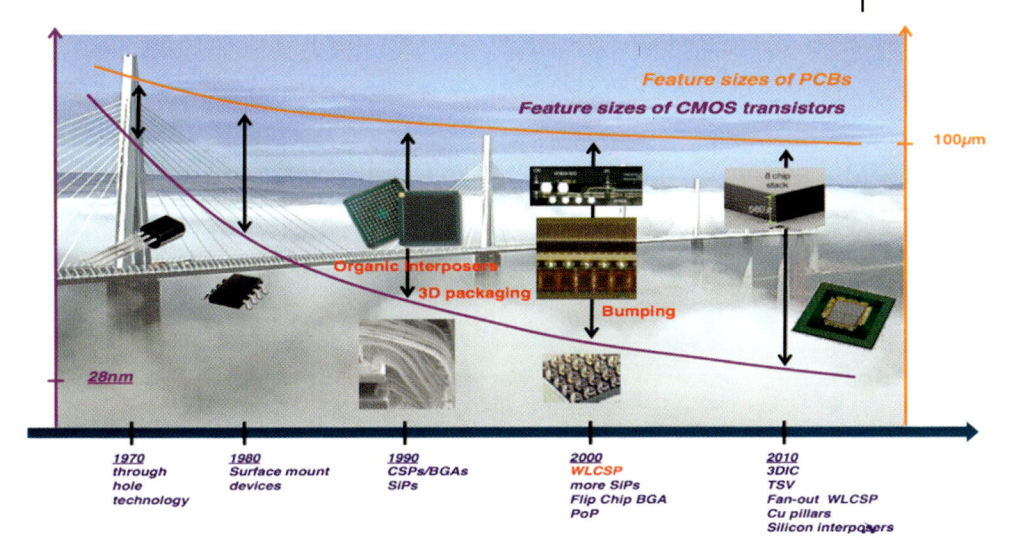

Figure 2.4 I/O interconnect gap between CMOS transistors and printed circuit board (PCB) features [4]. Courtesy of Yole Developpement.

ball grid arrays (BGAs), and SiPs in the 1990s; wafer-level chip-scale packaging (WLCSP), flip-chip BGAs, and more SiPs and package-on-packages (PoPs) developed around 2000; and more recently, 3D IC and through-silicon vias (TSVs), fan-out WLCSPs, Cu pillars, microbumping, silicon interposers, and embedded technologies. Advanced packaging is like a flower that bloomed over the years into a multicolor array of pastels. Many of these technologies, although developed several decades ago, are still available today and successfully supporting various packaging platforms.

Advanced packaging, especially wafer-level packaging, technologies are gaining more significance and importance within the semiconductor industry. They have the potential for notable growth in this industry. In 2012, 13 million wafers, which is approximately 16% of semiconductor IC wafers, were manufactured using various packaging technologies such as bumping, redistribution layers (RDLs), and TSV interconnect technologies. By 2017, the growth of advanced packaging is forecasted to reach approximately 23% penetration rate; that means 35 million wafers (in 300 mm equivalent wafers) out of the forecasted 148 million IC wafers will be using packaging technologies. The advanced packaging industry is growing significantly, outpacing the growth expected of the total semiconductor industry. With a 21% compound annual growth rate (CAGR), the advanced packaging industry is growing twice as fast as the semiconductor industry, which is forecasted to have a 10% CAGR (as illustrated in Figure 2.5).

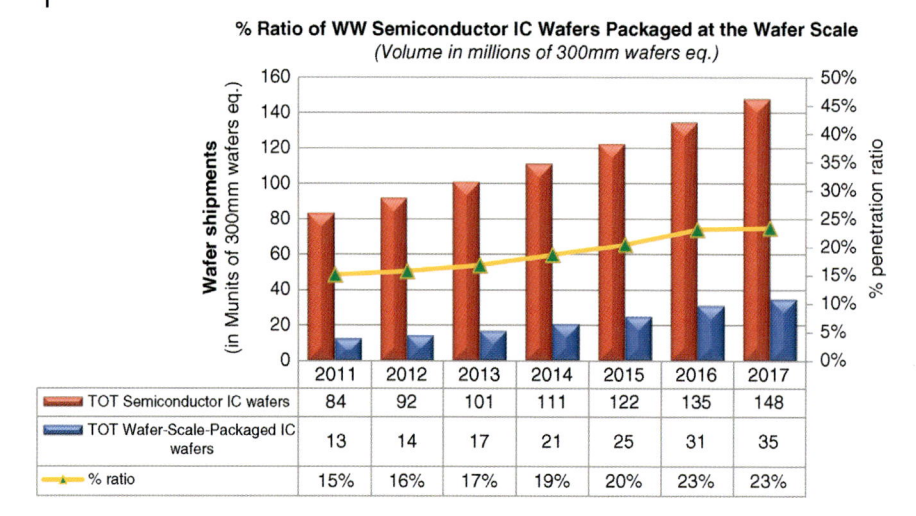

Figure 2.5 Wafer-level packaging in the semiconductor IC processing industry [4]. Courtesy of Yole Developpement.

2.3
3D Integration-Focused Activities – The Global IP Landscape

3D integration not only will bring diversification, but will also continue to drive the evolution of packages for several decades to come. Therefore, today 3D integration is considered a new paradigm for the semiconductor industry. There are various ways of vertically packaging devices, but the latest and most advanced technology for 3D stacking is TSV technology. Within wafer-level packaging, the platforms using TSV vertical interconnects are 3D WLCSPs, 2.5D interposers, and 3D ICs (Figure 2.6).

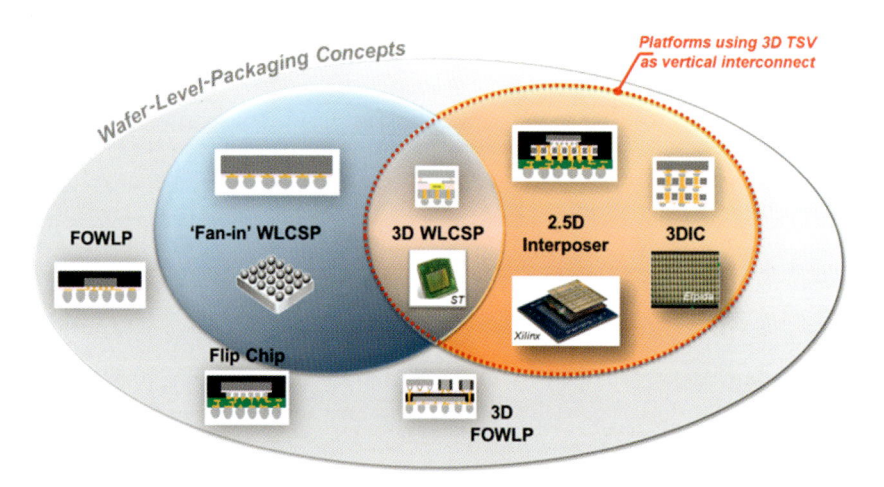

Figure 2.6 Wafer-level packaging technologies using TSV vertical interconnects [4]. Courtesy of Yole Developpement.

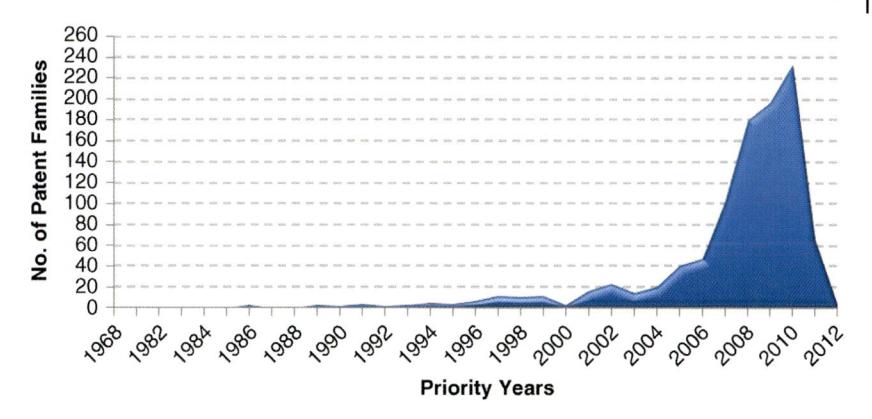

Figure 2.7 Trend of patent filing for 3D IC [6]. Courtesy of Yole Developpement.

3D integration is not a new concept. A recent IP landscape analysis performed by Yole Developpement, focusing on 3D ICs and 2.5D interposers, revealed 1969 as the year when the first 3D patent was applied. That IBM was the first assignee of this patent should not surprise anyone. The majority (82%) of the patents in this area were actually filed starting with 2006 (as shown in Figure 2.7).

A total of 1013 patent families were filed between 1969 and 2012. The United States (56%) followed by Korea (18%) were found to be the priority countries with the highest number of patents. The United States, China, Korea, Taiwan, and Japan, as illustrated in Figure 2.8, are the main countries where patents, once filed, are extended.

Even though the first patent was filed more than 40 years ago, for 30 years there was almost no activity in this domain. 3D IC is still a relatively young technology, also reflected by the high number of pending patents in this domain. As illustrated in Figure 2.9, by the end of 2012, only 39% of the patents were granted and almost half (45%) are still pending.

There were 260 players found to be involved with 3D IC. The top 10 (Figure 2.10) assignees represent 48% of the patents filed in 3D IC domain. IBM leads with respect to the number of patents filed, followed by Samsung, Micron, Taiwan Semiconductor Manufacturing Company (TSMC), SK Hynix, STATS ChipPac Ltd., Intel, Amkor, Elpida Memory, and Industrial Technology Research Institute (ITRI).

3D IC IP activities were found across different academic institutions with ITRI being the top institute with 21 patents, followed by CEA-Leti (France), Fraunhofer-Gesellschaft München (Germany), KAIST (Korea), IMEC (Belgium), University of Beijing (China), ASTRI (Hong Kong), University Tsinghua (China), Chinese Academy of Science (China), and ETRI (Korea). The top 10 academic assignees represent over 11% of the total 3D IC filed patents (Figure 2.11).

A more in-depth analysis of the patents with respect to processing showed TSV isolation, filling, bonding and debonding, and barrier and seed deposition to be the processing steps most patented. IBM, for example, initially focusing on using TSV

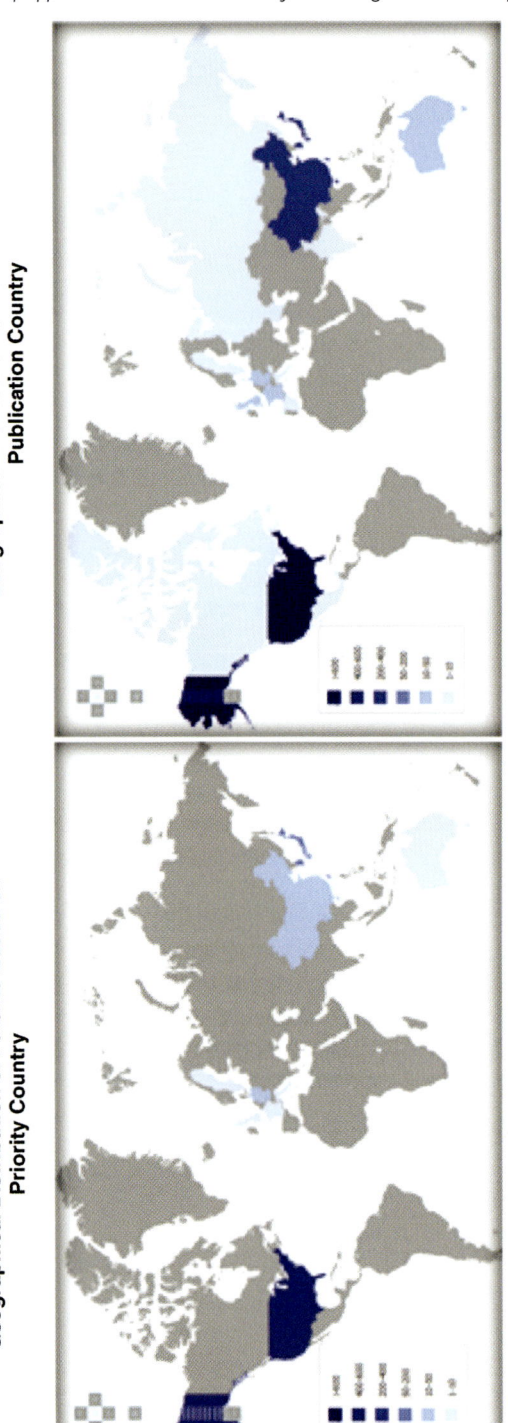

Figure 2.8 Geographical distribution of patent filing for 3D IC technology [6]. Courtesy of Yole Developpement.

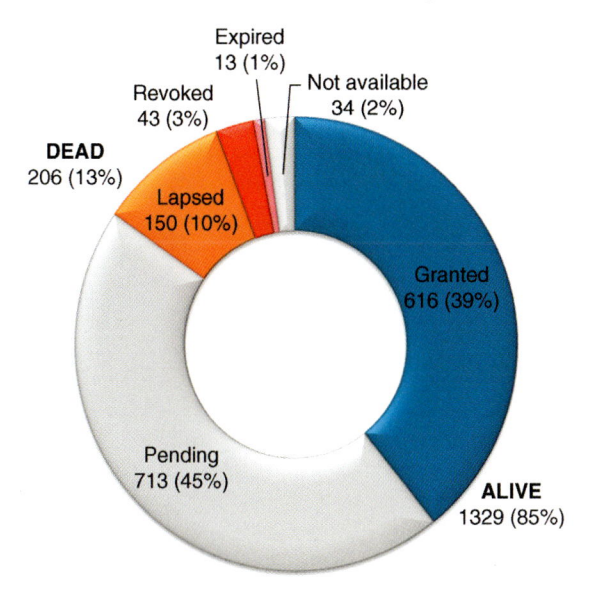

Figure 2.9 Legal status of 3D IC patents [6]. Courtesy of Yole Developpement.

for power amplifiers, has spent a lot of effort in trying to improve the electrical performance, especially at high frequencies; therefore, a large number of its patents (32%) were filed on the TSV isolation step. At IBM alone, there were more than 150 inventors and 200 patents found related to 3D IC technology.

In conclusion, the United States was the early player, increasingly involved in 3D IC since 1969, but new players, such as China and Korea, have entered the industry since 2005. If IBM and Micron (both US companies) are the main assignees and

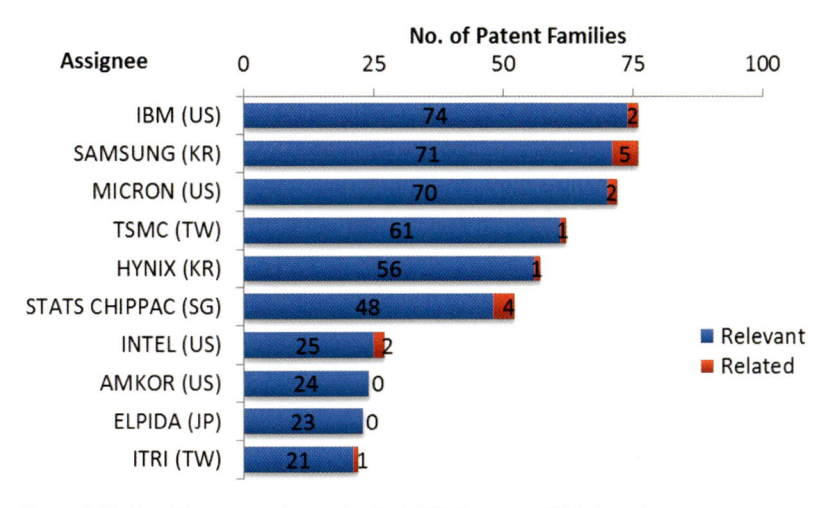

Figure 2.10 Top 10 patent assignees for 3D IC [6]. Courtesy of Yole Developpement.

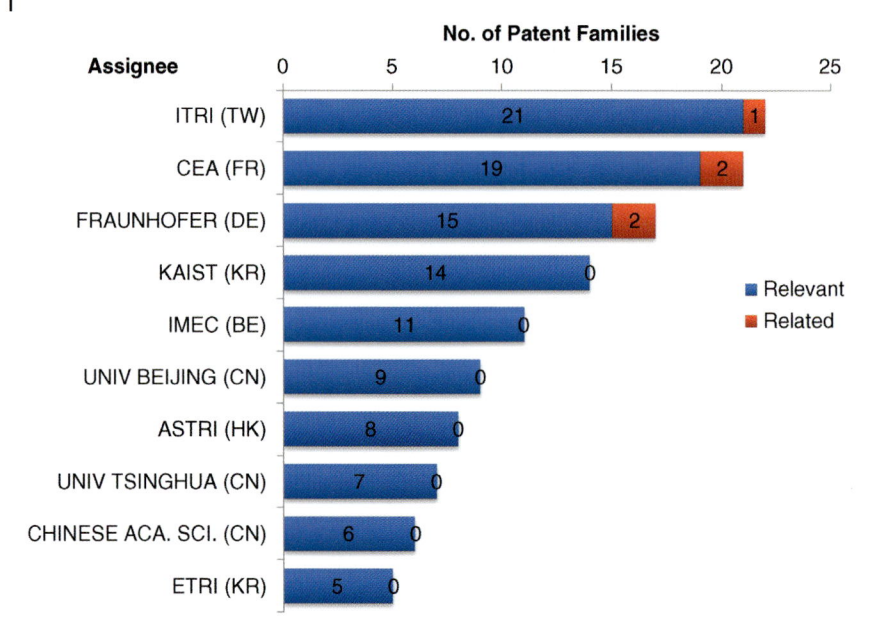

Figure 2.11 Top academic assignees for 3D IC patents [6]. Courtesy of Yole Developpement.

still active, SK Hynix (Korea) and STATS ChipPac Ltd. (Singapore) emerged as new players during the last 5 years.

2.4
Applications, Technology, and Market Trends

Several applications (Figure 2.12) have been identified where 3D TSV integration can be beneficial, in terms of performance, miniaturization, cost, or functionality.

Prior to 2012, we had mainly three applications using 3D IC and TSV technology, most of them processed on 150 mm type wafers:

- CMOS image sensors (CISs) with front-side imagers packaged in 3D WLCSP and backside imagers (BSIs) with TSV.

The main drivers for CIS are increased performance and integration:

  - The BSI "backside illuminated" image sensor continues to be a very hot product today. BSI is expected to increase image sensor performance (low light sensitivity) while relaxing back-end-of-line (BEOL) design requirements as the photodiodes will be at the top side of the CIS design.
  - It will also enable 3D integration with the possibility of having "more intelligence per pixel" for higher sensing performance and management of the information at

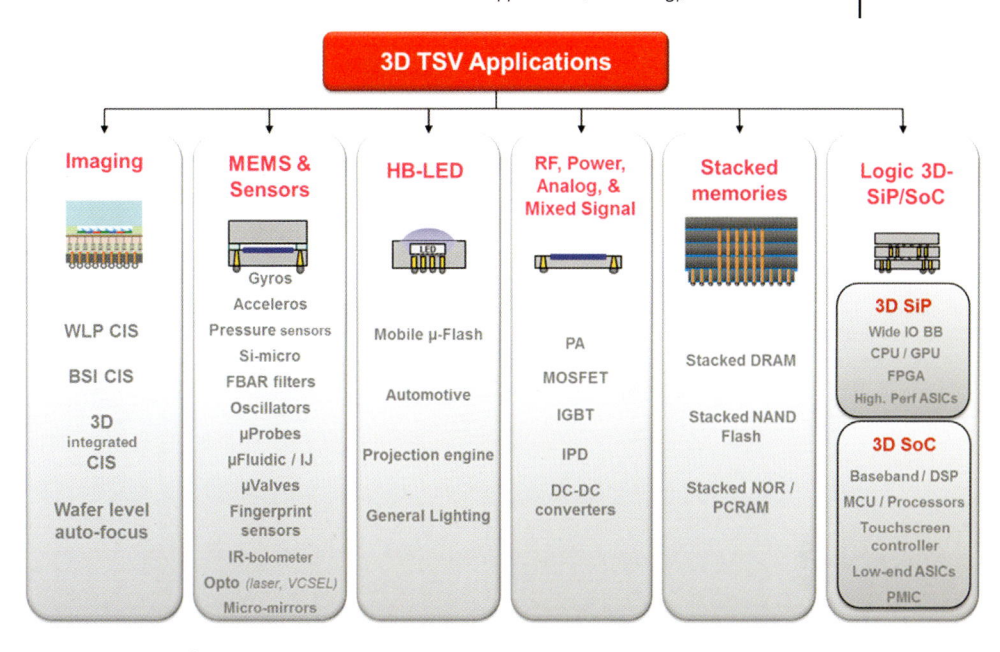

Figure 2.12 Six applications using 3D TSVs [6].

the pixel level, for recognition capability features, automotive sensors (safety and security applications), and gaming applications (human body/sensor interface with the computer).

High-end BSI sensors are already in production at Sony (Japan), and mostly dedicated to DSC and DSLR video cameras. BSIs are also expected to progressively replace CCDs in high-end applications. Other players include Omnivision, Panasonic, Aptina Imaging Corporation, and Samsung (Figure 2.13). Consumer BSI sensors (driven by cellphone applications) penetrated the market in 2011 on 1.1 μm pixel generation sensors (see Figure 2.20; top track).

Mobile phones and tablets will be driving future innovation in imaging. Of the imaging and optoelectronic 3D devices that will be shipped in 2017, 77% are forecasted to be integrated in mobile phones and tablets.

- *MEMS applications*: accelerometers and gyroscopes, fingerprint sensors, pressure sensors (TPMS, gas, etc.), RF-MEMS (FBAR, resonators, switches), microfluidics (microvalves, POC, etc.), microprobes, and optical MEMS (micromirrors, IR bolometers).

The driving forces for adopting TSVs in MEMS are form factor, cost, and integration:

- MEMS and sensor package costs are relatively high (often 40–60% of the overall cost) in addition to the large size of the package.

24 *2 Key Applications and Market Trends for 3D Integration and Interposer Technologies*

Figure 2.13 TSV/WLP reality in high-end, BSI CMOS image sensors from Samsung, Sony, Omnivision, and Toshiba [6]. Courtesy of Omnivision, SystemPlus Consulting, and Chipworks.

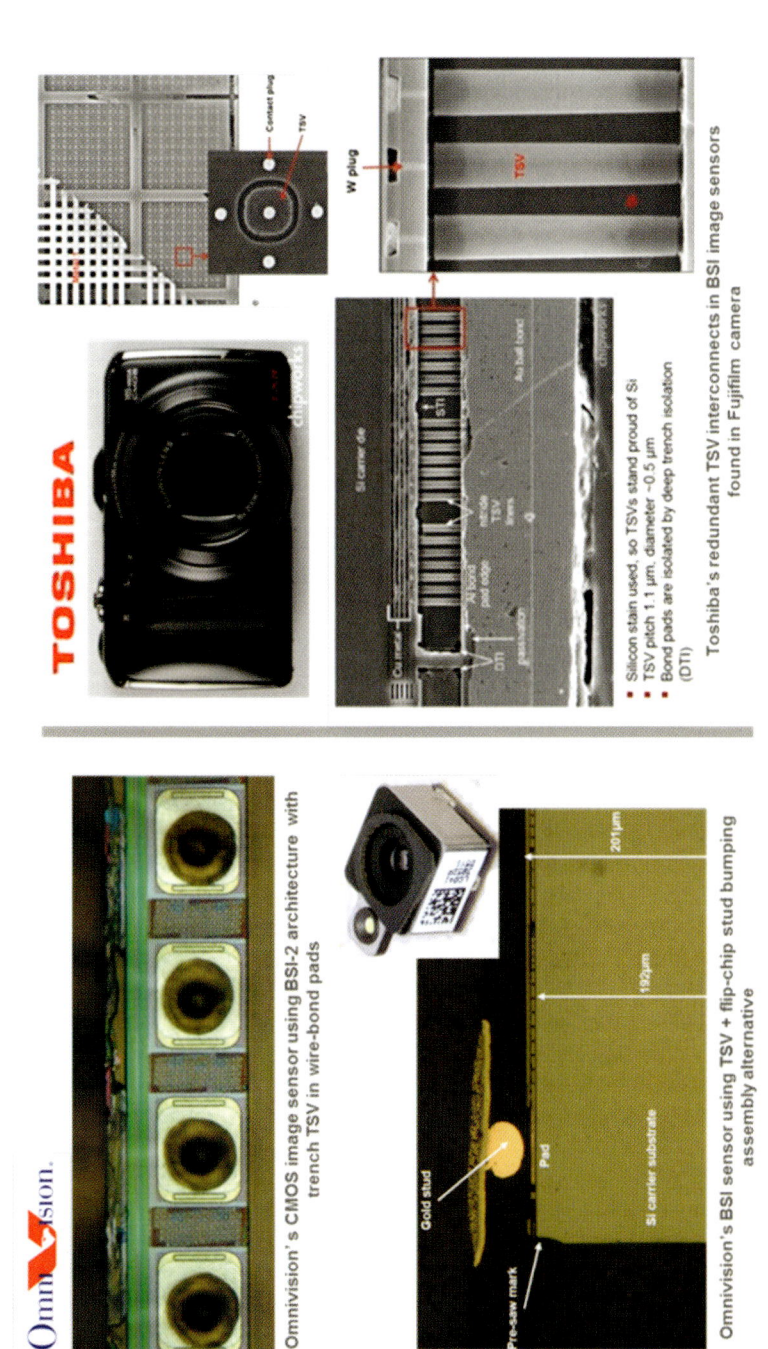

Figure 2.13 (*Continued*)

- Some sensor modules are becoming quite complex (TPMS and IMU modules), driving the need for a higher level of integration.

3D integration of TSVs has become a reality in MEMS and sensor applications. Avago is in production for their FBAR filters and VTI and ST Microelectronics are in production for MEMS inertial sensors (accelerometers and gyroscopes). Other applications are entering the market (silicon microphones from Sonion, fingerprint sensors from IDEX Corporation, and more) and a high level of activity is currently running on 150 mm in MEMS foundries such as Silex, Teledyne Dalsa, Touch Microsystems Technology, Innovative Micro Technology (IMT) MEMS, but 200 mm facilities are ramping up. There is also a high interest from giant CMOS foundry players (TSMC, UMC, Chartered, Semiconductor Manufacturing International Corporation (SMIC)); they have the ability to recycle their aging 200 mm fabs, since they are already deeply involved in the MEMS ASIC business. With respect to IDMs, ST, Bosch, TI, Avago, Omron, and others are also very active (Figure 2.14).

- *RF and power, analog, and mixed signal power applications*: IGBT, MOSFET for medical and automotive electronics, high performance MMIC components, and DC/DC converters. Examples are illustrated in Figure 2.15.

The main motivation for TSVs for these applications are cost, form factor, and performance.

- *Power applications*: With 3D integration, the wire bonding can be replaced with "ground" TSVs, preventing arcing effects by placing all the electrical connections on the backside of the device, reducing packaging size, and increasing reliability.
- Integrated SiPs operating at high power often incorporate high-power vertical components such as IGBT and MOSFETs.
- Power "GaN on sapphire" and "GaN on silicon" components benefit from TSV packages in terms of performance because wire bonding is not able to provide the performance requirements for the interconnects. IR, TI, Maxim, ST Microelectronics, and Panasonic are developing such types of components.
- *3D integrated passive devices (IPDs)*: IPDs offer a wide range of system architecture and partitioning opportunities as well as cost reduction because the 3D IPDs can be manufactured at wafer level as well. Miniaturization is also enabled through 3D integration because the size of the IPD is no longer constrained by the size of the package substrate. Due to shorter connections between the passive and the active components, and the external interface, the level of parasitics is also significantly reduced.
- *Analog ICs*: The use of TSVs is a performance enabler for high-speed analog devices, reducing the bond length for critical nodes; the length of the connections to the chip's bond pad can be reduced from millimeters to microns.

The mobile phone industry is the primary driver for using RF and analog 3D integrated devices and will continue to drive the growth of such packages in the

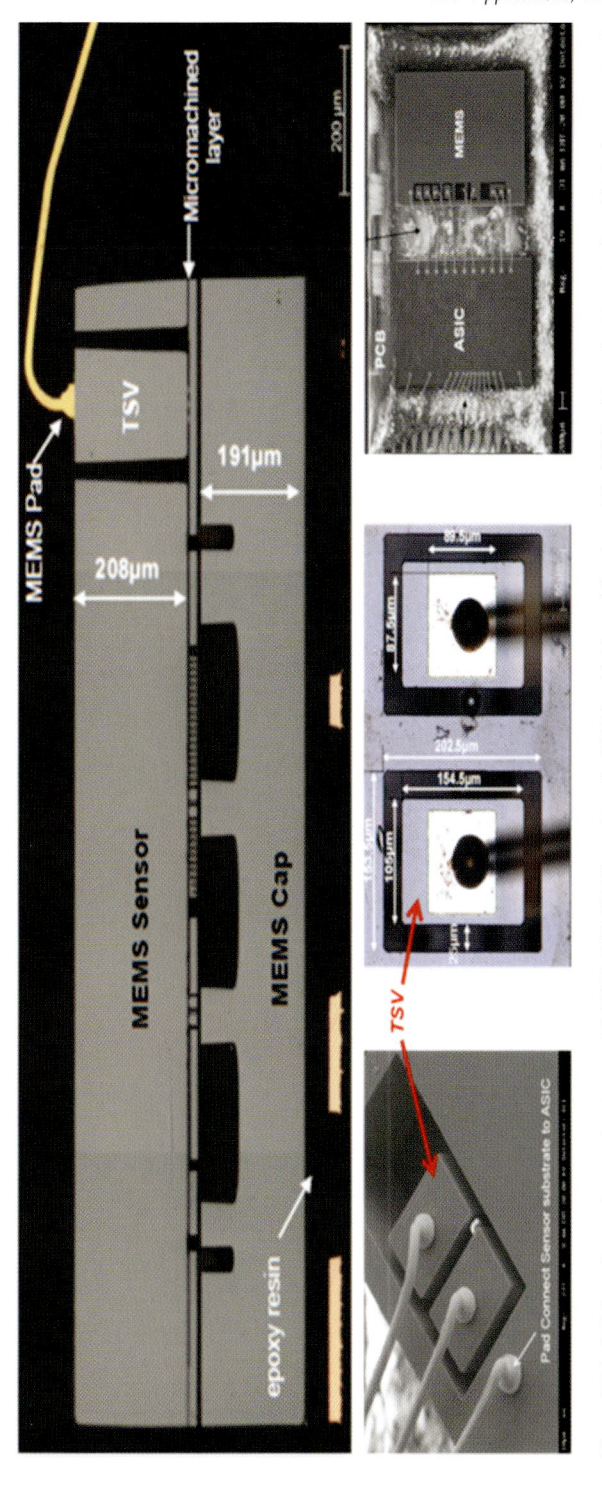

Figure 2.14 ST Microelectronics accelerometer with TSV in MEMS IC introduced in 2011 in a Nokia mobile phone. The TSV in this package was isolated from the MEMS with an air gap [6]. Courtesy of SystemPlus Consulting.

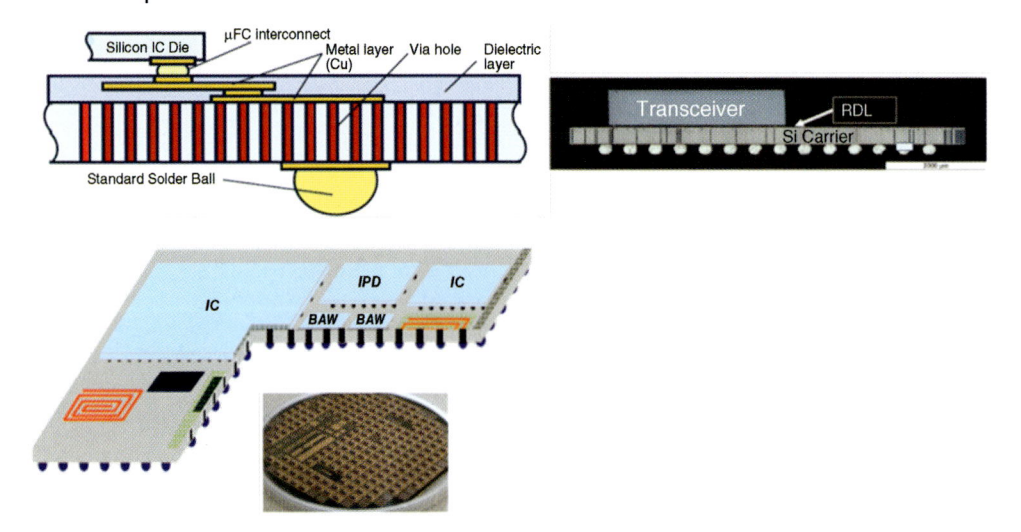

Figure 2.15 3D power/RF/analog passive interposers – 3D IPD at Infineon [6]. Courtesy of Infineon Technologies.

future. By 2017, it will still be the main application by far, followed by electrical and hybrid vehicles with power devices such as IGBT with TSV (ground via).

Newer activities in the industry are integration of 2.5D interposers and TSV technology in high-end applications. Large-die field-programmable gate array (FPGA) devices and ASICs have been commercialized; Xilinx has taken the lead with their first shipment targeting network applications such as smart TVs and set-top boxes. 2.5D glass/silicon interposers have actually emerged as key substrate elements for connecting the nanometer to millimeter worlds in future semiconductor chip packaging assembly (Figure 2.16).

2.5D interposers are viewed as the first generation of 3D ICs. To further reduce costs, many companies are considering alternatives to silicon interposers, such as glass interposers.

A large driver for using 2.5D interposers is system partitioning and ability to integrate at least one logic IC with one or several memory ICs and even mixed signal or analog ICs. It involves "four slices" instead of a single-die 3D SoC repartitioned logic design. This increases CMOS manufacturing yield because of the smaller die size and high-density wiring at the surface of the four-layer copper damascene silicon interposer wafer – leading to a breakthrough in cost versus power consumption versus performance. This type of packaging is expected to progressively replace monolithic SoCs. As illustrated in Figure 2.17, interposer use will be dominated by system and logic partitioning interposers.

For the heterogeneous integration of stacked memories, logic devices, or memory on logic without the use of an interposer, there is still much work left to

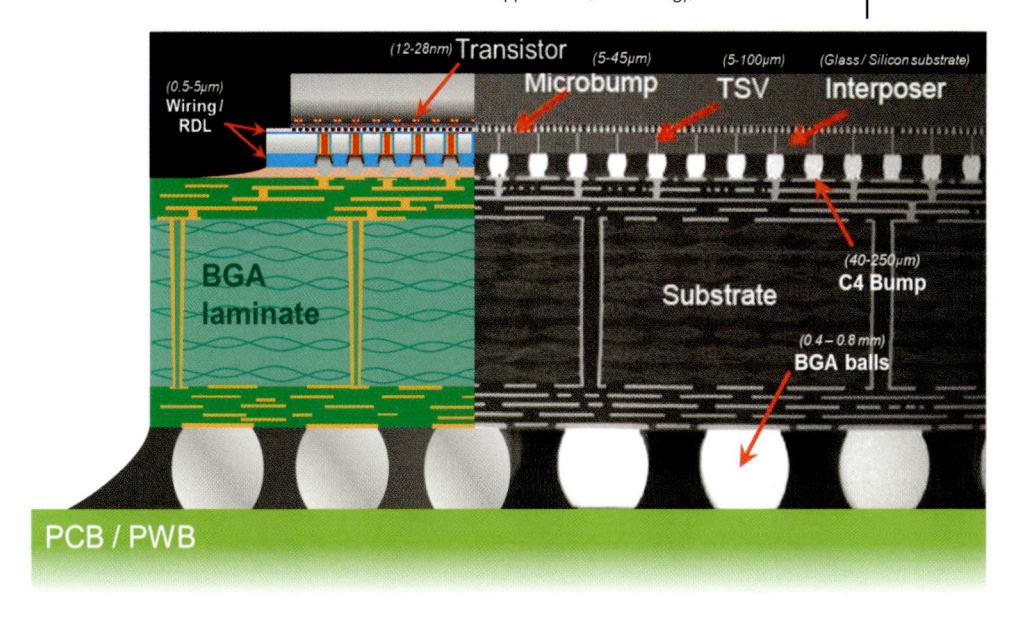

Figure 2.16 2.5D interposer – key enabler for connecting the nanometer to millimeter worlds [6].

do. The industry still needs design tools, thermal management, more work on bonding and debonding solutions, and accepted test methodologies. Once these challenges are overcome, significant growth in 3D IC technology adoption in high-volume manufacturing (HVM) is expected, especially in 3D stacked DRAM and 3D logic SoC applications, followed by CMOS image sensors, power devices, and MEMS (Figure 2.18).

In 2011, the market value of all of the devices that were using TSVs in 3D ICs or 3D WLCSP platforms – including CMOS image sensors, ambient light sensors,

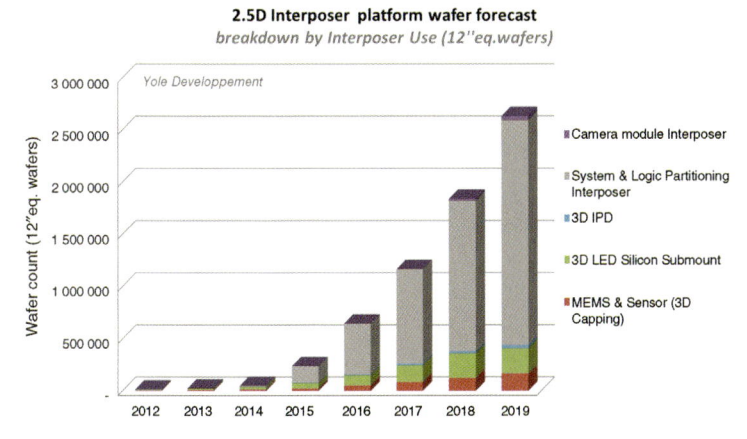

Figure 2.17 Interposer forecast by application.

30 | *2 Key Applications and Market Trends for 3D Integration and Interposer Technologies*

3DIC Platform Middle-End Revenues by 2019 (M. US$)
Breakdown by IC type

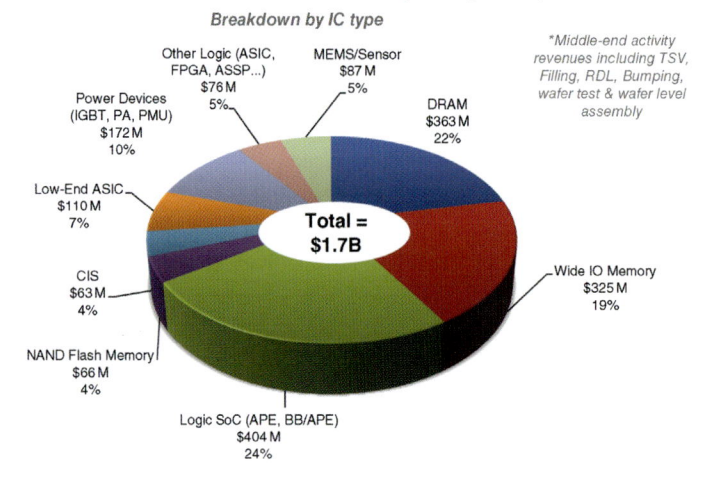

- Other Logic (ASIC, FPGA, ASSP...) $76 M — 5%
- MEMS/Sensor $87 M — 5%
- Power Devices (IGBT, PA, PMU) $172 M — 10%
- Low-End ASIC $110 M — 7%
- CIS $63 M — 4%
- NAND Flash Memory $66 M — 4%
- Logic SoC (APE, BB/APE) $404 M — 24%
- Wide IO Memory $325 M — 19%
- DRAM $363 M — 22%

Total = $1.7B

*Middle-end activity revenues including TSV, Filling, RDL, Bumping, wafer test & wafer level assembly

Figure 2.18 The future 3D IC market is driven by stacked memory and logic SoC applications. The chart shows 3D IC platform middle-end revenues by 2019 in US dollars. Courtesy of Yole Developpement.

power amplifiers, and radio frequency and inertial MEMS – was worth $3.2 billion, according to Yole Developpement. By 2019, the 3D TSV market value is expected to have a fast grow and is estimated to reach close to $40 billion. By 2017, if no further delays in adoption of this technology take place, the 3D TSV market is expected to reach a 9% penetration rate by 2017 (Figure 2.19). 3D IC, which typically uses via-middle TSVs for memory and logic IC stacking, is expected to grow the fastest in wafers as well as in overall value, whereas 3D WLCSP will continue growing at a 18% CAGR. The key factor leading to 3D IC integration is the growing demand for memory-enhanced applications.

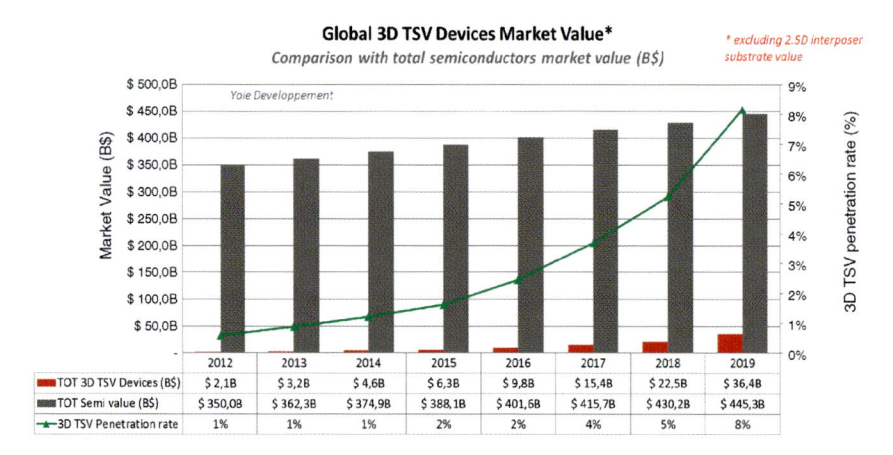

Global 3D TSV Devices Market Value*
Comparison with total semiconductors market value (B$)
*excluding 2.5D interposer substrate value
Yole Developpement

	2012	2013	2014	2015	2016	2017	2018	2019
TOT 3D TSV Devices (B$)	$2,1B	$3,2B	$4,6B	$6,3B	$9,8B	$15,4B	$22,5B	$36,4B
TOT Semi value (B$)	$350,0B	$362,3B	$374,9B	$388,1B	$401,6B	$415,7B	$430,2B	$445,3B
3D TSV Penetration rate	1%	1%	1%	2%	2%	4%	5%	8%

Figure 2.19 3D chip penetration of the IC semiconductor market.

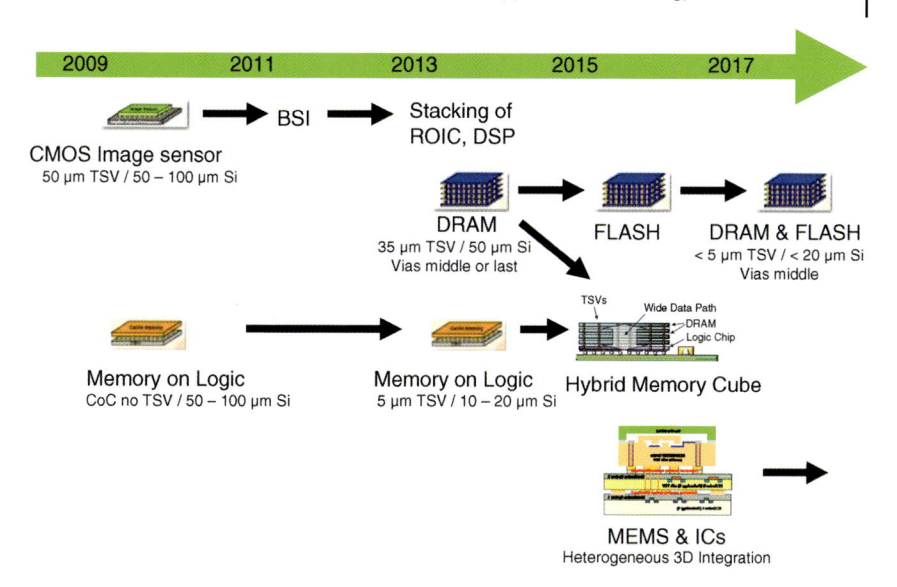

Figure 2.20 Application development for 3D integrated products.

As shown in Figure 2.20, the application roadmap for memory-enhanced products can be outlined in two 3D TSV tracks: stacked memories (homogeneous integration of identical dies of DRAM, flash, etc.) and memory on logic. End of 2013, Hynix started the production of DRAM stacks with copper TSVs and Tezzaron got in the up-ramp phase to high-volume production for 3D ICs, consisting of DRAMs and processors with tungsten TSVs.

The 3D TSV integration of memory stacks with logic – combining these two memory tracks – is announced by the "hybrid memory cube (HMC)" consortium to come to HVM production in 2016 at the latest (www.hybrid.memorycube.org). The leading-edge industrial group was founded by Micron Technology and Samsung Electronics. The HMC will combine high-speed logic process technology with a stack of TSV bonded memories and is expected to deliver dramatic improvements particularly in bandwidth (compared, for example, to DDR3 modules).

Beyond these memory-enhanced applications, 3D integration is recognized as a key technology for heterogeneous products, demanding smart system integration rather than extremely high interconnect densities. According to surveys on revenue forecast by Yole Developpement, MtM products and heterogeneous MEMS/IC systems could become the main driving markets for 3D TSV integration [7]. Many R&D activities worldwide are focusing on heterogeneous integration for novel functionalities [8]. Corresponding 3D integration technologies of such sophisticated heterogeneous products are still in evaluation at several companies and research institutions, such as the European e-BRAINS consortium, with a dedicated focus on reliability issues (www.e-BRAINS.org).

There are significant advances for integrated MEMS systems using 3D integration technologies. By 2020, Yole Developpement predicted: "It is likely that MEMS fabs will have developed internal standard process blocks but it will be fab-specific standard tools." [9]. There have been many publications describing the combination of CMOS with MEMS [10,11]. CMOS MEMS is likely to be restricted to specific applications where MEMS arrays will need very close electronic processing. For all other cases, it will depend on MEMS product cycle time, flexibility, cost, integration, market demand, and power consumption. However, with the development of TSV technology, for a few applications ". . . users are willing to pay the higher cost to get the better performance and smaller size from the shorter connections." Yole Developpement projects that the demand for MEMS with 3D TSV will reach several hundred thousand wafers a year by 2015 [7]. With the maturity of heterogeneous 3D technologies and corresponding low-cost fabrication, a diversity of 3D integrated CMOS-MEMS products will be developed, leading to a further key application track (Figure 2.20). There are high growth areas expected, in particular the Internet of Things (IoT), with the need for such 3D heterogeneous systems.

References

1 Colwell, B. (2013) Moore's Law Dead by 2022. EE Times magazine. Available at http://www.eetimes.com/document.asp?doc_id=1319330 (accessed February 28, 2014).

2 Nowak, M. (2011) Chapter 1: Introduction to high density through silicon stacking technology, in *3D IC Stacking Technology Book*, vol. 2, McGraw-Hill, p. 1.

3 ITRS (2011) Emerging Research Materials, ITRS Interconnect 2011 Edition, p. 11. Available at http://www.itrs.net/Links/2011itrs/2011Chapters/2011ERD.pdf. (accessed February 28, 2014)

4 Yole Developpement (2012) 3DIC & TSV Interconnect Report.

5 Ramm, P., Wolf, J., and Wunderle, B. (2008) Wafer-Level 3D System Integration. *Handbook of 3D Integration*, Wiley-VCH Verlag GmbH, Weinheim, p. 300.

6 Yole Developpement (2013) 3DIC & 2.5D Interposer IP Landscape Report.

7 Mounier, E. (2010) MEMS technology roadmap: demand for smaller, lower cost devices drives major technology trends for next decade; www.yole.fr.

8 Fernández-Bolaños, M. and Ionescu, A. (2010) Heterogeneous integration for novel functionality. Proceedings of the IEEE International 3D System Integration Conference – 3DIC 2010, Munich (eds P. Ramm and E. Beyne).

9 Yole Developpement (2011) MEMS: Trends in MEMS Manufacturing & Packaging, Technology Report.

10 Lapisa, M., Stemme, G., and Niklaus, F. (2011) Wafer-level heterogeneous integration for MOEMS, MEMS, and NEMS. *IEEE Journal of Selected Topics in Quantum Electronics*, 17 (3), 628–644.

11 Ramm, P., Klumpp, A., Weber, J., and Taklo, M.M.V. (2010) 3D system-on-chip technologies for More than Moore systems. *Microsystem Technologies*, 16 (7), 1051–1055.

3
Economic Drivers and Impediments for 2.5D/3D Integration

Philip Garrou

3.1
3D Performance Advantages

3D performance advantages can be summarized by looking at Figure 3.1 in which Samsung compares a conventional package-on-package (PoP) solution and an equivalent 3D solution and concludes that the 3D solution offers a 35% reduction in package size, a 50% reduction in power consumption, and an 8× increase in bandwidth [1]. These are significant performance advantages especially for mobile devices.

In volume 1, Chapter 2 of the *Handbook of 3D Integration*, we took a look at the basic technical drivers for 3D integrated circuits (ICs) [2]. While all of those justifications remain, a closer look at the economic motivations and roadblocks for 2.5D and 3D ICs is warranted at this time.

3.2
The Economics of Scaling

For years the IC industry has been driven by Moore's law – the notion that by regularly reducing the minimum dimensions of its manufacturing processes, the functionality that fits on a single die will double every 18–24 months. As minimum dimensions have shrunk from 45 to 28 and now to 20 nm, it has become increasingly difficult to design with those processes, and increasingly expensive to manufacture them. The shift to 20 nm is introducing new technical difficulties such as a shift to nonplanar finFET transistor structures and more complex double-patterning lithography for critical layers.

While scaling can and will continue to at least the 7 nm node, it will not be easy or cheap. Iyer of IBM has pointed out that cost per transistor has begun to saturate; Figure 3.2 shows that simply moving to the next node is not delivering the economic gains that it used to [3].

Handbook of 3D Integration: 3D Process Technology, First Edition.
Edited by Philip Garrou, Mitsumasa Koyanagi, and Peter Ramm.
© 2014 Wiley-VCH Verlag GmbH & Co. KGaA. Published 2014 by Wiley-VCH Verlag GmbH & Co. KGaA.

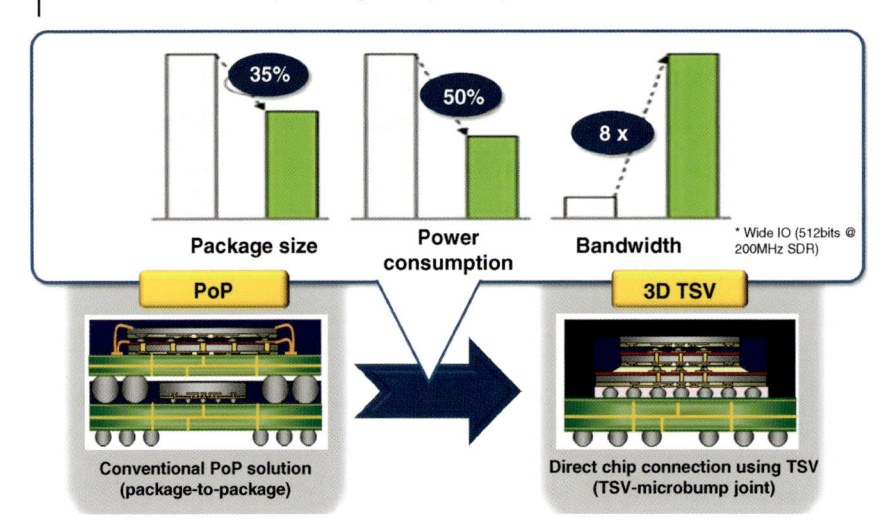

Figure 3.1 Comparison of PoP to 3D packaging solutions for mobile devices [1].

Similarly, Figure 3.3, reported by Jones of IBS, indicates that the lower costs resulting from scaling (i.e., more chips per wafer and more transistors per chip) cross over around the 22 nm node and because of the excessive costs at that node actually produce a higher transistor cost than the previous generation [4].

3.3
The Cost of Future Scaling

Everyone understands the "Moore's law coming to an end" arguments. Exactly when it will happen is less relevant than the fact that it is happening or has already happened for many mid-tier IC fabricators.

While some top-tier fabs/foundries will find a way to move forward past 22 nm to 14 nm and beyond, the vast majority will not. This is not because the technology will not be available, but rather because it will be too expensive. Figure 3.4 shows

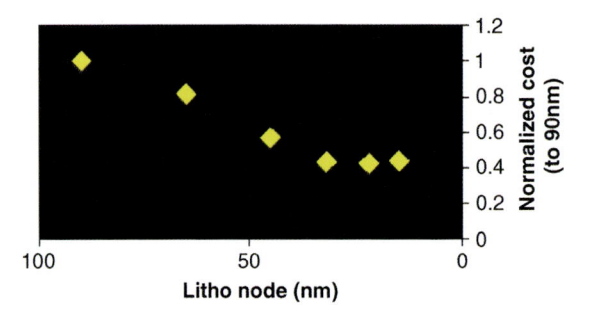

Figure 3.2 Cost per transistor [3].

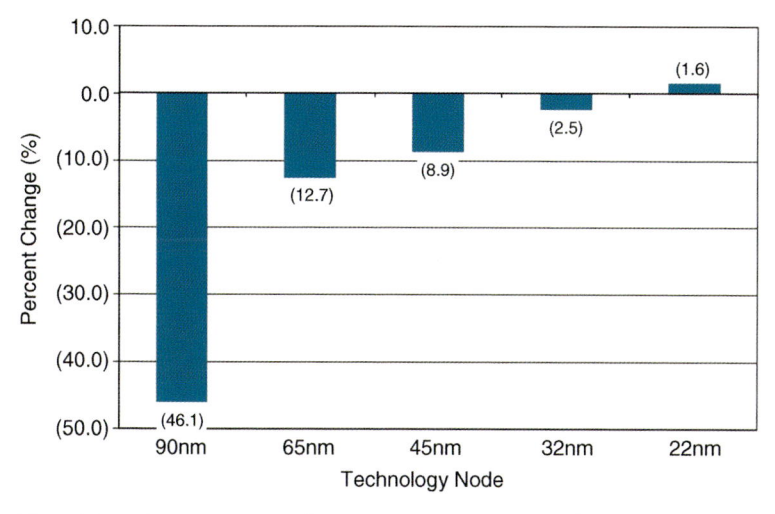

Figure 3.3 Reduction in semiconductor costs versus previous node [4].

GlobalFoundries' account of the rising costs incurred by building a state-of-the-art IC fab.

As R&D costs exceed $1 billion and construction costs exceed $6 billion, there are not many integrated device manufacturers (IDMs) or foundries that can afford to participate. If we look at IC sales for 2012 and eliminate the fabless companies (denoted by **), we find few companies left that have the financial wherewithal to undertake building a fab costing $6 billion or more (Table 3.1).

Gartner has looked at design costs by node and concluded that a 22 nm design will cost upward of $150 million (Figure 3.5). The volume of production needed to absorb >$150 million nonrecurring engineering (NRE) design costs will severely limit the number of HVM products that can be manufactured at that node.

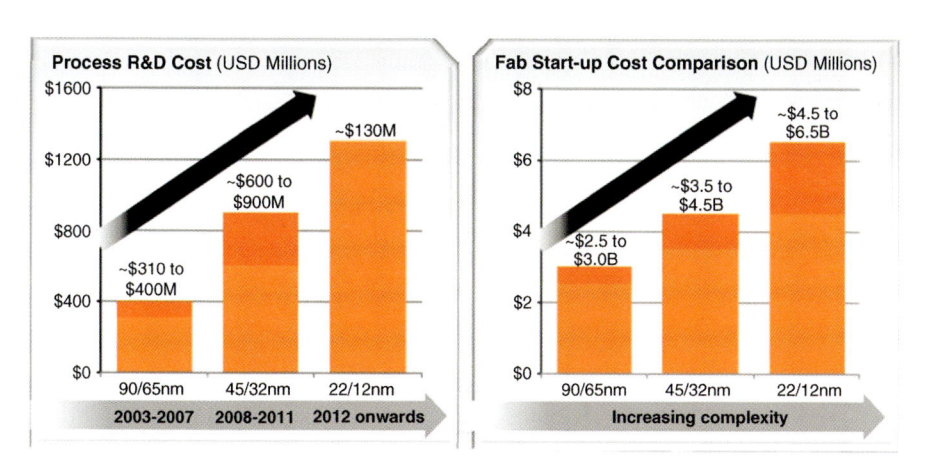

Figure 3.4 The cost of future scaling [5].

3 Economic Drivers and Impediments for 2.5D/3D Integration

Table 3.1 IC sales, 2011/2012 [6].

Company	Headquarters	1Q12 Tot Semi
Intel	United States	11 874
Samsung	South Korea	7067
TSMC[a]	Taiwan	3526
Qualcomm[b]	United States	3059
Toshiba	Japan	3255
TI	United States	2934
SK Hynix	South Korea	2115
Micron	United States	2102
ST	Europe	1999
Broadcom[b]	United States	1770
Renesas	Japan	2363
GlobalFoundries[a]	United States	1170
Infineon	Europe	1292
AMD[b]	United States	1585
NXP	Europe	969
Sony	Japan	1514
Nvidia[b]	United States	935
Freescale	United States	910
UMC[a]	Taiwan	804
Fujitsu	Japan	1216

a) Foundry.
b) Fabless.

A look at logic fab construction by IDMs, in Figure 3.6, supports these conclusions. Over the last decade, we have seen a steady drop in the number of participants at the next node.

It is expected that in the future systems will, in a lot of cases, not be differentiated by the latest node chip, but rather by how they are interconnected and packaged.

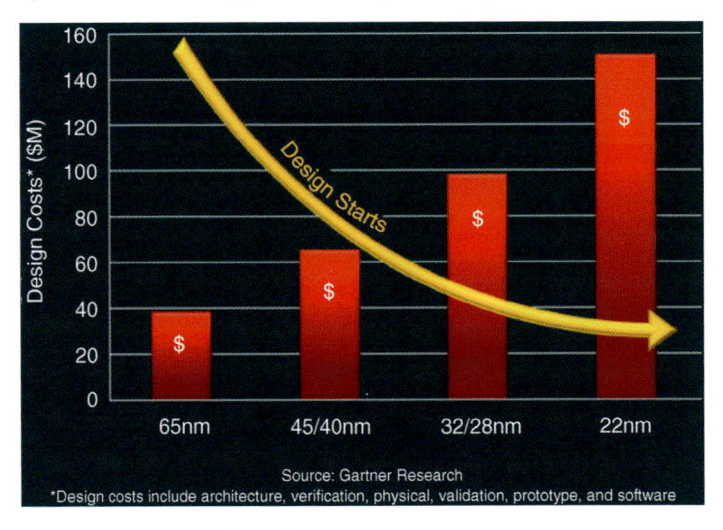

Figure 3.5 Design costs by node [7].

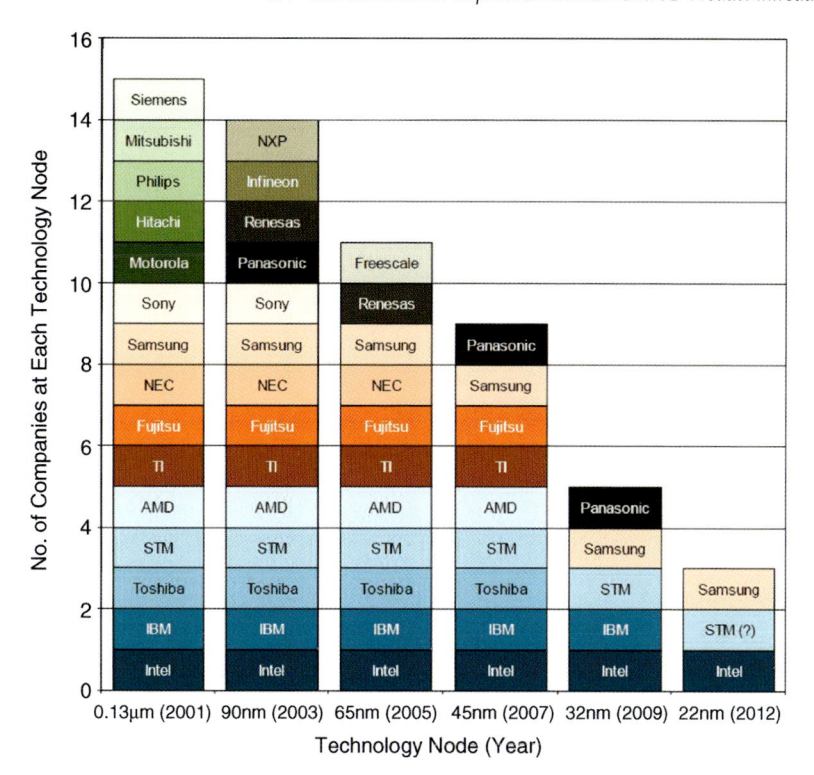

Figure 3.6 Logic IDM fab construction [4].

So called "More than Moore" [8] or "heterogeneous integration" will develop 2.5D SiP and 3D vertical structures, which have not only memory and logic incorporated into the stacks but also DSPs, sensors, displays, power sources, Rf, and so on. Such technologies will all revolve around through-silicon via (TSV) technology.

3.4
Cost Remains the Impediment to 2.5D and 3D Product Introduction

While the cost of the latest node fab or the latest node design will preclude many potential customers, the cost of 2.5D interposers or full 3D IC stacked chip sets is currently acting as a barrier to their introduction into today's proposed products.

All of the first-generation 2.5D/3D products contain memory, either stacked or in conjunction with another function. As of late 2013, the memory stacks and/or the silicon interposers remain too expensive for incorporation into low-cost products.

As shown in Table 3.2, Ma of SPIL points out that cost remains the major concern for product launches for both 2.5D and 3D ICs [9]. Ma contends that the entry of OSATs such as SPIL (who announced their entry as a high-density silicon

Table 3.2 Cost is the major concern for 2.5D/3D IC product launches [9].

2.5D/3D IC	Package content	Application	Major concerns for product launching
2.5D	Logic/(on) TSI (homogeneous integration)	FPGA/high-speed logic	Cost
2.5D	Logic + analogs + · · ·?/TSI (heterogeneous integration)	System partitioning (and aggregation)	Cost and TSI design
2.5D	Logic (processors) + DRAM stacks/TSI	Computing	Cost, DRAM stack availability
3D	Memory stack	Wide IO DRAM, HMC	No consensus on mainstream (versus DDR4) yet, cost
3D	Wide IO DRAM on logic	Mobile communication	Cost/design/thermal issues, DRAM availability

interposer supplier in the summer of 2013) will help drive down prices, which up to now have been controlled by the foundries.

3.4.1
Required Economics for Interposer Use in Mobile Products

Qualcomm's Nowak has cautioned that interposers would add substantial cost and as such probably would not be a broadly accepted solution for low-cost mobile products, which would prefer straight 3D stacking. In response to the Qualcomm statement, TSMC's Yu stated that while the addition of an interposer adds cost to the overall product, " . . . this [2.5D] solution also offers cost savings by reuse of IP and separating digital and analog circuitry and allowing partitioning of costly SoC" and "that this could make it the lowest-cost solution" [10].

3.4.2
Silicon Interposer Pricing

At the 2012 International Interposer Conference, Nowak of Qualcomm and Vordharalli of Altera both pointed to the necessity for interposer costs to reach 1 cent mm^{-2} for wide acceptance in the high-volume mobile arena. Nowak proposes that a standard 200 mm^2 interposer should cost \$2 [11].

Ramaswami of Applied Materials has published a cost analysis that shows 300 mm interposer wafer costs of \$500–650 per wafer. His cost analysis indicates

3.4 Cost Remains the Impediment to 2.5D and 3D Product Introduction

that the major cost contributors are damascene processing (22%), front pad and backside bumping (20%), and TSV creation (14%). Ramaswami notes that the dual damascene costs have been optimized for front-end processing, so there is little chance of cost reduction there, and the cost reduction focus should be on backside bumping and TSV creation.

Since one can produce ~286 dies (200 mm^2) on a 300 mm wafer, Ramaswami's $575 per wafer cost (midpoint cost) results in the required $2/200 mm^2 silicon interposer, giving hope that such pricing is possible.

Yole Developpement has published cost modeling calculations for the Xilinx Vitrex 7 2000T interposer, the first 2.5D product to be available in the market [12]. This 100 μm thick silicon interposer is 31 × 31 mm^2, has 12 μm TSV, 3 Cu damascene layers, and uses 65 nm (top layer) design rules for routing. Yole calculates a 300 mm "Xilinx interposer" wafer cost of $683. Fifty-six good dies per wafer results in a manufacturing cost of $12 per interposer. Yole breaks out manufacturing cost as shown in Figure 3.7.

Using these assumptions, if we again assume 286 dies (200 mm) per 300 mm wafer, this would result in $2.38 per interposer (at 100% yield).

So are we close to the needed mass production costs? Is the problem really price and not cost? Are the yields simply not good enough? Or are these calculations simply way off base? Hopefully, in the next few years we will better understand these economics.

Some remain convinced that only panel-size formats (i.e., flat panel glass or laminates) can deliver the economics necessary to make 2.5D packaging mainstream; most agree that while glass panels and even possibly advanced laminates present interesting possibilities for low-cost future products, they currently cannot meet density requirements and are in the earliest stages of R&D (see Chapter 4 for a complete discussion of alternative interposer materials).

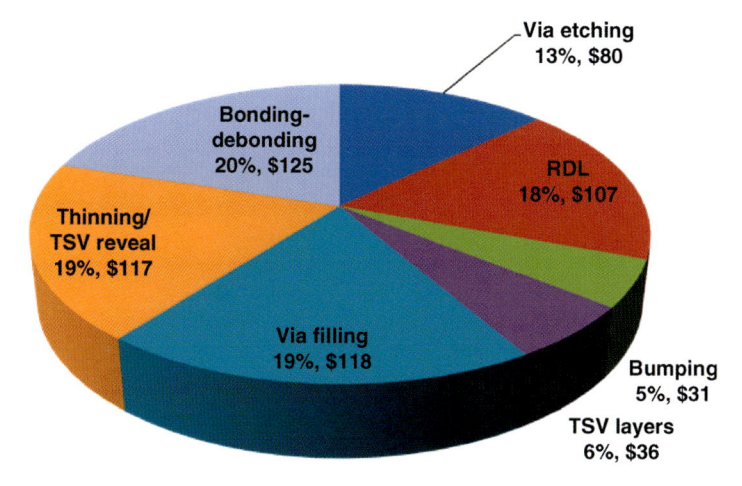

Figure 3.7 Yole Developpement cost breakout for Xilinx interposer [12].

References

1 Samsung (2012) 3D TSV technology & wide IO memory solutions. Design Automation Conference. Available at http://www.samsung.com/us/business/oem-solutions/pdfs/Web_DAC2012_TSV_demo-ah.pdf (accessed December 31, 2013).
2 Garrou, P., Vitkavage, S., and Arkalgud, S. (2008) Chapter 2: Drivers for 3D integration, in *Handbook of 3D Integration*, vol. 1 (eds P. Garrou, C. Bower, and P. Ramm), Wiley-VCH Verlag GmbH, Weinheim.
3 Iyer, S.S. (2013) Orthogonal scaling to fill today's fabs in the future. ConFab Conference, Las Vegas, NV.
4 Jones, H. (2010) Cost of participation in semiconductor industry increasing: impact after 32/28nm. SEMI ISS, Half Moon Bay, CA.
5 Walker, J. (2011) The market challenge of manufacturing convergence: going vertical. SEMICON Taiwan, 3D IC Technology Forum.
6 IC Insights (2013) Eleven companies move up in 1Q13 top 20 Semi supplier ranking. Available at http://www.icinsights.com/news/bulletins/Eleven-Companies-Move-Up-In-1Q13-Top-20-Semi-Supplier-Ranking/ (accessed December 31, 2013).
7 Garrou, P. (2013) IFTLE 148: The future of Packaging: a look from 50,000 ft., Solid State Technology, Insights from the Leading Edge, May 25, 2013. Available at http://www.gslb.cleanrooms.com/index/packaging/packaging-blogs/ap-blog-display/blogs/ap-blog/post987_2555779367382983596.html (accessed December 31, 2013).
8 Arden, W. *et al.* (2010) More than Moore. White paper, ITRS. Available at http://www.itrs.net/Links/2010ITRS/IRC-ITRS-MtM-v2%203.pdf (accessed December 31, 2013).
9 Ma, M. (2013) The expanding role of OSATs in the era of system integration. ConFab Conference, Las Vegas, NV.
10 Solid State Technology (2012) Interposer supply/ecosystem examined at IMAPS device packaging. Available at http://www.electroiq.com/articles/ap/2012/03/interposer-supply-ecosystem-examined-at-imaps-device-packaging.html (accessed December 31, 2013).
11 Solid State Technology (2012) Will the $2 interposer be silicon or glass? Available at http://www.electroiq.com/articles/ap/2012/11/will-the-2-dollar-interposer-be-silicon-or-glass.html (accessed December 31, 2013).
12 Micronews Media (2012) Are silicon interposers "luxury" solutions? Available at http://www.i-micronews.com/upload/pdf/3D%20Packaging_May2012_AC.pdf (accessed December 31, 2013).

4
Interposer Technology

Venky Sundaram and Rao R. Tummala

4.1
Definition of 2.5D Interposers

In microelectronic packaging, an interposer is an electronic substrate that facilitates interconnection between the fine-pitch I/Os at the die level on the top side of the interposer to the coarser dimensional features on the package on the bottom side of the interposer. An interposer has through-via connections, sometimes referred to as through-silicon vias (TSVs) or through-package vias (TPVs), which provide electrical connections between the components on the top side to those at the bottom. The interposer has multiple wiring layers called redistribution layers (RDLs), with one or more layers of RDLs on each side of the interposer substrate.

Figure 4.1 shows a schematic cross section of a traditional package on the left and a double-sided interposer with through-vias mounted on a package on the right. The wiring layers, along with the TSVs or TPVs, help interconnect 2D and 3D ICs to the ball grid array (BGA) package or printed wiring board (PWB) on the bottom side. Traditionally, the chip-package community defined a package as the bridge between the chip pitch and the board pitch. Now the interposer has become the bridge between the chip and package. Thus, an interposer can facilitate chips side by side with interconnections along the same plane of the interposer, or structures wherein the components are attached on both sides of the interposer. The 2.5D interposers bear a strong resemblance to the large multichip modules (MCMs) of the 1990s, but are differentiated by their pitch, I/O density, and TSV. The typical thickness of the interposer is 50–300 μm with a through-via pitch of 10–200 μm.

The "side-by-side" configuration of a high wiring density interposer is shown in Figure 4.2a, where multiple ICs are placed side by side and interconnected by wiring on the top side of the interposer. An alternate structure is shown in Figure 4.2b, where ICs are placed on both the top and bottom sides of the thin interposer and interconnected by TSV in the interposer to achieve high-bandwidth interconnections.

Handbook of 3D Integration: 3D Process Technology, First Edition.
Edited by Philip Garrou, Mitsumasa Koyanagi, and Peter Ramm.
© 2014 Wiley-VCH Verlag GmbH & Co. KGaA. Published 2014 by Wiley-VCH Verlag GmbH & Co. KGaA.

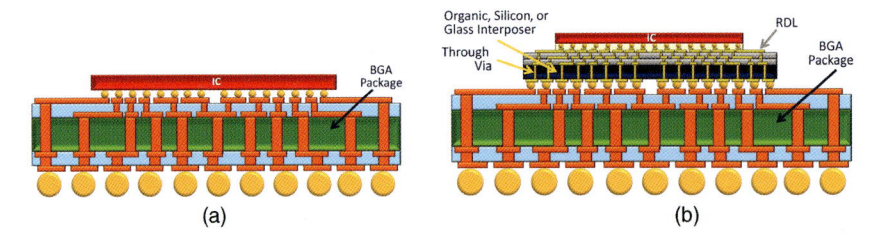

Figure 4.1 Comparison of (a) conventional IC package interconnected to PWB and (b) interposer with ICs on the top side and BGA package on the bottom side for interconnection to PWB. Courtesy of GT PRC.

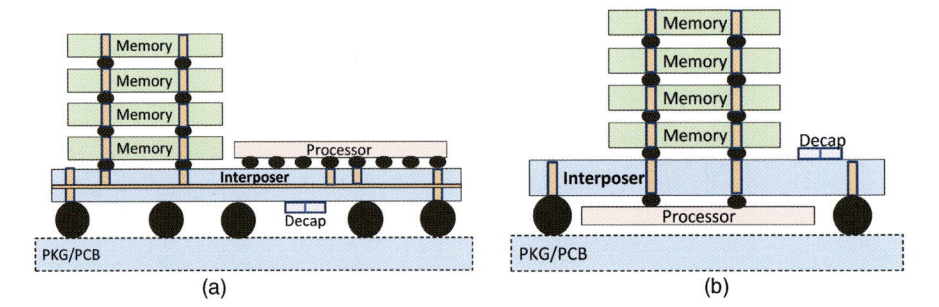

Figure 4.2 Schematics illustrating the different interposer structures.

The interposer can be either attached to a BGA package, common for larger interposer sizes, or directly mounted to a motherboard or PWB, for small interposer sizes.

4.2
Interposer Drivers and Need

The primary driver for interposer technology is the increased I/O density for chip-to-chip interconnections, driven by the scaling down of transistor size in accordance with Moore's law. The I/O pitch gap between ICs at 30–225 μm pitch and PWBs at 500–1000 μm pitch created the need for the original package. Similarly, the advancement of IC nodes to below 32 nm, with a roadmap down to 7 nm front-end-of-line (FEOL) design rules, has created a gap between the on-chip I/O pad pitch need of 10–50 μm and the current BGA package pitch at 100–150 μm. This is the primary driver for the development of high-density and fine-pitch interposers as shown in the interconnections roadmap in Figure 4.3.

The escalating need for high-bandwidth data consumption in both mobile devices and cloud computing servers is the single biggest driver for I/O density and fine-pitch interconnections between logic and memory chips. Although 3D IC stacking with TSVs offers the potential for high bandwidth and low power

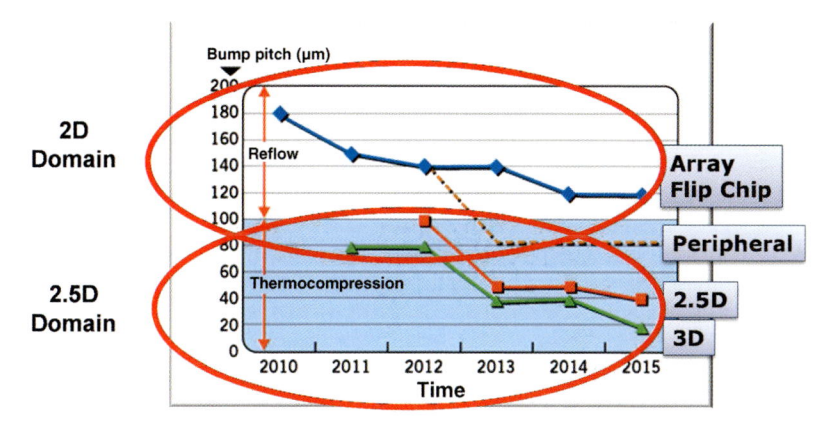

Figure 4.3 Fine-pitch interconnections needed for 2.5D and 3D interposers (McCann, D. (2012) Interconnection Roadmap, private communication.).

consumption, other challenges such as thermal dissipation, test, yield, and standardization of I/O positions in ICs have fueled the use of 2.5D interposers with logic and memory ICs on high-density silicon or other interposers.

Limits of organic packages: The challenges in standard organic packages include

- I/O density limitation due to poor dimensional stability,
- warpage during fabrication and assembly due to low elastic modulus,
- poor heat dissipation due to low thermal conductivity, and
- on-chip interlayer dielectric (ILD) and off-chip interconnection reliability due to high coefficient of thermal expansion (CTE).

There are three major types of applications driving the development of high-density interposers (Figure 4.4); (a) logic to memory bandwidth improvement; (b) "die breakup" where a single large IC is broken into multiple smaller ICs and reconnected through wiring layers on the interposer, leading to potential higher yield at smaller IC nodes; and (c) the integration of multiple heterogeneous ICs to form complex multifunctional subsystems, similar to current SiP packages.

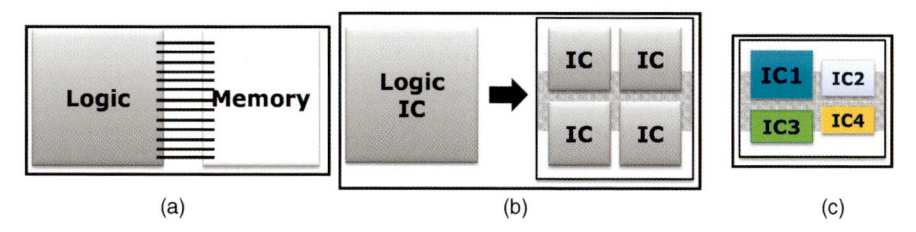

Figure 4.4 Three major application drivers for 2.5D interposers: (a) high-bandwidth logic–memory interconnections, (b) die breakup for higher yield, and (c) heterogeneous integration.

4.3
Comparison of Interposer Materials

Table 4.1 provides a qualitative comparison of interposer materials for electrical, mechanical, thermal, and physical properties as well as for ease of through-via processing, and relative cost. Besides providing high I/O interconnections exhibiting good signal and power integrity, they need to have mechanical robustness, chemical inertness, and high thermomechanical reliability.

Interposer materials have evolved from ceramics in the 1980s to organics. Ceramic substrates are limited in I/O density and suffer from high processing cost. Organic substrates were developed to replace ceramics, and are widely used; however, they are limited in both line and space density and via density due to their high CTEs. Silicon interposers provide excellent dimensional stability leading to fine-pitch and small TSVs for package and board-level interconnections. Their disadvantage, however, lies in the high process cost that arises from required back-end-of-line (BEOL) processing. In addition, the lower electrical resistivity of silicon makes it susceptible to signal loss and leakage in applications such as RF. To address these challenges, Georgia Tech and others have examined alternative interposer materials [1–3].

Glass, as shown in Table 4.1, has high electrical resistivity, large panel availability similar to organics, and the excellent dimensional stability and surface finish similar to silicon. The key advantages of using glass are (a) high electrical resistivity, leading to lower signal insertion loss and smaller crosstalk; (b) excellent dimensional stability, resulting in better layer-to-layer via registration with smaller capture pads; (c) tailorable CTE (3–10 ppm/°C), providing flexibility in mechanical

Table 4.1 Material property comparison between different interposer materials.

Characteristic	Ideal properties	Glass	SC Si	Poly Si	Organic	Metal	Ceramic
Electrical	• High resistivity • Low loss						
Physical	• Smooth surface finish • Large area availability • Ultrathin						
Thermal	• High conductivity • CTE matched to Si						
Mechanical	• High strength • High modulus						
Chemical	• Resistance to process chemicals						
Manufacturable high TPVs	• Low-cost via formation and metallization						
Cost/mm²	• At 25 μm pitch						

Legend: Good / Fair / Poor

design to minimize stress at various interfaces within the system; (d) chemical inertness; and (e) high thermal stability.

4.4 Silicon Interposers with TSV

Silicon interposer processing is described in various chapters throughout this book and will not be discussed in detail in this chapter. Silicon interposers were initially developed by IBM and Bell Labs in the 1980s using Cu-polyimide thin-film wiring on silicon wafers (Figure 4.5) [4,5]. These non-TSV-containing silicon modules (thin-film MCMs) saw limited commercial use due to their high cost compared to the ceramic and organic alternatives at that time.

The first commercial implementation of high-density silicon interposers is based on wafer silicon with TSVs of 10 μm diameter and 100 μm depth using the TSMC-developed chip-on-wafer-on-substrate (CoWoS) technology for Xilinx field-programmable gate array (FPGA) modules [8].

In general, Bosch etching is the standard process used for forming blind vias in thick silicon wafers. This is followed by liner, barrier, and seed layer deposition by chemical vapor deposition (CVD) or other dry processes, and finally copper via fill by electrochemical deposition (ECD). After chemical–mechanical polish (CMP) for planarization, high-density wiring layers are fabricated on the front side of the silicon interposer using BEOL SiO_2 Cu damascene processes. A carrier wafer is bonded to the front side using a temporary bonding adhesive and TSV backside reveal is performed by backgrinding and final polish. Backside RDLs are then fabricated either by BEOL processing or more commonly by wafer-level Cu-polymer RDL processes. Since TSV materials and process technologies have been extensively covered elsewhere in this book, this section only highlights the application of silicon wafer with TSV for passive interposers. Figure 4.6 shows a cross-sectional schematic and typical design rules for BEOL Si interposers.

The main benefit of using TSV silicon interposers (versus 3D stacking) for multidie integration is that TSVs are only needed in the silicon interposer, thus minimizing the cost and manufacturing complexity of TSV integration in advanced

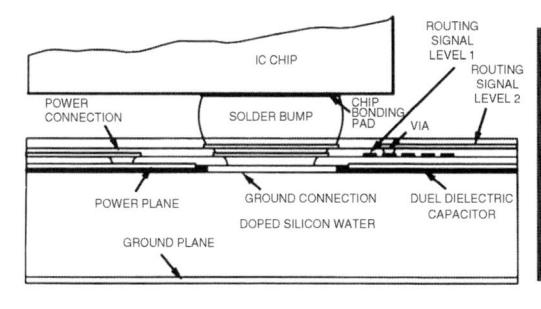

Figure 4.5 Historic development of silicon interposers without TSVs [6,7].

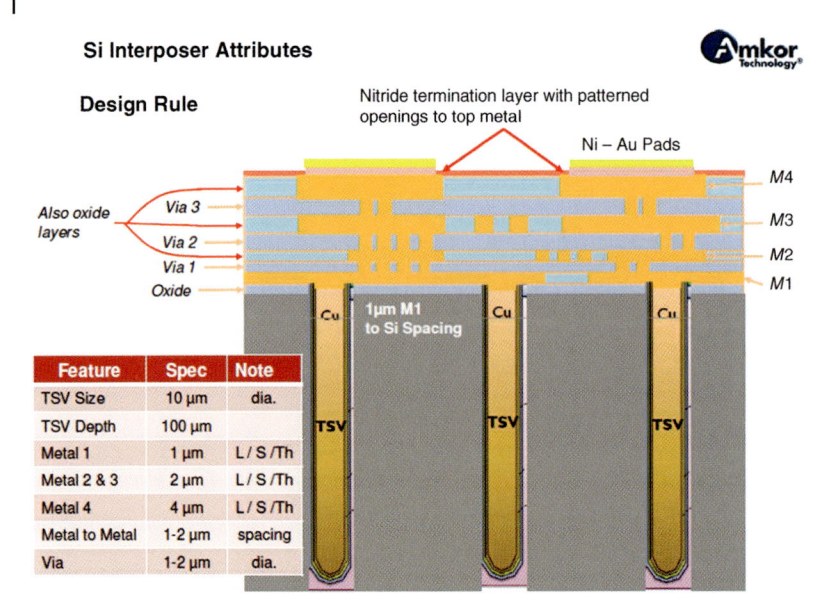

Figure 4.6 Typical design rules and cross section of TSV silicon interposer [9].

logic wafers and associated thinning challenges. Figure 4.7 illustrates the first commercial example of the application of 2.5D silicon interposers: a schematic and a scanning electron microscope (SEM) cross section of Xilinx Virtex 7, which is made of multiple slices (die) of FPGAs using a TSMC 28 nm HPL process with 6.8B transistors, stacked onto a passive silicon interposer fabricated using a TSMC 65 nm process with four layers of metal. Only the silicon interposer, which is manufactured using a mature 65 nm process, goes through TSV processing. The 65 nm design rules result in a factor of 100–1000× smaller features than conventional package substrates, thus allowing for orders of magnitude higher interconnect density, dramatically reducing latency and power consumption [10].

There are many other applications for wafer-based silicon interposers with TSV and fine-pitch RDL. The first and most compelling application is the high-

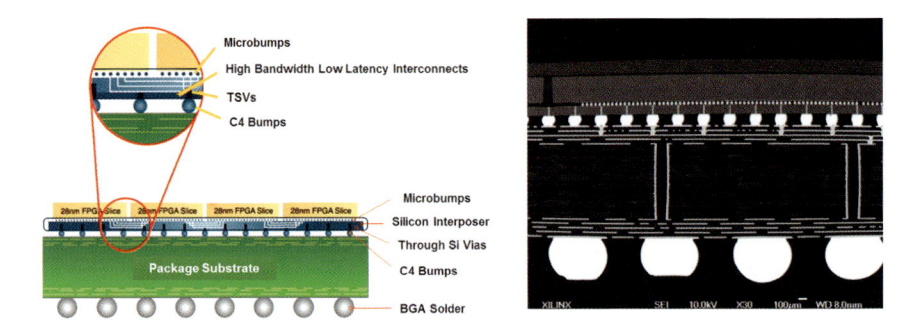

Figure 4.7 Advanced logic onto silicon interposer with 2.5D configuration. *Source*: Xilinx.

bandwidth logic-to-memory interconnections needed for smartphones and tablets. Currently used wire-bonded package-on-packages are limited in bandwidth ($<8\,\text{GB s}^{-1}$) because of low I/Os and long interconnection paths that increase signal delay. This approach will not significantly scale to higher bandwidth without significant increase in power consumption. The 2.5D interposer approach scales to much higher bandwidth, due to the higher number of I/O channels within a small form factor and significant reduction in signal delay per channel. As an example, with 2–4 μm lines and spaces for RDL, combined with 50 μm pitch die-to-interposer interconnections, bandwidth in excess of $25\,\text{GB s}^{-1}$ can be achieved. Such a high bandwidth is achieved at significant power savings (up to 50%) over a wide-I/O bus configuration. Another application is in the high-performance computing and networking environments, where bandwidth capacities of the order of a few TB s^{-1} are driven by memory intensive multicore processors, or multiple logic devices such as ASICs and FPGAs. Bandwidth optimization, taking into account crosstalk and resistance at high frequencies, may indicate that 2 μm line RDLs would be appropriate to maximize performance. Thus, 2.5D interposers built with low-loss substrates with TSVs and RDLs can be designed to achieve very high bandwidth, perhaps as high as $2.5\,\text{TB s}^{-1}$. In summary, the 2.5D interposer at 2 μm ground rules is expected to achieve high bandwidth at low cost and at low power consumption without requiring TSV in the logic ICs in the short term.

In addition to the high cost, a significant challenge for silicon interposers is the high electrical loss of TSVs with thin oxide liners. Recent activity has focused on alternate liner materials and structures to decrease the electrical loss and improve the reliability of TSVs in silicon interposers. The main focus has been on thicker liners using low-modulus polymer dielectrics by spin coating or vacuum lamination of dry film polymers. Figure 4.8 shows the approach taken by ST Microelectronics to apply polymer liners to TSVs, demonstrating lower electrical loss and higher reliability.

An interesting alternate approach to reducing the TSV loss is the use of air as dielectric in through-silicon hole (TSH) structures, developed by ITRI Taiwan, as shown in Figure 4.9. The copper interconnection between the two chips on the top and bottom of the silicon interposer is accomplished by a modified wire-bonding technique.

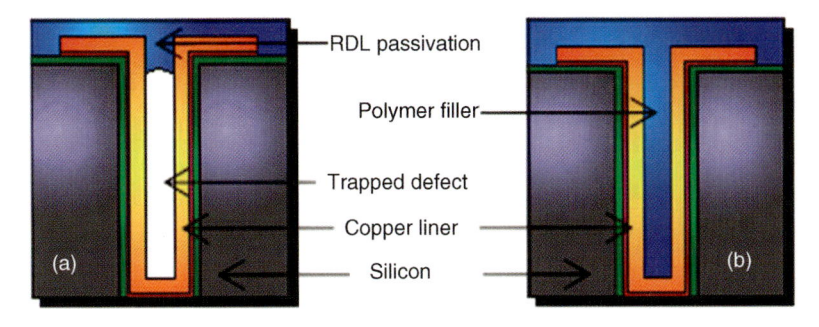

Figure 4.8 Polymer-lined TSV for silicon interposers [11].

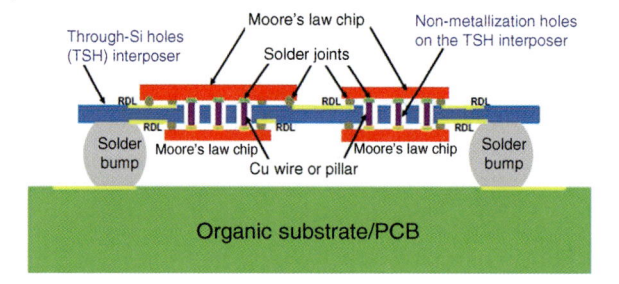

Figure 4.9 TSH with air as dielectric liner demonstrating lower electrical loss than TSVs [12].

4.5
Lower Cost Interposers

The increased cost focus driven by smart mobile devices and Internet of things (IoT) applications has resulted in the proposal of several low-cost interposer alternatives to silicon, primarily based on large panel processing as the primary cost reduction driver, as shown in Figure 4.10.

4.5.1
Glass Interposers

Glass has excellent dimensional stability, high electrical resistivity, excellent chemical inertness, CTE matched or tailored to silicon die and PWB, and availability in thin (30 μm), large panel formats [14]. Glass has two major fundamental limitations, however: its brittleness and its low thermal conductivity. It also has major engineering challenges: (a) handling of ultra-thin glass panel substrates, (b) small TPVs produced at high throughput and fine pitch, and

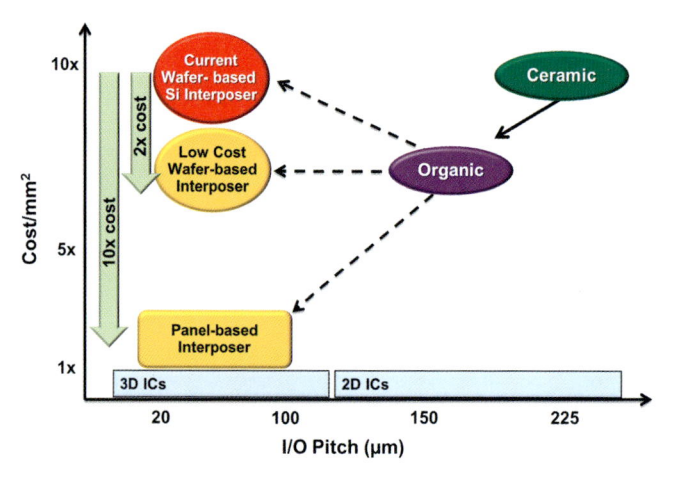

Figure 4.10 Cost-reduction potential using panel-based processes [13].

(c) reliability of copper TPVs in the glass interposer due to large CTE mismatch and the brittleness of glass.

4.5.1.1 Challenges in Glass Interposers

There are few fundamental and engineering challenges that need to be addressed in forming glass interposers. This section outlines four key challenges and discusses potential solutions to overcome these technical challenges.

a) Formation of ultrasmall via holes and thin glass handling without damaging the glass surfaces.
b) TPV metallization with copper and fine-line RDL.
c) Copper TPV reliability.
d) Thermal dissipation.

4.5.1.2 Small-Pitch Through-Package Via Hole Formation and Ultrathin Glass Handling

The high Young's modulus and chemical inertness of glass make it difficult to machine through-holes at small pitch. DRIE based on SF_6 plasma has been known and used to form vertical feedthroughs in glass (Figure 4.11) at larger pitch (>300 μm) for MEMS applications. The etch rates in glass are slower (~1 μm/min) compared to TSV formation in CMOS grade silicon. Wet etching using HF-based chemistry or KOH yields higher etch rates (~10/min); however, the isotropy of etch profile makes it difficult to machine TPVs at aspect ratios greater than five [16]. Other potential approaches that can be employed include (a) mechanical machining, (b) photo-via formation in photosensitive glass, (c) sand blasting, and (d) laser ablation.

Mechanical machining, using proprietary technology, has been used to form via diameters down to 50 μm in 100–300 μm thin glass [17]. It is a serial process, that is, TPVs are drilled sequentially one at a time. The mechanical machining, by its very nature, may induce damage on via wall surfaces, especially around the periphery of the through-holes. Photo-via formation is an attractive technology that was developed by Corning glass in the 1950s to form glass first and then photo expose to form regions that crystallize [18]. So it has the potential to form all the

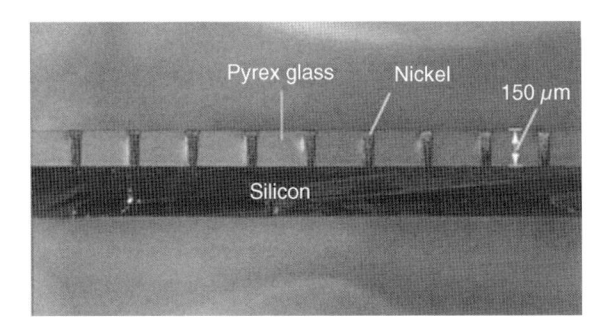

Figure 4.11 Vertical feedthrough in Pyrex glass using DRIE [15].

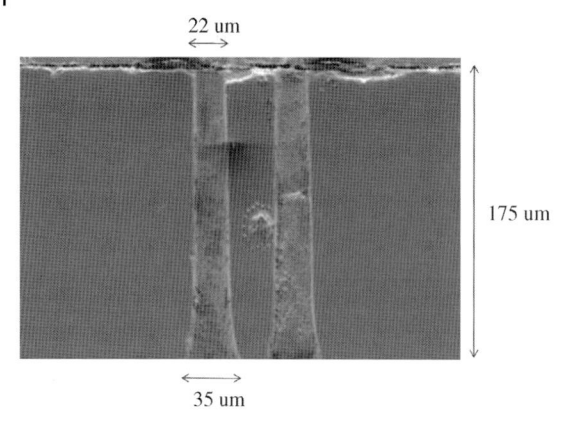

Figure 4.12 SEM cross section of TPV using excimer laser. *Source*: Georgia Tech PRC.

TPVs simultaneously. Such glasses are exposed to ultraviolet (UV) radiation and heated to high temperatures (>500 °C) to crystallize the exposed areas. Subsequently, the crystalline regions are preferentially etched using HF-based wet chemistry. High aspect ratio vias can be achieved using this method; however, it is necessary to develop glass compositions that can be drawn into thin glass sheets and that would also crystallize into crystals that will have preferential etching characteristics compared to the glass that is not crystallized. This is a complex technology. Sandblasting is a technology that can be scaled to parallel processing, which can help achieve via diameters down to 100 μm.

Laser ablation has been investigated as a potential high-throughput via formation method. Because glass is transparent in the visible region, maximum absorption lies in the deep UV (<266 μm) or infrared (>9 μm) wavelengths. The mechanism of laser interaction in glass is a function of the incident wavelength, frequency, and fluence. An example of TPV drilled using excimer laser is shown in Figure 4.12. A slight taper is observed with TPVs, the entrance diameter being slightly greater than that of the exit side. TPVs down to 30 μm diameter and 50 μm pitch can be easily machined in 175–200 μm glass thicknesses using excimer lasers.

Emerging lasers such as picosecond and femtosecond lasers can also be used to ablate vias in glass. The pulse width of such lasers operating in the UV or IR wavelengths is in the femtosecond or picosecond range, and the mechanism of material removal is based on nonlinear absorption.

Figure 4.13 shows the cross-sectional image of TPVs drilled in thicker glass substrates using different laser methods.

It is evident that thinner glass will facilitate smaller vias at finer pitch at higher throughput. Although thin glass can be readily manufactured, handling without breakage is a key challenge. Glass is a hard material with high modulus of elasticity (>70 GPa), yet the fracture toughness of glass is very low. The brittle nature of glass makes handling very difficult, especially as ultrathin large panels (<100 μm). Three potential handling techniques have been developed recently that include (a) glass carrier, (b) polymer-laminated glass, and (c) roll-to-roll processing.

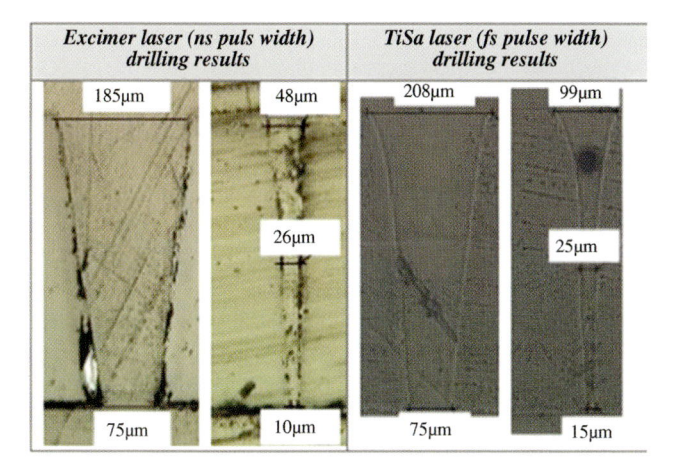

Figure 4.13 Via formation in glass using different lasers [19].

The use of carrier glass has been proposed by Asahi Glass Corporation (AGC), wherein the thin glass substrate (<0.1 mm) is attached to a thicker carrier substrate (0.5 mm) [20].

Another promising handling method is to apply a thin layer of polymer on both sides of thin glass (Figure 4.14). The polymer conforms to the surface and protects glass surfaces from getting damaged, thus preventing crack nucleation and propagation that may eventually lead to failures. Thin polymer films can be applied by any number of processes such as dry film lamination and spin or spray coating.

4.5.1.3 Metallization of Glass TPV

Metallization of small-diameter and small-pitch TPVs is important for achieving electrical interconnections between the RDL on the top and bottom sides. Copper is the most desirable conductor due to its high electrical conductivity ($\sim 6 \times 10^7$ S m^{-1}) and the ease with which it can be electroplated. Copper vias and RDLs can be metallized either using BEOL-based wafer processes used by

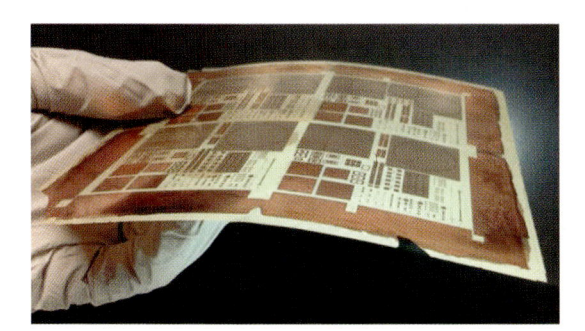

Figure 4.14 Ultrathin glass substrate (30 μm) laminated with polymer and after double-sided metallization. *Source*: Georgia Tech PRC.

foundries, or panel processes typically used by package substrate manufacturers. Three approaches have been reported to achieve metallization of glass interposers, namely, (a) sputter seeding followed by electroplating, (b) electroless plating followed by electroplating, and (c) flowing molten glass over high-temperature metal plugs such as tungsten.

Wafer-based sputtering consists of depositing thin layers of chromium and copper, typically in the order of few hundred to few thousand nanometers. The substrate is head in a vacuum chamber, followed by CVD or physical vapor deposition (PVD) to deposit thin seed layers. The microvias and lines are typically plated using bottom-up electroplating. The wafer-based processes on glass are similar to processing silicon interposers, and the wafer fabrication infrastructure can be applied to form RDLs on glass in the BEOL; the process is limited to single-side processing, requiring carrier wafers to process on both sides.

By using panel-based wet processing approach to plating, one can realize double-side metallization of thin glass interposers. In the electroless approach, electroless copper is used as the seed layer, followed by copper electroplating. Direct plating on glass, using electroless plating is challenging, requiring the functionalization of the glass surfaces using silanes. Alternatively, the glass surfaces can be textured to achieve sufficient roughness, thereby providing mechanical anchoring sites to the metal ions. Laser-assisted metallization using nonlinear absorption can also be used to define metal traces on glass. Subtractive etching can be used to define coarser dimensions ($>30\,\mu m$) for lines, while semi-additive plating (SAP) is conducive for achieving ultrasmall lines ($<15\,\mu m$). An example of metallized vias using electroless plating is shown in Figure 4.15. The TPV pitch in this example is $50\,\mu m$, and complete metallization of the vias is observed.

The third approach to realize metallized feedthroughs in glass is to flow molten glass over metal plugs; via diameters down to $100\,\mu m$ have achieved using this approach [21].

4.5.1.4 Reliability of Copper TPVs in Glass Interposers

Thermomechanical reliability is a critical requirement for any electronic package. Mechanical stresses arise due to the difference in CTE between the different materials within the glass interposer. Copper is typically used to metallize the

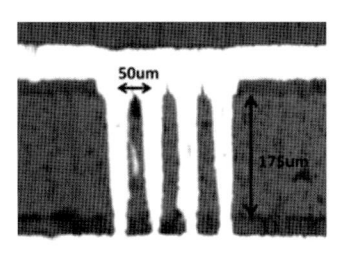

Figure 4.15 Metallized glass vias using electroless and electrolytic plating process. *Source*: Georgia Tech PRC.

TPVs and fine-line RDLs. The interface of interest is between the low-CTE glass (3 ppm/°C) and the high-CTE (17 ppm/°C) copper. This differential expansion causes failures by either delamination or cracking. A number of test vehicles that are fabricated with connected TPV arrays and fine-line structures in a daisy-chain configuration were subjected to thermal cycling tests (−55 to 125 °C) and highly accelerated stress tests (HAST) based on JEDEC standards. The daisy-chain resistance was monitored after a definite number of thermal cycles. Initial tests conducted on glass TPVs showed promising results, with stable daisy-chain resistance until 1500 thermal cycles, with no cracks in the glass.

4.5.1.5 Thermal Dissipation of Glass

Glass has lower thermal conductivity (approximately 1 W/(mK)), compared to silicon (approximately 150 W/(mK)), and hence heat dissipation through glass interposers is a significant issue. Proposed solutions include (1) the use of heat sinks on top of the logic die and (2) the use of thermal vias to provide a low thermal resistance path to the PWB and the ambient.

4.5.1.6 Glass Interposer Fabrication with TPV and RDL

Glass interposers can be fabricated using either wafer- or panel-based approaches. A typical panel-based process flow is shown in Figure 4.16. The glass surface is initially cleaned, a thin layer of polymer is deposited onto both sides of the glass, followed by through-via formation using lasers. Subsequently, metallization is carried out using electroless plating. Buildup layer fabrication for RDL is performed similar to organic substrates, wherein a thin dielectric material is laminated over the first metallization layer. Microvias are then formed using laser ablation to provide electrical contact between the two metal layers separated by the thin dielectric.

Figure 4.17 shows the optical image of a four-metal-layer glass substrate with through-vias. The glass substrate thickness is 175 μm and through-via diameter is 60 μm.

4.5.2
Low-CTE Organic Interposers

The desire for ever thinner packages has led to the development of ultralow-CTE core materials. Low-CTE organic core materials address two of the most important limitations of traditional organic packages for interposer use, namely, mismatch-related reliability and warpage during chip assembly. A thinner package made of lower CTE materials performs similar to a high-CTE thicker core in warpage [22].

The current approach to low-CTE organic cores uses advanced resins and inorganic fillers. Organic cores with low CTE <10 ppm/°C, high T_g, and high modulus have been recently developed by Hitachi Chemical, Panasonic, Sumitomo, Mitsubishi, and other material suppliers. For example, Hitachi Chemical has produced a core with a CTE of less than 6 ppm/°C [23].

The ultralow CTEs of the organic cores, as expected, were found to reduce the warpage after assembly. The warpage of ultralow-CTE cores was found to be

54 *4 Interposer Technology*

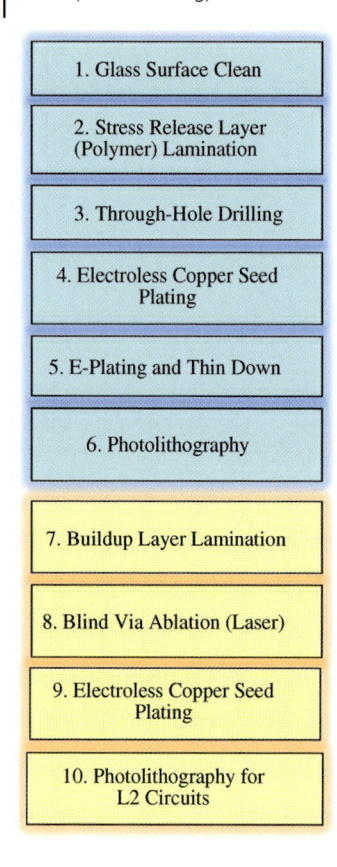

1. Glass Surface Clean
2. Stress Release Layer (Polymer) Lamination
3. Through-Hole Drilling
4. Electroless Copper Seed Plating
5. E-Plating and Thin Down
6. Photolithography
7. Buildup Layer Lamination
8. Blind Via Ablation (Laser)
9. Electroless Copper Seed Plating
10. Photolithography for L2 Circuits

Figure 4.16 Typical process flow for fabricating glass interposer using panel approach. *Source*: Georgia Tech PRC.

significantly less at both 25 and 260 °C in comparison with high-T_g FR-4, with a core thickness of 100 μm (Figure 4.18).

A list of currently available low-CTE organic laminate materials with their key properties is shown in Table 4.2.

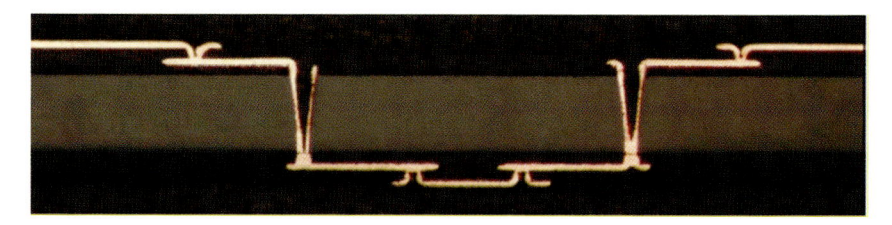

Figure 4.17 Four-metal-layer glass substrate with TPVs and double-side RDL. *Source*: Georgia Tech PRC.

4.5 Lower Cost Interposers | 55

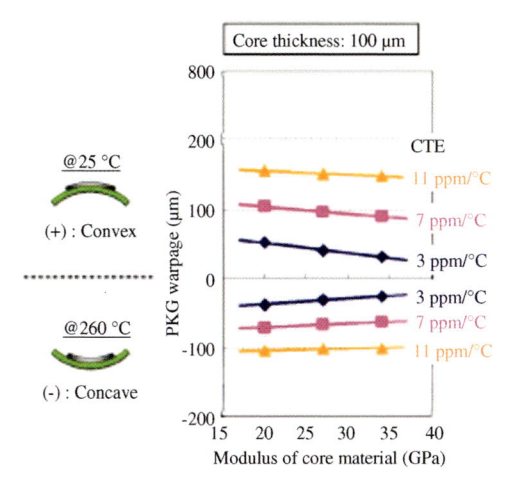

Figure 4.18 Warpage improvement using low-CTE organic cores.

4.5.3
Polycrystalline Silicon Interposer

Polycrystalline silicon interposer technology is being developed by Georgia Tech to address both the cost and performance issues of traditional wafer-based single-crystalline Si. Polycrystalline Si cost is lower by an estimated factor of 10× from single-crystal silicon wafers. The materials, processes, and tools are very different in fabricating polysilicon interposers. A comparison of the polycrystalline silicon interposer with conventional wafer silicon interposers is shown in Table 4.3. The polysilicon silicon interposer starts with low-resistivity polycrystalline thin silicon panels down to 100 μm thickness with an unpolished surface finish. The TPV hole is formed by UV lasers (lower than DRIE). The TPV liner is made of a thick

Table 4.2 Low-CTE organic cores from different manufacturers.

Manufacturer	Trade name	T_g (TMA, °C)	CTE (xy) (ppm/°C)	Elastic modulus (GPa)
Hitachi Chemical	E-700G#l	229	8.5	35.1
	E-700G (R)	221	9	33
	E-705G	260	6 (E-glass)	33
			3.3 (S-glass)	36
	E-800G	230	5 (E-glass)	20
			2.8 (S-glass)	23
MGC	HL-832X2	210	7	30
	HL-830X3	220	6.5	33
Panasonic	R1515αl	225	9.5	33
Sumitomo Bakelite	ELC9963B	240 (DMA)	9	30
	ELC9970A	240 (DMA)	7	—

Table 4.3 Comparison of traditional wafer Si interposer and low-cost silicon interposer.

Technology	Traditional Si interposer	GT low-cost silicon wafer and panel interposer
Raw material	CMOS Si	Polycrystal Si
Size	200–300 mm wafer	300 wafer to 700 mm panel
Thickness	50–200 µm with polish	100–200 µm w/o polish
TSV hole	DRIE	Laser
TSV liner	Thin oxide	Thick polymer
TSV metal	PVD/CVD barrier/seed/Cu	Electroless seed/Cu
RDL	Single-side process	Double-side process
	Cu-oxide	Cu-polymer buildup
Dielectric	Thin SiO_2	Dry film polymer
Via	Dual damascene	Laser
Relative cost		
• Wafer	High	2×
• Panel	N/A	10×

polymer by a low-cost process technology, in contrast to thin SiO_2 liner by CVD or high-temperature oxidation. The RDLs are fabricated, not with liquid dielectrics, but with dry films.

4.5.3.1 Polycrystalline Silicon Interposer Fabrication Process

The process flow for polysilicon interposers is shown in Figure 4.19. TPVs with 80 µm diameter on 150–200 µm pitch were formed in 150 mm × 150 mm, 200 mm thick polysilicon by 355 nm UV laser ablation. The thick polymer liner was fabricated by a "double-laser" process. The through-vias in silicon were first filled with polymer using dry film vacuum lamination, followed by a second UV laser ablation process to form a 40 µm diameter inner via within the polymer fill. This resulted in a via sidewall liner of approximately 20 µm thickness to isolate the lossy silicon substrate, which has a resistivity of 0.5–0.6 Ω cm. An electroless plating process was used to form a thin Cu seed layer on the polymer liner both on the top

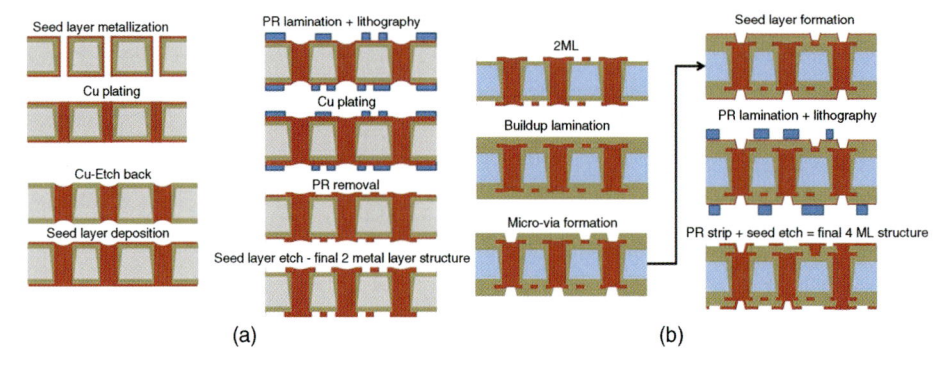

Figure 4.19 Process flow to fabricate (a) two-metal-layer and (b) four-metal-layer structures [24].

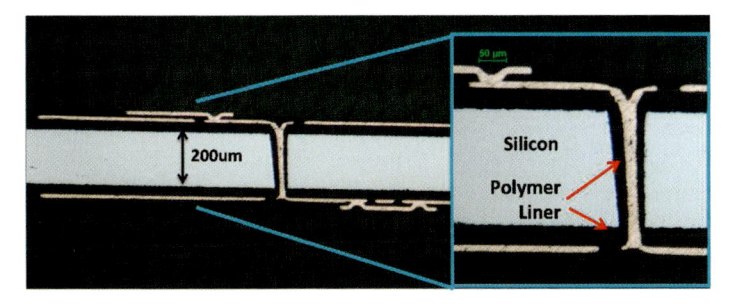

Figure 4.20 Cross section of the final four-metal-layer polycrystalline silicon interposer.

and the bottom of the silicon and in the via. The through-vias were filled using a void-free electrolytic plating process. Copper overburden on the top and bottom surfaces was removed by a double-sided chemical etching process. A SAP process was used for double-sided copper pattern formation. A double-side dry film lithography process was then used to pattern the traces, followed by electrolytic plating to achieve a two-metal-layer RDL on polycrystalline silicon interposers.

Figure 4.19b shows the process flow used to fabricate a four-metal-layer structure. The buildup polymer dielectric layers were deposited by a double-sided vacuum lamination process. The blind microvias were formed by 355 nm UV laser ablation in the cured polymer. These microvias and outer circuit layers were then metallized by SAP process using double-sided dry film photoresist lamination. The finished structure of the four-metal-layer interposer is shown in the microsection photograph in Figure 4.20.

The high-frequency characterization results of a four-metal-layer polycrystalline silicon interposer with smaller TPVs were reported by Georgia Tech [25].

4.6
Interposer Technical and Manufacturing Challenges

Interposer integration offers a lot of benefits for system integration, form factor miniaturization, and performance. However, interposers also present a number of technology and manufacturing challenges. The key challenges are as follows:

Thermal management: 2.5D interposers result in side-by-side high-performance multichip modules leading to higher thermal density. Hence, efficient and higher heat flux thermal dissipation techniques are critical to maintain junction temperatures and skin temperatures within specified limits.

Testability and test cost: Interposer testability, and its cost and complexity must be addressed to achieve high yield in high-volume manufacturing.

Manufacturing supply chain readiness: Silicon interposers from wafer foundries are now available in limited volumes from a select list of suppliers. But a widespread availability of interposers, especially low-cost interposers, remains a major hurdle to successful adoption of interposer technology into products.

4 Interposer Technology

EDA tools to design with interposers: Although design tools are widely available for IC and SiP packages, as well as for on-chip interconnections, there is a lack of a set of EDA tools optimized for high-density 2.5D and 3D interposers. This is an area of intense development activity in the design tool community today.

Interposer cost: The single biggest challenge to the pervasive use of interposer technology is their cost, especially for large body size interposers. Large panel processing with thin and low-cost materials, such as ultrathin fusion drawn glass, as well as high-throughput tools for TPV formation and for copper metallization offer great potential to reduce the cost of interposers for high-volume adoption.

4.7
Interposer Application Examples

This section highlights some of the initial application examples of interposers for electronic systems in consumer and high-performance products.

IBM high-performance 2.5D silicon interposer: Researchers at IBM recently demonstrated a high-performance system combining SOI-based Cu-45HP technology with 8HP BiCMOS SiGe technology interconnected through a single wiring layer on an interposer, which supports a bandwidth of greater than 1.3 Tbps as seen in Figure 4.21 [26]. The silicon interposer combines cost-effective 90 nm

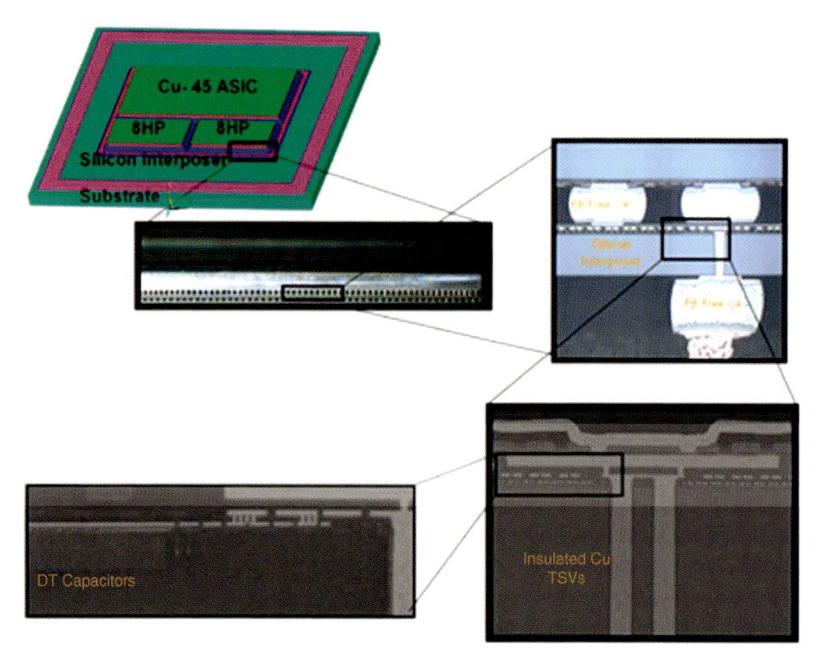

Figure 4.21 Logic–memory integration on 2.5D silicon interposer. Courtesy of IBM.

4.7 Interposer Application Examples | 59

DRAM die size	10.7 mm × 13.3 mm
DRAM die thickness	50 μm
TSV count in DRAM	1,560
DRAM capacity	512 Mbit/die × 2 strata
CMOS logic die size	17.5 mm × 17.5 mm
CMOS logic die thickness	200 μm
CMOS logic bump count	3,497
CMOS logic process	150 nm CMOS
DRAM-logic FTI via pitch	50 μm
Package size	33 mm × 33 mm
BGA terminal	520 pin/1mm pitch

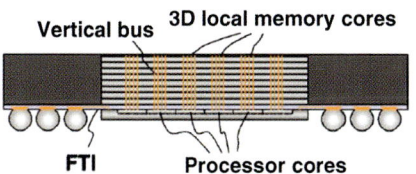

Figure 4.22 High-bandwidth logic–memory integration via 3D SMAFTI interposer. Courtesy of NEC/Renesas Electronics.

BEOL wiring levels with Cu TSVs to eliminate the complexity of integrating TSVs in the active ICs. Memory integration using hybrid memory cube technology from Micron is also being developed.

NEC high-bandwidth SMAFTI 3D interposer: A team from NEC demonstrated the first implementation of a 3D interposer using a multilayer Cu-polyimide coreless substrate called smart feedthrough interposer (SMAFTI). This 3D interposer interconnected a stacked memory on the top side with a logic IC on the bottom side of the interposer to achieve ultrashort connection length and high bandwidth as shown in Figure 4.22 [27]. This technology has also been demonstrated by Renesas Electronics and Elpida memory using silicon interposer with TSV, and more recently by a team of TSMC, ITRI Taiwan, and Rambus [28].

Glass interposers for RF applications: High-frequency applications require low signal insertion loss and crosstalk. Glass-based integrated passive devices (IPDs) for mobile phone and WLAN applications have been developed by ST Microelectronics [29]. The integration of through-vias in thin glass to significantly reduce the form factor and cost of the glass interposers for RF modules has been recently demonstrated by a team of Georgia Tech, ST Microelectronics, and Asahi Glass Company [30].

3D glass interposer for MEMS integration with ASIC: Glass caps have been used for hermetic sealing of MEMS devices at wafer level for several years [31]. Glass interposers are an ideal solution for MEMS packaging due to their moisture insensitivity. VTI has used glass interposers with TGV to interconnect gyro-MEMS devices to ASIC chips as shown in Figure 4.23 [32].

Silicon interposer for high-brightness LED submounts: The high thermal conductivity of silicon lends itself to interposer submounts and packaging of high-power HB-LEDs. VisEra (a subsidiary of TSMC) recently started production of silicon wafer interposers with plated TSVs for LED layers assemblies as shown in Figure 4.24.

- **WLP / TGV process flow:**

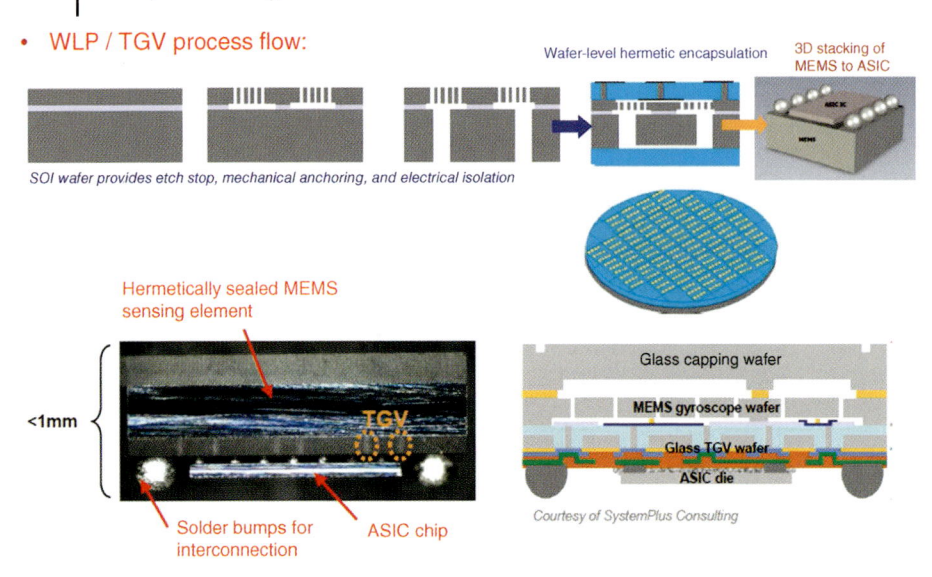

SOI wafer provides etch stop, mechanical anchoring, and electrical isolation

Wafer-level hermetic encapsulation 3D stacking of MEMS to ASIC

Figure 4.23 Glass interposer with through-glass via for hermetic MEMS 3D integration [33].

4.8 Conclusions

High-density silicon interposers with TSVs are the initial interposer technology moving into commercial production by TSMC and others. For applications where these thin-film interposers are too expensive, glass, polysilicon and new low-CTE organic interposers are being pursued because of their lower cost potential and

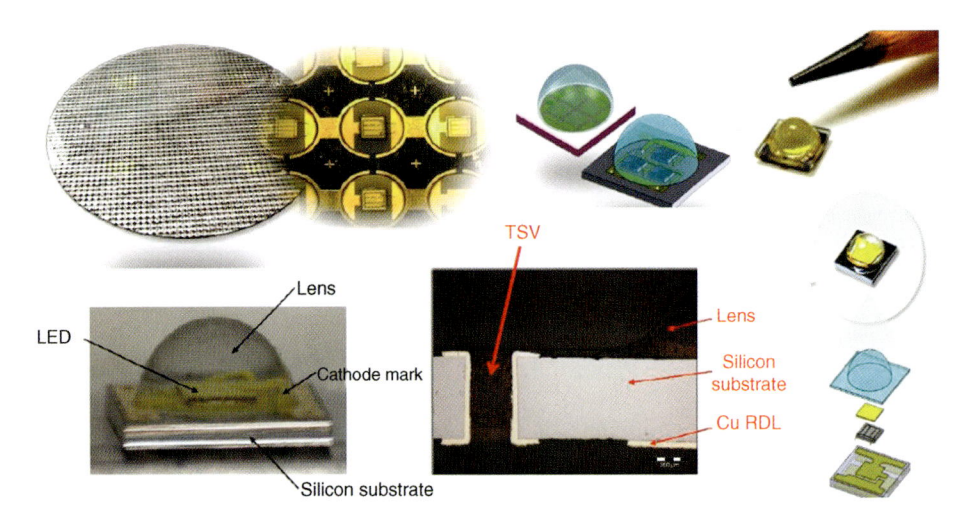

Figure 4.24 HB-LED submount integration on silicon interposer with conformal TSVs [34].

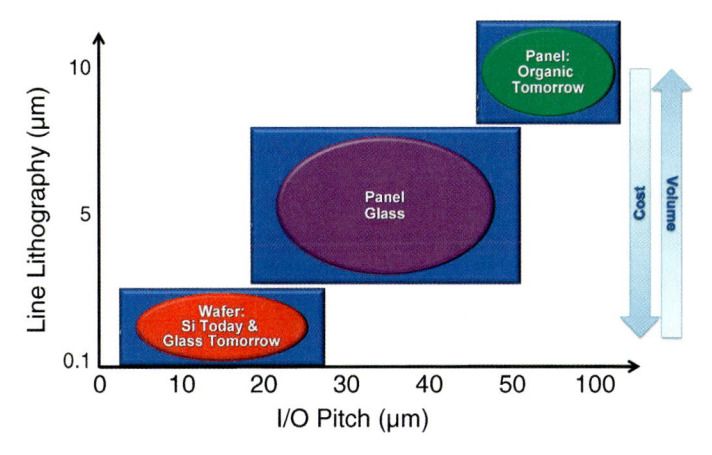

Figure 4.25 Summary of interposers in three regimes.

ability to be panel processed. Figure 4.25 is a good summary of the I/O pitch and cost regimes for wafer-based silicon below 1 μm at high cost, organics at or near 10 μm at low cost and glass covering the space in between.

References

1 Sukumaran, V. *et al.* (2010) Through-package-via formation and metallization of glass interposers. IEEE 60th Electronic Components and Technology Conference (ECTC 2010), June 1–4, 2010, Las Vegas, NV, pp. 557–563.

2 Topper, M. *et al.* (2010) 3D thin film interposer based on TGV (through glass vias): an alternative to Si-interposer. IEEE 60th Electronic Components and Technology Conference (ECTC 2010), pp. 66–73.

3 Yamanaka, K., Kobayashi, K., Hayashi, K., and Fukui, M. (2009) Advanced surface laminar circuit packaging with low coefficient of thermal expansion and high wiring density. IEEE 59th Electronic Components and Technology Conference (ECTC 2009), pp. 325–332.

4 Ho, C.W., Chance, D.A., Bajorek, C.H., and Acosta, R.E. (1982) The thin-film module as a high-performance semiconductor package. *IBM Journal of Research and Development*, **26** (3), 286–296.

5 Tai, K.L. (1987) Integrated circuit chip-and-substrate assembly. US Patent 4,670,770.

6 Ho, C.W., Chance, D.A., Bajorek, C.H., and Acosta, R.E. (1982) The thin-film module as a high-performance semiconductor package. *IBM Journal of Research and Development*, **26** (3), 286–296.

7 Tai, K.L. (1987) Integrated circuit chip-and-substrate assembly. US Patent 4,670,770.

8 Banijamali, B., Ramalingam, S., Nagarajan, K., and Chaware, R. (2011) Advanced reliability study of TSV interposers and interconnects for the 28nm technology FPGA. IEEE 61st Electronic Components and Technology Conference (ECTC2011), pp. 285–290.

9 Lee, C. (2011) Presentation at the 1st IEEE Global Interconnect Technology Workshop (GIT 2011), Atlanta, GA, November 2011.

10 Kim, N., Wu, D., Kim, D., Rahman, A., and Wu, P. (2011) Interposer design optimization for high frequency signal transmission in passive and active interposer using through silicon via (TSV). IEEE 61st Electronic Components and Technology Conference (ECTC 2011), pp. 1160–1167.

11 Chausse, P., Bouchoucha, M., Henry, D., Sillon, N., and Chapelon, L.L. (2009) Polymer filling of medium density through silicon via for 3D-packaging. IEEE 11th Electronics Packaging Technology Conference (EPTC'09), pp. 790–794.

12 Wu, S.-T., Lau, J.H., Chien, H.-C., Hung, J.-F., Dai, M.-J., Chao, Y.-L., Tain, R.-M., Lo, W.-C., and Kao, M.-J. (2012) Ultra low-cost through-silicon holes (TSHs) interposers for 3D IC integration SiPs. IEEE 62nd Electronic Components and Technology Conference (ECTC 2012), pp. 1618–1624.

13 Rao, T. and Sundaram, V. (2011) Impact of 3D ICs with TSV is profound but complex and costly – is there a better way? *Chip Scale Review*, **8**, 31–32.

14 SCHOTT, Ultra-thin glass: not just flexible, but even rollable. Available at http://www.us.schott.com/advanced_optics/english/products/wafers-and-thin-glass/glass-wafer-and-substrates/ultra-thin-glass/index.html. Accessed: 2013

15 Li, X. et al. (2002) Fabrication of high-density electrical feed-throughs by deep-reactive-ion etching of Pyrex glass. *Journal of Microelectromechanical Systems*, **11**, 625–630.

16 Iliescu, C. et al. (2007) Deep wet etching-through 1mm Pyrex glass wafer for microfluidic applications. IEEE 20th International Conference on Micro Electro Mechanical Systems (MEMS 2007), pp. 393–396.

17 Egashira, K., Mizutani, K., and Nagao, T. (2002) Ultrasonic vibration drilling of microholes in glass. *CIRP Annals – Manufacturing Technology*, **51** (1), 339–342.

18 Stookey, S.D. (1953) Chemical machining of photosensitive glass. *Industrial & Engineering Chemistry*, **45** (1), 115–118.

19 Brusberg, L. et al. (2009) Photonic system-in-package technologies using thin glass substrates. IEEE 11th Electronics Packaging Technology Conference (EPTC'09), pp. 930–935.

20 AGC, AGC's carrier glass technology enables current production processes to handle ultra-thin glass. Available at http://www.agc.com/english/news/2012/0530e.pdf. Accessed: 2012

21 NEC SCHOTT, SCHOTT HermeS® substrate. Available at http://www.nec-schott.co.jp/english/auto/others/hermes.html. Accessed: 2010.

22 Yim, M.J., Strode, R., Brand, J., Adimula, R., Zhang, J.J., and Yoo, C. (2011) Ultra-thin PoP top package using compression mold: its warpage control. IEEE 61st Electronic Components and Technology Conference (ECTC 2011), pp. 1141–1146.

23 Satoru, A., Shimizu, H., and Hanawa, A. (2012) Prepreg and its application products for low thermal expansion and low dielectric tangent. US Patent 8,115,105.

24 Chen, Q., Bandyopadhyay, T., Suzuki, Y., Liu, Fuhan, Sundaram, V., Pucha, R., Swaminathan, M., and Tummala, R. (2011) Design and demonstration of low cost, panel-based polycrystalline silicon interposer with through-package-vias (TPVs). IEEE 61st Electronic Components and Technology Conference (ECTC 2011), pp. 855–860.

25 Sundaram, V., Chen, Q., Suzuki, Y., Kumar, G., Liu, Fuhan, and Tummala, R. (2012) Low-cost and low-loss 3D silicon interposer for high bandwidth logic-to-memory interconnections without TSV in the logic IC. IEEE 62nd Electronic Components and Technology Conference (ECTC 2012), pp. 292–297.

26 Baez, F.M., Cranmer, M., Shapiro, M., Audet, J., Berger, D., Sprogis, E., Collins, C., and Iyer, S. (2012) Electrical design and performance of a multichip module on a silicon interposer. IEEE 21st Conference on Electrical Performance of Electronic Packaging and Systems (EPEPS 2012), pp. 303–306.

27 Kurita, Y., Matsui, S., Takahashi, N., Soejima, K., Komuro, M., Itou, M., Kakegawa, C., Kawano, M., Egawa, Y., Saeki, Y., Kikuchi, H., Kato, O., Yanagisawa, A., Mitsuhashi, T., Ishino, M., Shibata, K., Uchiyama, S., Yamada, J., and Ikeda, H. (2007) A 3D stacked memory integrated on a logic device using SMAFTI technology. IEEE 57th Electronic Components and Technology Conference (ECTC'07), May 29–June 1, 2007, pp. 821–829.

28 Zhan, C.-J., Tzeng, P.-J., Lau, J.H., Dai, M.-J., Chien, H.C., Lee, C.-K., Wu, S.-T. et al. (2012) Assembly process and reliability

assessment of TSV/RDL/IPD interposer with multi-chip-stacking for 3D IC integration SiP. IEEE 62nd Electronic Components and Technology Conference (ECTC 2012), pp. 548–554.
29 Titz, D., Bisognin, A., Ferrero, F., Luxey, C., Jacquemod, G., Laporte, C., Ezzeddine, H., Valente, M., and Brachat, P. (2012) 60 GHz patch antenna using IPD technology. Antennas and Propagation Conference (LAPC 2012), Loughborough, pp. 1–4.
30 Sridharan, V., Min, S., Sundaram, V., Sukumaran, V., Hwang, S., Chan, H., Liu, Fuhan, Nopper, C., and Tummala., R. (2010) Design and fabrication of bandpass filters in glass interposer with through-package-vias (TPV). IEEE 60th Electronic Components and Technology Conference (ECTC 2010), pp. 530–535.
31 Esashi, M. (2008) Wafer level packaging of MEMS. *Journal of Micromechanics and Microengineering*, **18** (7), 073001.
32 Kuisma, H. (2008) Wafer level packaging of MEMS. Keynote at Smart Systems Integration Conference, Barcelona, Spain, April 9–10, 2008.
33 I-Micronews (2011) 3D wafer-level-packaging of MEMS gyroscope sensor with VTI's CMR3000, System Plus Consulting, Issue No. 20, September 2011.
34 Zinck, C. (2010) 3D integration infrastructure & market status. IEEE International 3D Systems Integration Conference (3DIC 2010), pp. 1–34.

5
TSV Formation Overview

Dean Malta

5.1
Introduction

Through-silicon vias (TSVs) are the foundation of today's emerging 3D integration and 2.5D Si interposer technologies. These interconnects extend through the silicon substrate, enabling vertical integration and shortened interconnect lengths for reduced size, weight, and power consumption [1,2]. It is expected that 3D and 2.5D integration will play a major role in the continued advancement of semiconductor technology, as the scaling of 2D Si microelectronics slows down and becomes increasingly more difficult [3,4]. Examples of emerging 3D integrated circuit (3D IC) applications, as demonstrated by companies such as Tezzaron and IBM, are illustrated in Figure 5.1 [5,6]. An assembled 2.5D wide I/O field-programmable gate array (FPGA) architecture, recently demonstrated by Xilinx (with TSMC and Amkor), is shown in Figure 5.2 [7].

The TSV design and process considerations differ slightly for 3D IC applications versus 2.5D Si interposers. For TSV implementation in 3D IC wafers, TSV dimensions typically will tend to be small, most commonly in the range of 1–5 μm in diameter and 10–50 μm in depth (as determined by the final substrate thickness). The thermal budget for TSV process steps must be compatible with the IC layers already fabricated, generally limiting processing to approximately 400 °C or less, depending on the wafer technology. Also, device proximity effects must be taken into account, as TSVs placed near active circuitry may impact IC performance due to induced stress in the region around the vias [8]. For Si interposers, TSVs can be slightly larger, possibly 10 μm or greater in diameter, with depths of 50–100 μm or more. The device proximity concern does not apply for passive Si interposer substrates, and there are no restrictions on the process temperatures for TSV formation, since the TSVs can be formed in bare Si wafers.

The early development of TSVs has been covered in detail in Ref. [9]. Much of the initial focus was on the establishment of unit process capabilities such as deep Si etching, conformal TSV liner depositions, and via-fill metallization approaches. Another key issue was to decide how TSV formation would be integrated with the IC fabrication and assembly, mainly when the TSVs are to be formed and on which

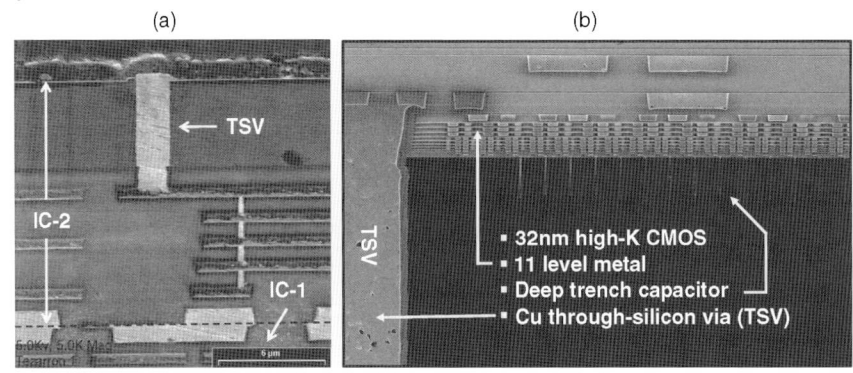

Figure 5.1 Examples of TSVs in 3D IC stacks. (a) Tezzaron 3D processor wafer stack with tungsten TSVs. (b) IBM 3D embedded DRAM module fabricated in a stackable 32 nm high-k/metal gate technology with Cu TSVs.

side of the supply chain (IC foundry versus OSAT). Significant progress has been made in these areas since the earlier volumes of this book were published. Recently, TSV reliability and manufacturability (cost, throughput, and yield) have seen increased focus, as these ultimately will determine the potential for TSV technology to be implemented commercially on a widespread basis.

This chapter will provide an overview of TSV process approaches and key fabrication steps, along with a discussion of yield and reliability issues. In particular, new developments and progress made since the prior volumes of this book will be noted. Subsequent chapters will provide greater detail on the specific fabrication methods for TSV formation.

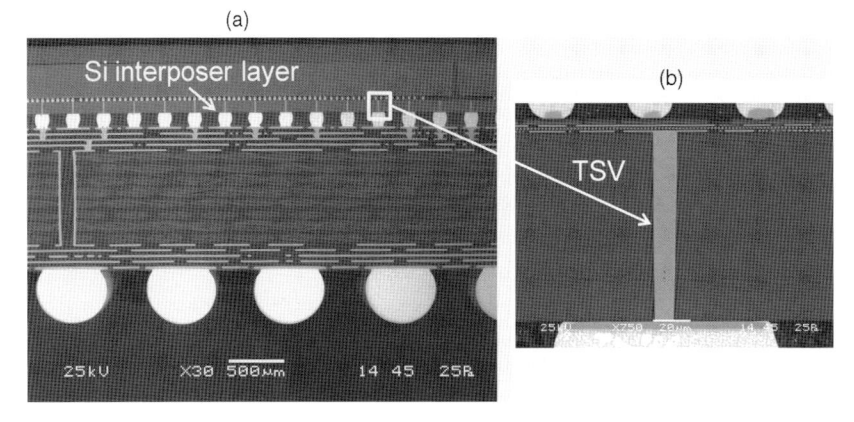

Figure 5.2 TSVs in 2.5D Si interposer application. (a) 2.5D wide I/O FPGA from Xilinx, showing multiple die assembled on TSMC Si interposer, attached to organic laminate. (b) Close-up image of Cu TSV in Si interposer layer.

5.2
TSV Process Approaches

Historically, there have been several different process approaches under consideration for TSV formation in IC wafers. These include TSV-first, TSV-middle, backside TSV-last, and front-side TSV-last. Of the two TSV-last approaches, backside vias are the more common approach. These flows vary in terms of how the vias are fabricated, when the vias are made (relative to the IC process), and potentially where the vias are made (IC foundry or OSAT facility). The concept of each primary process approach is indicated in Figure 5.3.

Since the previous volumes of this book were released, the viewpoint of the 3D community has been converging on TSV-middle as the preferred manufacturing approach for most commercial 3D IC applications. This approach seems to have the most favorable balance in terms of TSV process requirements, integration with the IC process, and overall manufacturability. The backside TSV-last approach has been employed in the production of CMOS image sensors, which use backside contacts because of front-facing sensor and optics layers. The TSV-first approach is no longer viewed as a viable option due to inherent limitations, mainly via resistance. This approach requires the use of poly-Si as the TSV conductor due to front-end-of-line (FEOL) process compatibility requirements and the high temperatures the TSV must be able to withstand during the IC process. The front-side TSV-last approach is technically viable, though generally not a preferred commercial manufacturing process. Still, this approach may enable custom 3D integration for specialized applications in which TSV-middle wafers are not offered and backside TSVs are not practical. The 3D integration approaches based on TSV-middle and TSV-last are discussed in further detail below.

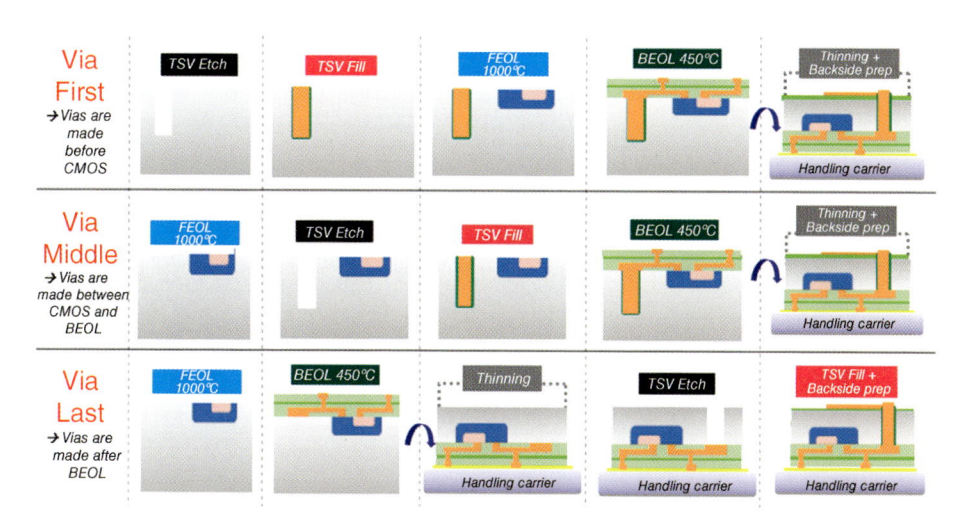

Figure 5.3 Process flow diagrams illustrating the three primary alternatives to TSV implementation in IC wafers for 3D IC applications. Courtesy of Yole Développement.

5.2.1
TSV-Middle Approach

In the TSV-middle approach, the TSVs are made at the IC fabrication facility following the FEOL active device layers and prior to the back-end-of-line (BEOL) interconnect wiring layers. Because the TSVs are processed before the BEOL sequence, it is possible to route subsequent metal layers over the TSVs in this flow. The TSVs are initially formed as blind vias, which are later revealed from the back during wafer thinning, for backside interconnection and routing. Generally, solder bumps or microbumps are added for interconnection to another IC layer or substrate. The TSV reveal and backside metallization steps could be done at an OSAT facility, although some of the larger IC foundries have indicated plans to provide the full TSV process, including thinning and backside processing. The TSV-middle approach has many advantages, including the lowest consumption of real estate, since the IC fabrication facility has capability for making small TSVs with high placement accuracy and BEOL metal layers are routed over the TSVs. The remaining steps, including precise wafer thinning, TSV reveal, and backside processing of thinned IC wafers, are all considered viable for manufacturing, thanks in part to recent progress in temporary carrier wafer bonding technology.

5.2.2
Backside TSV-Last Approach

The backside TSV-last approach involves etching vias from the back surface of a completed IC wafer after it has been thinned, landing the vias on front-side metal. This approach has been adopted for production use in CMOS image sensors, which use relatively low aspect ratio periphery TSVs, often with sloped sidewalls and a conformal metallization layer (unfilled TSVs) [10,11]. These image sensors, which have widespread use in mobile devices, represent the first volume commercial product utilizing TSVs.

This TSV approach becomes more challenging for higher density 3D IC applications requiring small, high aspect ratio TSVs. Etching vias from the bottom of the wafer requires back-to-front lithography alignment capability. The via etch terminates on the IC metal at the bottom of the TSV, exposing the metal surface. In the process of insulating the TSV, this metal surface is covered by the deposited via dielectric. A specialized etch step, sometimes called a "bottom-clear etch," is required to reopen the metal at the base of the via, while not removing the deposited insulator from the TSV sidewalls. For high aspect ratio TSVs, forming a low-resistance connection to the front-side landing metal can be challenging. This metal surface can be affected by exposure to the various etch plasmas and is difficult to restore using standard sputter etch techniques, since it is at the bottom of a deep via. For these reasons, this process flow is unlikely to see widespread use for higher density, higher aspect ratio TSV applications.

5.2.3
Front-Side TSV-Last Approach

Most high-volume commercial 3D applications, including stacked memory, logic–memory, and CMOS image sensors, will likely be covered by the TSV approaches discussed previously. However, there may be certain 3D applications that fall outside of these approaches for a variety of reasons. These may include defense and aerospace applications, such as focal plane arrays or heterogeneous integrated microsystems, which involve analog, mixed signal, or RF ICs. The ICs used in these applications may not be made available with TSV-middle, since some of the wafer suppliers do not offer a TSV production capability, or offer it only for select wafer technologies.

For many defense and aerospace applications, there is a strong driver for the performance benefits of 3D integration, while the product volumes are lower and somewhat less cost sensitive than consumer electronics. The same may be true for other applications, such as advanced medical imaging. In such cases, custom 3D approaches using front-side TSV-last may find implementation, provided that the wafers can be designed and fabricated with "TVS exclusion" areas, free of metal and active circuitry. The exclusion areas allow for TSVs to be etched through the interlayer dielectric (ILD) stack and the Si layer, without obstruction from metal or damage to active circuitry. In this flow, TSV processing would likely be done by a third-party 3D integration foundry, or perhaps by the system integrator, which is the user of the 3D components. Once the blind vias are formed from the front of the wafer, the remainder of the process flow can be similar to the TSV-middle approach, including backside TSV reveal and metallization. An example of a functional 3D IC demonstration using a front-side TSV-last flow is shown in Figure 5.4, which illustrates a 3D infrared focal plane array with a TSV density of $>10^5 \, \text{cm}^{-2}$ [12]. In this example, completed IC wafers of different types, obtained

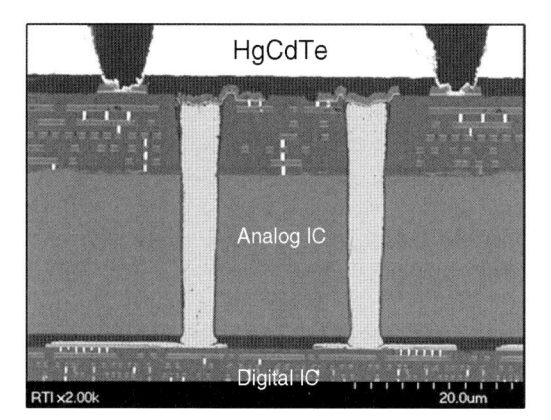

Figure 5.4 3D integrated IR focal plane array, including analog and digital readout ICs interconnected using front-side TSV-last process. Courtesy of RTI International/DRS Technologies.

from two separate ASIC foundries, were integrated using a front-side TSV-last 3D IC approach.

5.3
TSV Fabrication Steps

The specific sequence of steps required for TSV fabrication and interconnection will vary with each process flow outlined above, or for 2.5D interposers versus 3D IC. Many of the individual steps, however, are common to the different TSV approaches. Because TSV-middle is emerging as the preferred manufacturing process for many commercial 3D electronics applications, the fabrication steps described in this section will be largely based on that flow. This section will provide an overview of the TSV processing techniques, their basic requirements, and challenges. The specific process steps will be described in greater detail in subsequent chapters on TSV formation.

5.3.1
TSV Etching

The method used for etching TSVs is known as deep reactive ion etching (DRIE). It was first introduced by Bosch in the mid-1990s and is widely used in MEMS fabrication [13]. This technique is based on the principle of alternating steps of etching and passivation to enable etching of deep features in Si with vertical sidewalls and high selectivity to photoresist or SiO_2 mask layers. Several equipment manufacturers have licensed the Bosch process and sell DRIE systems based on this approach. Others have developed alternative proprietary approaches to achieve similar deep silicon etching capabilities.

The process generally involves an inductively coupled plasma (ICP) source, with Si etching using SF_6 and sidewall passivation using C_4F_8. In each etching step, a small amount of Si is removed by the SF_6 plasma. Next, a passivation layer is applied over all of the surfaces. In the subsequent etch step, the passivation is removed from the bottom of the feature, allowing the Si etch to continue downward, while remaining on the sidewall and protecting it from further etching. The process is repeated many times and the result is a generally vertical feature sidewall, made up of a series of small ridges, known as "scallops."

DRIE techniques and equipment are now being tailored to address the challenges of TSV fabrication, which requires etching of small vias with high throughput and minimal sidewall roughness. The reduction of scallops is necessary to enable sufficient coverage of TSV insulation and seed metal in high aspect ratio vias. In recent years, much progress has been made in achieving high etch rates for TSV cost reduction, while greatly reducing scallops. This has been done through the use of advanced plasma sources, along with shortened process cycles enabled by high-speed control modules. Equipment companies, such as SPTS and Applied Materials, are now offering TSV etch processes that are indicated to be essentially scallop-free. TSMC reported scallop-free TSVs in a recent paper on

| (a) | (b) | (c) |

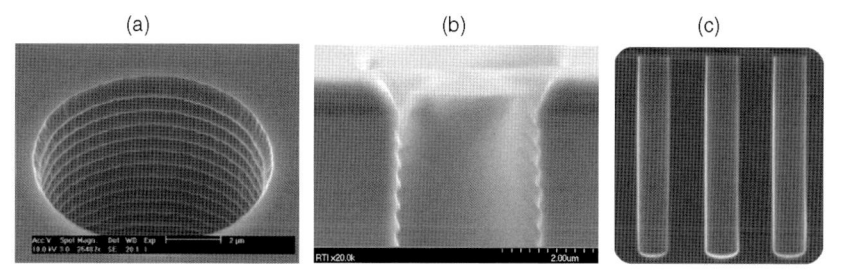

Figure 5.5 DRIE features etched in Si. (a) Top view of TSV with sidewall scallops. Courtesy of University of Arkansas. (b) Cross-sectional view of scallops in Si trench. Courtesy of RTI International. (c) "Scallop-free" TSV etch. Courtesy of SPTS.

their 28 nm 3D IC process [14]. Representative images of DRIE features, illustrating sidewall scallops and "scallop-free" TSVs, are shown in Figure 5.5.

5.3.2
TSV Insulation

The insulation of TSVs is generally accomplished by deposition of a highly conformal SiO_2 layer. Because this layer is deposited after active devices have been formed, it is necessary to limit the deposition temperature, usually within the range of 200–400 °C or less. With high aspect ratio vias, vertical profiles, sidewall roughness, and limited thermal budget, achieving conformal insulator coverage can be a challenge for many of the low-temperature deposition techniques. In addition, the TSV insulator films must have adequate electrical insulation properties, including dielectric constant, leakage current, and breakdown voltage. For maximum conformality in TSVs, the preferred approach is subatmospheric chemical vapor deposition (SACVD) using O_3/TEOS. This thermally activated process has a deposition temperature near 400 °C. Sometimes, lower temperatures are required, for example, in the processing of memory wafers, or for TSV-last flows on temporary carrier wafers. In these cases, certain plasma-enhanced chemical vapor deposition (PECVD) processes can be used. With specially developed PECVD techniques, sufficient conformality in TSVs may be achieved at temperatures below 200 °C.

5.3.3
TSV Metallization

In the previous volumes of this book, multiple options for the TSV conductor were discussed. These included electrochemically deposited (ECD) Cu, CVD W or Cu, and poly-Si. At this time, most TSV approaches are based on ECD Cu metallization, although CVD W is still used in some cases. Film stress and wafer bow are the main potential concerns for W TSV filling. This limits the thickness that can be deposited; therefore, W is only used for small-diameter holes or annular vias. As mentioned earlier, poly-Si is no longer commonly considered for use in TSVs due to concerns over via resistance.

For Cu TSV metallization schemes, it is necessary to add a barrier material between the Cu and the SiO_2 layers. Common TSV barrier choices include TiN or TaN, which are often deposited by the ionized metal plasma (IMP) sputtering process. For TiN barrier layers, metallo-organic (MO) CVD processes are also used [15]. Prior to ECD Cu metallization, a thin Cu seed layer is deposited, typically by IMP sputtering. It is important that this layer has continuous coverage along the via walls, to enable uniform and electrochemical deposition and void-free fill.

In recent years, there have been significant advances in Cu ECD techniques for TSVs, enabled by highly engineered TSV plating chemistries and process equipment. Currently, most TSV approaches are based on ECD Cu filling, using a "bottom-up" growth process to ensure there are no voids in the Cu fill [11,16]. Plating voids are a major TSV reliability concern and can also lead to yield problems during subsequent fabrication steps. The chemistries used for Cu TSV filling involve organic additives, which act as accelerators, suppressors, or levelers. The balance of these additive levels is critical in determining the relative plating rates at various points along the via profile and wafer surface. By adjusting these, it is possible to slow plating at the surface and top portion of the via, while promoting plating near the bottom of the via, resulting in a bottom-up TSV fill profile [11,16]. Once optimized, the levels of these additives, along with other chemical components of the makeup solution, must be maintained very precisely for consistent TSV filling. This requires bath monitoring by chemical analysis methods and regular adjustment of additives and makeup chemicals, since some of these are depleted during use.

In addition to void-free filling, ECD Cu TSV processes must be able to achieve fast fill time, in order to make the process throughput and cost viable for commercial production use. Similarly, the Cu overburden thickness should be minimized, in order to keep the throughput and cost of the subsequent chemical–mechanical polish (CMP) step practical. These concerns are addressed by reduction in TSV size, increases in growth rate, and use of a bottom-up fill process, which enables lower overburden than conformal plating.

The use of MOCVD Cu has some advantages as a plating seed layer or metallization for lined (unfilled) TSVs, since it can produce nearly 100% conformal deposition in high aspect ratio features. It can also be used to fill small TSVs [12,17]. By itself, it is not very practical for filling large TSVs, due to limited deposition rates. It does not appear that MOCVD Cu will be a preferred choice for commercial TSV production, based on process cost and throughput. Also, production equipment options are limited, due to the much wider adoption of ECD Cu by the semiconductor industry. At this point, it appears that ECD Cu fill will likely remain as the primary option for commercial TSV manufacturing.

5.3.4
Overburden Removal by CMP

After TSV metallization, which is most often Cu, the overburden layer is typically removed by CMP. The polishing processes for Cu are fairly well established due to the emergence of Cu dual-damascene IC metallization. However, the larger size of TSVs compared to BEOL interconnects generally means thicker overburden layers and longer polishing times. This has created concern regarding throughput and cost for commercial TSV production. However, reductions in TSV size and the development of bottom-up Cu plating approaches have helped to minimize Cu overburden thickness and reduce polishing time. Also, some high-rate Cu CMP slurries have been developed specifically for TSV processing. These may allow further gains in throughput and help reduce the cost of CMP, particularly for larger TSV dimensions.

5.3.5
TSV Anneal

It is widely understood that the difference in coefficient of thermal expansion (CTE) between Cu and Si is a potential concern for the implementation of Cu TSVs in Si IC wafers. This is generally addressed by keeping the TSV dimensions as small as possible, while also placing TSVs far enough from active regions to avoid stress-induced changes in the device operation [8]. In recent years, an additional element of concern has emerged during 3D process demonstrations and has received considerable attention since the earlier volumes of this book. Tezzaron first reported in 2009 that the plated Cu TSV conductor tends to demonstrate a permanent extrusion at the TSV surface, when heated during subsequent process steps [18]. The surface of the TSV, which is planar after CMP, can become domed or form hillocks during heating, resulting in a permanent surface topography [14,18–20]. This effect has been called "Cu protrusion," "Cu extrusion," or "Cu pumping." This Cu protrusion at the TSV surface can induce stress in metal and dielectric layers above the TSVs, even causing mechanical failures [14,18]. The amount of Cu protrusion scales with via size, which is another reason why small TSVs are recommended. Studies have shown that the ECD Cu undergoes microstructural changes during heating, resulting in increased grain size, while there is also a tendency for plating impurities from the bath chemistry to migrate out of the TSVs to the surface [11,20]. Representative images of Cu protrusion and its effect on subsequent device layers are shown in Figure 5.6.

IMEC first proposed stabilization of this effect through annealing of the TSVs after plating and CMP, before depositing additional layers [21]. This anneal, which is now a common practice, typically involves heating of the TSVs to temperatures approaching 400 °C or more. After some time, the extrusion of the TSV surface stops and the TSV is therefore "stabilized" for further processing. If the topography is significant, a brief "touch-up" CMP may be used to restore planarity. Improvements to Cu ECD chemistries have also been shown to limit Cu protrusion by reducing the amount of impurities incorporated into the plated Cu material [11]. TSMC reported an optimized 3D IC process that addresses the Cu protrusion issue through both ECD Cu improvements and a TSV anneal step [14].

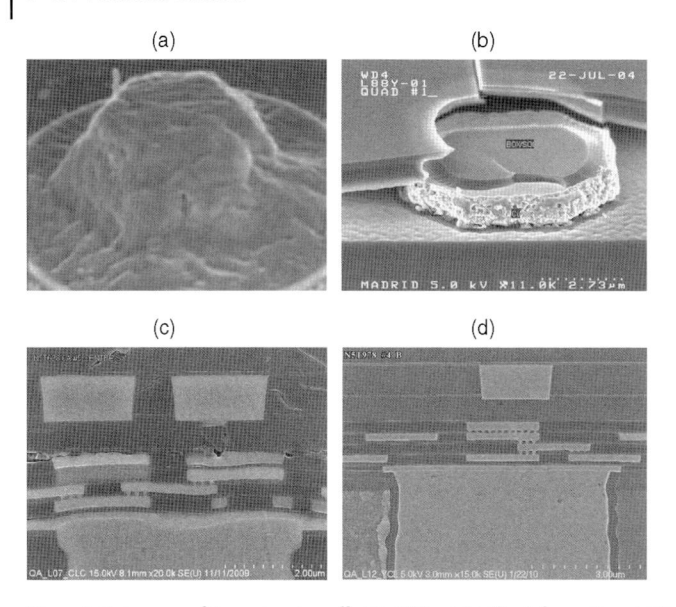

Figure 5.6 Images of Cu protrusion effect in TSVs. (a) Hillock formation at TSV surface. Courtesy of IMEC. (b) Tezzaron image of dielectric failure over TSV. (c) TSMC image showing deformation of BEOL layers over TSV. (d) TSMC's optimized process with improved plating and TSV anneal.

5.3.6
Temporary Carrier Wafer Bonding and Debonding

After the TSVs have been formed as blind vias from the front-side of the wafer, the wafer is thinned to its final target thickness and the TSVs are revealed from the backside for later 3D interconnection. These processes require temporary bonding of the device wafer or interposer substrate to a carrier wafer, for mechanical support during thinning and processing of backside layers. There has been much progress in this area in recent years. The materials and techniques used for temporary carrier wafer bonding and debonding will be covered in detail in subsequent chapters.

5.3.7
Wafer Thinning and TSV Reveal

Following temporary carrier wafer bonding, the device wafer is thinned using a combination of backgrind and CMP. In order to manage device wafer total thickness variation (TTV), it is important that the TTV of the carrier wafer and adhesive are minimized at the bonding step. Careful measurements must be made of each layer and the full stack prior to thinning, so that the TTV induced by the thinning processes can be determined and controlled. Typically, the CMP step ends below the TSVs, and then a plasma etch process is used to remove remaining Si and reveal the TSVs, which are still encapsulated in the via dielectric layers. This approach helps to ensure that the TSV insulator and barrier layers are not damaged by CMP. A dielectric layer is deposited over the backside wafer surface and TSVs, and then polished by CMP until the metal TSV conductor is exposed for interconnection. Further detail on wafer thinning and TSV reveal will be provided in subsequent chapters.

5.4 Yield and Reliability

The main yield considerations and challenges for TSV fabrication have been mentioned in the prior sections regarding individual process steps. The etching of TSVs is relatively well established. Once the process has been optimized for rate and desired sidewall characteristics, this step will be very repeatable, provided that the chamber is properly maintained for etch consistency. For the TSV liner steps, the main yield considerations are related to achieving continuous, conformal film coverage while maintaining consistent electrical insulation and barrier properties. These processes should also be reliable and consistent, once qualified. Filling of TSVs by Cu ECD is a process that requires careful controls to maintain consistency. This involves regular bath chemical analysis, additive adjustments, and bath replenishment. The main yield risk with the plating step is the formation of voids in the Cu via fill due to pinch-off in the Cu plating near the top of the TSV. Trapped void defects are somewhat difficult to detect using nondestructive methods, although some techniques are emerging, including high-resolution X-ray imaging and microwave acoustic microscopy. Once the TSVs are successfully formed, the primary concern is the control of TTV in the temporary carrier bonding and thinning processes, so that the device wafer is thinned uniformly to its target thickness.

Some of the primary reliability concerns for TSVs have been mentioned in the fabrication sections and relate mainly to the CTE mismatch between Cu and Si. This can result in stress around the TSVs, creating potential device proximity effects that can impact the performance of active devices. The problem is addressed through the implementation of keep out zones (KOZ) around TSVs, creating a safe distance between TSVs and active devices [8]. The size of the KOZ required to eliminate proximity effects is dependent on the TSV diameter and the particular device technology. The Cu protrusion effect, which was described in the TSV anneal section, is also an important reliability concern, creating the potential for thermomechanical failures in subsequent layers. This is generally addressed by limiting TSV size, optimization of ECD chemistries, and performing a stabilization anneal prior to building up additional layers of metal and dielectric over the TSVs. Plating voids can be a major reliability issue, as these can trap chemicals and later lead to contamination or corrosion concerns. These must be completely eliminated through the use of optimal ECD techniques and process monitoring, ideally using nondestructive void-detection methods. The remaining potential reliability concerns are generally related to insulator integrity or forms of interconnect degradation. The primary insulator concern would be leakage current resulting from insufficient sidewall coverage or damage induced during later process steps, such as TSV reveal or die bonding. As with any interconnect, electromigration is another potential area of concern that must be addressed through materials selection and process controls.

Acknowledgments

The author would like to gratefully acknowledge the contributions made by various RTI International technical staff members and by Dr. Philip Garrou, a lead editor for this book.

References

1 Beyne, E. (2006) 3D system integration technologies. International Symposium on VLSI Technology, Systems, and Applications, Hsinchu, Taiwan, April 2006, pp. 1–9.
2 Shapiro, M. (2008) 3-D silicon technology: applications and requirements. 5th Annual 3D Architectures for Semiconductor Integration and Packaging Conference, Burlingame, CA, November 2008.
3 Eloy, J. (2008) 3-DICs & TSV: a market analysis. 5th Annual 3D Architectures for Semiconductor Integration and Packaging Conference, Burlingame, CA, November 2008.
4 Garrou, P. (2013) 3DIC & TSV market update. Presented at Suss MicroTech's 3D Integration Workshop at *Semicon West*, San Francisco, CA, June 2013.
5 Patti, R.S. (2006) Three-dimensional integrated circuits and the future of system-on-chip designs. *Proceedings of the IEEE*, 94 (6), 1214–1224.
6 Golz, J. *et al.* (2011) 3D stackable 32nm high-k/metal gate SOI embedded DRAM prototype. Proceedings of 2011 Symposium on VLSI Circuits, Honolulu, HI, June 2011, pp. 228–229.
7 Banijamali, B., Ramalingam, S., Nagarajan, K., and Chaware, R. (2011) Reliability study of TSV interposers and interconnects for the 28 nm technology FPGA. Proceedings of 61st Electronic Components Technology Conference, Lake Buena Vista, FL, May 2011, pp. 285–290.
8 Mercha, A. *et al.* (2010) Comprehensive analysis of the impact of single and arrays of through silicon vias induced stress on high-k/metal gate CMOS performance. Proceedings of 56th IEEE International Electron Devices Meeting (IEDM), San Francisco, CA, December 2010.
9 Garrou, P., Bower, C., and Ramm, P. (2008) *Handbook of 3D Integration: Technology and Applications of 3D Integrated Circuits*, vols 1 and 2, Wiley-VCH Verlag GmbH, Weinheim.
10 Yoshikawa, H. *et al.* (2009) Chip scale camera module (CSCM) using through-silicon-via (TSV). Proceedings of 2009 IEEE international Solid-State Circuits Conference, San Francisco, CA, February 2009.
11 Beica, R., Siblerud, P., and Erikson, D. (2010) Advanced TSV copper electrodeposition for 3D interconnect applications. Proceedings of IMAPS 6th International Conference and Exhibition on Device Packaging, Scottsdale, AZ, March 2010, p. 45.
12 Temple, D., Bower, C., Malta, D., Robinson, J., Coffman, P., Skokan, M., and Welch, T. (2006) High density 3-D integration technology for massively parallel signal processing in advanced infrared focal plane array sensors. Proceedings of 52nd IEEE

International Electron Devices Meeting, San Francisco, CA, December 2006.
13. Laermer, F. and Schilp, A. (1996) Method of anisotropically etching silicon. US Patent 5,501,893.
14. Lin, J.C. *et al.* (2010) High density 3D integration using CMOS foundry technologies for 28nm node and beyond. Proceedings of 56th IEEE International Electron Devices Meeting (IEDM), San Francisco, CA, December 2010.
15. Riedel, S. *et al.* (2000) Investigation of the plasma treatment in a multistep TiN MOCVD process. *Microelectronic Engineering*, 50, 533–540.
16. Liu, Z., Chiu, J., Keigler, A., and Zhang, J. (2008) Reliable and economical TSV copper filling. 5th Annual 3D Architectures for Semiconductor Integration and Packaging Conference, Burlingame, CA, November 2008.
17. Ramm, *et al.* (2002) Interchip via technology by using copper for vertical system integration. Proceedings of Materials Research Society Advanced Metallization Conference 2001, vol. 17, p. 159.
18. Garrou, P. (2009) Researchers strive for copper TSV reliability. Semiconductor International, December 3, 2009.
19. Huyghebaert, C. *et al.* (2010) Enabling 10 μm pitch hybrid Cu–Cu IC stacking with through silicon vias. Proceedings of 60th Electronic Component and Technology Conference, Las Vegas, NV, June 2010, pp. 1083–1087.
20. Malta, D. *et al.* (2011) Characterization of thermomechanical stress and reliability issues for Cu-filled TSVs. Proceedings of 61st Electronic Component and Technology Conference, Lake Buena Vista, FL, May 2011, pp. 1815–1821.
21. Okoro, C. *et al.* (2010) Elimination of the axial deformation problem of Cu-TSV in 3D integration. Proceedings of the 11th International Workshop on Stress-Induced Phenomena in Metallization, Dresden, Germany, April 2010, *AIP Conference Proceedings*, 1300, 221–228.

6
TSV Unit Processes and Integration

Sesh Ramaswami

6.1
Introduction

After decades of conceiving "integrated circuits" in two dimensions, the industry is extending integration to the third dimension with through-silicon via (TSV) technology. The use of TSVs in devices for power, signal, and ground circuits increases interconnect bandwidth, reduces latency (delay), decreases form factor (size), and improves power efficiency [1–6]. It also offers the freedom to integrate different device types (memory, RF, sensors, etc.) supplied from different technology nodes and platforms.

The first few applications of TSVs are in field-programmable gate arrays (FPGAs) [7,8], DRAM memory stacks [9], and logic–memory stacks for mobile technologies. DRAM devices are challenging to scale per Moore's law [10] and greater densities are obtained not only by lithography scaling, but also by 3D stacking. This 3D approach satisfies both performance and form factor needs for future end products [11,12]. Longer term, stacked DRAM products can significantly benefit from the transition from chip-to-chip (C2C) or chip-to-wafer (C2W) bonding [13] to wafer-to-wafer (W2W) bonding technologies [14,15]. Similar architectures can be envisioned, with stacks of logic–memory, logic–RF, and others.

In true 3D stacking, TSVs run through the active chips. The parallel "2.5D" approach based on interposer technology enables die to be laid side by side (adjacent) and attached to a larger piece of silicon (the interposer) with high-density interconnects between them. Silicon interposer is gaining momentum in the industry for homogeneous and heterogeneous integration. This technology creates new product opportunities for FPGAs as well as other high-performance applications. Early indications suggest that integration of logic and memory for applications such as GPUs, wireline communications, and high-performance computing will be the leading high-volume application of silicon interposer. GPUs or multicore CPUs and DRAMs placed at close proximity and connected with tight pitch copper interconnects reduce data latency and power consumption. In this technology, microbumps connect the die to the interposer. Multilevel copper lines running horizontally through the interposer connect the die to each other with dense connectivity. TSVs running vertically through the silicon thickness connect to the substrate. (Since TSVs do not run through the active die, the dies are left untouched and remain as they were in the wafer fab.) All the TSV and redistribution layer (RDL) processing takes place on the simpler (no transistors or devices) interposer.

TSV diameter, depth (desired die thickness), pitch, and pattern density are the key areas where end-product design, process integration, process window, unit process robustness, and cost intersect. Since TSVs extend through the thickness of the die, the TSV depth (wafer thickness) can be determined by several factors: (a) lowest die thickness that can be successfully handled after the wafer is diced, (b) structural integrity (interlayer delamination, warpage), (c) cracking during dicing or assembly, (e) thermal management (spreading hot spots and overall heat dissipation), (f) form factor (thickness) of multi-die stacks, (g) capacitance, and (h) post-assembly yield. Considering these factors, the "sweet spot" for wafer thickness can range from 30 to 120 μm and TSV diameters range between 4 and 10 μm, with roadmaps extending down to 2 μm. This puts the aspect ratio (depth/diameter) of the via in the range of 5 : 1 to 15 : 1.

6.2
TSV Process Overview

This chapter provides an overview of unit process and process integration technology to enable the realization of TSVs for 3D chip stacking and 2.5D interposer implementations. In both architectures, the central element is the TSV itself, whose fabrication is subject to many unit process and process integration challenges [16,17]. Typically, the primary steps in forming the via are deep silicon etching, chemical vapor deposition (CVD) of an insulating oxide liner, physical vapor deposition (PVD) of a metal barrier and seed layer, electrochemical deposition (ECD) of copper to fill the via hole, and chemical–mechanical planarization (CMP).

TSV processes are done on full-thickness wafers or on thinned wafers as shown in Figure 6.1. In the via-middle flow, vias are created after transistor formation but before the formation of all or some of the back-end-of-line (BEOL) damascene interconnects. Via reveal is a companion process to via middle and is done on the backside of wafers after thinning. In the via-last flow, vias are created from the backside of thinned wafers. These are explained in more detail later in the chapter.

Optimization of etch rate, profile, and related parameters is well understood for typical aspect ratios ranging from 5 : 1 to 15 : 1 (typically vias 4–10 μm in diameter and 40–100 μm deep). A CVD process for dielectric liner deposition (from 0.25 to 1.0 μm on the sidewall) has demonstrated approximately 100% sidewall step coverage in such vias. PVD of titanium or tantalum barriers and copper seed layers has been co-optimized with ECD to ensure void-free metal filling. Copper, barrier, and oxide films on the top surface are removed by CMP using end-point and process controls to achieve uniform transition between layers and preserve surface

6.2 TSV Process Overview

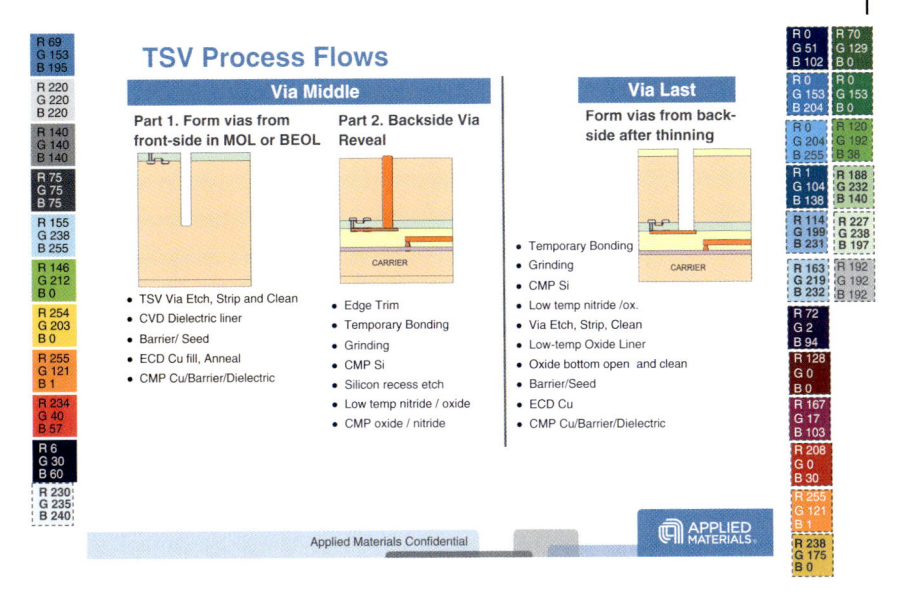

Figure 6.1 TSV flows for via middle, via reveal, and via last.

topography. TSV structures have been characterized for capacitance–voltage (C–V), leakage (I–V), time-dependent dielectric breakdown (TDDB), and impact on stress and performance of nearby transistors. These are discussed later in the chapter.

In the case of interposer fabrication, via-middle TSVs are created on a full-thickness silicon wafer (with no active components) followed by approximately three layers of copper interconnect (damascene) to enable high-density inter-die connectivity. Such interposers typically have 10 μm diameter, 100 μm deep TSVs and 1 μm line/space (L/S) copper interconnects with an aspect ratio of approximately 2. Interposers use the same process and integration learning as developed for the mainstream via-middle and via-reveal applications discussed in this chapter. Table 6.1 details the TSV creation process for via middle and interposer.

Table 6.1 Process steps comprising the TSV creation for via middle and interposers.

Via-middle formation unit process steps	Key process attributes
Lithography	Resist thickness, CD uniformity, selectivity for etching
Silicon etching	Etch rate, via profile (scallops, striations, tapers), selectivity, via depth uniformity
Via cleaning	Post-etch residue removal and surface treatment
Dielectric oxide liner deposition	Step coverage, dielectric constant, leakage, breakdown voltage
Metal barrier/seed deposition	Step coverage, continuity, Cu barrier properties, Cu seed layer for filling
ECD via filling	Void-free filling, stability, overburden thickness minimization
Annealing	Film stability, protrusions, microvoids
CMP of Cu/barrier/oxide	Uniformity, flatness, topography

Table 6.2 Process steps comprising via-reveal flow.

Via-reveal formation unit process steps	Key process attributes
Edge trimming	Trim the edge of the product wafer around the circumference to allow for bonding to a 300 mm carrier
Temporary bonding of TSV wafer (face down to silicon or glass carrier)	Bond temperature/method, bond quality, total thickness variation, adhesive outgassing, shrinkage, carrier-wafer edge engineering
Wafer thinning (grinding and CMP)	Total thickness variation of product wafer, defects, contamination
Silicon etching to recess and reveal via	Silicon etch rate, selectivity (to oxide, barrier, copper), uniformity, defects
Wet cleaning	Surface cleanliness for good adhesion
Low-temperature CVD oxide and/or nitride or polymer deposition	Temperature (compatible with bonded wafer), adhesion, stress, hermeticity
CMP of oxide/nitride/polymer	Coplanarity, defects, uniformity

In the via-reveal flow (Table 6.2), the via-middle TSVs described above (in Table 6.1) are opened or revealed from the back of the wafer after thinning. This integration is more completely described later in the chapter.

In via-last TSVs, the vias are created from the back of the wafer after thinning (Table 6.3). Just as in the via-reveal scheme above, this process occurs on thinned wafers that are temporarily bonded to carriers.

6.3
TSV Unit Processes

6.3.1
Etching

After lithography, vias are etched through the silicon using a deep reactive ion etching (DRIE) process. Via etching sets the foundation for successful integration. It involves a combination of dielectric etching (since a pre-metal dielectric layer at least is typically present on the silicon wafer surface) and deep silicon etching. Key TSV etch requirements are smooth sidewall, minimal re-entrance or undercut between layers, etching of dielectric and silicon in the same chamber, high resist selectivity, and depth uniformity across the wafer. The etch reactor alternates between gases that etch the silicon and gases that deposit a protective passivation coating using polymerizing chemistries. This process creates a series of etched "scallops" in the silicon (sidewall), which can vary in size depending on process parameters. With advanced reactor design, source and wafer bias power, power modulation, and gas switching techniques, scallops can be virtually eliminated, producing substantially smooth sidewalls at high etch rates.

Table 6.3 Process steps comprising via-last flow.

Via-last formation unit process steps	Key process attributes
Temporary bonding of TSV wafer (face down to a silicon or glass carrier)	Bond temperature/method, bond quality, TTV, adhesive outgassing, shrinkage, carrier-wafer edge engineering
Wafer thinning (grinding and CMP)	TTV of product wafer, defects, contamination
Lithography	Resist thickness, CD uniformity, selectivity for etching
Silicon etching and stripping	Etch rate, via profile (scallops, striations, tapers), selectivity, undercut, via depth uniformity (stopping on Cu interconnect lines), resist stripping (compatibility with adhesive system)
Via cleaning	Compatibility of post-etch residue removal with exposed Cu and adhesive system, residue-free sidewall
Low-temperature dielectric oxide liner deposition	Step coverage, mechanical properties, leakage, breakdown voltage, capacitance, temperature compatibility with adhesive
Bottom oxide-open etching	Oxide etching (directional) to remove the liner oxide from the via bottom to enable metal–metal contact
Resist stripping	Low-temperature adhesive- and carrier-compatible photoresist stripping
Wet cleaning	Adhesive- and metal-compatible wet cleaning
Metal barrier/seed deposition	Step coverage, Cu barrier properties, Cu seed layer for filling, temperature compatibility with adhesive, wafer chucking compatibility with carrier type
ECD via filling	Void-free filling, stability, uniformity
Annealing	If used, temperature compatibility with adhesive
CMP of Cu/barrier/oxide	Flatness, topography, uniformity

For via-last etching, the etching process (on thinned product wafers temporarily bonded to silicon or glass carriers) and post-etch resist stripping (temperature and chemistry) need to be compatible with the adhesive system and the underlying metal (tungsten or aluminum) on which the via lands. Additionally, the post-etch wet cleaning should not attack the exposed metal nor the exposed surface of the adhesive at the carrier-wafer edge.

6.3.2
Insulator Deposition with CVD

After etching and post-etch cleaning, a CVD process is used to line the via holes with a SiO_2 dielectric. Because the dielectric liner functions as electrical insulation and affects the capacitance of individual vias, variation in dielectric thickness can lead to current leakage, inconsistent reliability, and inconsistent signal delays through the TSV arrays. Studies have shown that both subatmospheric (SA) and plasma-enhanced (PE) CVD processes can be suitable for TSV dielectric liner applications.

SACVD is ideally suited for schemes in which temperatures of approximately 400 °C or higher are permitted. SACVD, a thermally activated process, is insensitive to mass transfer effects in high aspect ratio structures. SACVD technology uses production-proven O_3/TEOS chemistry. Its films exhibit sidewall coverage of 100% in high aspect ratio structures. This enables continuous PVD metal barrier and seed layers, which in turn are critical for subsequent void-free copper filling. SACVD film stacks with 2 to 10 000 kÅ sidewall thickness meet capacitance requirements of typical TSV structures.

The temperature limitations of the via-last approach tend to favor PECVD or low-temperature CVD processes, which operate below 200 °C.

6.3.3
Metal Liner/Barrier Deposition with PVD

The most common approach for forming a conductive pathway through the etched via involves depositing a PVD or sputtered barrier/seed layer followed by ECD copper filling. The barrier layer acts as a diffusion barrier layer for copper, while the copper seed layer provides the conductivity for the subsequent ECD filling. The key requirement is to create a continuous barrier layer (typically tantalum nitride/tantalum (TaN/Ta) or titanium (Ti)) and a continuous copper seed. When used in a via-last application, the wafer temperature must be kept below the safe limit for the bond adhesive (<180 °C). Also, the PVD system and process need to be compatible with silicon and glass carriers. Since via-last TSVs land on an underlying metal (in the interconnect), it is important for the barrier/seed deposition to occur on a clean surface. Typically, sputter precleaning using argon or a reactive preclean chemistry is needed prior to the PVD. It is also important to ensure that the adhesive between the device wafer and carrier is fully cured, since any residual solvent or moisture will outgas in the 10^{-9} Torr high-vacuum environment of the PVD equipment, thereby contaminating the device. From a roadmap perspective, it is foreseen that as TSV dimensions shrink and the via aspect ratio approaches 20:1, CVD or ALD barrier/seed deposition techniques may need to be used, though ongoing efforts are made to extend PVD further.

6.3.4
Via Filling by ECD of Copper

The main requirements for ECD of TSV structures are (a) void-free filling, (b) stable copper microstructure, (c) minimal contamination, (d) high filling rate, (e) minimal overburden, and (f) good thickness uniformity. Void-free filling of high aspect ratio structures requires bottom-up, superconformal filling technology, in which deposition at the bottom of the feature is accelerated while that at the top and sidewall is suppressed. The electrolytic deposition process is controlled primarily by the electrical potential between wafer and electrolyte, and the current applied to the electrolyte. Hence, the ECD reactor must optimize the current density uniformity across the wafer. Copper and additives are transported inside the via more by diffusion and migration than by fluid flow. Since the metallic ion concentration is higher at the via mouth than at the via bottom, there is a propensity for pinch-off at the top, which creates an enclosed void inside the via. The process is more completely described in Ref. [18].

6.3.5
CMP of Copper

The polishing process removes copper, barrier metal, and dielectric layers while maintaining tight control over topography and defects. Although the post-ECD copper overburden in TSV is an order of magnitude thicker than conventional back-end-of-line copper, copper slurries for TSV operate at a high polishing rate, thereby enabling lower cost. Process criteria [19] for effective TSV CMP include the following:

a) High planarization efficiency, equating to a high copper/barrier/oxide removal rate while meeting post-CMP profile requirements.
b) End-point capability and profile control, enabling within-wafer and wafer-to-wafer control with tunable post-CMP topography.
c) High throughput, enabled by high removal rate.
d) In-line cleaning capability.
e) Low cost of ownership, co-optimized with prior CVD/PVD/ECD steps.

Real-time process control (RTPC) enables a uniform cross-wafer copper thickness profile to be maintained during the bulk copper removal step. Advanced CMP tools are equipped with sophisticated end-point and process control technologies, and a variety of sensors. Using a three-platen system as an example, a sensor on platen 1 continuously monitors across-wafer copper film thickness and automatically updating polishing head zone pressures to achieve and maintain a flat thickness profile. Once the desired thickness profile has been achieved, the sensor automatically triggers an end to the platen 1 polishing step, thus preventing overpolish. Once the wafer has moved to platen 2, a different sensor monitors the copper clearing step in real time, ensuring minimal dishing with no copper residue. On platen 3, a third sensor monitors barrier clearing and senses the stopping point when the dielectric is revealed. This end-point technology can stop the polish either within a dielectric layer or at the interface of an underlying stop layer. For backside processing, CMP of thinned, TSV device wafers (typically 20–100 μm thick) bonded to glass or silicon carriers can be damaged at the wafer edge if the wafer and carrier are misaligned in the temporary bonding system.

6.3.6
Temporary Bonding between Carrier and Device Wafer

Device wafers are typically bonded to silicon or glass carriers with a temporary adhesive before being taken through thinning and backside processing. The glass or silicon carrier mechanically stabilizes the device wafer, preventing bending or warping, and protects the wafer edge. Some of the basic requirements of this module include dimensional tolerance of the bonded pair (diameter, thickness), alignment between the product wafer and carrier, edge treatment and profile, (minimal) exposed adhesive, and adhesive compatibility with post-bonding/post-thinning vacuum, plasma, and wet processes.

Other techniques that do not involve bonding (but involve a device wafer being held by a support structure during processing) have been considered. These include vacuum chucks, electrostatic chucks, Bernoulli chucks, and so on. However, while these techniques may be used to temporarily hold a device wafer, none of them has demonstrated the compatibility with wafer processing described in this chapter.

A typical process for temporary bonding with spin-on adhesives starts with fully processing the device wafer on the front-side. Then, either (or both) the carrier or device wafer is (are) coated with an adhesive and moved to the bond chamber, where vacuum bonding occurs at a specified temperature. The adhesive must fully encase the microbumps after coating and curing (e.g., for a 30 μm microbump, a 40 μm adhesive coating is typically used). The temporary adhesive must be thermally and chemically stable and must resist decomposition and outgassing during processes such as via etching and liner deposition by CVD or PVD. The material must also be resistant to the chemicals used for wafer cleaning, ECD, and CMP to preserve the bond strength and edge integrity. The bonded wafer–wafer pair must withstand backside processing steps, including grinding (thinning), etching and cleaning, dielectric liner deposition, barrier/seed deposition, via filling, redistribution layer deposition, and solder bumping. After processing is completed, the device wafer is separated from the carrier, transferred to dicing tape, diced, and then taken through the die-to-wafer stacking process. When the device wafer is debonded, it is important that the microbump surface (especially at the corners) and the wafer surface be free of any residual adhesive.

6.3.7
Wafer Backside Thinning

Wafers are thinned from 775 μm (typical prime silicon thickness for 300 mm wafers) to 30–150 μm while bonded to a carrier wafer, as described above. The key requirements of this process step are as follows:

a) Uniform thickness, 2 mm to the wafer edge, with a total thickness variation (TTV) of <1 μm.
b) Minimal damage to residual silicon.

c) Negligible (nanometer scale) roughness, free from silicon residue, metallic contamination, and defects.
d) Minimal residual stress in the bulk silicon that could lead to post-singulation die breakage.
e) No edge breakage or cracking.
f) No edge/bevel damage.

Mechanical thinning is the most common technique to start the wafer thinning sequence. A two-step process is commonly used in which a coarse grinding step removes most of the backside silicon and a subsequent fine grinding step reduces surface roughness and removes damage caused by coarse grinding. Even after fine grinding, a defect band may remain near the surface. Grinding is followed by CMP. CMP Si and post-CMP cleaning provide stress relief, consistent wafer–wafer thickness and within-wafer thickness control, and lower defects.

6.3.8
Backside RDL

RDLs are made using a semi-additive or subtractive process. The former uses photodefined organic dielectrics (e.g., polyimide), while the latter uses inorganic (oxide, nitride) dielectrics. After the dielectric layer is patterned, the next steps are barrier/seed sputter deposition, electrochemical deposition, etching, and wet cleaning.

6.3.9
Metrology, Inspection, and Defect Review

The purpose of automated metrology, defect inspection, and review is to provide the information needed for measurement, process control, and quality control at various stages in the fabrication process, some of which are listed below:

a) TSV critical dimension (CD), top opening.
b) TSV silicon etch depth and uniformity in bulk silicon (via middle and interposer).
c) TSV silicon etch depth and uniformity on bonded wafers, with metal (via last).
d) Edge and bevel inspection of the device wafer.
e) Alignment (notch-to-notch, carrier-to-device wafer prior to bonding).
f) Adhesive thickness and TTV and defects.
g) Wafer carrier thickness and TTV.
h) Bond quality (voids, adhesive texture/smear, bubbles).
i) Post-grinding thinned wafer thickness and TTV.
j) Post-grinding inspection (particles, scratches, cracks, roughness).
k) Edge bead removal (EBR) measurement of the adhesive.
l) Edge trim measurement.
m) Silicon thickness over the copper TSV (post-thinning, from the backside).

n) TSV pillar height (when exposed from the backside with recess etching).
o) Post debonding clean inspection (residue on microbump/pillar).
p) Post via-reveal TSV pillar CD.
q) Post via-reveal TSV pillar height coplanarity and surface defects.
r) Microbump/copper pillar coplanarity and defects.

6.4
Integration and Co-optimization of Unit Processes in Via Formation Sequence

TSV etching is typically tuned to achieve the highest etch rate to maximize productivity, while maintaining the required via profile with minimal scallops and no defects. In via-middle flows, after via cleaning, O_3/TEOS oxide deposition occurs with equipment and process optimized for high step coverage across dense and isolated structures, with demonstrated extendibility down to 2–3 μm TSVs. Sidewall thicknesses can range from 2 kÅ to 1 μm, based on the end application (Figure 6.2a). Recipes are tuned to optimize capacitance, leakage, breakdown field, and TDDB. Previously published modeling results have shown the impact of the liner's intrinsic properties and cohesive strength on TSV coupling capacitance, RC delay, and TSV reliability [20].

A PVD barrier layer of TaN/Ta or Ti is deposited to achieve adequate step coverage, especially at TSV bottom corners. Ti is preferred to Ta-based barriers in terms of cost. PVD copper seed deposition processes are optimized for step coverage, symmetry, and film morphology. Improved ECD processes exhibit enhanced bottom-up filling, significantly lower overburden, and more stable microstructure (Figure 6.2b). Overburden is defined as the thickness of the copper in the field region after the via is filled. Early plating processes (typically associated with conformal electroplating deposition) deposited overburden equal to 0.5–0.75

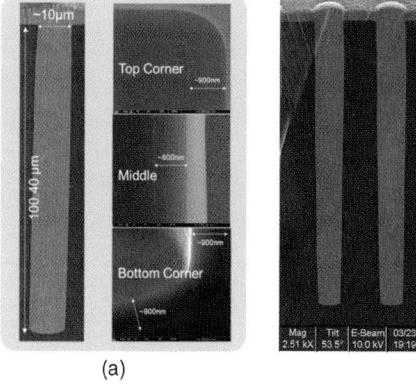

Figure 6.2 (a) Conformal sidewall oxide in 10 μm × 100 μm interposer TSVs. (b) Result of enhanced bottom-up filling process of 10 : 1 aspect ratio structure with low copper overburden in the field region.

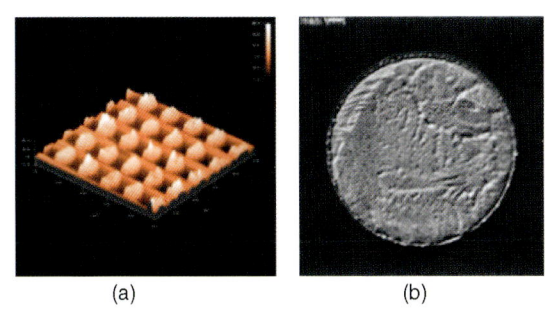

Figure 6.3 (a) Topography resulting from optimized ECD filling process. (b) Top view of a TSV (post-CMP) with low copper protrusion.

of the TSV diameter, or up to several microns in thickness for large vias. This high thickness added stress to the wafer and caused excessive wafer bowing, which interfered with post-ECD steps (especially CMP) and/or induced breakage. A thick copper overburden also increases the cost of copper CMP. Enhanced bottom-up ECD filling processes outlined above have reduced overburden and thus the cost of the subsequent CMP process needed to remove the copper and barrier material from the field region is lower. The selection of the CMP process and slurry is governed by three key requirements: (a) clean copper and barrier removal, (b) no attack or corrosion of the copper in the via or the barrier metal on the sidewall, and (c) no divot or attack of the oxide liner on the inner circumference of the via.

Copper is typically annealed between the ECD and CMP processes at 410–450 °C in forming gas for approximately 30 min to stabilize its microstructure and film composition. This step is critical for reducing copper protrusion height (Figure 6.3) to achieve stable copper microstructure and interconnect structural integrity during post-CMP annealing or thermal cycling steps. All subsequent processes, such as BEOL damascene processing and final annealing, must be done at or below this temperature to minimize protrusion and avoid the risk of dielectric cracking and intermetal shorts. Copper protrusion (sometimes referred to as "copper pumping" or "copper pistoning") is affected by ECD chemistry composition and process [18], TSV geometry (shape and volume) [21], copper microstructure, post-ECD annealing process conditions [22], and the mismatch between the coefficients of thermal expansion (CTEs) of copper and silicon. Since copper protrusion can scale with volume, smaller TSVs are preferred for via-middle applications (Figure 6.3b). Several details regarding process integration can be found in Refs [23,24].

6.5
Co-optimization of Unit Processes in Backside Processing and Via-Reveal Flow

Vias created in the middle of line need to be exposed from the backside of the wafer so that connections can be made. This process is known as "via reveal,"

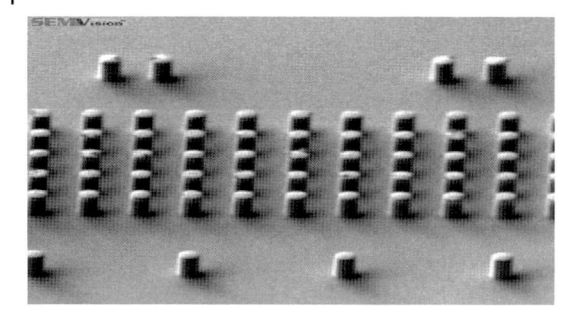

Figure 6.4 SEM image of via- of the backside of a wafer showing revealed TSV's.

"tip reveal," or "backside contact," and is the companion process to via middle, which means that every wafer processed through via middle has to be processed through via reveal. Wafers are first bonded face down on a temporary carrier, and then mechanically thinned by grinding. The first challenge is managing the TTV of the bonded stack and the thinned device wafer. This variation can emanate from the uniformity of one or all of the following: (a) carrier thickness, (b) adhesive thickness, (c) via-middle etch depth, and (d) grind thickness. Hence, successful backside via integration requires that the process chosen accounts for these variations. CMP of silicon can be used to improve TTV and create a smooth, defect-free surface. More commonly, backside wafer thinning (grinding followed by silicon CMP) stops short of the TSV; the TSV remains below the surface of the thinned silicon wafer to ensure that copper is not exposed. After CMP, highly selective silicon recess etching is used to expose the vias without damaging the liner oxide encasing them (Figure 6.5).

After the vias are revealed, low-temperature CVD nitride/oxide layers are deposited for isolation/passivation. The nitride serves as a copper diffusion barrier and stress compensator, and the oxide provides strength to avoid potential pillar knock-off during the downstream dielectric CMP step. It is important for the CVD process to have high step coverage so that a seam is not formed at the base of the protruding via (Figure 6.5).

In an optimized via-reveal sequence, the dielectric CMP process planarizes the resultant pillar and exposes the copper TSVs with minimal defectivity (scratches, pillar knock-off, contaminants, etc.) and smooth tunable topography leaving a robust matrix of dielectric insulation between them (Figure 6.6) for the subsequent RDL/bump process. The above process is more completely described in Ref. [17].

As shown in Figure 6.7, poor step coverage of conventional CVD passivation film could cause divots (circular recesses in dielectric caused by seam). These can potentially lead to failures at electrical or reliability testing.

Conventional PECVD process

Seams formed at corners acting as leakage paths for moisture

Improved PECVD process

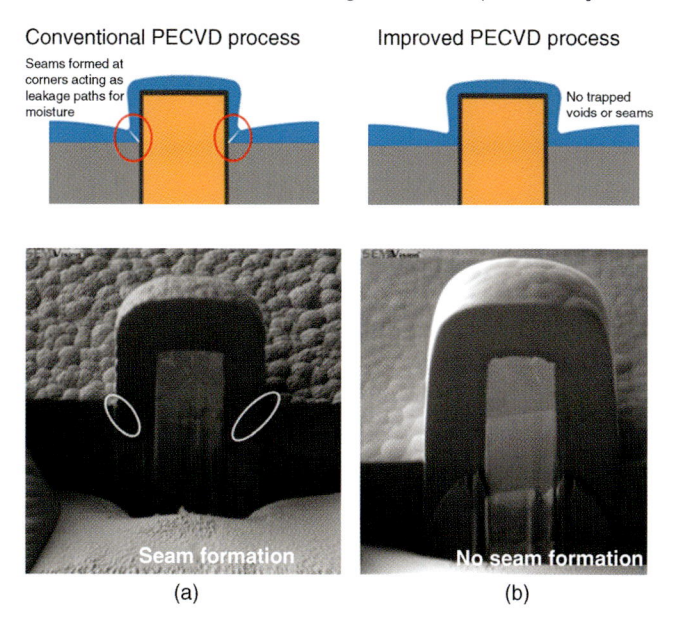

No trapped voids or seams

(a) Seam formation

(b) No seam formation

Figure 6.5 (a) Conventional PECVD process with poor coverage resulting in seam at pillar bottom/corner. (b) Improved nitride/oxide process with no seam at pillar bottom corner.

6.6
Integration and Co-optimization of Unit Processes in Via-Last Flow

After thinning, a low-temperature CVD silicon nitride (SiN) is deposited to seal the backside or stress-compensate the wafer. Thermal budget is a significant concern, since device wafers are processed while temporarily bonded to carriers; all

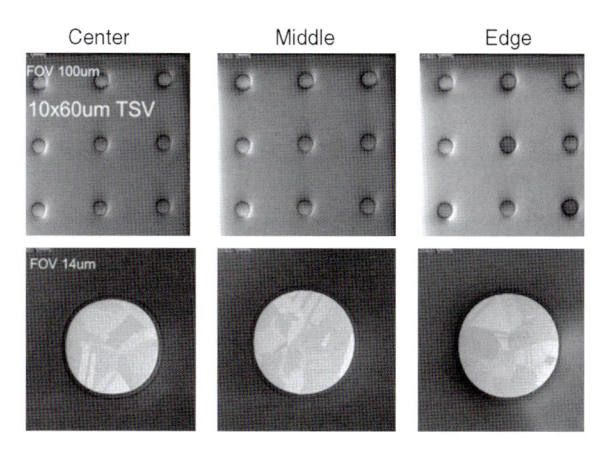

Figure 6.6 Top-down images of post-CMP revealed copper vias, with silicon nitride/oxide between them.

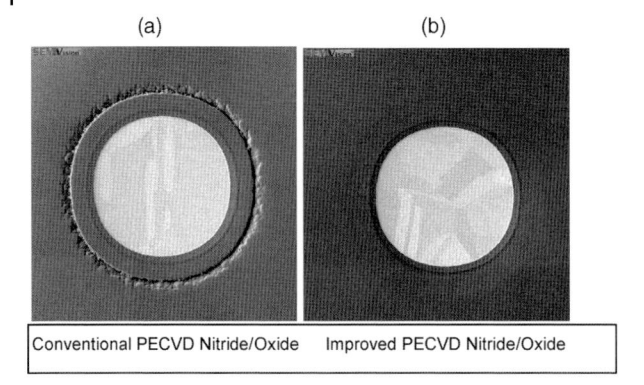

Figure 6.7 Top-down SEMs following final CMP show (a) circular divot footprint of conventional PECVD compared with (b) smooth copper and dielectric surface achieved with optimized CVD nitride/oxide process.

processing conducted on temporarily bonded wafers needs to be done at low temperature (e.g., below 200 °C), compatible with the adhesive used to temporarily bond the wafer to the carrier. The TSVs are created from the wafer backside so that the vias land on an underlying metal layer in the device wafer interconnect, thereby creating a back-to-front connection. While via-middle TSVs are subjected to 10 or more thermal heat cycles in a typical logic damascene flow, the only heat cycles the via-last TSVs are exposed to are the ones associated with final in-fab annealing with solder reflow or copper-to-copper bonding during final packaging.

6.7
Integration with Packaging

Other advanced packaging concerns listed below need to be qualified with the insertion of the TSV module.

a) Integration with ultralow dielectric constant films, thermal cycles, and package stress.
b) Stand-off heights (between chips) and underfill material.
c) Die thickness and its impacts.
d) Die warpage.
e) Die cracking during dicing.
f) Heat dissipation and hot spots: for DRAM and low-power logic + DRAM chip stacks.

6.8
Electrical Characterization of TSVs

Several key TSV parametric attributes can be evaluated electrically at the wafer level [25]. In a 3D or 2.5D stack, it is important for the capacitance of the TSV to be low

6.8 Electrical Characterization of TSVs

so that little power is lost to capacitive charging and little crosstalk occurs between neighboring TSVs. It is likewise vitally important for the leakage current from the TSV to the silicon body to be low. Finally, the breakdown voltage (V_{bd}) of the oxide must be sufficiently high for the device to perform reliably with a long service life.

The paragraphs that follow describe a study in which TSVs were formed in p-doped silicon wafers with a doping concentration of $10^{15}\,\text{cm}^{-3}$. Before TSV patterning, 2 kÅ of CVD oxide was deposited to simulate the pre-metal dielectric layer on a device wafer. Vias 5 μm in diameter and 50 μm deep were etched as described earlier. A proprietary SACVD oxide liner of 220 nm sidewall thickness was deposited. PVD barrier and seed depositions were picked to ensure a minimum thickness of 50 Å barrier and seed at the thinnest point. The vias were filled by copper ECD in a chamber with special agitation and field-shaping features, using a programmed controlled-current waveform. After plating, the wafer was annealed at 400 °C for 30 min in forming gas. CMP was performed by a three-platen sequence with end-point control.

A comb pattern was made with aluminum traces 9 μm wide above the copper TSVs, leading to 210 μm × 180 μm probe pads. Figure 6.8 shows schematics of the aluminum traces and the TSVs connected by them, as well as an optical micrograph of a portion of the structure at probe test. Three different comb-pair structures were used, each pair consisting of one comb with TSVs descending from it and a corresponding "blank" comb without TSVs. Each of the three pairs had a different pitch or center-to-center spacing between the TSVs: 10, 25, and 50 μm, respectively. During measurement, the terminal pad was contacted by one probe tip, and the grounded metal chuck holding the wafer provided the return path to

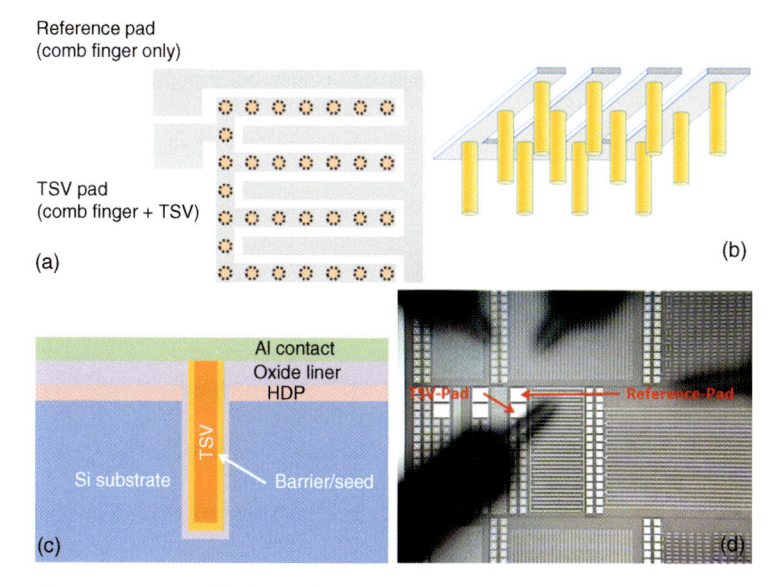

Figure 6.8 Structure for electrical testing: (a) top-view schematic, (b) 3D-view schematic, (c) cross-sectional schematic, and (d) optical micrograph.

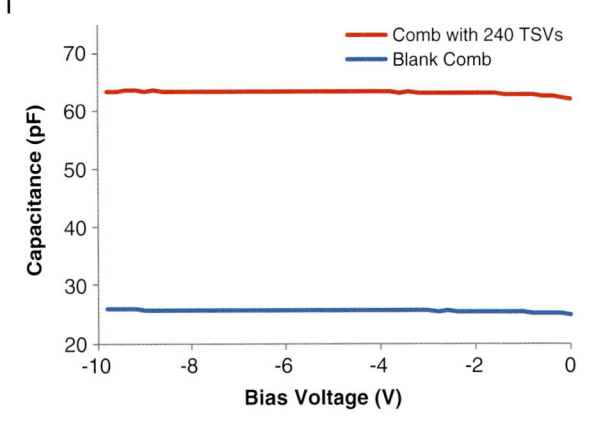

Figure 6.9 QSCV measurement of accumulation capacitance of combs with and without TSVs.

the LCR (inductance–capacitance–resistance) meter. Capacitance measurements were made on the TSV-populated comb and its corresponding blank comb so that the net capacitance of the vias could be estimated by the difference between the two values.

TSV capacitance was measured by sweeping the bias voltage from -10 to 10 V with a small-signal AC component at 1 MHz. TSVs were located in three regions where they were 10, 25, and 50 µm apart. The comb structures were designed such that for each of these three pitches, the total number of comb fingers and the comb finger length were all different. Corresponding TSVs totaled 522, 256, and 240 for the 10, 25, and 50 µm pitches, respectively. Quasi-static capacitance–voltage (QSCV) technique is frequency independent and has high measurement sensitivity [26]. We used this method to extract accumulation capacitance C_{ox}. Figure 6.9 shows the QSCV measurement results for TSV-populated combs and blank combs for 50 µm pitch TSVs in the accumulation region. By subtracting the two and dividing by the number of TSVs, we obtained a C_{ox} of 157 fF per TSV.

It has been suggested that operating within the minimum depletion capacitance regime can effectively reduce the total TSV capacitance, which is desirable for 3D integrated circuits [27]. Considering that the TSV bottom region does not contribute much to the total capacitance, a TSV can be modeled as a cylindrical metal–oxide–silicon structure. Thus, the minimum depletion capacitance is

$$C_{\text{dep min}} = \frac{2\pi \varepsilon_{Si} L_{TSV}}{\ln(R_{max}/R_{ox})}, \tag{6.1}$$

where ε_{Si} is permittivity of silicon, L_{TSV} is TSV depth, and R_{ox} is outer radius of the oxide liner. The maximum depletion radius, R_{max}, is obtained by solving [27]

$$\frac{qN_a}{4\varepsilon_{Si}}(R_{ox}^2 - R_{max}^2) + \frac{qN_a R_{max}^2}{2\varepsilon_{Si}} \ln\left(\frac{R_{max}}{R_{ox}}\right) = 2\frac{k_B T}{q} \ln\left(\frac{N_a}{n_i}\right), \tag{6.2}$$

where k_B is the Boltzmann constant, T is absolute temperature, q is charge of an electron, N_a is doping concentration, and n_i is intrinsic carrier concentration of silicon. From Equation 6.2, R_{max} is 3.45 μm. Substituting into Equation 6.1, we obtained a $C_{dep\,min}$ of 119 fF.

The minimum TSV capacitance is series combination of C_{ox} and $C_{dep\,min}$:

$$C_{TSV\,min} = \frac{1}{(1/C_{ox}) + (1/C_{dep\,min})}. \qquad (6.3)$$

Using the above C_{ox} and $C_{dep\,min}$ values, the minimum per TSV capacitance is calculated to be 67 fF.

TSV capacitance at different pitches is illustrated in Figure 6.10. Nine dies across the wafer were checked for each of the three pitches. For 10, 25, and 50 μm pitch TSVs, the minimum average capacitance per TSV is 64, 66, and 64 fF, respectively. This is in good agreement with the calculated value from Equation 6.3. It is also worth noting that in all cases the capacitance shows very little variation. Similar findings that TSV capacitance is essentially independent of inter-via spacing have been reported in the literature [28]. Repeat C–V measurements were performed at 10 and 100 kHz in both series and parallel modes. The results, while not shown here, are consistent with the reported data.

Figure 6.11 shows the leakage current distribution in accumulation and depletion modes, where the TSVs are biased at ±5 and ±10 V. It can be seen that under low bias (±5 V), the leakage currents per TSV in accumulation and depletion are close. Also, in accumulation mode the TSV leakage is not as sensitive to bias voltage as in depletion. In both cases, the leakage current is below 1 pA [29].

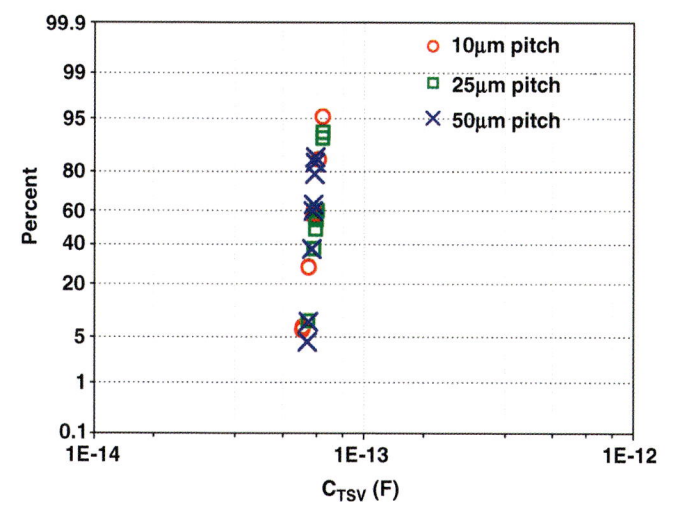

Figure 6.10 Distribution of minimum capacitance per TSV for different via spacings.

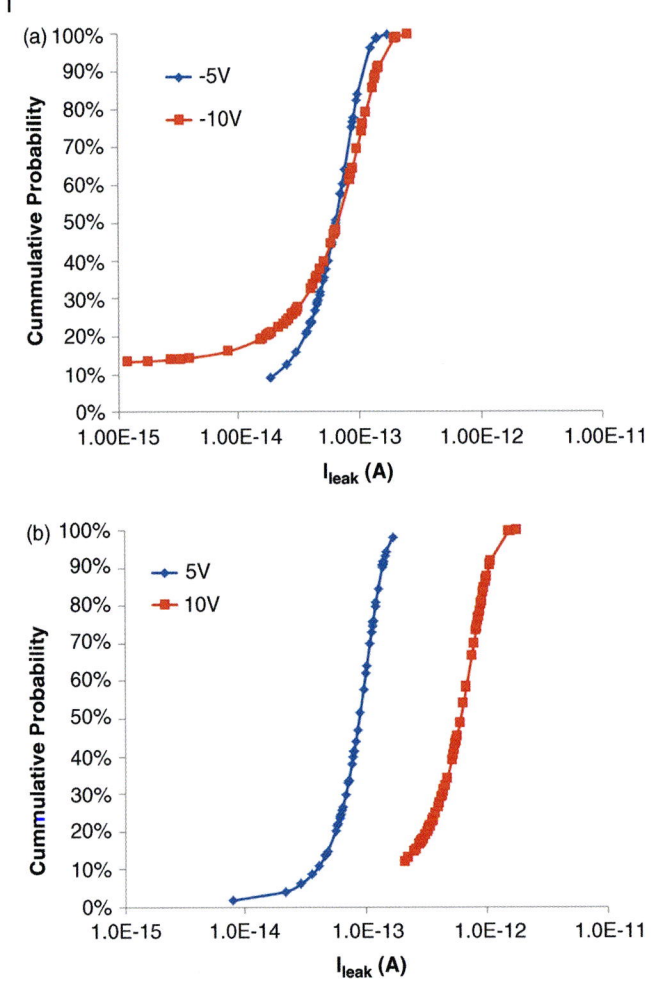

Figure 6.11 Distribution of leakage current in (a) accumulation and (b) depletion.

6.9
Conclusions

Unit processes have been optimized and integration is well advanced, with "best known methods" for typical interposer, via-middle, and via-last flows. Extensive co-optimization between unit processes in etching, CVD, PVD, ECD, and CMP has led to a better understanding of process trade-offs early in the technology life cycle. Collaboration across the industry ecosystem in wafer thinning, bonding/debonding, metrology, and TCAD has enabled rapid learning. Future work will focus on extending these processes to smaller TSVs (2–5 μm diameter) to meet device scaling requirements and will continue to enable smaller, faster, more functionally sophisticated, cost-efficient, and more energy-efficient consumer and industrial electronics.

Acknowledgments

The author acknowledges the contributions from etching, CVD, PVD, ECD, and CMP process engineering and members of the mask, lithography, integration, and characterization teams for all their contributions in enabling TSV technology.

References

1 Iyer, S.S. et al. (2009) Process design considerations for three dimensional memory integration. Proceedings of the VLSI Technology Symposium, p. 60.
2 Knickerbocker, J.U. et al. (2008) Three-dimensional silicon integration. *IBM Journal of Research and Development*, **52** (6), 553–569.
3 Farooq, M.G. (2011) 3D integration in silicon technology. *ECS Transactions*, **35** (2), 83.
4 Savidis, L. et al. (2010) Electrical modeling and characterization of through-silicon vias (TSVs) for 3-D integrated circuits. *Microelectronics Journal*, **41** (1), 9–16.
5 Koyanagi, M. et al. (2009) High-density through silicon vias for 3-D LSIs. *Proceedings of the IEEE*, **97** (1), 49–59.
6 Koester, S. et al. (2008) Wafer-level 3D integration technology. *IBM Journal of Research and Development*, **52** (6), 583–597.
7 Xilinx (2010) Xilinx Stacked Silicon Interconnect Extends FPGA Efficiency: Technology to Deliver 'More than Moore' Density, Bandwidth and Power. Available at pres.xilinx.com/phoenix.zhtml?c=212763&p=irolnewsArticle.
8 Altera (2012) Altera and TSMC Jointly Develop World's First Heterogeneous 3D IC Test Vehicle Using CoWoS™ Process. Available at http://www.altera.com/corporate/news_room/releases/2012/corporate/nrcowos.html?GSA_pos=1&WT.oss_r=1&WT.oss=cowos.
9 Kang, U. et al. (2009) 8 Gb 3-D DDR3 DRAM using through-silicon-via technology. Proceedings of the ISSCC, pp. 130–131.
10 Moore, G.E. (1965) Cramming more components into integrated circuits. *Electronics*, **38** (8), 114–117.
11 Koyanagi, M. (1989) Roadblocks in achieving three-dimensional LSI. Proceedings of the 8th Symposium on Future Electron Devices, pp. 50–60.
12 Ramm, P. et al. (2001) Interchip via technology for vertical system integration. Proceedings of the IEEE International Interconnect Technology Conference, pp. 160–162.
13 Chen, Q. et al. (2011) Chip-to-wafer (C2W) 3D integration with well-controlled template alignment and wafer level bonding. Proceedings of the IEEE Electronic Components and Technology Conference, pp. 1–6.
14 Diehl, D. et al. (2012) Formation of TSV for the stacking of advanced logic devices utilizing bumpless wafer-on-wafer technology. *Microelectronic Engineering*, **92**, 3–8.
15 Ohba, T. et al. (2010) 3D large scale integration technology using wafer-on-wafer (WOW) stacking. Proceedings of the IEEE International Interconnect Technology Conference, pp. 1–3.
16 Wu, B., Kumar, A., and Ramaswami, S. (2011) *3D IC Stacking Technology*, McGraw-Hill.
17 Kumar, N. et al. (2012) Robust TSV via-middle and via-reveal process integration accomplished through characterization and management of sources of variation. Proceedings of the IEEE Electronic Components and Technology Conference, Las Vegas, NV, May 2012, pp. 787–793.

18 Papanu, J.S. *et al.* (2011) Processing and integration considerations for successful copper electrodeposition in 3D IC applications. Proceedings of the 220th Meeting of the Electrochemical Society.
19 Gage, M. *et al.* (2010) CMP for copper TSV applications. Materials Research Society Symposium Proceedings, vol. 1249.
20 Karmarkar, A.P. *et al.* (2010) Material, process and geometry effects on through-silicon via reliability and isolation. Materials Research Society Symposium Proceedings, vol. 1249.
21 Andry, P.S. *et al.* (2006) A CMOS-compatible process for fabricating electrical through-vias in silicon. Proceedings of the 56th IEEE Electronic Components and Technology Conference, p. 831.
22 Malta, D. *et al.* (2011) Characterization of thermo-mechanical stress and reliability issues for Cu-filled TSVs. Proceedings of the IEEE Electronic Components and Technology Conference, pp. 1815–1821.
23 Ramaswami, S. *et al.* (2009) Process integration considerations for 300mm TSV manufacturing. *IEEE Transactions on Device and Materials Reliability*, **9** (4), 524–528.
24 Ramaswami, S. (2011) A holistic approach to process co-optimization for through-silicon via. Proceedings of the International Reliability Physics Symposium, pp. 5D.1.1–5D.1.3.
25 Li, Y. *et al.* (2012) Electrical characterization method to study barrier integrity in 3D through-silicon vias. Proceedings of the 62nd IEEE Electronic Components and Technology Conference, San Diego, CA, June 2012, pp. 304–308.
26 Stucchi, M. *et al.* (2012) Capacitance measurements of two-dimensional and three-dimensional IC interconnect structures by quasi-static C–V technique. *IEEE Transactions on Instrumentation and Measurement*, **61** (7), 1979–1990.
27 Katti, G. *et al.* (2011) Temperature-dependent modeling and characterization of through-silicon via capacitance. *IEEE Electron Device Letters*, **32** (4), 563–565.
28 Stucchi, M. *et al.* (2012) Test structures for characterization of through-silicon vias. *IEEE Transactions on Semiconductor Manufacturing*, **25** (3), 355–364.
29 van der Plas, G. *et al.* (2011) Design issues and considerations for low-cost 3-D TSV IC technology. *IEEE Journal of Solid-State Circuits*, **46** (1), 293–307.

7
TSV Formation at ASET

Hiroaki Ikeda

7.1
Introduction

ASET is a research-focused consortium that transcends barriers between corporations and different industries (Figure 7.1). ASET has been focused on a project aiming to develop "Functionally Innovative Three-Dimensional Integrated Circuit (Dream Chip)" technology that overcomes the limits of semiconductor scaling. This technology will realize very high-speed and low-power consumption chips that use a through-silicon via (TSV). It will be "High-Density Three-Dimensional Integration Technology for Multifunctional Devices" satisfying the specifications needed for its assumed practical applications.

The project will demonstrate the functions and practical effectiveness of the developed chip through producing a 3D integration SiP prototype utilizing the developed integration technology.

For TSV/3D stacking technology developments, memory cubes (such as 4-high stack DDR3 SDRAM, wide I/O DRAM, and hybrid memory cube (HMC) have been leading the progress of the technologies [1,2].

They will be served by new infrastructure (design systems, process development, testing/evaluation methodologies, failure analysis, and reliabilities) for TSV/3D stacking technologies (Figure 7.2). With these breakthroughs, we will be ready for developing more sophisticated heterogeneous 3D stacking systems.

Data traffic growth predictions indicate a +78% CAGR in the period 2011–2016 [3]. This rapid growth will be driven not only by data downloading but also by the uploading. Bidirectional data exchange must contain the sensor data of each terminal. Sensor terminals having hard-sensing and soft-sensing capabilities will serve the key contents of the data [4].

This growth of data traffic of the network (mainly wireless) requires more performance and headroom for the future. The fusion system by hard sensing and soft sensing will be the key contents of the data on traffic. It will be the "sensor network" and each network node (terminals including on automobile) requires dynamic data allocation by cognition aids (Figure 7.3). This sophisticated system

Handbook of 3D Integration: 3D Process Technology, First Edition.
Edited by Philip Garrou, Mitsumasa Koyanagi, and Peter Ramm.
© 2014 Wiley-VCH Verlag GmbH & Co. KGaA. Published 2014 by Wiley-VCH Verlag GmbH & Co. KGaA.

100 | *7 TSV Formation at ASET*

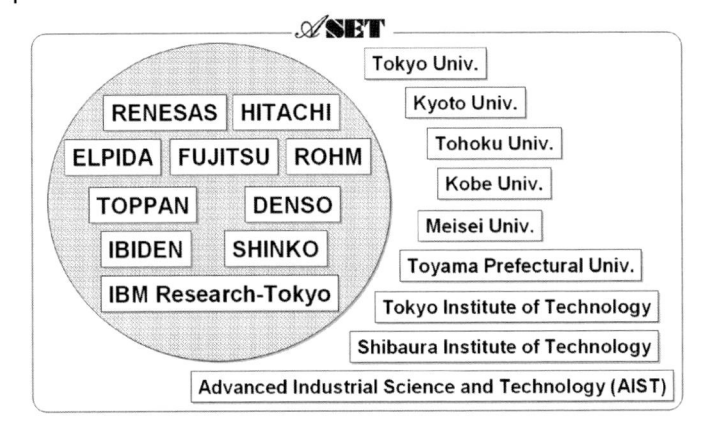

Figure 7.1 ASET organization for the project.

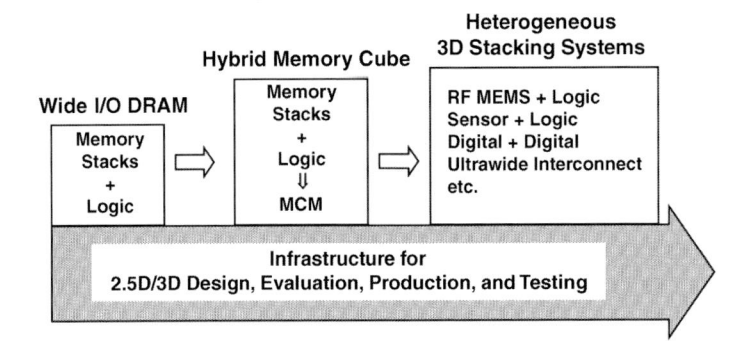

Figure 7.2 Infrastructure movements for 2.5D/3D stacking.

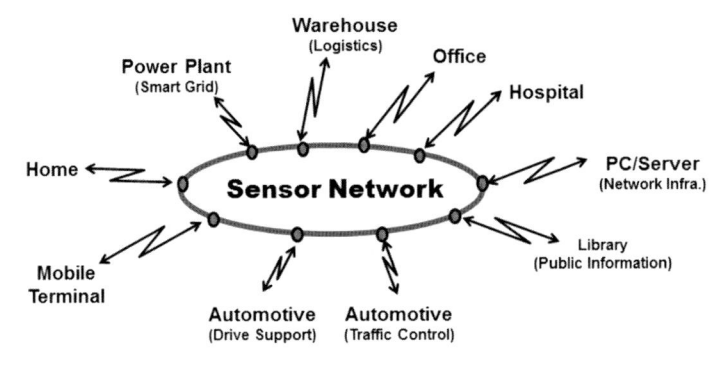

Figure 7.3 Sensor network.

demand must be the motivation to have the heterogeneous 3D stacking system after forerunners.

For the continuous expansion of data traffic, we recognized that the data processing system for the movement (such as memory + logic) requires a 3D stacking structure if it is going to cover more than $10\,\text{GB}\,\text{s}^{-1}$ bandwidth by one unit. Wide I/O DRAM and HMC are the solutions for higher bandwidth per unit.

They will be produced commercially by several suppliers within a few years. Such devices require new supply chain infrastructure for designing, manufacturing, testing, and evaluating those suitable for producing 3D structured devices. We could expect the infrastructure will be adopted for more sophisticated heterogeneous 3D system developments for the rapid expansion of data traffic.

ASET has been developing TSV/3D process technologies suitable for heterogeneous 3D stacking system manufacturing. These systems (structured by analog/digital devices such as memory, logic, image sensors, field-programmable gate arrays (FPGAs), high-speed I/O devices, RF-MEMS, and testing and monitoring devices) must have multiple supply chains for production. They have different process choices and are not always structured by the common "via-middle" process. We recognized that "via-last" TSV process can be more flexible and suitable for heterogeneous 3D stacking system production. Additionally, a heterogeneous 3D system must have lower power consumption than a conventional system under the same performance, but high-density packaging may have larger power density. The power integrity (PI) of the 3D system is the key issue to be resolved, and thus the power distribution network (PDN) design using TSV is also a key consideration.

Via-last processes have good capability for larger diameter TSVs. We see power transmission through these large TSV dimensions as a solution for future requirements of these heterogeneous systems. Of course, TSV pitch should have room to be shrunk. The coexistence of fine-pitch TSV for signal and larger TSV structures for power lanes is an issue ASET is researching.

For heterogeneous 3D stacking system developments, there are several basic technologies.

1) Cooling and stacking and bonding technology.
2) Thin wafer technology.
3) 3D integration technology.
4) Ultrawide bus SiP 3D integration technology.
5) Digital and analog 3D integration technology.
6) Heterogeneous 3D integration technology.

There are six working groups at ASET covering these themes (Figure 7.4):

1) Cooling and Stacking/Bonding Technology Working Group:
    - Simulation and evaluation of 3D stacked system's stress and thermal behavior.
    - Cooling system design and prototyping.
    - 10 µm pitch microbump interconnection.

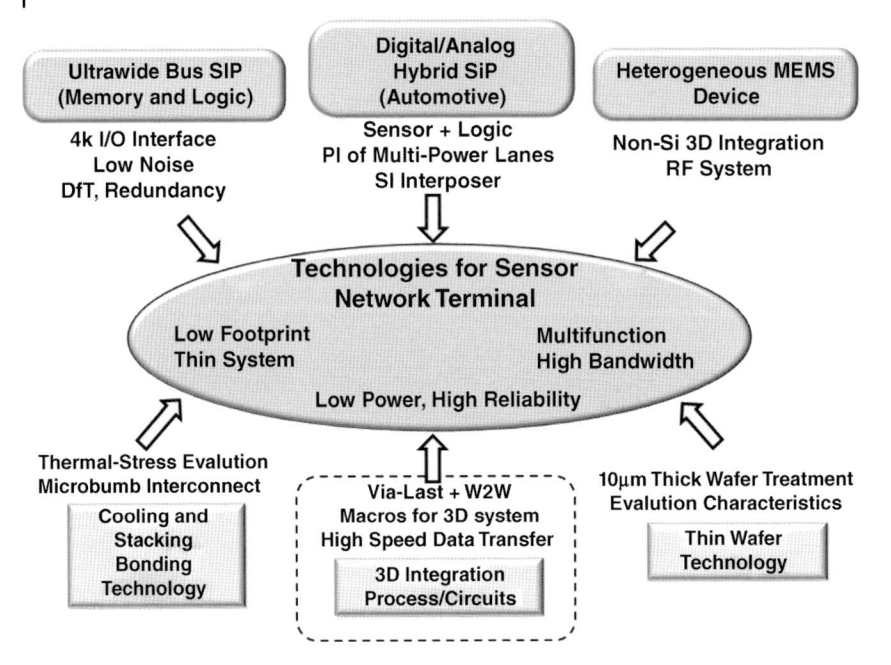

Figure 7.4 Six working groups of ASET.

2) Thin Wafer Technology Working Group:
   - Fabrication and evaluation of thin wafer characteristics (thickness is down to 10 μm).
   - Thin wafer handling (thinning, gettering layer forming, dicing).
   - Methodology for evaluating diffusion characteristics of metal contamination.
3) 3D Integration Technology Working Group:
   - Cost-effective TSV/3D process and circuit macros for 3D system.
   - Via-last TSV-forming technology with wafer-to-wafer (W2W) and die-to-die (D2D) stacking.
   - Synchronization among layers in 3D structure.
   - Evaluation of electromigration of interconnects and keep-out zone around TSV.
4) Ultrawide Bus SiP 3D Integration Technology Working Group:
   - High-bandwidth data processing (more than $100 \, \text{GB s}^{-1}$) with low power consumption.
5) Digital and Analog 3D Integration Technology Working Group:
   - Analog plus digital and/or sensor plus logic system design and evaluation.
   - PI design technology.
6) Heterogeneous 3D Integration Technology Working Group:
   - Non-Si hybrid RF-MEMS systems.

The rest of this chapter will focus on the activity of the "3D Integration" working group whose goal was to develop 25 μm pitch TSV by a via-last approach, W2W stacking process.

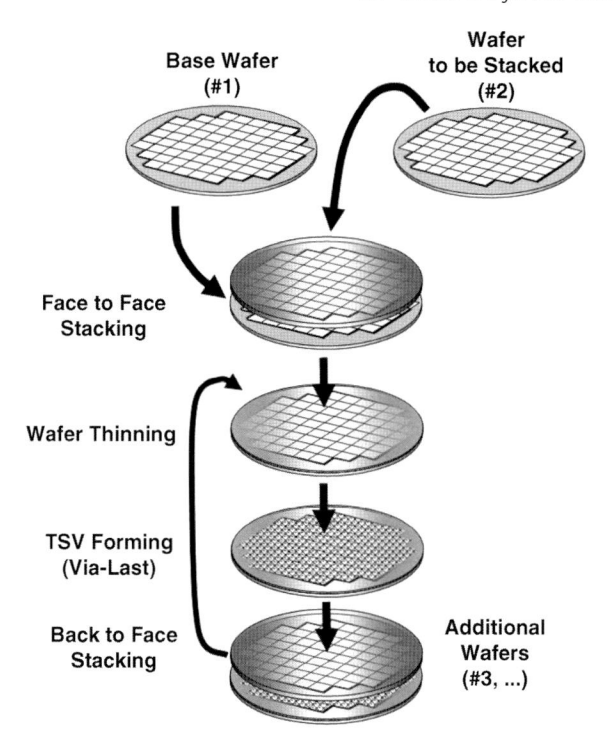

Figure 7.5 Concept of W2W stacking with via-last process.

ASET is mainly developing a W2W stacking process because of its headroom to improve the throughput of stacking processes. We also have been making use of D2D stacking for quick evaluation.

ASET's W2W stacking process is started by microbumping onto the surface of finished wafers using patterned polymer and Cu plating followed by chemical–mechanical polishing for flattening the surface. The first stacking step is face-to-face wafer stacking (by #1 and #2 wafers). The second step is wafer thinning for a backside of the wafer (e.g., #2). The third step is TSV forming from the backside of the thin wafer (Figure 7.5).

7.2
Via-Last TSV for Both D2D and W2W Processes in ASET

ASET decided to develop both D2D and W2W stacking technologies, which each have their own advantages. W2W is an attractive method for back-end-of-line (BEOL) interconnects. It can eliminate a requirement of the wafer support system (WSS) for tentative support of the thin wafer process (e.g., thinning, TSV forming, bumping) in middle-end-of-line (MEOL) interconnects. WSS is a major part of the TSV fabrication cost structure when the support glass wafer/support Si wafer

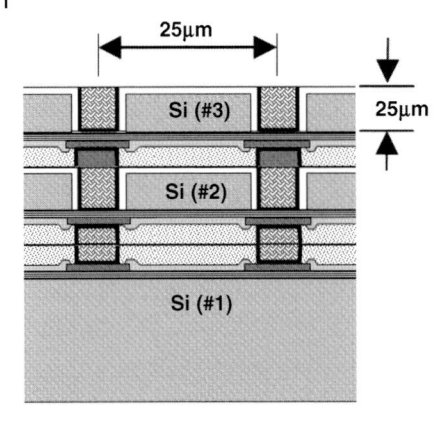

Figure 7.6 Cross section of stack (W2W, not in scale).

cannot be reused. The W2W stacking structure itself can support a thin wafer by using a thick-bottom Si wafer (Figure 7.6). For TSV/microbump pitch shrinkage, the W2W process has better potential by its more precise alignment capability. To make heterogeneous 3D stacking systems, W2W stacking has the restriction that equal size die be stacked. But even in the heterogeneous 3D stacking systems or 3D/2.5D systems, memory cubes will be mandatory for data processing performance. BEOL cost reduction and throughput improvement by W2W processing of memory cubes will be quite important. According to concerns of W2W yield degradation, we know there are studies for selecting wafers having similar distribution of failure dice for W2W stacking (i.e., wafer matching) realizing yield improvements [6]. Because memory has long been designed to have redundant cells, words, and bits, optimum redundancy density design must be the solution for W2W stacking yield improvements.

D2D stacking development activities in ASET have several motivations. The first is heterogeneous 3D stacking. Even if our current demonstrators have common-sized slices, there must be requirements to stack differently sized dice on heterogeneous 3D structures. Optimized designs for sensors, AD/DA converters, power delivery and clock distribution layers, and so on must have their own aspect ratio and size. So, ASET decided to include D2D stacking activities in parallel with W2W. Additionally, D2D stacking can be supported by outsourced semiconductor assembly and tests (OSATs) in Japan. We can expect quick evaluation of our TEGs and demonstrators after MEOL process for stacking and packaging.

Following is a detailed explanation of the TSV forming processes (for both D2D and W2W) under development at ASET.

Our TSV target dimension is 25 μm pitch. Related dimensions, such as wafer thickness and via diameter, are determined by usage of physical vapor deposition (PVD) for the barrier metal and seed layer. The aspect ratio of via (TSV) thickness and its diameter is around 2–3 by this choice.

This target range is determined by consideration of the JEDEC standard for wide I/O DRAM. Its 40 μm × 50 μm pitch TSV/bump metrics will be long lasting, and

Table 7.1 Target characteristics of TSV.

R (TSV)	$\leq 0.2\,\Omega$ (include bump to bump interconnect)
C (TSV)	$\leq 50\,fF$ (exclude fringe capacitance)
Via density	40k–100k via cm^{-2}
TSV pitch	25 μm
TSV diameter	8–12 μm
Via depth	25 μm

Table 7.2 Bump structure for D2D and W2W.

	D2D	W2W
Front-side bump	SnAg/Cu	Cu
Backside bump	Au/Ni	Cu
Stacking underfill	Capillary UF	Hybrid

the next movement of pitch reduction will be in the range of around 25 μm. Die thickness target is mainly determined by W2W handling, but it also works for D2D stacking.

Our target dimension and characteristics of TSV are indicated in Tables 7.1 and 7.2.

For consideration of TSV usage for both signal paths and power lanes, resistance of TSV for power lanes should be low enough compared with the surface routing of power lines. Conversely, the signal path is not sensitive to its resistance since the response time of signal TSV is determined by its capacitance and drive circuit's equivalent resistance. Capacitance of the TSV should be low enough for signal paths. TSV diameter range is around one-third to one-half of its pitch.

Bump structure for W2W and D2D are different. In D2D stacking, one has to consider warpage of slices during the stacking process. So the tolerance for different standoffs between slices should be covered by wet metal. Sn–Cu-based soldering is our choice for microbump structure for D2D stacking. For W2W, more accurate (flat) surface conditions can be supported and thus we use Cu–Cu interconnects for our W2W.

7.3
TSV Process for D2D

For quick evaluation of small device counts of TEGs and demonstrators, the D2D stacking process has advantages over W2W stacking. Since D2D stacking requires separated thin dice, thinning and TSV forming are done on the wafer support system. We had selected glass-wafer-based WSS for our purposes. The via-last process does not require specific via preparation in font-end-of-line (FEOL)

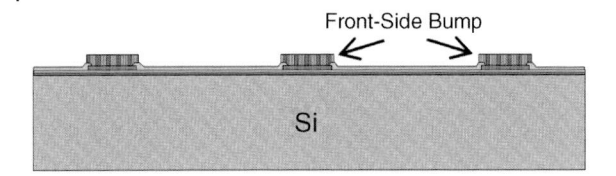

Figure 7.7 Front-side bump forming.

interconnectors. We can start the MEOL process by making microbumps on the finished wafer.

7.3.1
Front-Side Bump Forming

By our choice for D2D stacking, front-side bumps have a Cu pillar and SnAg solder layer. They are formed by a conventional electroplating process (Figure 7.7). After cleaning the wafer, as a preparation process for electroplating the bump, a barrier metal (titanium) layer (around 70 nm thickness) and Cu seed layer (around 600 nm thickness) are deposited using PVD. Resist coating and development make an opening in the metal basement (the top metal layer of the device is aluminum). Electroplating of Cu followed by SnAg plating makes the Cu pillar with a SnAg cap on the top. The height of the Cu pillar is 8–12 µm and the thickness of SnAg cap is 2–4 µm. After plating, the resist mask, seed layer, and barrier metal layer beneath the resist are removed. The seed and barrier metal layers are removed by wet process. This gives us a wafer with microbumps on the front side.

7.3.2
Attach WSS and Thinning

The finished thick wafer with front-side bumps is coated by tentative adhesive for MEOL process treatments (Figure 7.8). The thermal budget of the adhesive is the key consideration to designing the MEOL process. The adhesive of our choice has a thermal budget less than 200 °C. So successive process treatments, such as CVD and PVD, should be done within this thermal budget. The wafer is attached onto a glass-based WSS. The surface of the glass wafer is precoated by a specific layer for

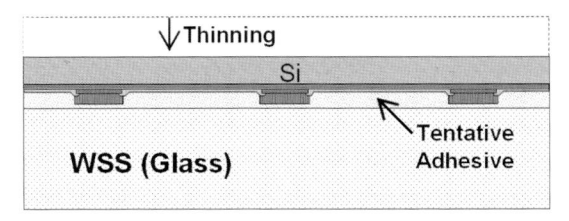

Figure 7.8 Attach WSS and wafer thinning.

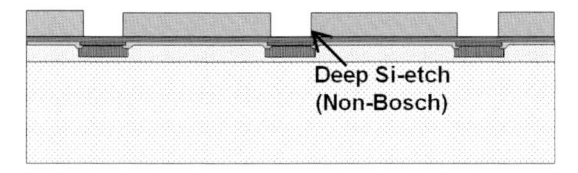

Figure 7.9 Deep Si etch.

detaching. Then the wafer is thinned by backgrinding (BG) down to a 25 μm thickness.

7.3.3
Deep Si Etching from the Backside

The thin wafer is coated by resist and patterned by lithography. Since alignment marks are located on the front side of the wafer, infrared (IR) monitoring from the backside of the wafer is used for lithography alignment (Figure 7.9). After patterning, the resist pattern is a mask for deep Si etching. For our purposes, we chose a non-Bosch type etching system. Etch-stop is done at the interface between the bulk Si and the dielectric layer beneath M1 (premetal dielectric (PMD)). The via diameter is 8–12 μm and its depth is 25 μm.

7.3.4
Liner Deposition

The sidewall and bottom of the blind via and thin wafer backside are coated by SiO_2 using low-temperature chemical vapor deposition (CVD) (Figure 7.10). The liner thickness of the sidewall is around 0.5 μm, and the wafer backside is 2 μm. This deposition is a common way to make TSV liner and the dielectric layer for the backside of a thin wafer.

7.3.5
Removal of SiO_2 at the Bottom of Via

Since the thickness of the dielectric layer at the bottom of blind via is thinner than that of the backside surface, dry etching for the whole backside results in removal

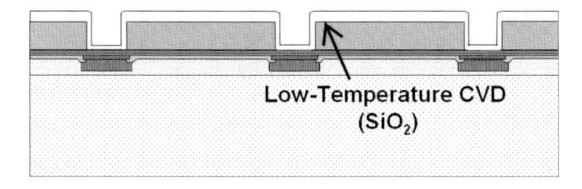

Figure 7.10 TSV liner and backside isolation.

7 TSV Formation at ASET

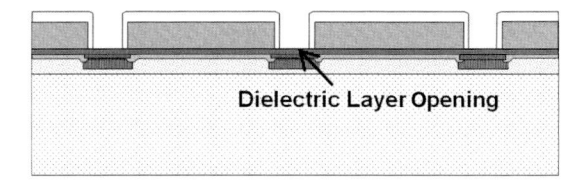

Figure 7.11 Cross section at the stage of SiO$_2$ removal.

of the dielectric layer at the bottom of the blind via and makes the opening for the interconnection between M1/M2 and TSV (Figure 7.11).

By using chemical–mechanical polishing (CMP) on the Cu-wiring process (wide and flat Cu pattern), one can reduce any concern over "dishing." Cu layers close to the TSV boundary (M1/M2) are patterned as stripes having spacing between the lines. The spaces are filled by SiO$_2$. They act as windows for etching. The removal process of the dielectric layer at the bottom of the blind via also removes the SiO$_2$ layers among the M1 spaces and the M1/M2 isolation layer. To eliminate unexpected penetration of etching, there M1 and M2 layout must be considered. This is a specific situation of the via-last process for Cu-wired devices (Figure 7.12).

ASET examined several positions for M1 and M2, such as placing M1 and M2 by "parallel with offset" or "orthogonal." "Parallel with offset" placing of M1 lines and M2 lines effectively act as etch-stoppers that prevents penetration of SiO$_2$ etching beyond M2 (Figure 7.13). The spacing between M1 lines and M1 and M2 are coated by a barrier metal (Ti) and seed layer (Cu). Then it is filled by Cu electroplating. Thus, the interface between the TSV and the device wiring is done by connection with both M1 and M2.

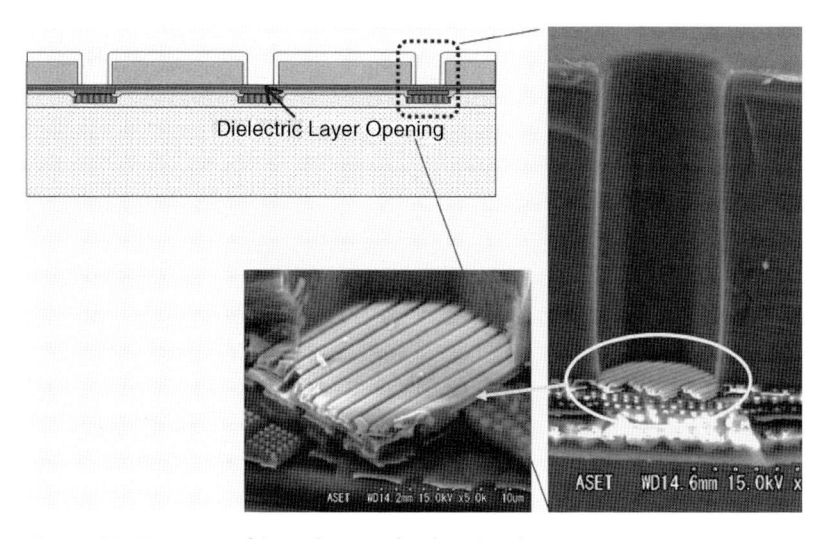

Figure 7.12 SEM image of the via bottom after deep Si etching.

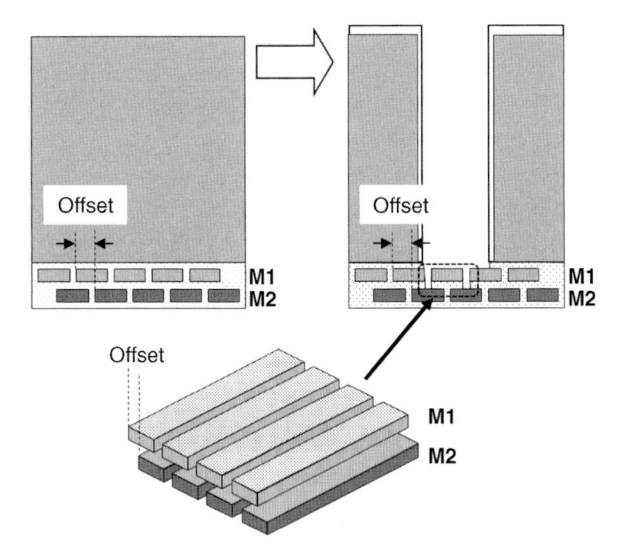

Figure 7.13 M1 and M2 parallel with offset.

If the metal lines are aluminum, a wide and flat pattern is adoptable for M1 design. In this way, interconnection with TSV can be done by M1 only.

Isolation layer removal between M1 and M2 followed by Cu plating makes a more robust interconnection of TSV and M1/M2. Daisy-chain resistance evaluation by M1/M2 positioning, "parallel with offset" placing makes better connectivity than "orthogonal" placing (Figure 7.14). At ASET we determined "parallel with offset" M1/M2 placing to be the standard design and the layout rule for our TEGs and demonstrators.

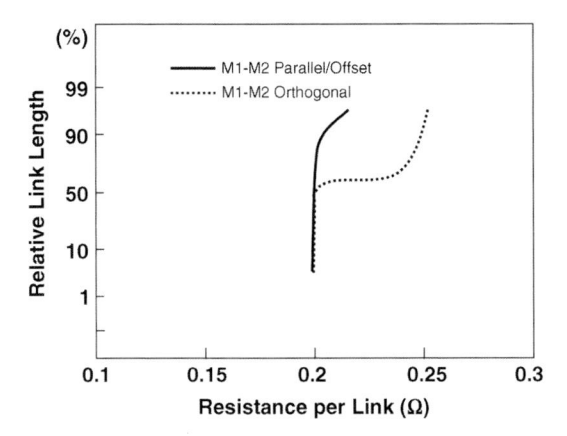

Figure 7.14 Daisy-chain link resistance by layout.

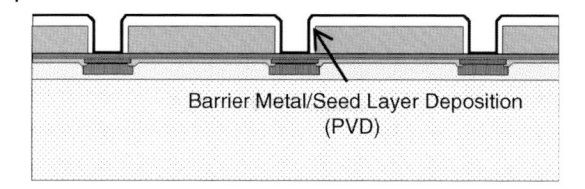

Figure 7.15 Barrier metal and seed layer deposition.

7.3.6
Barrier Metal and Seed Layer Deposition by PVD

After opening the bottom liner of the via, barrier metal (60 nm thick Ti) and seed layer (600 nm at sidewall Cu) are deposited by PVD for via-fill electroplating (Figure 7.15). For PVD coverage at the bottom of the via and sidewall, the aspect ratio of the via (via depth/via diameter) should be less than 3–5. It is also critical to keep the thermal budget (less than 200 °C) through the process by keeping the adhesive layer intact.

7.3.7
Cu Electroplating

A thin wafer on a WSS with a Cu seed layer is ready for Cu plating (Figure 7.16). Conventional conformal plating is not suitable for filling these blind vias because there is high probability that a void or seam might be included in the via filling. We chose a bottom-up plating bath for seamless plating of the via. In bottom-up plating, the plating speed at the bottom of via is much faster than that of the surface. This reduces the overburden thickness of the backside surface. Because the via open area covers only 5% or less of the entire surface, the rest of the backside surface covered by Cu overburden has to be removed. The thickness of the overburden is critical to the successive CMP process TAT.

7.3.8
CMP

The Cu overburden and the barrier metal on the backside of the thin wafer are removed by CMP (Figure 7.17). First, Cu CMP is done. Torque control and CMP

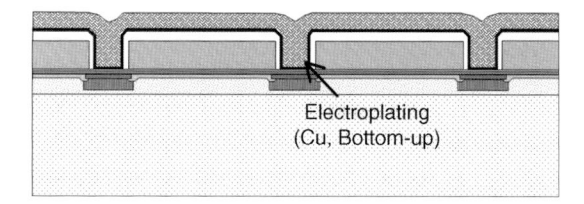

Figure 7.16 Electroplating of Cu.

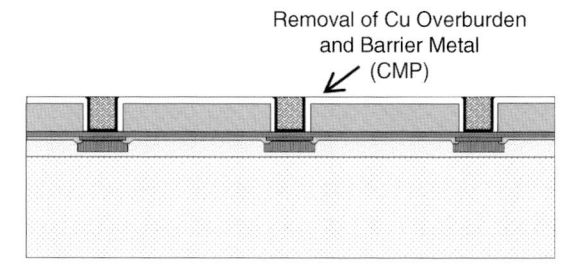

Figure 7.17 CMP to remove overburden and barrier metal.

monitoring allow full removal of the Cu layer down to the barrier metal (Ti) layer. Since the CMP stopper is SiO_2, after the removal of Cu, CMP is continuously done to reach the SiO_2 layer of the backside removing the barrier metal. The appearance of the Cu via top can be an additional alignment mark for the next lithography for bumping.

7.3.9
Backside Bump

The backside surface of the wafer is deposited by barrier metal (Ti) and seed layer (Cu) again by PVD for plating of the backside bump (Figure 7.18). Ni/Au electroplating follows a resistance mask coat and development process for bump patterning. After the Ni/Au bump forming, the resistance mask is removed and the seed layer and barrier metal layers are removed by wet etching.

7.3.10
Detach WSS

The backside of the thinned wafer is attached to a dicing sheet (Figure 7.19). For detaching WSS, an ultraviolet (UV) laser scan is done from the backside of the glass WSS. It makes a specific detach layer on the WSS surface instantaneously heat up to evaporate the adhesive near the detach layer. Then, the WSS is removed. Any remaining adhesive layer on the device surface is also peeled off.

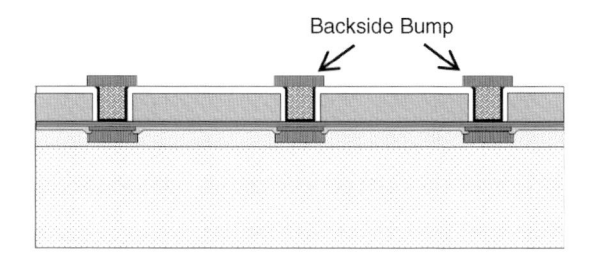

Figure 7.18 Backside bump forming.

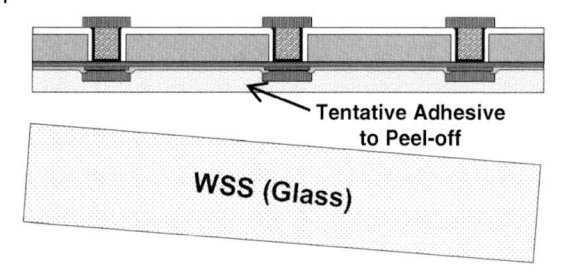

Figure 7.19 Detach WSS.

7.3.11
Dicing

The backside of the thinned wafer is attached to a dicing sheet (face up) with the front side exposed for the next stage—dicing and dice pickup (DPU) stage (Figure 7.20). The choice of dicing sheet for a thin wafer (25–50 μm thickness) is important. We evaluated several available dicing sheets and selected one. After dicing, the thin dice are ready for DPU and D2D stacking (BEOL).

An example of the via-last process for D2D stacking is shown in Figure 7.21. This photomicrograph is a cross section of 50 μm thick die with 50 μm pitch TSV.

The front-side bump is structured by Cu pillar and SnAg cap. The backside bump is structured by Ni/Au. In this case, the backside bump is mainly for interconnection to an organic substrate, so the diameter of the bump is larger than normal (20 μm).

ASET has a demonstrator using this TSV process for D2D stacking. It is the "ultrawide-bus SiP." It has a three-layered structure. The top layer is SRAM (face down) and the logic chip is located on the bottom. Between them, there is an Si interposer as a redistribution layer (RDL). They are connected to each other by 4k I/O terminals (Figure 7.22). Its bandwidth is $100\,\text{GB}\,\text{s}^{-1}$ and power efficiency for interconnection is $0.5\,\text{pJ}\,\text{bit}^{-1}$. The TSV dimension for this device is 50 μm pitch, 20 μm diameter, and 50 μm deep. Since the Si interposer is facing both the front side of the SRAM and logic, the interposer has an Ni/Au microbump in both the front side and backside.

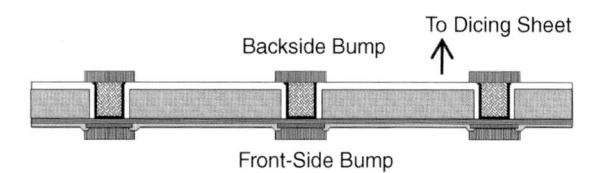

Figure 7.20 Dicing after peeling off tentative adhesive.

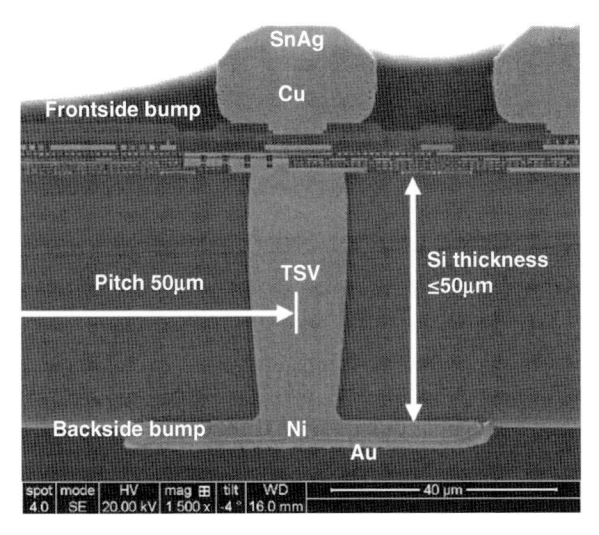

Figure 7.21 Cross section of thin die before dicing.

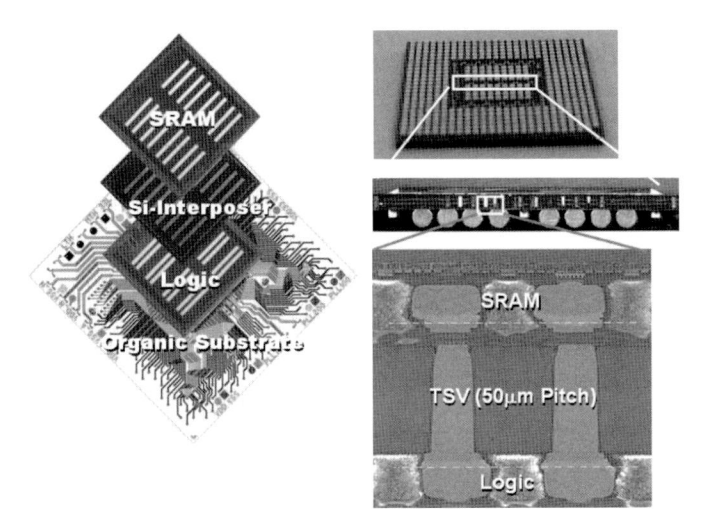

Figure 7.22 Stacked structure of 4k I/O memory, Si interposer, and logic SiP.

7.4 TSV Process for W2W

The TSV process for W2W differs from that of the D2D. The key difference is

a) The front-side and backside bumps are structured by Cu only for Cu-Cu bonding.

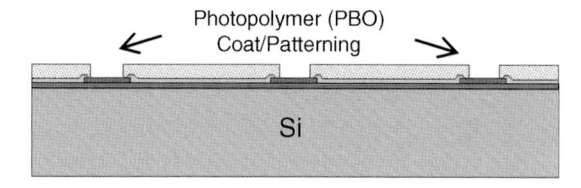

Figure 7.23 Photopolymer coat and patterning on the initial wafer.

b) The front side of the wafer is covered by polymer.
c) The MEOL process has a W2W stacking stage (Figure 7.5).

Since polymer for hybrid bonding is a permanent layer (not tentative), the material should have enough thermal budget for Cu–Cu bonding (thermal compression bonding). The bonding condition is around 300 °C and 30 min. It is also expected to have photolithograph capability to eliminate additional resist coat/develop steps. These requirements resulted in our choice of polybenzoxazole (PBO) as a solution for the hybrid bonding.

7.4.1
Polymer Layer Coat and Development

Photopolymer (PBO) is coated over the wafer surface and patterned by lithography. The opening pattern of PBO is for front-side microbumps (Figure 7.23).

7.4.2
Barrier Metal and Seed Layer Deposition

Using PVD, the barrier metal (Ti) and seed layer (Cu) are deposited over the wafer surface (Figure 7.24).

7.4.3
Cu Plating

Cu plating is done by electroplating. A bottom-up plating bath is not mandatory at this stage (Figure 7.25).

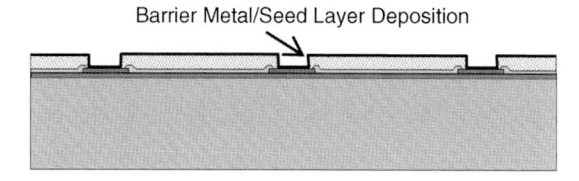

Figure 7.24 Barrier metal and seed layer depositions.

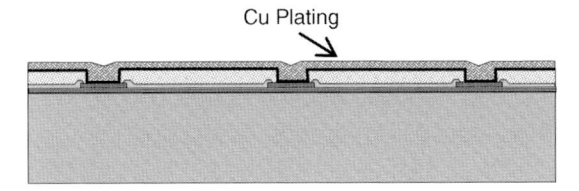

Figure 7.25 Cu plating for front-side microbumps.

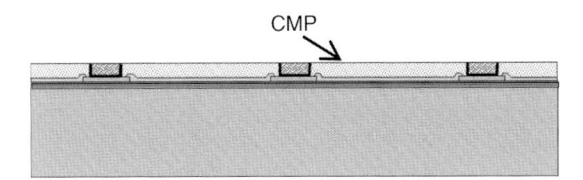

Figure 7.26 CMP for the front side.

7.4.4
CMP

The Cu overburden and barrier metal are removed by CMP (Figure 7.26). Bump height and polymer thickness are 5 μm after CMP. The Cu surface is slightly higher than that of polymer (50–100 nm) for bonding. Then the wafer is ready for first W2W stacking.

A bird's-eye view of the Cu bumps after CMP is shown in Figure 7.27.

Polymer shrinkage in the cure process after patterning results in a tapered cross section of the Cu bump. For the 5 μm thickness of PBO, the bump diameter of its bottom is around 6 μm and top of it is around 10–15 μm.

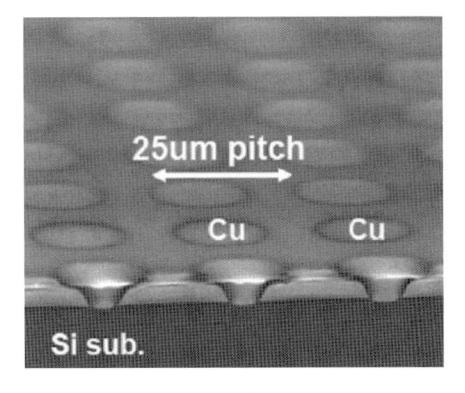

Figure 7.27 SEM image after CMP.

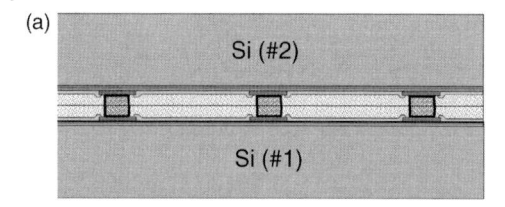

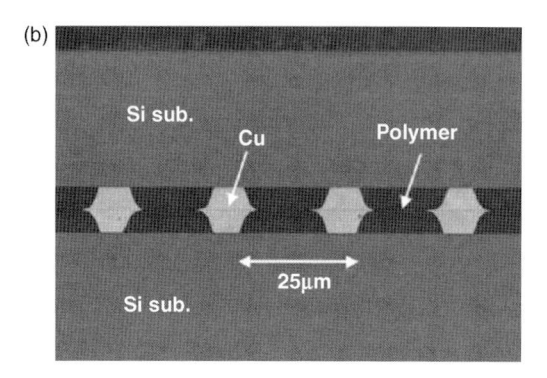

Figure 7.28 (a) First W2W stacking (F2F, hybrid). (b) SEM image of 2-high stack W2W cross section.

7.4.5
First W2W Stacking (Face to Face)

After cleaning the surface of both wafers (#1 and #2), they are attached by face-to-face (F2F) hybrid W2W bonding (thermocompression) (Figure 7.28a). The bonding condition is around 300 °C and 30 min. A scanning electron microscope (SEM) image of the cross section of W2W-F2F stacking is shown is Figure 7.28b. The bump pitch is 25 μm in this case.

The alignment of the two wafers is done at ordinary temperature and pressure. Then thermocompressive stacking is done under reduced pressure circumstances.

After stacking, the wafer edge is not always completely aligned and wafer edge trimming is required for a wafer to be thinned (in this case, #2). Edge trimming is also effective to eliminate a fragile knife-edge shape at the bevel after thinning.

7.4.6
Wafer Thinning and Deep Si Etching

The backside of the second wafer is thinned by BG. Resist is coated onto the backside of the second wafer so that TSV location is determined by resist mask after developing. IR is used for alignment because the alignment mark is located on the front side of the wafers. Deep Si etching is carried out by resist mask and the etching is stopped at the dielectric layer beneath the metal wiring (Figure 7.29).

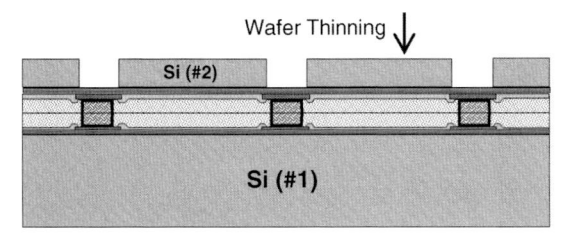

Figure 7.29 Wafer thinning and deep Si etch.

7.4.7
TSV Liner Deposition and SiO$_2$ Etching of Via Bottom

The SiO$_2$ liner and the dielectric layer of the wafer backside are deposited by CVD-SiO$_2$. Dielectric layer thickness of the via sidewall (liner) is around 0.5 μm thick, and the backside is 2 μm thick. Including PMD layer thickness, it is still thinner than that of the backside surface. By this difference, overall dry etching of SiO$_2$ results in a selected removal for the via bottom (Figure 7.30). To make a good interconnect between Cu wiring and TSV, the same layout consideration is important (M1 and M2 routing is in "parallel with offset").

7.4.8
Barrier Metal and Seed Layer Deposition and Cu Plating

Barrier metal (Ti) and seed layer (Cu) are deposited by PVD followed by Cu plating (Figure 7.31).

7.4.9
CMP

Cu overburden and barrier metal on the backside of the thin wafer are removed by CMP. CMP is stopped at the SiO$_2$ layer. The backside of second wafer is ready for the next W2W stacking (Figure 7.32).

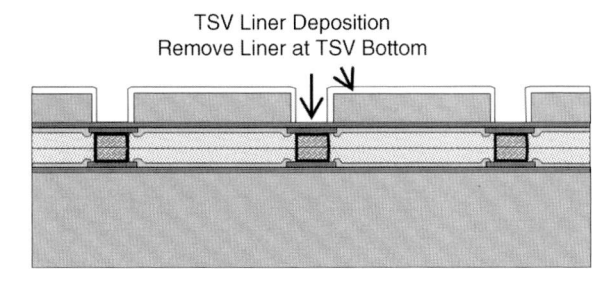

Figure 7.30 Liner deposition and its removal at the bottom of via.

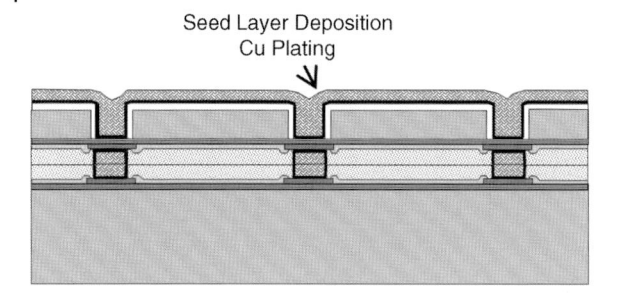

Figure 7.31 Barrier metal and seed layer deposition and Cu plating.

7.4.10
Next W2W Stacking

The front side of the third wafer is attached to the backside of the second wafer (back-to-face (B2F)) by hybrid W2W stacking (#2 and #3) (Figure 7.33). In this case, the interface between #2 and #3 wafers is different from that of #1 and #2 (polymer–polymer to polymer–SiO_2). This difference does not make problems for our developments. If the 3D stacked system is structured by these three slices, there should be terminals for external interconnection. The backside of either wafer (#1 or #3) can be thinned and made via to have TSVs by via-last for external terminals (not indicated in the figures).

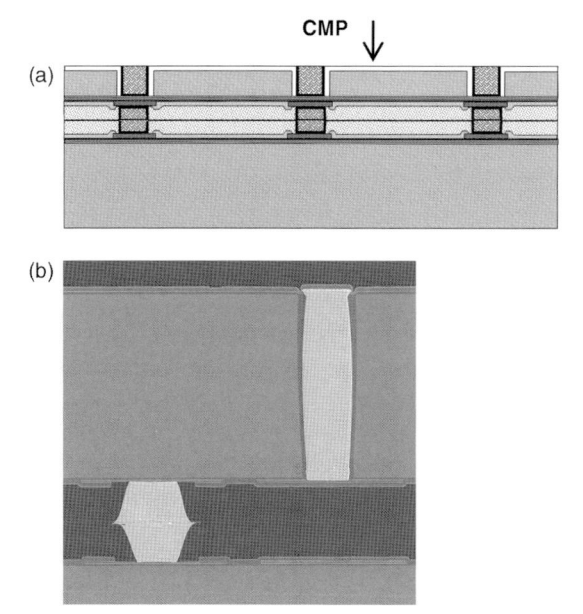

Figure 7.32 (a) CMP for second wafer backside. (b) Cross section of the stage (2-high F2F).

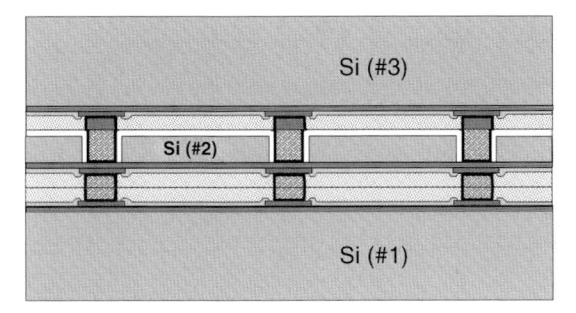

Figure 7.33 Third wafer stacking (B2F, hybrid).

7.5 Conclusions

The TSV process of ASET is aiming to be a solution for a heterogeneous 3D stacking system, which will be demanded by the rapid growth of data traffic. The 3D systems for the traffic are for both terminal and infrastructure sides. Our process is flexible to adopt a variety of supply chains for the system production.

Acknowledgments

This project is entrusted by NEDO "Development of Functionally Innovative 3D-Integrated Circuit (Dream Chip) Technology." The project is based on the Japanese government's METI "IT Innovation Program."

References

1 http://www.elpida.com/en/news/2011/index.html (June 27, 2011).
2 http://hybridmemorycube.org/news.html (October 6, 2011).
3 CISCO, http://www.cisco.com/en/US/solutions/collateral/ns341/ns525/ns537/ns705/ns827/white_paper_c11-520862.html (February 2012).
4 Rattner, J. (2010) Intel Developer Forum Day3 Keynote. http://download.intel.com/newsroom/kits/idf/2010_fall/pdfs/Day3_IDF_Keynote_Rattner.pdf (September 2010).
5 Hall, D.L., McNeese, M., Llinas, J., and Mullen, T. (2008) A framework for dynamic hard/soft fusion. Proceedings of the 10th Annual International Conference on Information Fusion, Fusion 2008, Cologne, Germany, June 30–July 3, 2008.
6 Singh, E. (2012) Impact of Radial Defect Clustering on 3D Stacked IC Yield from Wafer to Wafer Stacking, International Test Conference, Intel.

8
Laser-Assisted Wafer Processing: New Perspectives in Through-Substrate Via Drilling and Redistribution Layer Deposition

Marc B. Hoppenbrouwers, Gerrit Oosterhuis, Guido Knippels, and Fred Roozeboom

8.1
Introduction

3D stacking of semiconductor devices is on the roadmap of every major player working on further miniaturization. To interconnect them in the vertical dimension, "through-silicon via" (TSV) technology is under development worldwide. At least 15 pilot lines have been identified for 3D TSV production in various R&D centers and fabs worldwide [1]. The market forecast for 3D TSV packaging will amount to about $4 billion in 2015 and the forecast for the total available market for TSV creation is $37 million in 2014 [2].

TSVs are, in essence, conductive through-substrate interconnects with diameters ultimately down to 1 µm but currently still more than 10 µm [3] in wafers that are thinned down to 50 µm or less. Thus, TSVs can typically have relatively high aspect ratios up to 20 or more. Various methods, both traditional and novel, exist to create TSVs. In this chapter, we present two types of laser processes that can be used in 3D wafer integration: laser drilling of TSVs with medium-sized diameter (5–10 µm and up) and laser-assisted deposition of redistribution layers (RDLs).

We will show that laser drilling is not only technologically feasible but also a cost-effective technology option compared to competing techniques, of course under certain design and process conditions. Next, we will show that with laser-induced forward transfer (LIFT), it is possible to create conductive redistribution layers with a feature size that allows high-density interconnect applications.

8.2
Laser Drilling of TSVs

8.2.1
Cost of Ownership Comparison

Originally, the creation of vias with noncritical dimensions or layout in silicon has been done by photoelectrochemical etching with KOH or tetramethylammonium

Handbook of 3D Integration: 3D Process Technology, First Edition.
Edited by Philip Garrou, Mitsumasa Koyanagi, and Peter Ramm.
© 2014 Wiley-VCH Verlag GmbH & Co. KGaA. Published 2014 by Wiley-VCH Verlag GmbH & Co. KGaA.

hydroxide, or by powder blasting. Both approaches, however, have serious drawbacks. With wet etching, the rate of silicon removal is strongly dependent on the Si crystal orientation [4]. For Si(100) substrates this produces vias that have the well-known inverse pyramid shape, rather than the cylinder shape with the high aspect ratio that is required. With a photoelectrochemical variant of wet etching of Si(100), one can reach high densities of pores with ultrahigh aspect ratio, yet this process has large restrictions in terms of substrate choice and TSV layout [5]. With powder blasting (or abrasive jet machining) [6,7], the via diameter that can be obtained has a lower limit of about 50 μm [6,7], which is too large. Moreover, this process is likely to cause damage, that is, microcracking in the surrounding silicon due to its violent nature.

That leaves us with two techniques: *deep reactive ion etching* (DRIE) [8] and *laser drilling* [9]. In the end, cost of ownership (CoO) will be the deciding factor in the choice between these two technologies.

DRIE is a parallel plasma etch process using a pre-etched mask pattern on the semiconductor wafer: all vias are etched at the same time. So via depth and via shape determine the process rate. Laser drilling, conversely, is a serial process; hence, the process rate follows directly from the number of vias per wafer. This implies that DRIE will always be more cost effective for high via-density applications.

In order to compare the total process costs of TSVs obtained by DRIE versus laser drilling, we have developed a CoO model that includes the equipment-related cost (operation, maintenance, uptime, cleanroom space, tool depreciation, margin), consumables (utilities, resist, coatings, chemicals), and scrap cost. Data have been taken from internal and external databases. The model is quite similar to CoO models that have been published [10].

Figure 8.1 shows a plot of the CoO in $ per wafer as a function of the via density in vias mm^{-2}. Here we assumed 300 mm wafers with an effective diameter of 290 mm.

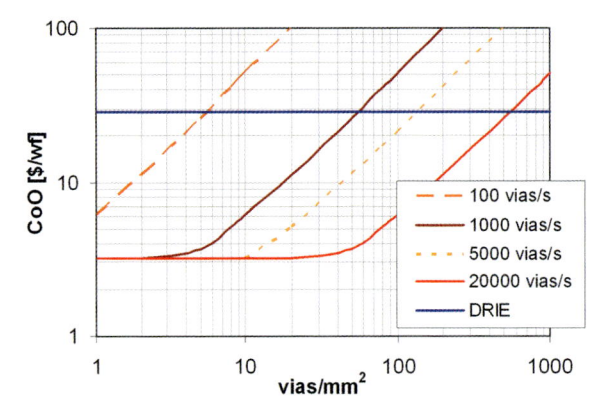

Figure 8.1 Cost of ownership of laser drilling for four different via laser drilling speeds. The figure shows the dependence of cost on the TSV density and the drill speed, while DRIE CoO is independent of via density.

We can see that the cost per wafer is not dependent on the via density if DRIE is used. This implies that for any given process speed, there will always be a via density above which DRIE is the more cost-effective process. Consequently, laser drilling is the favorable option for low- and medium-density applications, typically 10–100 TSVs mm^{-2}, provided that a drilling speed can be achieved of more than 5000 TSVs s^{-1}. Thus, for modest TSV dimensions and densities, laser drilling is very attractive from a cost point of view. We will focus on this option for the remainder of this paragraph. To some extent, the cost advantage is counterbalanced by a few disadvantages. These have been outlined before [11], and include a thermally affected zone due to the high-fluence laser, redeposited debris of ablated Si material, and a limited sidewall smoothness. However, most setbacks have been greatly reduced in recent years. For example, the laser-induced damage can be reduced by intermediate etching [12] or by lowering the average laser fluence by applying multibeam laser technology (see, for example, www.alsi-international.com (accessed December 12, 2012)). The latter solution will also smoothen the sidewalls. Finally, the wafers can be protected from the debris by applying an easily removable protective organic layer on top.

8.2.2
Requirements for an Industrial TSV Laser Driller

From the CoO calculation and other internal and external resource data such as the TechSearch roadmap [2], we compiled a minimum set of requirements that a laser drilling tool will have to meet to be cost effective over DRIE. Table 8.1 shows these requirements.

Apart from these hard requirements, a set of lower priority and softer requirements can also be identified for a successful market introduction. We briefly mention yield, via pattern, industry acceptance, via shape, via sidewall roughness, heat-affected zone, contamination, and drilling in other materials than Si (e.g., GaAs, GaN, Ge, SiC, etc.).

Table 8.1 Requirements for a laser drilling tool for TSV creation in silicon.

Specification	Current requirements	Requirements by 2016
Wafer size (mm)	100 and 150 (MEMS) 200 and 300	100 and 150 (MEMS) 200 and 300
Via density (vias mm^{-2})	10–100	20–200
Drill speed (vias s^{-1})	5000	10 000
Throughput (wafers h^{-1})	3–15	3–15
Via diameter (μm)	10	5
Via depth (μm)	50	25–50
Aspect ratio (height/diameter)	5	5–10
Resolution (μm)	1	1
Accuracy (μm)	1	1

8.2.3
Drilling Strategy

Drilling speed is one of the key requirements for the successful application of laser drilling of TSVs. This speed is partly determined by the physical drilling process itself and partly by the accumulated time that is needed for the combined mechanical and optical stepping from one via location to the next. The former part is determined by the properties of the via itself (diameter, depth, material) in combination with the fluence, wavelength, pulse length, and repetition frequency of the laser. For the latter, various strategies can be used. Fixing the wafer in place and scanning only the laser beam would require telecentric optics of at least the size of the wafer diameter. In addition, one would need an optical scanner with a large range, very high resolution, and angular speed. At the other extreme, having a fixed laser beam would require a stage to move the wafer to all via locations. This would imply a very slow process. To illustrate this, if we assume a density of $10\,\text{vias}\,\text{mm}^{-2}$ and a typical stage acceleration of $7\,\text{m}\,\text{s}^{-2}$, already about 3 h accumulated time would be needed to address all vias on a wafer. This does not include the time needed for stabilization of the stage and for actually drilling each via.

The solution is to define a so-called *field of view* (FoV) for an optical scanning technique to target the vias and address these FoVs by moving the wafer by means of a stage. Figure 8.2 shows this two-step approach.

8.2.3.1 Mechanical
Various mechanical scanning techniques have been considered:

1) *Step-and-scan*: Each time the wafer is moved to a new position and provides a new FoV to the optical scanner. Typical throughput is 1 wafer h^{-1}.

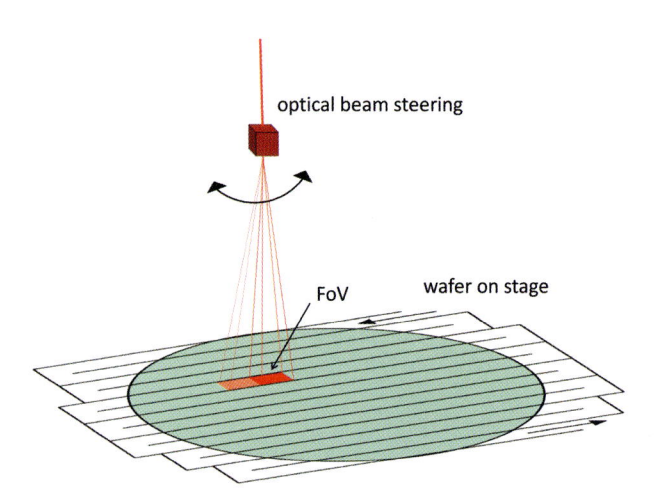

Figure 8.2 FoVs are selected mechanically by linearly moving a stage holding a wafer. The vias located within the selected FoV are addressed by an optical beam steering device.

2) *Line-by-line scan*: The wafer is scanned one line at the time. Of the total scanning time, 24% is lost in ramping the wafer up and down at the beginning and end of a line. Typical throughput here is 21 wafers h^{-1}.
3) *Vortex scan*: This is similar to writing a DVD. If rotation frequency and stage speed are optimized, the wafer throughput amounts to 27 wafers h^{-1}.

For the throughput calculations, we assumed an FoV of 1 mm^2, a 300 mm diameter wafer, and a stage with a maximum velocity of 500 mm s^{-1} and an acceleration of 7 m s^{-2}. It is obvious that the *step-and-scan* approach cannot meet the throughput requirement. The best choice between the final two options is not immediately clear. If the stage is used only for drilling, the vortex option could be chosen to optimize for speed. If the stage is also used, for example, for dicing, the obvious choice would be the line-by-line approach for versatility.

8.2.3.2 Optical

Once the optimal stage scanning strategy has been chosen, the optimal beam positioning technology has to be determined. Taking again a maximum stage velocity of 500 mm s^{-1} and a via density of 100 vias mm^{-2}, the time left for drilling the via after mechanically addressing the FoV is 20 µs. Within this time, the beam needs to be steered to the via location to within the required accuracy and, subsequently, the actual drilling needs to take place.

With a traditional line-scanning approach, a polygon mirror would, for example, need to have a pointing resolution equal to the accuracy with which a via needs to be addressed. This would result in a very low ratio between possible and needed via positions. This ratio would be as low as 0.01% given the required resolution of 1 µm.

Advanced beam steering approaches are, therefore, vector-based. Next we compare four different vector scanning methods:

1) *Galvano mirrors*: The typical response time for a 1% of range step is 250 µs. This already exceeds the available time by a factor of 10.
2) *Acousto-optical deflector (AOD)*: This method uses a sound wave traveling through a birefringent TeO$_2$ crystal. When operated in transverse mode, the response time is 1.6 µs per mm beam diameter. This is well within the time that is permitted for processing one FoV.
3) *Electro-optical deflector (EOD)*: Prisms using the Pockels effect to change the index of refraction and thus the beam path. Response time can be lower than 1 µs.
4) *Digital micromirror device (DMD)*: This method switches typically 10^6 mirrors within 200 µs on a 256 mm^2 area. Here, all vias within the FoV can be addressed at once. However, since the ratio of possible to needed positions is only 0.01%, 99.99% of the laser power is not used and it will be hard not to damage the DMD.

Summarizing, we can state that a line-by-line mechanical wafer scanning approach in combination with an optical deflector (AOD or EOD) for laser beam steering will meet the requirements listed in Table 8.1.

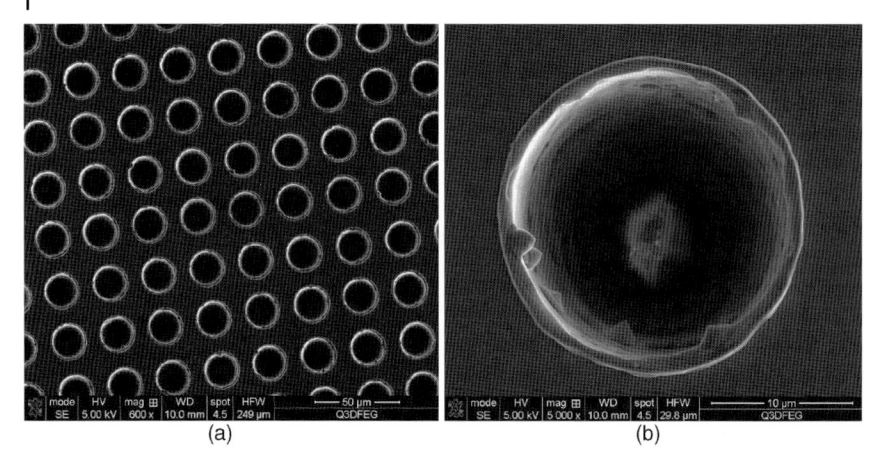

Figure 8.3 SEM images of TSVs with a diameter of about 18 μm, created by laser drilling at ALSI.

8.2.4
Experimental Drilling Results

Figure 8.3 shows scanning electron microscope (SEM) micrographs of TSVs created in silicon by laser drilling with the following process parameters: laser wavelength $\lambda = 355$ nm, laser fluence $= 90$ J cm^{-2}, pulse length $= 200$ ns, and number of pulses per via $= 4$. No AOD was used in this experiment. With laser drilling processes like this, we have been able to create vias with smooth sidewalls and rounded bottoms that meet most of the requirements of Table 8.1.

8.3
Direct-Write Deposition of Redistribution Layers

8.3.1
Introduction on Redistribution Layers

The ongoing miniaturization and 3D integration in semiconductor packaging has led to the use of so-called RDLs. The application of an RDL is a means to adapt the chip interconnect layout from a wire-bond scheme to flip-chip mounting. This adaptation reduces the effective chip package size and allows further 3D integration using interposer technology. Figure 8.4a shows the 2D layout of a typical redistribution layer. Figure 8.4b depicts a cross-sectional layout of a redistributed connection as it occurs in a wafer level chip scale package (WLCSP).

Traditionally, the RDL consists of alternating layers of insulator material and metal, where both are lithographically patterned. Each lithographic step in itself is a multistep process. Typically, for a complete RDL, the entire process adds up to 19 steps, as is shown in Figure 8.5.

8.3 Direct-Write Deposition of Redistribution Layers | 127

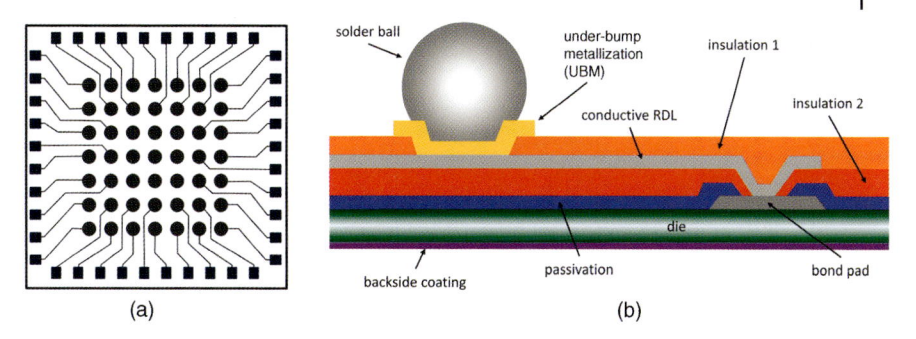

Figure 8.4 (a) Layout of a typical RDL. (b) Cross-sectional layout of a redistributed vertical connection in a WLCSP.

8.3.2
Direct-Write Characteristics

Direct-write (D-W) methods can replace each of these subtractive steps by a single-step additive deposition of the material at the desired locations only. Thus, D-W has the potential to reduce processing cost as well as startup costs by eliminating the mask steps. For this new proposed process architecture, D-W techniques are needed for the insulation/passivation layers, for the conductive tracks and pads, and for the under-bump metallization (UBM).

For the passivation and insulation layers, we propose the application of microstereolithography (µSLA). The basic process of µSLA is depicted in Figure 8.6.

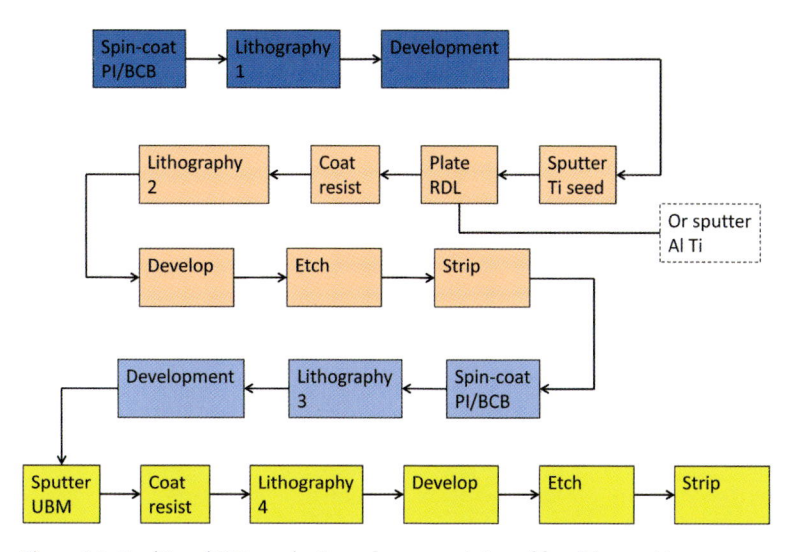

Figure 8.5 Traditional RDL production scheme consisting of four lithographic steps.

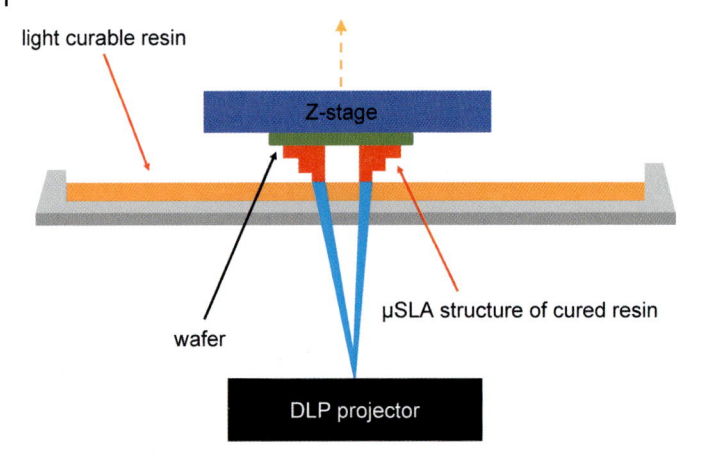

Figure 8.6 A schematic representation of the basic principle of microstereolithography.

A maskless projector (i.e., digital light projector (DLP)) projects a series of images through a glass plate onto the bottom surface of a photocurable resin. The images correspond to consecutive intersections of a computer-aided design (CAD) representation of the object to be built. Triggered by the incident radiation, the resin locally polymerizes to form a layer on the building platform (in this case a wafer). After each projection, the z-stage (holding the object) moves up to separate the wafer with the cured resin layer from the glass plate, and then moves down again to form a new thin layer of liquid resin. The glass plate is covered by a proprietary coating to facilitate easy release of the object. Current μSLA equipment is capable of producing parts with details down to 5–10 μm.

8.3.3
Direct-Write Laser-Induced Forward Transfer

In addition to using lasers for TSV drilling and wafer dicing, lasers can be used to deposit metals. Laser-induced deposition has been widely investigated [13]. Traditional pulsed laser deposition (PLD) is known as an inherently slow blanket layer deposition process. An alternative technology option with good industrial potential for fast and *direct-write* deposition of metal layers is offered by the so-called LIFT technique. Many varieties of this process exist [14–17]. In our experiments, we used a configuration where a pure metal (e.g., copper) donor on a transparent carrier is transferred by ultrashort (fs–ns) laser pulses onto a substrate as illustrated in Figure 8.7.

The substrate moves with respect to the stationary pulsed laser beam so as to continuously provide fresh donor material. The laser beam is focused onto the donor–carrier interface. The donor layer is positioned face down, close to the substrate surface. The air gap between the donor and substrate is typically 10–50 μm. The high local fluence generated during each laser pulse will partially melt and/or evaporate the donor layer and transfer a fraction of the donor material

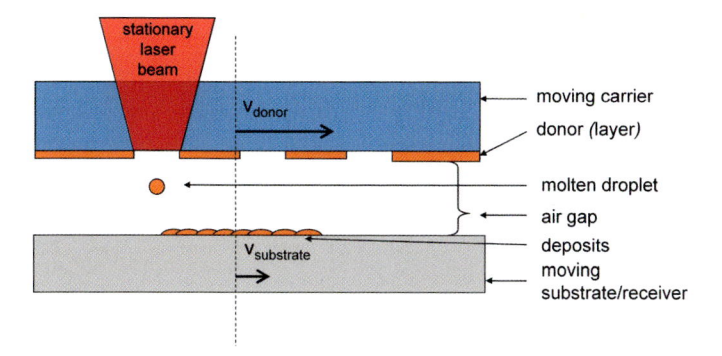

Figure 8.7 With LIFT, a layer of donor material that has been coated onto a transparent carrier (e.g., by sputter deposition) is heated by a pulsed laser beam to its melting point. Depending on the laser fluence, one or more droplets are launched forward and deposited on a receiving substrate. The carrier is moved after each laser shot to provide fresh donor material. In order to write conducting lines, the substrate is moved by a fraction of the deposit size while keeping the laser beam stationary. This ensures good overlapping of the deposits and thus good conductive contact.

forward toward the receiving substrate. The physical state of this piece of donor material depends on the energy levels involved and may range from almost fully solid to a "cloud" of clustered particles. This way, metal deposits in the μm range can be formed using a picosecond laser with a spot size of 5–20 μm focused onto the metal layer. To form 2D and 3D structures, overlapping droplets need to be deposited (see Figure 8.7) while scanning the substrate. The extent of overlapping deposits in the metal pattern is determined by the scan speed $v_{substrate}$ of the moving substrate.

When considering LIFT of conductive materials, one has to distinguish between two types of materials: bulk metals and metallic inks. Both are described extensively in the literature. For bulk metals, for example, copper, gold, chromium, zinc, aluminum, and nickel have been reported. With respect to metallic inks, gold nanoparticle (NP) ink, silver NP inkjet ink, and silver NP paste have been studied [18].

For comparison, the cross section and the production process needed for this new architecture based on additive techniques is shown in Figure 8.8. The passivation layers and the RDLs are straightforward applications of LIFT and μSLA as described earlier. For the UBM, an alternative process cannot be proposed without specifying the interconnect technology used. However, based on nanoparticle inks, new UBM-interconnect formulations can be developed. Also, LIFT can be used to transfer a multilayer or metal composite to form the UBM.

We calculated that the D-W approach will reduce the CoO by a factor of 2 over the conventional approach for a typical 9 mm^2 WLCSP with a 6 × 7 ball grid array (BGA).

Obviously, there are many technical challenges when introducing a new disruptive architecture and technology as described earlier. The main challenge in

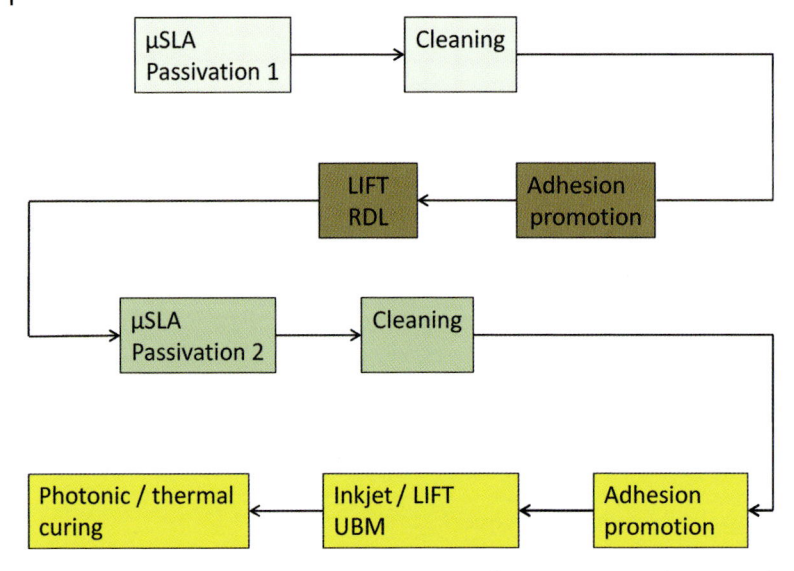

Figure 8.8 Alternative RDL production process based on direct-write techniques only. The number of steps is significantly reduced over the traditional approach.

the end is to qualify on package reliability. This will require application-specific research to further optimize the processing conditions.

8.3.4
LIFT Results

Focusing on the LIFT process, the main challenges are on the adhesion and the mechanical strength of the RDL as well as on electrical conductivity. Positioning accuracy and contamination (satellite depositions) are sufficiently under control to meet the geometrical requirements for the RDL. Given the many wafer-level laser machining tools that currently exist, industrialization is straightforward once the process is mature.

Figure 8.9 shows an optical micrograph (a) and a confocal image (b) of a conductive line feature produced by means of a LIFT process. The metal deposited here was copper. The carrier was a PET film with a thickness of 100 μm. The line has a full width of 20 μm and a resistivity that is 5.6 times that of bulk copper. The LIFT parameters were as follows: $\lambda = 515$ nm, pulse length $= 6.7$ ps, fluence $= 0.25$ J cm^{-2}; donor spacing $= 10$–20 μm; donor layer thickness $= 200$ nm; deposit overlap $= 80\%$.

In separate experiments, we were able to deposit lines with a thickness down to 8 μm in width, as shown in Figure 8.10.

LIFT has many process parameters that affect the results. These should all be optimized so as to create the best result for a certain application with a process

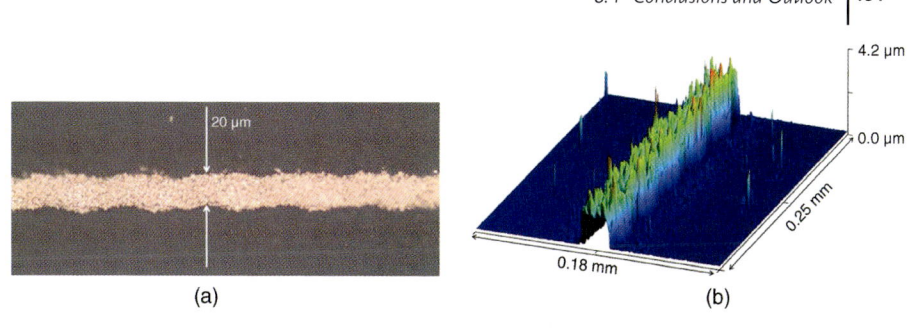

Figure 8.9 Conductive line produced by LIFT: (a) microscope image, (b) confocal image. See the text for process parameters.

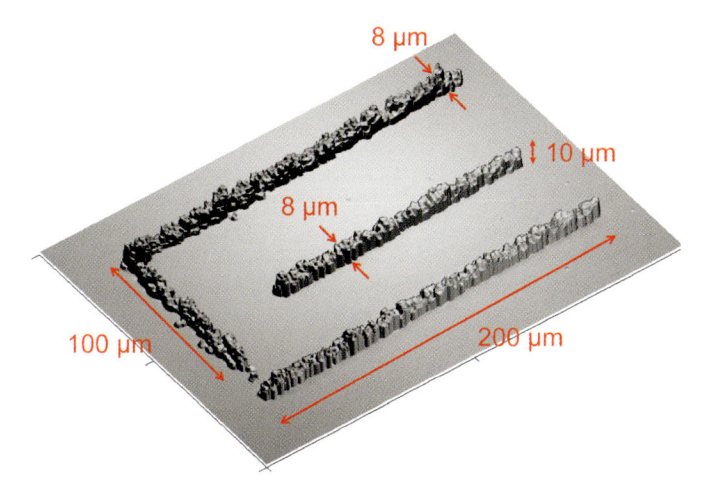

Figure 8.10 Confocal image of copper structures deposited on a glass substrate by means of LIFT.

window to allow for a robust production. This way, resistivity values of ∼3 times that of bulk copper have been obtained recently on 6 μm wide lines.

8.4
Conclusions and Outlook

Until recently short-pulsed laser processing has been used almost exclusively for die singulation, or "dicing." In this field, there are developments toward subsurface wafer dicing, which produces less debris, less recast, and an improved die strength, which have not been covered in this chapter but can be found elsewhere [12] (see, for example, www.alsi-international.com (accessed December 12, 2012)).

Besides dicing, laser-assisted techniques can also provide alternative steps in wafer processing. Recent work in laser microdrilling of TSVs (in silicon and other semiconductor substrates) has shown that medium-sized diameter (5–10 μm, and up) can be obtained. Another technology under development is the laser-assisted (LIFT) maskless deposition of RDLs. In addition, this additive, direct-write technique is currently under investigation by us for the application of TSV filling. Another potential application of this additive, direct-write technique is currently under investigation by us and aims at filling TSVs with copper.

References

1 Yole Développement (2010) 3D IC integration & TSV interconnects: market analysis. Webinar extracts, February 12, 2010.
2 Garrou, P.E. and Vardaman, E.J. (2008) Through silicon via technology: the ultimate market for 3D interconnects. TechSearch International.
3 Van der Plas, G., Limaye, P., Loi, I., Mercha, A., Oprins, H., Torregiani, C., Thijs, S., Linten, D., Stucchi, M., Katti, G., Velenis, D., Cherman, V., Vandevelde, B., Simons, V., De Wolf, I., Labie, R., Perry, D., Bronckers, S., Minas, N., Cupac, M., Ruythooren, W., Van Olmen, J., Phommahaxay, A., De Potter de ten Broeck, M., Opdebeeck, A., Rakowski, M., De Wachter, B., Dehan, M., Nelis, M., Agarwal, R., Pullini, A., Angiolini, F., Benini, L., Dehaene, W., Travaly, Y., Beyne, E., and Marchal, P. (2011) Design issues and considerations for low-cost 3-D TSV IC technology. *IEEE Journal of Solid-State Circuits*, **46**, 293–307.
4 Bean, K.E. (1978) Anisotropic etching of silicon. *IEEE Transactions on Electron Devices*, **ED-25**, 1185–1193.
5 Lehmann, V. (1993) The physics of macropore formation in low doped n-type silicon. *Journal of the Electrochemical Society*, **140**, 2836–2843.
6 Slikkerveer, P., Bouten, P.C.P., and de Haas, F.C.M. (2002) High quality mechanical etching of brittle materials by powder blasting. *Sensors and Actuators*, **85**, 296–303.
7 Wensink, H., Berenschot, J.W., Jansen, H.V., and Elwenspoek, M.C. (2000) High resolution powder-blasting micromachining. Proceedings of the 13th Annual International Conference on Micro Electro Mechanical Systems, Miyazaki, Japan, January 23–27, 2000, pp. 769–774.
8 Roozeboom, F., Blauw, M.A., Lamy, Y., van Grunsven, E., Dekkers, W., Verhoeven, J.F., van den Heuvel, F., van der Drift, E., Kessels, W.M.M., and van de Sanden, M.C.M. (2008) Deep reactive ion etching of through-silicon vias, in *Handbook of 3-D Integration: Technology and Applications of 3D Integrated Circuits* (eds P. Garrou, C. Bower, and P. Ramm), Wiley-VCH Verlag GmbH, Weinheim, pp. 47–91, and references therein.
9 Rodin, A.M., Callaghan, J., and Brennan, N. (2008) High throughput low CoO industrial laser drilling tool. Proceedings of the 4th IMAPS International Conference and Exhibition on Device Packaging, Scottsdale, AZ, March 17–20, 2008.
10 Velenis, D., Stucchi, M., Marinissen, E.J., Swinnen, B., and Beyne, E. (2009) Impact of 3D design choices on manufacturing cost. Proceedings of the IEEE International Conference on 3D System Integration (3DIC2009), San Francisco, CA, September 28–30, 2009, pp. 1–5.
11 Lo, W.-C. and Chang, S.M. (2008) Laser ablation, in *Handbook of 3-D Integration: Technology and Applications of 3D Integrated Circuits* (eds P. Garrou, C. Bower, and P. Ramm), Wiley-VCH Verlag GmbH, Weinheim, pp. 93–105, and references therein.
12 Li, J., Hwang, H., Ahn, E-C., Chen, Q., Kim, P., Lee, T., Chung, M., and Chung, T. (2007) Laser dicing and subsequent die strength enhancement technologies for ultra-thin wafer. Proceedings of the 57th IEEE International Conference on Electronic Components and Technology, San Francisco, CA, pp. 761–766.

13 Banks, D.P. (2008) Femtosecond laser induced forward transfer techniques for the deposition of nanoscale, intact, and solid-phase material. PhD thesis, University of Southampton, UK.

14 Piqué, A. and Chrisey, D.B. (eds) (2002) *Direct-Write Technologies for Rapid Prototyping Applications: Sensors, Electronics, and Integrated Power Sources*, Academic Press, San Diego, CA.

15 Bera, S., Sabbah, A.J., Yarbrough, J.M., Allen, C.G., Winters, B., Durfee, C.G., and Squier, J.A. (2007) Optimization study of the femtosecond laser-induced forward-transfer process with thin aluminum films. *Applied Optics*, **46**, 4650–4659.

16 Ko, S.H., Pan, H., Grigoropoulos, C.P., Luscombe, C.K., Fréchet, J.M.J., and Poulikakos, D. (2007) All-inkjet-printed flexible electronics fabrication on a polymer substrate by low-temperature high-resolution selective laser sintering of metal nanoparticles. *Nanotechnology*, **18**, 345202 1–8.

17 Narazaki, A., Kurosaki, R., Sato, T., Kawaguchi, Y., and Niino, H. (2009) Nano- and microdot array formation by laser-induced dot transfer. *Applied Surface Science*, **255**, 9703.

18 Nagel, M. and Lippert, T. (2012) Laser-induced forward transfer for the fabrication of devices, in *Nanomaterials, Processing and Characterization with Lasers* (eds S.C. Singh, H.B. Zeng, C. Guo, and W.P. Cai), John Wiley & Sons, Inc., New York, pp. 255–316, and references therein.

9
Temporary Bonding Material Requirements

Rama Puligadda

9.1
Introduction

All of the myriad structures and architectures for 3D integration require thinning of the device wafers to a thickness of 50 µm or less and connecting the layers in the stack by means of through-silicon vias (TSVs). Thinning the device wafer after all front-side processing is complete can be quite risky without the use of front-side protection. Handling the thinned wafer through the backside process without support is virtually impossible without severe breakage, chipping, or warping.

One of the ways to protect the device wafer through the stacking process is to permanently bond the wafer to another device wafer to create a stack and then thin the top device wafer to the desired thickness [1,2]. This approach, however, has several drawbacks: (a) it limits bonding options to only the face-to-face type; (b) the entire stack is exposed to the backside process of creating redistribution layers and interconnects on the thinned wafer, which typically includes harsh thermal and chemical steps; and (c) the wafer bonding scheme is limited to wafer-to-wafer only, thus making it difficult to stack dissimilar device wafers.

A current preferred approach is to temporarily bond the wafer face down to a rigid carrier, keeping the backside of the wafer completely exposed for thinning, via reveal and creating redistribution layers and interconnects needed for stacking. The temporary bonding material is usually designed to planarize the topography on the device side, providing a relatively flat surface for bonding to the carrier wafer. This bonding material also serves to protect any delicate features on the device wafer as it undergoes subsequent processing. Thinning to thicknesses less than 100 µm is typically achieved using a grinding process followed by a polishing process. The stack composed of the thinned wafer and carrier is not significantly different in thickness and size from a standard full-thickness wafer and therefore does not require any modification to lithography, deposition, etching, and other equipment. After the thinning and backside processes are accomplished, the device wafer can be separated from the carrier wafer using a specially designed process termed "debonding" and is then cleaned and transferred to a continuous support such as a film frame. Alternatively, the thin side of the temporary wafer stack may be attached to a film frame before debonding and cleaning.

9.2
Technology Options

Several different approaches and technology options [3–7] for bonding and debonding are available today, and several more are in development.

9.2.1
Tapes and Waxes

Tapes and waxes are commercially available and are used industry-wide for a limited number of applications today. However, they typically do not survive high-temperature processes and harsh chemical environments. Additional limitations include high total thickness variation (TTV) and the presence of residues after release.

9.2.2
Chemical Debonding

Commonly used by low-volume manufacturers, this method utilizes a perforated rigid carrier for support. To achieve debonding, a suitable solvent is allowed to penetrate through the holes in the carrier wafer and dissolve the adhesive that is holding the wafers together.

9.2.3
Thermoplastic Bonding Material and Slide Debonding

This approach uses carefully designed thermoplastic materials that can be heated to bond the wafers together and heated again to a temperature above the softening point of the material to allow sliding the carrier wafer across the device wafer to effect debonding. The material is designed to remain mechanically stable through the backside process. The bonding material solution is applied to either the carrier or the device wafer by spin coating. The solvent remaining in the film is then eliminated using a baking process. The device wafer and carrier are bonded in a vacuum chamber at a temperature above the softening point of the bonding material and at pressures ranging from 5 to 20 psi. This technology can be used with silicon, glass, or sapphire carriers of the same size as the device wafer.

After completing the backside process, debonding is achieved by a slide-off process where the thin device wafer is completely supported on a vacuum chuck in the debonding equipment. Both the device wafer and the carrier are then cleaned on a specially designed chuck using an appropriate remover. Optionally, the thin device wafer is transferred to a film frame before cleaning. A schematic of the

9.2 Technology Options | 137

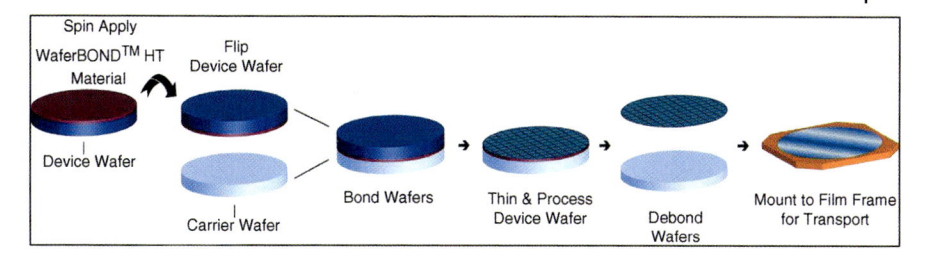

Figure 9.1 Slide-off approach.

[Figure 9.1 process flow: Spin Apply WaferBOND™ HT Material on Device Wafer → Flip Device Wafer → Bond Wafers (with Carrier Wafer) → Thin & Process Device Wafer → Debond Wafers → Mount to Film Frame for Transport]

slide-off approach is shown in Figure 9.1. When automated, this approach requires an equipment set consisting of a bonder unit where the device wafer is coated and bonded and a debonder unit where the wafers are separated and cleaned.

9.2.4
Debonding Using Release Layers

This approach [8] utilizes a specially designed release layer that, when altered by mechanical force, chemistry, ultraviolet (UV) light, heat, or laser energy, can effect the separation of the device layer from the carrier. When release layers are used, the stress applied to the device wafer during the debonding process is much reduced, in contrast with slide-off debonding, where a mechanical force is used at high temperature to slide one wafer along the other. Several different schemes are available today for evaluation and manufacturing. All of these methods cause separation of the carrier wafer from the bonding layer, thus leaving the bonding layer–device wafer interface undisturbed during separation. The bonding material is then removed from the surface of the device wafer by chemical or mechanical methods. Methods that involve dissolving the bonding material in suitable solvents can be a more preferred approach than approaches that utilize a mechanical force, such as peeling of the bonding material from the device wafer, because mechanical methods can apply significant stress in the direction perpendicular to the surface of the wafer. Such stresses can be especially problematic on bumped device wafers. The ZoneBOND® process developed and commercialized by Brewer Science differs from all other approaches in two important aspects: (1) it utilizes a thermoplastic bonding material for easy and low-stress removal, and (2) it bonds the device wafer to a carrier wafer having two zones of very different adhesion strengths to achieve selective bonding at the edges. The ZoneBOND process is described in more detail in subsequent sections.

Another important aspect of methods that utilize release layers is that the thinned and processed device wafer can be securely supported on a film frame or other form of continuous support before it is goes through separation of the carrier wafer and cleaning of the temporary bonding material from the device wafer surface. In the case of chemical removal, the bonding material removers are designed to not attack the tapes on the film frame. Alternatively, the tape can be treated to render it resistant to removers and other solvents.

9.3
Requirements of a Temporary Bonding Material

The temporary bonding material or adhesive must meet the following key requirements to effectively withstand processing demands:

- *Ease of application*: The process for applying the temporary bonding agent should enable high throughput and preferably use standard and accepted application processes such as spin coating, lamination, spray coating, and so on.
- *Coating uniformity*: Because the nonuniformity in the coating is transferred to the thinned wafer during the thinning process, the coating thickness of the bonding material must be extremely uniform. TTV is a commonly used parameter for assessing the uniformity of the coating. Current industry processes require the TTV of the thinned wafer to be less than 2 µm on a 300 mm wafer.
- *Adhesion to a wide variety of surfaces*: The surface of a device wafer typically is comprised of a wide variety of materials such as metals, dielectrics, solder, and so on, which will come in contact with the bonding material. Poor adhesion to any of these surfaces may lead to dewetting in areas, which can appear as voids in the bond line.
- *Thermal stability and mechanical strength*: A typical backside process flow for a TSV wafer usually includes high-temperature processes such as dielectric deposition, annealing, metal deposition, permanent bonding, and solder reflow. Such processes can cause the highly stressed device wafer to curl or buckle. The bonding material must remain mechanically stable during these high-temperature processes, where it must hold the device wafer completely bonded to the carrier. When thermoplastic adhesives are used, the melt viscosity and shear modulus of the material at temperatures of use become highly important considerations. Melt rheology curves of bonding material compositions show change in viscosity and modulus of the polymer composition with increasing temperature.

   Typically, subjecting the bonded wafer pair to temperatures far beyond the softening point of the bonding material can result in delamination or bubbling of the thin wafer. Figure 9.2a shows an example of a rheology curve for a temporary bonding material used in a slide debonding scheme. In addition, the adhesive should not undergo any decomposition, outgassing, or cross-linking through the high-temperature processes. Thermogravimetric analysis (TGA) is commonly used to measure any weight loss in the sample when it is heated to temperatures typically experienced in deposition chambers. An example of TGA results is shown in Figure 9.2b, which shows a TGA plot of a typical temporary bonding material that exhibits thermal stability up to 400 °C.
- *Chemical resistance to process chemicals*: The temporary bonding material must be resistant to process chemicals such as etchants, cleaning solutions, and corrosive electroplating solutions.
- *No residues*: The temporary bonding agent must be easily and completely removed from the surface of the wafer, especially when high topography or

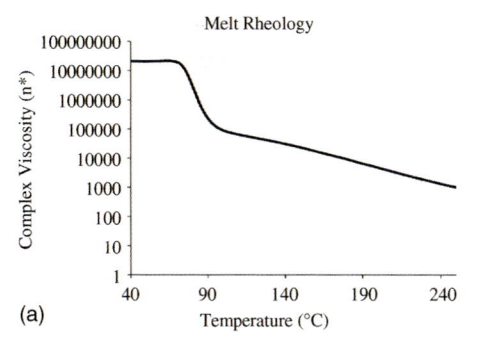

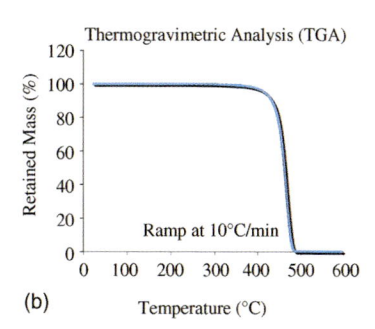

Figure 9.2 (a) Example of a rheology curve for a thermoplastic adhesive. (b) Example TGA plot of a temporary bonding material.

retrograde features exist on the device wafer. Residues left behind can cause difficulties in downstream processing and negatively affect the reliability of the device package.

9.4
Considerations for Successful Processing

9.4.1
Application of the Temporary Bonding Adhesive to the Device Wafer and Bonding to Carrier

Several parameters come into play during the coating and bonding steps, which must be optimized to ensure that the bonded pair goes through the downstream processes without any damage or failures. Each one of these parameters will be covered in the next few sections.

9.4.2
Moisture and Contaminants on Surface

Moisture and other contaminants on the surface of the wafer can cause outgassing during downstream processing. Figure 9.3a shows a scanning acoustic microscopy (SAM) image of a bonded wafer that showed delamination during a high-temperature deposition process due to release of moisture or outgassing of contaminants. Before the bonding material is applied, baking the device wafer at the highest allowable temperature for a period of 2–30 min can ensure complete removal of all contaminants. Figure 9.3b shows an SAM image of a bonded wafer that was subjected to a decontamination bake prior to coating.

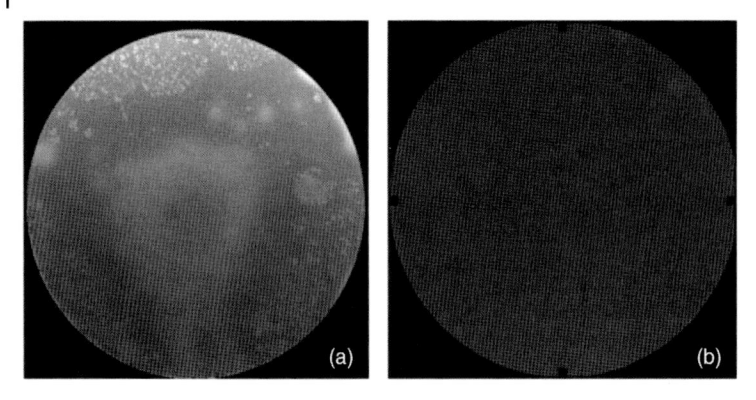

Figure 9.3 (a and b) SAM images demonstrating the need for a decontamination bake before coating the temporary bonding material.

9.4.3
Total Thickness Variation

Several factors contribute to the TTV of the thinned device wafer. The most important contributions come from the coating process, the bonding process, and the grinding process. Severe topography on the wafer, such as bumps and pillars, exacerbate the problem. An important consideration for the coating process is that the bonding material thickness must exceed the height of device features to ensure complete planarization. This requirement would mean that the bonding material thickness could be more than 100 μm to cover features such as bumps, which are commonly 80–90 μm in size. A common problem amplified while spin coating such a thick adhesive is a thick beadlike deposition at the edges called the "edge bead." Thickness uniformity of the adhesive after bonding can be measured using a spectral coherence technology offered by ISIS Sentronics. The effect of the edge bead on the TTV measured using ISIS techniques is shown in Figure 9.4. During the grinding process, the thicker areas in the adhesive layer get translated into thinner areas in the device wafers, causing nonuniformity in the wafer. Optimizing material and process factors such as spin and bake parameters, solvent type, and chemical composition of the temporary bonding material can reduce edge bead.

The bonding process can contribute to TTV as well. The uniformity of the platens in the bonding equipment directly affects the uniformity of the bond line. The bonding pressure used and the bonding program (center to edge or edge to center) can also affect the uniformity of the bond line.

9.4.4
Squeeze Out

When a thermoplastic material is used for temporary bonding, the bonding temperature is higher than the softening temperature of the material so that it

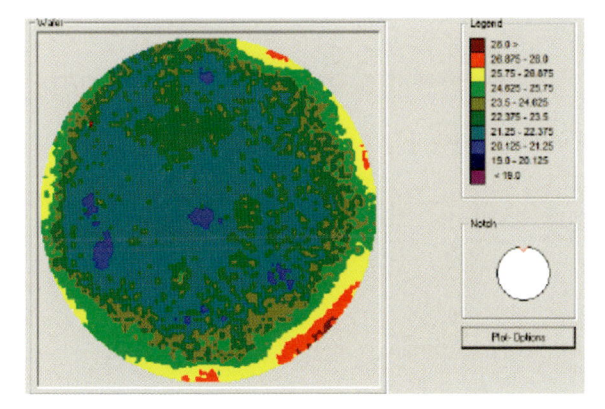

Figure 9.4 Effect of edge bead on thin wafer uniformity.

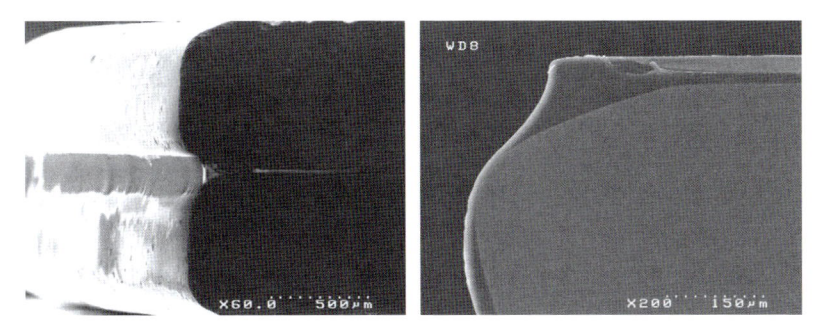

Figure 9.5 Image of "squeeze out" during bonding process.

can easily reflow to achieve a planar surface. Using nonoptimized temperature and pressure conditions during the bonding step, however, can cause bonding material to severely "squeeze out" from the edges of the bonded pair, as shown in Figure 9.5. This phenomenon not only affects the uniformity of the bond line, but also can contaminate the bonding equipment and other equipment used for further processing.

9.5
Surviving the Backside Process

Processing on the backside of the device wafer starts with grinding it to less than 50 μm in thickness to expose the copper nails, followed by patterning to create redistribution layers. Most of these steps expose the wafer pair to harsh conditions such as high plasma, high temperature, and corrosive chemicals. The following sections describe some important considerations for successful processing.

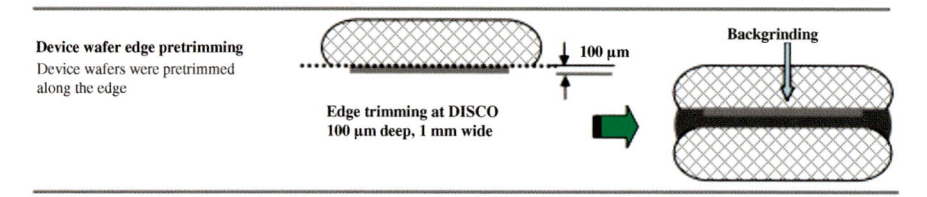

Figure 9.6 Pretrimming before coating and bonding.

9.5.1
Edge Trimming

One of the most common failure modes of the grinding process for wafer thinning is chipping and breakage at the edges. Grinding of the beveled edge of the wafer results in a knife edge that is easily prone to chipping that can further cause wafer cracking [9]. A study conducted by Brewer Science [10] compared the efficacy of the most commonly used edge protection schemes, including use of a larger wafer carrier, pretrimmed wafers, and prethinned carriers. The study showed that while all the edge protection schemes drastically reduce edge chipping, pretrimming was the most efficient. An illustration of the pretrimming process, in this case done at DISCO Corp., is shown in Figure 9.6.

9.5.2
Edge Cleaning

The difference in radii of the thinned wafer and the carrier wafer results in the presence of exposed temporary bonding material on the edge of the carrier wafer, as shown in Figure 9.7. This exposed material can undergo physical or chemical changes when subjected to the high-temperature and plasma processes and chemicals used in cleaning, electroplating, and so on. In the case of pretrimmed wafers, the difference in radii is much larger, and therefore, a larger area of adhesive on the carrier is exposed. Figure 9.8 shows a picture of exposed temporary adhesive on the edges of a carrier wafer after high-temperature treatment. The

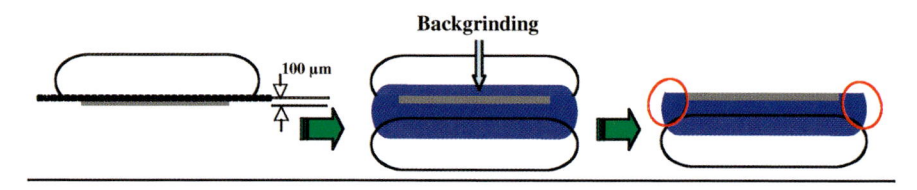

Figure 9.7 Schematic showing exposed temporary bonding material after backgrinding (circled areas).

Figure 9.8 Temporary bonding material on the edge of carrier wafer after high-temperature treatment.

deformed material at the edges can cause problems during downstream processing. Additionally, the material can outgas and contaminate equipment used for deposition. A thermoplastic temporary bonding material can advantageously be removed by dispensing a suitable solvent at the edge of the bonded wafer pair.

9.5.3
Temperature Excursions in Plasma Processes

The temperature observed at the backside surface of the device wafer during high-temperature plasma deposition processes, such as dielectric deposition, has been observed to be 20–50 °C higher than the set temperature. This difference occurs due to poor thermal conductivity of the polymeric adhesive and the carrier wafer that is in contact with the chuck in the deposition chamber. The temperature excursions are higher when glass carriers are used because thermal conductivity of glass is lower than that of silicon. Therefore, the material used for temporary bonding must have thermal and mechanical stability at temperatures at least 50 °C above the set processing temperatures.

9.5.4
Wafer Warpage due to CTE Mismatch

A mismatch between the coefficients of thermal expansion (CTEs) of the carrier and the device wafer can cause mechanical stresses to build up in the device wafer, causing it to warp and delaminate from the carrier wafer at high temperatures. The CTE mismatch is greater when a glass carrier is used and increases with increasing processing temperature, causing more severe issues. While a thermoset bonding material can provide superior mechanical support because of its rigidity, it can cause additional stress on the thin wafer by not being compliant enough to allow expansion of the thin wafer in all directions. Conversely, at high temperatures, a thermoplastic material allows uniform expansion of the wafer as it changes to a soft solid or high-viscosity liquid.

9.6
Debonding

The term "debonding" describes the process of separating the device wafer from the rigid carrier. The following sections describe some of the important considerations for the debonding process in order to meet the demand for fast and secure separation needed for high-volume manufacturing.

9.6.1
Debonding Parameters in Slide-Off Debonding

Thermoplastic adhesives enable this process by turning into viscous liquids when heated to temperatures above their softening points. The debonding process parameters [11] such as debonding temperature, debonding force, and debonding speed depend on the rheological properties of the bonding material, bonding material thickness, the topography on the wafer, and the sensitivity of the device to high temperatures. An optimized debonding process would use the lowest possible temperature and least debonding force while achieving the required throughput and causing no damage to the devices or interconnects.

9.6.2
Mechanical Damage to Interconnects

Damage to the bumps that are within the bond line during debonding is manifested in three main modes: flattening, displacement, and smearing. Damage of any of these types leads to yield and reliability issues. Bumps on the backside of the device pose additional challenges in separating the thin wafer from the carrier. While a slide-off process can be optimized to minimize damage to interconnects, debonding methods that use release layers and do not require heating the wafers are advantageous when processing bumped wafers.

Acknowledgments

The author would like to gratefully acknowledge the contributions made by Mark Privett, Jason Neidrich, Jeremy McCutcheon, and the advanced packaging project team members at Brewer Science, Inc. The author would also like to acknowledge the work done by our partners at Leti and IMEC. Finally, the author would like to acknowledge contributions made by Catherine Frank.

References

1 Garrou, P., Bower, C., and Ramm, P. (eds) (2008) *Handbook of 3D Integration: Technology and Applications of 3D Integrated Circuits*, Wiley-VCH Verlag GmbH, Weinheim, pp. 463–486.

2 Hosali, S. et al. (2008) Through-silicon via fabrication, backgrind, and handle wafer technologies, in *Wafer Level 3-D ICs Process Technology* (eds C.S. Tan et al.), Springer, pp. 85–115.

3 Puligadda, R. et al. (2007) High performance temporary adhesives for wafer bonding applications. Materials Research Society Symposium Proceedings, vol. 970, Paper No. 0970-Y04-09.

4 Hermanowski, J. (2009) Thin wafer handling – study of temporary wafer bonding materials and processes. IEEE International Conference on 3D System Integration.

5 Pargfrieder, S. et al. (2009) 3D process integration with TSV: temporary bonding and debonding. *Advanced Packaging*, April 2009.

6 Wolf, J.M. et al. (2008) 3D process integration – requirements and challenges, in *Materials and Technologies for 3-D Integration, Materials Research Society Symposium Proceedings* (eds F. Roozeboom et al.), Cambridge University Press, vol. **1112**, pp. 3–15.

7 Kessel, C.R. et al. (2004) Wafer thinning with the 3M wafer support system. Proceedings of the International Wafer-Level Packaging Conference (IWLPC).

8 Mathias, T. (2007) Ultrathin-wafer processing utilizing temporary bonding and debonding technology. Proceedings of the International Wafer-Level Packaging Conference (IWLPC).

9 Garrou, P., Bower, C., and Ramm, P. (eds) (2008) *Handbook of 3D Integration: Technology and Applications of 3D Integrated Circuits*, pp. 27–28.

10 Bai, D. et al. (2009) Edge protection of temporarily bonded wafers during backgrinding. *ECS Transactions*, **18** (1), 757–762.

11 Privett, M. et al. (2008) TSV thinned wafer debonding process optimization. Proceedings of the International Wafer-Level Packaging Conference (IWLPC) October 13–16, 2008, pp. 144–148.

10
Temporary Bonding and Debonding – An Update on Materials and Methods

Wilfried Bair

10.1
Introduction

Since the first edition of the *Handbook of 3D Integration* was published, much has changed in the world of temporary bonding and debonding. Spin-on materials are no longer "typically debonded by a slide-off method," and heat and solvents are no longer needed to initiate separation of the device wafer from the carrier wafer. Instead, the industry has turned to room-temperature mechanical lift-off debonding methods that do not add to thermal budget requirements and reduce the total cost of ownership.

The drivers for this change in debonding strategy have been further tightening of specifications and increased complexity of the device wafer: lower total thickness variation (TTV), smaller through-silicon vias (TSVs), high topography, and reduction in final wafer thickness. The primary challenges remain the same – surviving all processes and producing a thinned wafer clean enough (i.e., residue free) for further downstream processing. Eliminating heat from the equation is just the first step. Removing mechanically applied stress during the debonding process is also required and the latest commercially available temporary bonding adhesive systems are designed to release (fail) well before damage to the device wafer and carrier wafer can take place.

The reasons for temporary bonding have not changed. There is still a need for device wafers to be temporarily held flat so they can be thinned, typically to reveal TSVs, and survive backside redistribution layer formation processes such as photolithography and plating, before being sent off to assembly or bonded to another wafer. Bonding methods have also remained similar except for the inclusion of release layers and/or primer layers that facilitate debonding at a selected interface in the bonded stack and with a specified force.

See Figure 10.1 for a typical bond process flow using a release layer/primer and room-temperature mechanical lift-off debond. The engineering of surface properties is the fundamental change that has enabled room-temperature mechanical lift-off debonding methods to become a mainstream process option. Due to this revolutionary change in temporary bond strength management (stress/strain

---

Handbook of 3D Integration: 3D Process Technology, First Edition.
Edited by Philip Garrou, Mitsumasa Koyanagi, and Peter Ramm.
© 2014 Wiley-VCH Verlag GmbH & Co. KGaA. Published 2014 by Wiley-VCH Verlag GmbH & Co. KGaA.

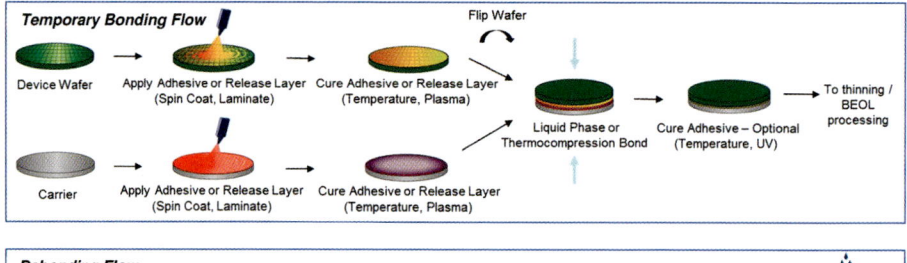

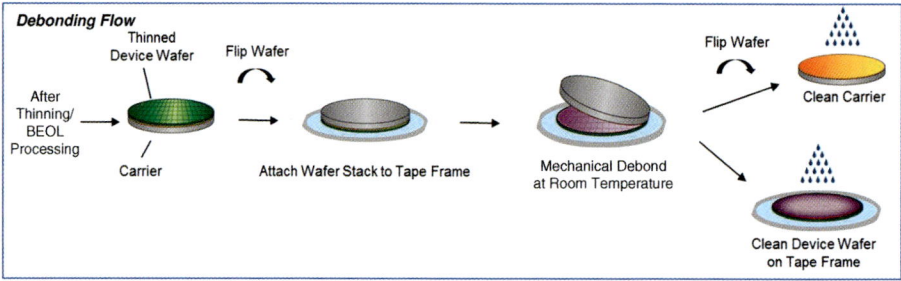

Figure 10.1 Typical process flow for temporary bonding and debonding that includes the applications of a release layer or primer layer and debonding at room temperature.

management), the range of chemistries that can be used for temporary bonding has increased significantly – silicone, BCB, acrylate, and polyimide to name a few. By employing the fundamentals of fracture mechanics to the temporary bonding adhesive interfaces, one can use many different materials with the same debonding tool. The debonder's function is simply to peel two wafers apart and as long as the adhesive system has been engineered correctly, the bonded wafers should come apart well before any damage can occur to either device or carrier wafer.

10.2
Carrier Selection for Temporary Bonding

Now that we understand why room-temperature lift-off debonding is currently the state-of-the-art debonding method, we can discuss the full process flow in more detail. And, since carrier wafers are an integral part of the temporary bonding/debonding process, let us start with them. Carrier wafer material selection has a significant influence as to what adhesive materials and type of bonding and debonding methods can be employed. Glass and silicon carriers are the two most commonly used solid carrier types. The biggest differences between glass and silicon as a carrier wafer material are thermal conductivity (Table 10.1), transparency, and trace element impurities, as well as the ability of a glass carrier to be made in nonstandard sizes. And, although pricing of high-quality glass and silicon carriers is similar today, there appears to be a path for glass wafers to be made at lower cost when scaled to mass production levels.

10.2 Carrier Selection for Temporary Bonding

Table 10.1 Comparison of the physical properties of silicon versus glass [1].

Mechanical properties	Glass	Silicon
Density (g cm^{-3})	2.38	2.33
Young's modulus (GPa)	73.6	129.5
Knoop hardness (kg mm^{-2})	453	1150
CTE (0–300 °C, $\times 10^{-7}$ °C^{-1})	31.7	31.5
Thermal conductivity (23 °C, W cm^{-1} K^{-1})	0.0109	1.31

Photosetting adhesives and laser ablation debonding require a carrier wafer that is transparent at the wavelength of light needed for curing and ablation – typically ultraviolet (UV) radiation. For most other aspects of the processing, glass and silicon are interchangeable. As mentioned earlier, glass is a much better insulator, and this needs to be taken into consideration when thermal processes are involved. Process chambers used for processing wafers while on a carrier may read a different temperature of the thinned silicon wafer when planed on an adhesive layer and mounted to a carrier, in particular with glass carriers and the aforementioned insulation properties.

The size of the carrier wafer needs to be carefully considered and may vary by material category or material type. The typical choices are a carrier wafer that is equal to or slightly greater in diameter to the device wafer. If, however, the device wafer undergoes edge trimming (see Figure 10.2) with the intent to reduce the likelihood of cracking and chipping of the wafer edge during backgrinding, the effect will likely be similar to attaching an oversized carrier with the adhesive slightly extending beyond the silicon wafer and providing support to the wafer edge. Since the thinned device wafer is very fragile at the edges, there is a handling advantage with an oversized carrier as carrier wafer protrudes beyond the silicon wafer and provides added edge protection.

The choice of carrier wafer materials also influences the inspection techniques that can be used in subsequent process steps such as bonding. After the silicon wafers

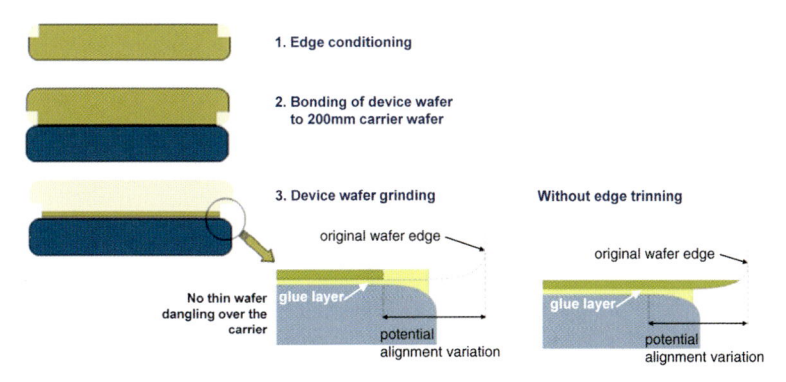

Figure 10.2 Edge conditioning (trimming) of the device wafer prior to bonding reduces the likelihood of cracking or chipping while also reducing the wafer diameter after backgrinding [2].

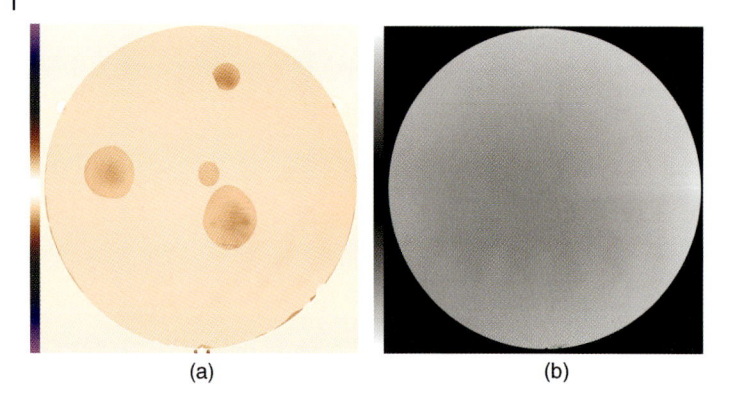

Figure 10.3 SAM scans of temporary bonded wafers with (a) and without (b) post-bond voids.

have been bonded to the carrier, an inspection of the bond quality is commonly used. There are several options available including scanning acoustic microscopy (SAM) imaging as shown in Figure 10.3 or a full field inspection using a system such SPARK by Nanometrics [3]. Since very small voids can potentially lead to delamination under high vacuum or during high-temperature processing, the detection resolution capabilities of the above-mentioned techniques are being continuously challenged. There is a delicate balance between the need for strong enough adhesion to prevent bubbles from growing and weak enough adhesion to permit debonding at the end of the processing steps while the wafer is on the carrier.

After determining that a void-free and fully bonded wafer to a carrier pair has been achieved, the next critical specification to check is the TTV of the bonded pair (Figure 10.4). TTV control in the low single-digit micron range is required primarily for the final TSV reveal process and the electrical connections to be made during the stacking process. If there is a great deal of variation in thickness, the high spots will be ground down too much, and more importantly, the low spots may not be ground down enough to expose the tops of the vias leading to open circuits. Although some thickness variation can be overcome by sophisticated

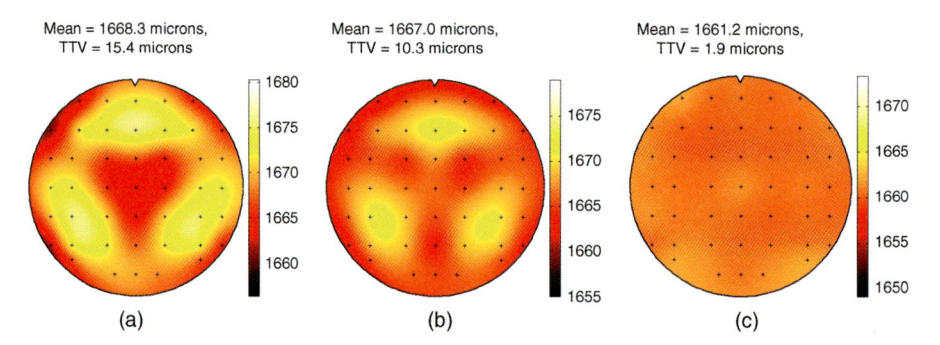

Figure 10.4 Typical TTV mapping result display format for a bonded wafer stack prior to thinning. Images (a) and (b) show TTV before process optimization while image (c) shows TTV after process optimization.

grinding equipment and process control during the via reveal process, it is critical to achieve a consistent and small TTV. To achieve the best bonding performance in terms of TTV control, low TTV coating is considered by many to be important; however, newer bonding techniques do not necessarily require good post-adhesive-coat TTV to achieve good post-bond TTV. Furthermore, TTV can be affected by post-bond processing, so caution must be taken depending upon the adhesive properties and the processes it is subjected to [4]. TTV is so critical in the temperature bonding processes that SEMI is currently reviewing material for establishing standards in metrology for measuring the TTV, bow, warp/sori, and flatness of bonded stacks (Draft Document 5409).

10.3
Selection of Temporary Bonding Adhesives

With the primary goals to achieve a void-free and low-TTV bonded pair with the proper adhesion strength for the adhesive and release layers, an approach to achieving these results has to be developed and factors influencing these performance parameters have to be identified. The right adhesive selection is therefore essential. There are many types of temporary bonding adhesives available today and more in various stages of making their way onto the market. There are three typical categories for the adhesives used in temporary bonding and debonding: thermoplastic, thermoset, and photoset.

a) Thermoplastic adhesives bond the wafer and carrier by using heat and pressure. What differentiates them from other adhesives is their ability to "flow" or be reset by applying more heat or pressure.
b) Thermoset adhesives typically use the same bond equipment as thermoplastic adhesives, but once the adhesive is set, it cannot be reworked. The cross-linking reactive groups have been activated and the process cannot be undone.
c) Photoset adhesives are polymerized by exposure to light and do not require elevated temperatures for the mounting process of the wafer to the carrier.

It is often very helpful to put together a matrix to sort out which adhesive(s) qualify on paper, such as cost versus thermal stability [5]. Other notable qualities to be considered are chemical resistance, vacuum stability, bonding speed, debonding speed, and final clean requirements. Thermal stability can be established with thermal gravimetric analysis (TGA) and thermal desorption spectroscopy (TDS). To what degree the adhesives need to be thermally stable is a function of the adhesion strength and process conditions. Some adhesives are very chemically inert, which is beneficial for resisting chemical attack during lithography or etch processes, but it may also result in more difficult and expensive cleaning steps to remove the adhesive or adhesive residue after debonding. A chemical rinse or dissolve of the adhesive residue may be required during the final cleaning process to ensure that the device wafer's surface meets the cleanliness requirements for subsequent process steps, such as soldering or permanent bonding, for final device stacking.

However, some very chemically resistant adhesives create films that can be easily removed by peeling techniques with the possibility of eliminating the need for costly and hazardous chemical stripping baths. A thorough evaluation and comparison of all material properties will be required to identify the most adequate adhesive material. Compromises may be necessary for this adhesive selection process.

10.4
Bonding and Debonding Processes

Once the adhesive and carrier type have been selected, the bonding process and recipe need to be created to meet high-volume process requirements. Bonding is typically achieved with UV light, heat, or contact force singly or in combination. Minimizing device wafer and carrier movement and potential stress on the wafers during bonding will significantly improve the quality of the temporary bond. In order to minimize void formation during bonding, it is advantageous to perform the bond process inside a vacuum chamber. Commercially available bond chambers, such as the SUSS LF300 temporary bond module, are able to pump down quickly to a chamber pressure that will reduce the voids to below detection levels of the metrology equipment. An additional benefit to bonding in vacuum is that any volatile solvents remaining in the adhesive films that could outgas and cause delamination in downstream processing will be drawn out by the low-pressure environment.

The bond chamber has to keep the wafers centered and rotationally aligned (notch aligned) during the bonding process until the adhesive has set. The difference between success and failure of the mounting process can be a matter of a few tens of microns shift or a tenth of a degree of rotation.

The device and carrier wafers are now being held firmly in place inside the vacuum chamber ready for bonding. The best strategy for putting the two wafers together depends on the state of the adhesive – liquid, gel, or solid. Certain adhesives require a significant amount of pressure to achieve good bond strength, whereas others might squeeze out during bonding and potentially contaminate the chuck or process chamber. This is undesirable in a high-volume manufacturing environment since cleanup will reduce the wafer throughput that can be achieved.

For this reason, the bonding force needs to be variable – from very low force for liquids to very high force for solids (pseudosolids). Other critical factors are the coefficient of thermal expansion (CTE) and thermal conductivity match. Bonding usually takes place at a predetermined steady-state temperature. However, for throughput reasons, the bonded pair is not extracted from the bond chamber at room temperature and the differences in CTE and thermal conductivity of the adhesive, device wafer, and carrier wafer may add stress to the bonded stack while it is cooling. If the cooling process is not done in a controlled manner, stress will build up and manifest itself in the form of bow, warp, and increased TTV. Temporary wafer bonding for the purposes of 3D integrated circuit (3D IC) manufacturing is very challenging and requires a significant amount of materials and equipment engineering to be successful.

Table 10.2 Debond method/material process compatibility matrix.

Debond method/material	Solvent Release	Thermal slide	Laser assisted release	Mechanical debonding at room temperature
High Throughput Capable	↘	→	↗	↗
Silicon Carrier (standard)	↘	↗	↘	↗
Glass Carrier (standard)	↘	↗	↗	↗
No Laser Induced or Thermal Stress / Carbonization	↗	↘	→a)	↗
Allows for Double Carrier Process / Selective De-bond	↘	↘	↗	→

a) Depends on absorption properties of adhesive/release layer.

As with bonding, the debonding method is largely dictated by the adhesive as well. A head-to-head comparison of the most common debonding methods is shown in Table 10.2.

a) Thermal slide debonding has been available for many years and utilizes thermoplastics that soften when heated [6]. Unfortunately, this property also limits the thermal capabilities of the mounted wafer stack while adding more stress during debonding.

b) Chemical dissolution is an almost stress-free process, but it takes a considerable amount of time and requires the use of expensive perforated carriers and lots of solvent.

c) Laser ablation to remove the carrier wafer has been successfully used in production, but prohibits the use of silicon carrier wafers.

d) Room-temperature mechanical lift-off debonding is the most recent innovation in separating the thinned bonded wafers from the carrier. The reason that this debonding technique is considered advantageous over existing methods is most easily explained by looking at Figure 10.5. By forming a straight debond line across the wafer, the force needed for separation can be minimized. The debond line length is directly proportional to the amount of force needed for debonding. By peeling the carrier from the device wafer utilizing a straight debond line, the most efficient debond wave front and thus the least amount of force applied can be achieved. This is independent of the type of adhesive used for bonding. Furthermore, by securing the device wafer to a vacuum chuck using a film frame, and only applying force to the carrier wafer during the

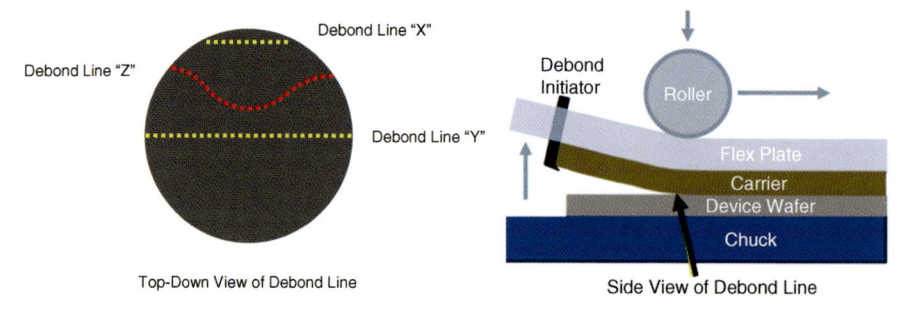

Figure 10.5 The debonding wave front lines "X" and "Y" (yellow dashed lines) are shortest possible. The debond wave line "Z" (red dashed line) represents a nonoptimal uncontrolled debond line.

debond process, the carrier wafer will see any stresses created during the debond process. The delicate device wafer will only see the absolute least amount of stress during carrier separation and is left protected and ready for the subsequent processes. Figure 10.6 shows the plot for debonding force versus roller displacement for 300 mm blanket and bumped wafer pairs. There is no difference in debonding force between blanket and bumped wafer pair.

Once the carrier wafer has been removed, the final requirement is to clean the device wafer surface that was exposed to the adhesive. Any residue remaining after debonding can interfere with downstream assembly processes. Since the wafer is very fragile at this point, it needs to be supported and the best method is to use tape on a film frame (Figure 10.7). Cleaning solvents used for some of the adhesives may damage the supporting tape, and in this case, it is important to protect the tape if solvents are used for final cleaning. Some equipment manufacturers are providing process options to protect the tape for debond clean. In parallel, tape manufacturers are developing new tapes specifically for these applications with the goal to withstand commonly used chemicals for this cleaning process.

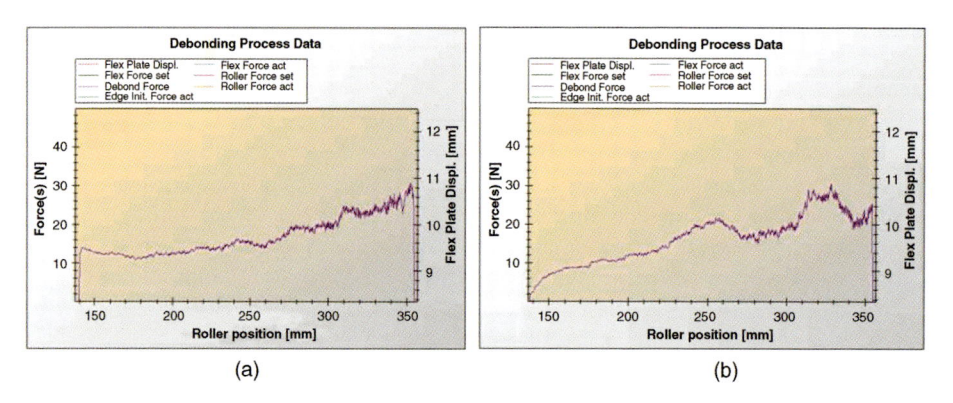

Figure 10.6 Debonding force as a function of displacement on the DB12T debonder module for a 300 mm blanket pair with 25 μm adhesive (a) and a bumped wafer with 60 μm adhesive (b). The debonding force stayed below 30 N in both cases.

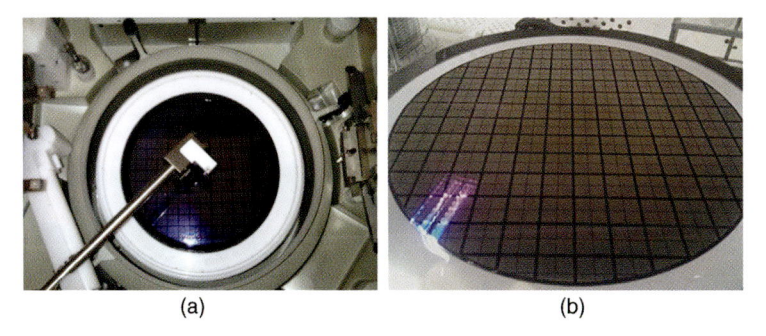

Figure 10.7 (a) Debonded thinned device wafer on a tape frame. (b) Integrated thin wafer cleaning inside a debonding cluster.

10.5
Equipment and Process Integration

The market need for stacked devices and interposer technologies requires temporary bond and debond technologies to enable low-cost and high-yield process integration schemes. With the new class of room-temperature mechanical lift-off adhesives and the latest generation of lift-off debonding equipment, the path to high-volume manufacturing of 3D ICs is well in sight [7]. It can be expected that high-performance 3D IC packages will be found in many of our consumer electronic devices and infrastructure systems.

From the very beginning of the temporary bonding and debonding process and equipment development, SUSS MicroTec's strategy was to break with the approach of equipment and materials tied together and focus on the development of a universal platform that would accept commonly available materials. For the debond processes, the focus was set on room-temperature lift-off processes versus the standard thermal slide-off processes at that time.

The XBS300 high-volume production temporary bonding platform supports coating and bonding the device wafer to the carrier as well as inline metrology for determining TTV for carrier, device wafer, and bonded stack (Figure 10.8). Built as an open platform for 200 and 300 mm wafers, it supports all major commercially available temporary bonding processes with notch and center alignment capability for same and different size carriers [8].

After completion of the wafer mounting process and metrology, the wafer will go through the thinning and backside processing steps. These backside processing steps vary by application (e.g., device wafer versus interposer processing) and the adhesives chosen have to be compatible to these specific chemical, thermal, and mechanical requirements.

With backside processing complete, the wafers are mounted on a film frame and dicing tape and presented to the XBC300 fully automated debonder and cleaner. The modular design of the XBC300 enables configurations for debonding cleaning that support the commonly used materials for temporary bonding. The lift-off

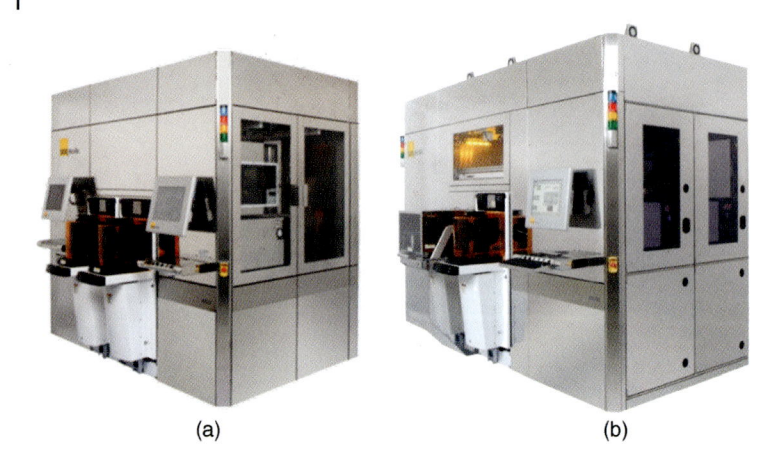

Figure 10.8 SUSS MicroTec's XBS300 temporary bonding platform (a) supports all key process steps for temporary bonding and is flexible enough to process all commercially available temporary bonding adhesives and (b) XBC300 Gen 2 debonder/cleaner is configurable for major debonding and cleaning processes via a wide selection of process modules.

debond process may be preceded by a laser release process. To minimize or eliminate any stress to the device wafer, the XBC300 debonder loads the device wafer with its face down on a proprietary chuck surface that keeps the thinned wafer in place without adding stress to the device wafer. Once the debond sequence has been initiated with a proprietary (patent pending) debond initiator, the debond front is precisely controlled with variable speed and debond force input. Debond force and speed are closed loop controlled and all parameters are logged during the mechanical debond process. This will guarantee a repeatable debond process no matter what the wafer topography and surface properties are. In addition to the laser release and debond modules, the XBC300 allows for cleaning of devices wafers while mounted on tape and a film frame as well as carrier wafer cleaning.

Acknowledgments

The authors would like to acknowledge contributions from Stefan Lutter, Peter Bisson, Greg George, Hale Johnson, Markus Gabriel, and other team members in the temporary bonding engineering, product development, and process development teams at SUSS MicroTec. Finally, the authors would like to acknowledge our partners at IMEC and temporary bonding material suppliers.

References

1 Trott, G.R. and Shorey, A. (2011) Glass wafer mechanical properties: a comparison to silicon. 6th International Microsystems, Packaging, Assembly and Circuits Technology Conference.

2 http://www.imec.be/ScientificReport/SR2009/HTML/1213307.html, January 10, 2014.
3 http://www.nanometrics.com/process_control/3D-packaging.html, January 10, 2014.
4 Tamura, K. (2010) Proceedings of the 60th Electronic Components and Technology Conference, June 2010, pp. 1239–1244.
5 Farrens, S., Sood, S., and Bisson, P. (2010) Thin wafer handling challenges and emerging solutions. *ECS Transactions*, **27** (1), 801–806.
6 Jourdain, A. *et al.* (2010) 300 mm wafer thinning and backside passivation compatibility with temporary wafer bonding for 3D stacked IC applications. 2010 IEEE International 3D Systems Integration Conference (3DIC), November 16–18, 2010, pp. 1–4.
7 Gabriel, M., Bisson, P., Sood, S., Bair, W., and Hermanowski, J. (2010) Equipment challenges and solutions for diverse temporary bonding and de-bonding processes in 3D integration. 2010 IEEE International 3D Systems Integration Conference (3DIC), November 16–18, 2010, pp. 1,5.
8 Lutter, S. (2013) Advances in wafer bonding technologies. Semicon Singapore 2.5D/3D IC Forum, May 2013.

11
ZoneBOND®: Recent Developments in Temporary Bonding and Room-Temperature Debonding

*Thorsten Matthias, Jürgen Burggraf, Daniel Burgstaller,
Markus Wimplinger, and Paul Lindner*

11.1
Introduction

While the introduction and adoption of 3D has been slower than originally anticipated, there has been significant progress made, and today multiple products are in production.

From the manufacturing point of view, one of the most challenging topics is the need to process ultrathin wafers.

11.2
Thin Wafer Processing

Figure 11.1 shows the basic concept of thin wafer processing. The device wafer with fully or partially processed front side is temporarily bonded to a rigid carrier wafer. The rigid carrier wafer enables thinning and backside processing of the thin device wafer. The carrier wafer gives mechanical stability to the wafer stack, which enables the usage of standard wafer processing equipment. Standard wafer cassettes, robot end effectors, and processing modules can be used for processing of the thin wafer. After backside processing, the thin wafer gets debonded from the carrier and is either mounted on dicing tape or directly bonded to another wafer. The general principles of thin wafer processing have been discussed in the first volume of this series [1].

Thin wafer processing has been demonstrated for wafers with thicknesses down to 20 µm. The biggest challenge for thin wafer processing is not temporary bonding and debonding, but rather the integration with all the processing steps in between. The fundamental concept of thin wafer processing is that by adding two process steps, temporary bonding to a carrier and debonding, every piece of equipment in the fab is usable for thin wafer processing. Today more than 10 adhesives from multiple suppliers are available. However, the process window for each of the different adhesives depends on the specific process flow to which the thin wafer is

Handbook of 3D Integration: 3D Process Technology, First Edition.
Edited by Philip Garrou, Mitsumasa Koyanagi, and Peter Ramm.
© 2014 Wiley-VCH Verlag GmbH & Co. KGaA. Published 2014 by Wiley-VCH Verlag GmbH & Co. KGaA.

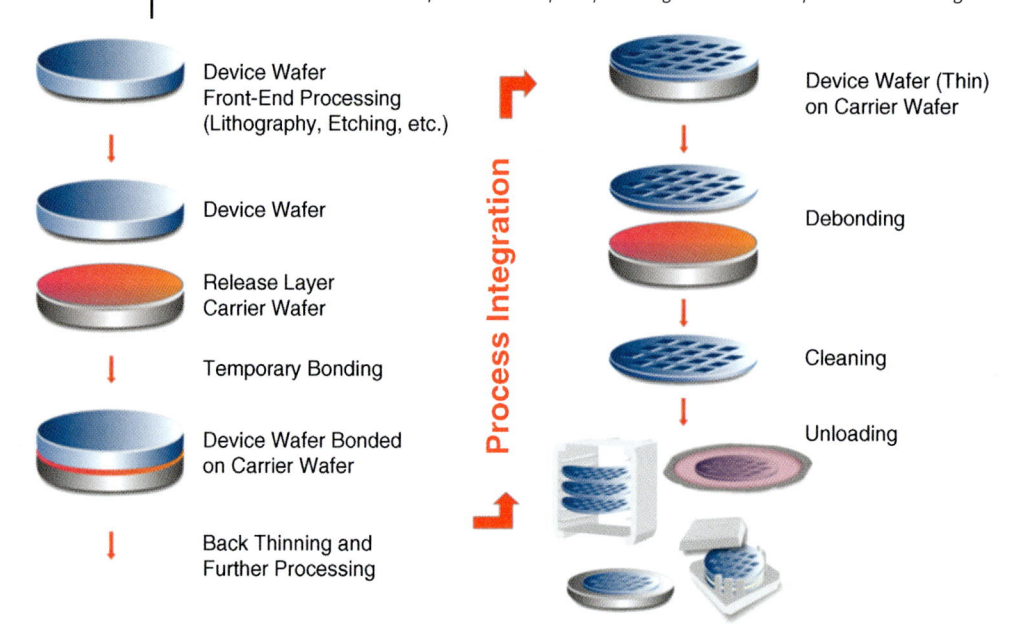

Figure 11.1 Basic process flow of thin wafer processing by temporary bonding to a rigid carrier and debonding.

exposed to. An important aspect is the stress of the thin wafer and stress management. The SEMI standard for the wafer thickness of a 300 mm wafer is 775 μm, while the actual device thickness is less than 1 μm. Such a thick wafer absorbs the stress created by all the deposited films, whereas in thin wafer processing, the thin wafer cannot absorb the stress. The temporary adhesive film transfers the stress to the carrier wafer. For the adhesive film to mechanically transfer the stress, it needs to be rigid and the adhesion between adhesive and (a) device wafer and (b) carrier wafer needs to be very high. This has important implications to the process window of many adhesives.

Technically, the maximum operating temperature of an adhesive is defined as the lower temperature of (a) the temperature where adhesive decomposition and outgassing starts or (b) the temperature at which the adhesive can keep a thin wafer flat on the carrier wafer. Thermoplastic adhesives are very interesting for thin wafer processing because they allow for easy solvent cleaning after debonding. However, since thermoplastic materials exhibit lower viscosity at elevated temperature, at some point, the adhesive gives in to the stress of the thin wafer and the thin wafer starts bending. This means that for a low-stress wafer, the maximum operating temperature is about equal to the decomposition/outgassing temperature, whereas for highly stressed wafers, the maximum operating temperature can be 50–100 °C lower.

Several new debonding technologies are based on mechanical release (also referred to as *peel-off* or *lift-off*) at room temperature. The underlying technical concept is that the thin wafer can be released from the carrier if the separation energy is higher than the adhesion. High stress of the thin wafer can result in delamination. An important technical improvement was achieved with Zone-BOND® technology [2], which is described in the following pages. The edge zone avoids unwanted delamination caused by wafer stress, but upon removal of the edge zone, the device wafer can be mechanically debonded at room temperature in a controlled way.

11.2.1
Thin Wafer Total Thickness Variation

It is of highest importance that the device wafer has a uniform thickness after thinning. For via-first/via-middle integration schemes, the uniform thickness ensures a smooth through-silicon via (TSV) reveal process. Poor total thickness variation (TTV) of the thin wafer would result in either grinding into the TSV or wafer areas that are too thick for the TSV reveal process. For via-last integration schemes, a poor TTV would increase the risk that the TTV does not electrically connect to the front-side metallization layers.

Backgrinding should result in a coplanarity between the backside of the carrier wafer and the backside of the device wafer. The carrier wafer and the device wafer themselves should have a TTV of less than 1 μm each. The potentially biggest contribution to thin wafer TTV stems from the adhesive layer. The adhesive layer TTV is impacted by the adhesive properties, the adhesive film thickness, and the coating, baking, and bonding processes. The TTV contribution from coating and baking depends on the film thickness. Table 11.1 shows the common thickness regimes for thin wafer processing. The topography on the front side of the device wafer, which needs to be embedded in the adhesive film, determines the thickness of the adhesive film. Coating, especially of thick films, can result in center high or low spots and in edge bead. The baking process can even increase the edge bead. There are multiple sophisticated technologies available to mitigate the TTV, such as dynamic dispense, area dispense, multiple film coating, and spray coating. The TTV contribution from the bonding process, conversely, is less dependent on the adhesive film thickness and can even improve the TTV.

Table 11.1 The topography on the front side of the device wafer, which needs to be embedded in the adhesive film, determines the thickness of the adhesive film.

Topography	Interconnect	Application	Adhesive thickness (μm)
<10 μm	Bonding pads	Die to die Various	20
~30–40 μm	Microbumps, Cu pillars	Die to die Die to interposer	40–50
~80–90 μm	Bumps	Die to substrate interposer to substrate	90–100

Figure 11.2 The EVG850TB temporary bonding system with nine process modules for wafer coating, baking, and bonding plus an integrated metrology module.

On the EVG850TB (Figure 11.2) temporary bonding system we use a setup with a rigid pressure plate, which allows us to planarize the edge bead as well as center high spots.

Table 11.2 shows achieved TTV results today and the roadmap for 2013. It is important to note that TTV is very much dependent on the adhesive properties. The table shows the results achieved with selected adhesives.

Backgrinding is a nonreversible process. Therefore, it is important to verify TTV of the adhesive layer prior to backgrinding, which allows defect correction before it potentially destroys the wafer. On the EVG850TB temporary bonding system, we have implemented an inline metrology module, which enables simultaneous measurement of carrier wafer TTV, device wafer TTV, and adhesive film TTV. The entire wafer is mapped with 282 000 data points at a cycle time of 80 s, which allows 100% inspection in production. In addition to TTV inspection, the system detects coating and bonding defects, as well as bow and warp of the wafer stack. Figure 11.3 shows an example for a wafer map in 2D and 3D plots. The details of the metrology system have been described in the literature [3].

Table 11.2 Achieved results and roadmap for selected materials: TTV data are based on full wafer mapping with 3 mm edge exclusion zone.

Bondline	TTV 2013	2015
20 μm	2 μm	1 μm
50 μm	4 μm	2 μm
100 μm	7 μm	4 μm

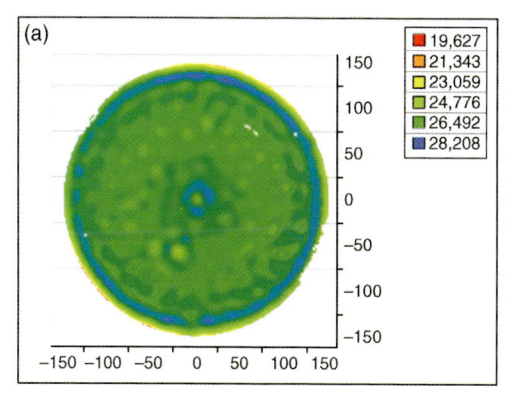

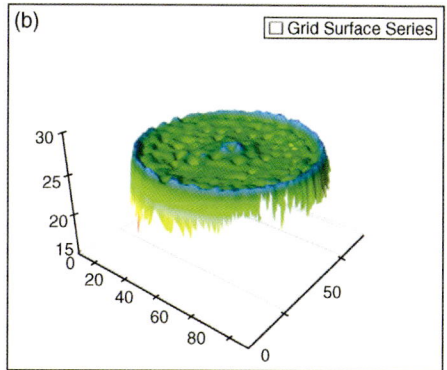

Figure 11.3 2D (a) and 3D (b) wafer maps showing adhesive film thickness.

11.2.2
Wafer Alignment

An important aspect of thin wafer processing is the alignment of the device wafer to the carrier wafer. In the past, it was believed that a coarse edge alignment of device wafer to carrier would be sufficient. However, while this is sufficient from the mechanical handling and edge protection point of view, it is not sufficient from a holistic manufacturing point of view for multiple reasons. (1) Empirical production data have shown that wafer-to-carrier alignment correlates positively with the yield of the edge dies. (2) If the wafer is not well aligned to the carrier, it increases the risk of arcing in plasma chambers. (3) For lithography steps, the wafer stack is pre-aligned based on the outer perimeter of the carrier wafer. Without a good center-to-center alignment, the alignment keys of the device wafer will not be within the field of view of photolithography steppers or scanners. If these systems, the most expensive in a wafer fab, have to search for the alignment keys, it reduces significantly their throughput. (4) For ZoneBOND debonding, the cycle time of the Edge Zone Release (EZR®) step depends very strongly on the width of the edge zone. A better alignment allows reducing the width of the edge zone and thereby speeds up the debonding cycle time.

Wafer alignment can be done by precision mechanical alignment, with a spec of 30 μm (3σ) or by optical edge alignment with a spec of 10 μm (3σ).

11.3
ZoneBOND Room-Temperature Debonding

ZoneBOND [2,4,5] breaks the link between debonding method and adhesive properties, with debonding now becoming a function of the carrier. Figure 11.4 shows the setup of the ZoneBOND carrier. The ZoneBOND carrier has two zones, which differentiate by the degree of adhesion between the adhesive and the carrier. The adhesion in the center zone is reduced, whereas full adhesion is at work in the

(a)

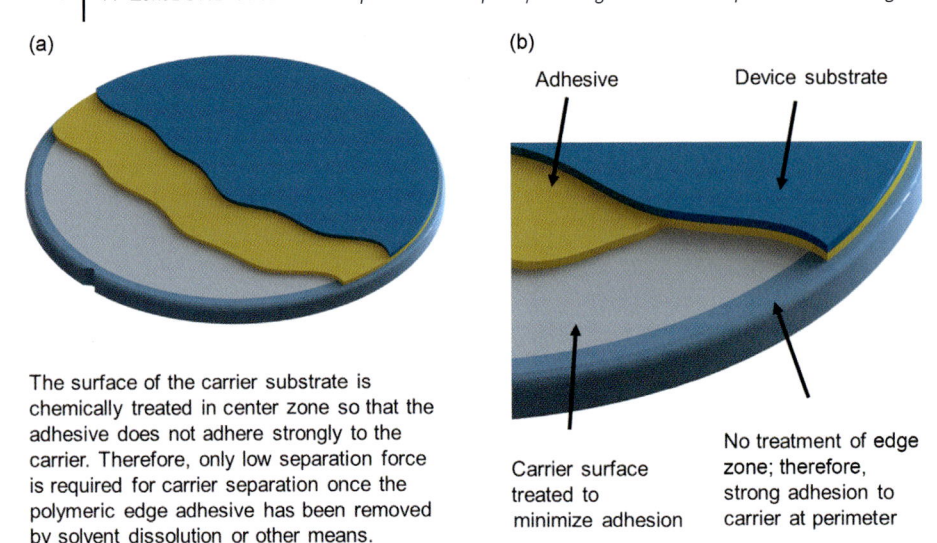

(b) Adhesive   Device substrate

The surface of the carrier substrate is chemically treated in center zone so that the adhesive does not adhere strongly to the carrier. Therefore, only low separation force is required for carrier separation once the polymeric edge adhesive has been removed by solvent dissolution or other means.

Carrier surface treated to minimize adhesion

No treatment of edge zone; therefore, strong adhesion to carrier at perimeter

Figure 11.4 (a) ZoneBOND carrier. (b) Cross section of a ZoneBOND carrier; the carrier surface of the center zone is treated such that the adhesion is reduced.

edge zone. It is important to note that the surface of the device wafer does not have to be treated at all for ZoneBOND, which makes the technology compatible with any kind of surface passivation.

The ZoneBOND debonding principle is shown in Figure 11.5. During the first step, EZR, the adhesive in the edge zone is dissolved. The center zone with the reduced adhesion is now the only connection between the thin device wafer and the carrier. The device wafer is separated from the carrier during the Edge Zone Debond (EZD®) step with a pure mechanical separation at room temperature. It is important to note that the actual separation happens between the adhesive layer and the carrier wafer. This means that the debonding process is independent from

1st Step : Edge Zone Release (EZR®)

Release adhesives at the wafer stack edge

2nd Step : Edge Zone Debond (EZD®)

Separation at the carrier; no vertical force to device wafer

Mechanical separation of the carrier wafer

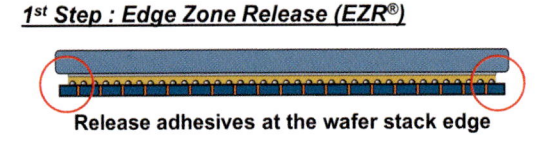

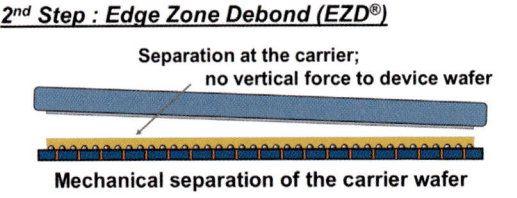

Figure 11.5 ZoneBOND debonding principle: During the EZR step, the adhesive edge zone is dissolved; the thin device wafer is separated from the carrier.

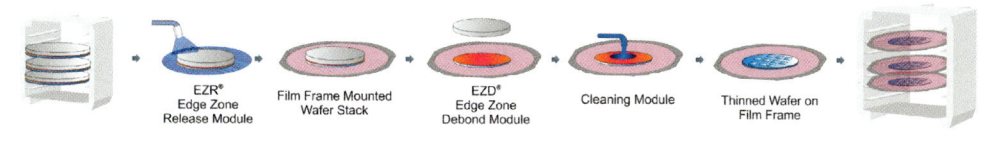

Figure 11.6 Full ZoneBOND debonding process sequence.

the top passivation layer of the device wafer and independent from the topography on the device wafer. During the EZD step, the bumps in the bond interface are embedded in the adhesive layer. No vertical or shear force is applied to the bumps during debonding, which eliminates the risk of bump damage. After debonding, the device wafer is cleaned in a dedicated thin wafer cleaning module.

Figure 11.6 shows the entire debonding process flow for ZoneBOND as it is implemented in the EVG850DB production debonding system. The bonded wafer stacks are delivered in a FOUP to the system. First the EZR step is performed as a single-wafer process. Then the wafer stack is mounted on a film frame. Performing the EZR process prior to film frame mounting allows using dicing tapes, which are not compatible with the solvents used for EZR. This gives foundries and outsourced semiconductor assembly and test (OSAT) companies full freedom of choice for the dicing tape. Then the thin wafer is debonded from the carrier with the EZD process step. After debonding, the adhesive is stripped from the thin wafer and the carrier by a solvent stripping process. The thin wafers on film frame are unloaded into a film frame cassette.

The analysis of the process sequence shows that the described debonding process is completely independent from the adhesives properties. It is a standardized process on standardized equipment independent of the specific adhesive being used. The best material for a specific application can be chosen, and multiple materials can be processed in parallel on one system [2].

11.4 Conclusions

Since the first volume of this series was published, significant progress has been made in temporary bonding and debonding. While in the past the focus was on the development of the unit processes, today the focus is on the holistic manufacturing with an emphasis on yield management. The temporary bonding process, particularly TTV control, wafer alignment, and inline metrology, has a huge effect on manufacturing yields and cost. The ZoneBOND process enables room-temperature debonding independent from the properties of the temporary adhesive, thereby enabling standardization of the debonding process.

Acknowledgment

ZoneBOND® is a registered trademark of Brewer Science, Inc., Rolla, MO, USA.

References

1 Garrou, P., Bowers, C., and Ramm, P. (eds) (2008) *Handbook of 3D Integration*, Wiley-VCH Verlag GmbH, Weinheim, pp. 240–248.

2 Matthias, T. *et al.* (2012) The material breakthrough towards standardized thin wafer processing. *Chip Scale Review*, **16**, 17–21.

3 Wimplinger, M. *et al.* (2012) Equipment and process solutions for low cost high volume manufacturing of 3D integrated, Proceedings of the Pan Pacific Microelectronic Symposium.

4 McCutcheon, J. *et al.* (2010) ZoneBOND thin wafer support process for wafer bonding applications. *Journal of Microelectronics and Electronic Packaging*, **7** (3), 138–142.

5 McCutcheon, J. and Bai, D. (2010) Advanced thin wafer support processes for temporary wafer bonding. *IMAPS 2010 – 43rd International Symposium on Microelectronics*, pp. 361–363.

12
Temporary Bonding and Debonding at TOK

Shoji Otaka

12.1
Introduction

Along with the technology of 3D integration, which has been in high demand for IC fabrication, the solution for handling and processing the thinned silicon wafers has become the most critical topic in development [1–3]. Normally, the silicon wafer is thinned down to less than 50 μm (200 or 300 mm; see Figure 12.1) and the backside of the thinned wafer is treated by normal IC fabrication types of process [4], such as plasma-enhanced chemical vapor deposition (PECVD), lithography, and so on. It is important that the silicon wafer is required to be supported by the carrier substrate before the thinning steps and the carrier substrate should support the thinned wafer during all the backside processes, in order to stabilize the wafer form and avoid wafer breakage or chipping. Also, after all of the backside processes are completed on the thinned wafer, the support carrier needs to be easily removed from the thinned silicon wafer.

TOK has developed temporary bonding and debonding technology called Zero Newton®, which allows us to handle the thinned wafer by bonding the silicon wafer to a glass carrier substrate, letting the bonded wafer go through all the IC processes, and finally debond the glass carrier from the thinned silicon wafer. Figure 12.2 is an example of a 200 mm thinned wafer with and without the bonded glass carrier. The thinned wafer without the carrier glass is so thinned that it is easily bent just by holding it. From an IC fabrication point of view, it is too fragile for process handling so that the wafer would be probably easily broken. Conversely, the thinned wafer with the glass carrier looks almost as same as the wafer before it is thinned down.

12 Temporary Bonding and Debonding at TOK

Figure 12.1 A 300 mm wafer thinned down to 30 μm.

Figure 12.2 A 200 mm wafer (a) without the glass carrier and (b) with the glass carrier.

12.2
Zero Newton Technology

Zero Newton is a total solution for thinned wafer handling, consisting of process steps and the temporary bonding and debonding (Figure 12.3). Zero Newton consists of the wafer bonder and debonder, the temporary adhesive, adhesive cleaning solvent, and glass carrier. The temporary bonding and debonding technology needs to cover the process capability of not only how to bond and debond but also how to stabilize the bonded wafer during the varieties of process steps. The key to optimizing performance is the perfecting of bonder and debonder equipment configuration with the temporary adhesive designs and the process method. TOK has focused on the total solution so that the key to optimization is clearly achieved by one supply solution.

12.2.1
The Wafer Bonder

The wafer bonder achieves temporary bonding of the wafer to the glass carrier by the process steps of coating adhesives, prebaking adhesives, and bonding. The thermal bonding step is performed under vacuum atmosphere, in order to eliminate voids in the bonding stack. The wafer bonder is equipped with alignment and total thickness variation (TTV) measurement as the general specs. For the Zero Newton wafer bonder, the bonding TTV is carefully

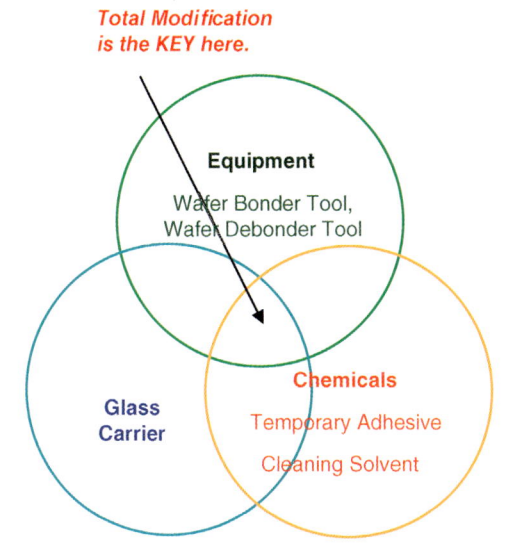

Figure 12.3 The Zero Newton concept: total integration of equipment, chemicals, and carrier.

optimized by the process condition along with the physical characteristics of the adhesive. Figure 12.4 shows an example of the Zero Newton bonding TTV. This bonding was performed on a 300 mm wafer with 80 μm height solder bump topography. The adhesive is coated to 120 μm thickness to cover the bump topography and bonded to the glass carrier. The total bonded stack

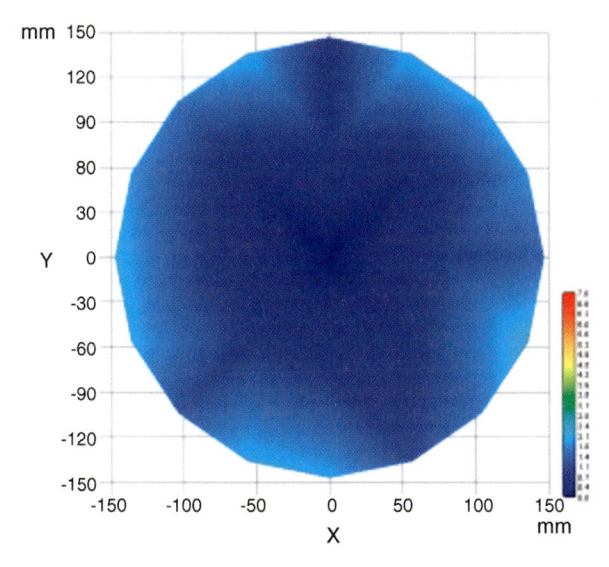

Figure 12.4 Post-bonded TTV of 300 mm sized 80 μm bump topography wafer, bonded with the glass carrier, using a 120 μm thickness adhesive.

TTV is 2.3 μm by 25 points and the edge exclusion area is 3 mm. For 3D integration, the TTV is very important especially when the wafer contains high topography; the TTV value often increases and keeping the bonded TTV is critically important [5,6] at the bonding area.

12.2.2
The Wafer Debonder

The wafer debonder removes the glass from the thinned wafer and chemically cleans the wafer surface by removing the temporary adhesive. Figure 12.5 shows the surface of the wafer after separation and chemical cleaning. This analysis was performed by XPS. We observed that the level of remaining adhesive, which is represented by the –C1s, the carbon substance, is almost the same as a bare Si wafer.

12.2.3
The Wafer Bonder and Debonder Equipment Lineups

Zero Newton bonder and debonder can be supplied with manual controls or fully automated for both 200 and 300 mm wafers (Figure 12.6).

12.2.4
Adhesives

The adhesive, which is placed in between the wafer and the glass carrier, needs to be capable of resistance and stability during the many variations of the thinned wafer backside processes, such as wafer grinding, chemical–mechanical polishing

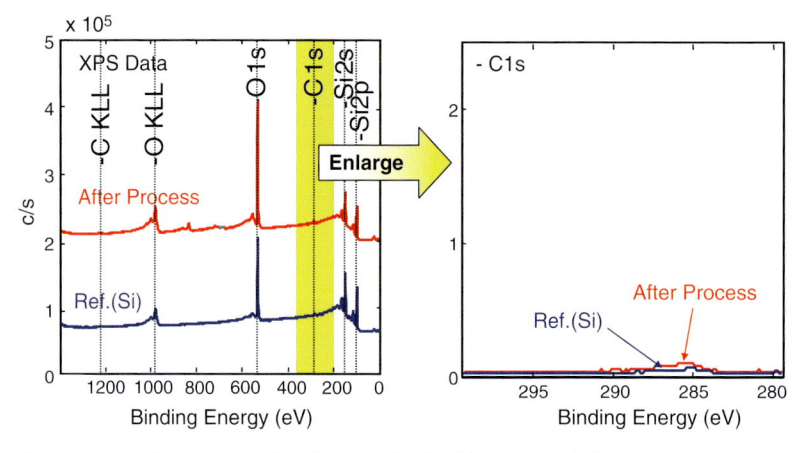

Figure 12.5 Wafer surface analysis by XPS after the debonding and cleaning process.

Equipment Lineup

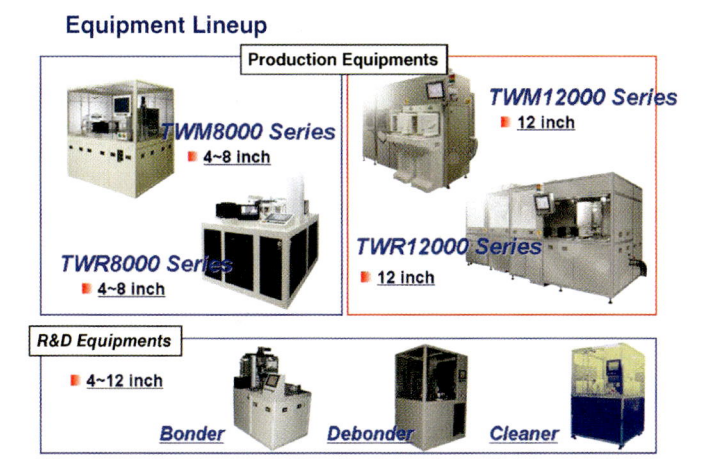

Figure 12.6 Zero Newton equipment.

(CMP), CVD, etching, plating, and so on (see Chapter 9). This is especially important for the processes that require high temperature, such as CVD or reflow, so that the adhesive thermal resistance is essential. Figure 12.7 shows the thermal characteristics of the adhesive by TG-DTA and TDS. It indicates that the adhesive has only $\Delta 1\%$ weight loss by 388 °C and no indication of outgassing to higher than 250 °C. Such high thermal stability and outgas free characteristics are needed during the backside processes, otherwise the adhesive would outgas in the bonding stack during processes such as CVD, and this degassing would cause voids and possible wafer delamination.

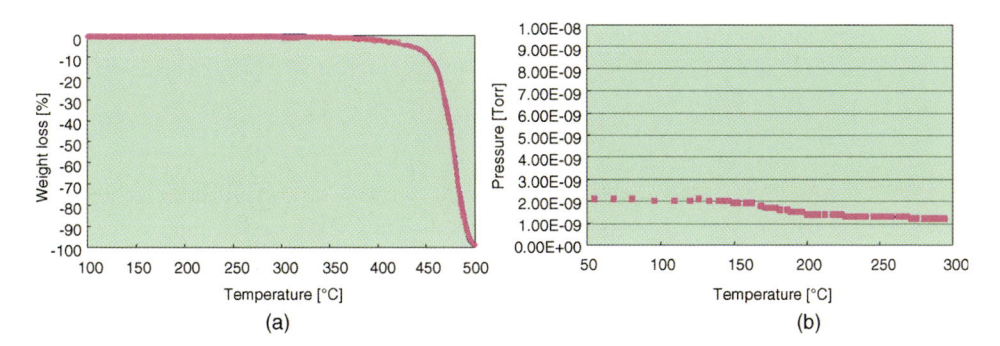

Figure 12.7 Thermal characteristics of adhesive: (a) TG-DTA by 30–500 °C (rump up 10 °C min^{-1}), N$_2$ flow with 115 μm adhesive thickness; (b) TDS by 30–500 °C with 115 μm adhesive thickness.

Table 12.1 Adhesive solubility test by the target solvent after the listed high temperature is applied.

Baking temperature (°C)	Baking time	
	30 min	60 min
250	Soluble	Soluble
300	Soluble	Soluble
350	Soluble	Soluble

Adhesive must also be easily removed after carrier separation. This is a significantly important point because the wafer surface needs to be recovered without any adhesive residue to avoid the device yield loss. To have good cleaning capability, TOK has selected a thermoplastic polymer as the base platform, instead of ultraviolet (UV) or thermoset polymers, which are more difficult to remove by chemical cleaning. Table 12.1 shows that the adhesive is still soluble by the solvent after processing at 350 °C/1 h in N_2. At same time, although the adhesive is removable by the target solvent, it should not be attacked by other chemicals, which are used in the IC process. Table 12.2 shows an example of the chemical resistance of the adhesive. It indicates that the adhesive has no solubility or absorption to these chemicals. This is also important as an adhesive characteristic because the adhesive damage by chemicals usually produce the delamination and wafer chipping.

12.2.5
Integration Process Performance

For 3D integration, the bonded wafer with the glass carrier is usually processed starting from the backgrind and CMP, and followed by many different processes. Usually, grinding capability is checked by the thinned wafer edge profile, whether or not the edge has chipping or crack by grinding. Figure 12.8 is an example of the wafer backside appearance after backgrinding to 50 μm. It shows no significant wafer chipping or cracking.

After backgrinding to 50 μm, the bonded wafer is taken to high temperature to see whether the bonded thinned wafer would survive. After applying a temperature of 220 °C for 15 min under 900 W RF power and 133 Pa pressure in N_2, no delamination or any abnormal phenomena are observed.

We performed a similar experiment on a 300 mm 80 μm solder-bumped wafer. Backgrinding this wafer to 50 μm thickness and applying the same 220 °C thermal treatment results in the same result, that is, no delamination or other wafer damage.

Table 12.2 Adhesive chemical resistance by different chemicals.

Solvents	Dipping condition		Film thickness		Adhesive weight change (mg)	Observation	Assessment
	Temperature (°C)	Time (min)	Post-coating (μm)	Post-dipping (μm)	Film thickness change (μm)		
1% HF	25	3	42.7	42.7	0.0	0	No change	Good
5% HF		3	42.7	42.7	0.0	0	No change	Good
31% H$_2$O$_2$		6	42.7	42.7	0.0	0	No change	Good
2.38% TMAH		10	42.7	42.7	0.0	0	No change	Good
PGME		10	42.7	42.7	0.0	0	No change	Good
PGMEA		10	42.7	42.7	0.0	0	No change	Good
IPA		30	42.7	42.7	0.0	0	No change	Good
ST104	60	20	42.7	42.7	0.0	0	No change	Good

Figure 12.8 Wafer appearance after backgrinding. No CMP polish.

12.3 Conclusions

Since 2005, TOK has been developing Zero Newton for 3D integration. It is a total solution for handling and processing the thinned silicon wafer. Zero Newton allows bonding, debonding, and backside processing, such as backgrinding and high-temperature and various chemical steps for thinned Si wafers.

References

1 Tamura, K. (2010) Novel Adhesive Development for CMOS-Compatible Thin Wafer Handling. 60th, IEEE-Electronics Components Technology Conference, June 2010, pp. 1239–1244.
2 Najash, N. (2013) Foundry Perspective of 2.5D & 3D Integration. Semicon Singapore 2.5D/3D IC Forum, May 2013.
3 Jonathan, J., Alvin, L., and Dongshun, B. (2013) Thin Wafer Debonding Mechanism. Semicon Europe, October 2013. pp. 1–6.
4 Dang, B., Andry, P., Tseng, C., Maria, J., Polastre, R., Trzcinski, R., Prabhanker, A., and Knickerbocker, J. (2010) CMOS Compatible Thin Wafer, Adhesive and Laser Release of Thin Chip/Wafers for 3D Integration. 60th, IEEE-Electronics Components Technology Conference, June 2010, pp. 1393–1398.
5 Jaesik, L., Justin, S., Guan Kian, L., Vincent, L., Hong Yu, L., Keng Hwa, T., Yubun, K., Atsushi., K., and Shan., G. (2011) High Temperature Thermo-Plastic Temporary Adhesive and Chemical Assisted Room Temperature Debonding for 3D Integration, International Conference on Materials for Advanced Technologies, June 2011.
6 Lutter, S. (2011) Equipment and Processes for Temporary (De-) Bonding, 3D Integration Workshop, Semicon West, July 2011.

13
The 3M™ Wafer Support System (WSS)

Blake Dronen and Richard Webb

13.1
Introduction

To create the ultrathin wafers that today's complex 3D architectures require, particular attention must be paid to supporting the wafer throughout backgrinding and subsequent process steps, in order to protect it from mechanical stresses. It is especially important to fully support the wafer between any raised topography on its surface and to support the edges that can be easily ground into a very thin "blade" shape, making them highly vulnerable to cracking or chipping. Following thinning and other processes, it is essential to be able to separate the wafer from its support system quickly and easily, without damaging or deforming it, and without time-consuming cleaning after debonding [1].

3M has developed a temporary wafer bonding system, enabling high-volume manufacturing of wafers down to 20 μm, while maintaining compatibility with high-temperature through-silicon via (TSV) processes and typical semiconductor process chemistries.

13.2
System Description

The 3M™ Wafer Support System (WSS) (used by wafer fabs around the world since 2008) combines proprietary 3M temporary bonding technologies with automated or semi-automated equipment, specifically designed to process wafers using 3M WSS materials.

The current 3M WSS consists of

- 3M™ Liquid UV-Curable Adhesive – a family of 100% solids acrylic adhesives, used for temporary bonding of the wafer to a glass carrier,
- 3M™ Light-To-Heat Conversion (LTHC) Coating – used to create a release layer on the glass carrier for debonding,

- 3M™ Wafer De-Taping Tape – used to peel the adhesive from the wafer after debonding,
- a glass carrier (customer-supplied), and
- automated bonding, debonding, and glass recycling equipment (supplied by 3M authorized equipment manufacturers).

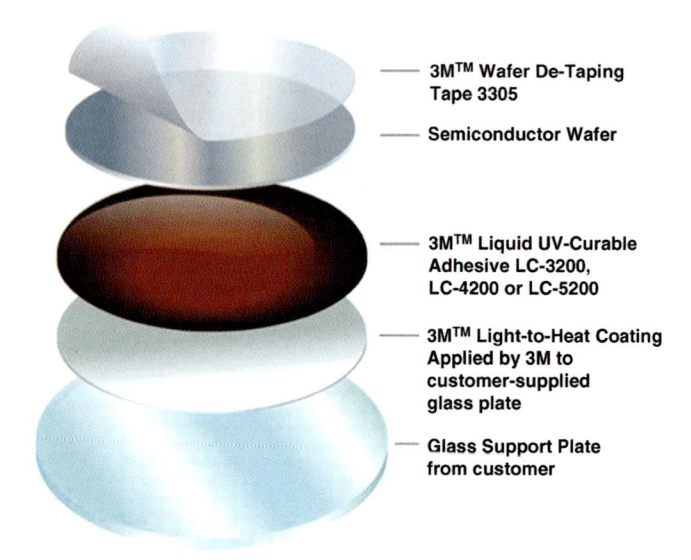

- 3M™ Wafer De-Taping Tape 3305
- Semiconductor Wafer
- 3M™ Liquid UV-Curable Adhesive LC-3200, LC-4200 or LC-5200
- 3M™ Light-to-Heat Coating Applied by 3M to customer-supplied glass plate
- Glass Support Plate from customer

The process steps for the current 3M WSS are illustrated in the left-hand column of Figure 13.1, and are briefly described here:

Bonding:

1) First, the UV-curable adhesive is spin coated onto the wafer. It is then vacuum bonded to the glass carrier, which has been treated with a release layer of LTHC coating, and quickly cured with UV light.
2) The fully supported wafer is thinned using standard backgrinding processes.
3) After thinning, the bonded wafer stack can be processed through standard semiconductor and TSV processes.
4) Following processing, standard dicing tape is applied to the back of the wafer, which is then placed in the debonding module.

Debonding:

1) A laser debonding process separates the glass carrier from the adhesive.
2) The glass carrier can now be separated from the adhesive layer with very low force.
3) The UV adhesive is peeled from the wafer using 3M Wafer De-Taping Tape. No post-peel cleaning is required.

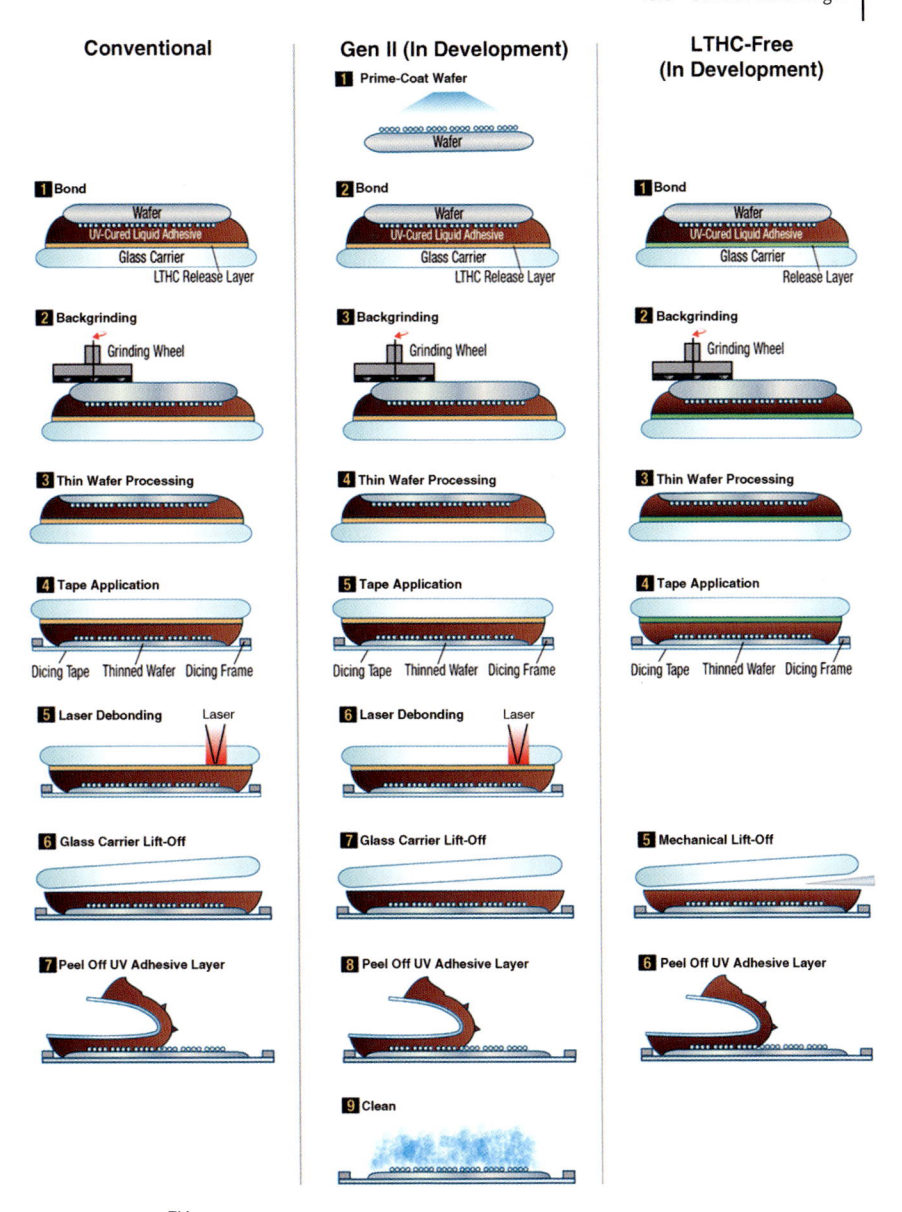

Figure 13.1 3M™ WSS process steps.

13.3
General Advantages

The 3M WSS system features fast and simple bonding and debonding processes, which in turn provide high throughput – typically 25 WPH for both bonding and debonding.

3M WSS adhesives are 100% solids, UV-curable materials. Because they contain no solvents, the adhesives can be spin coated at thicknesses up to several hundred microns in a single step, and there is no drying step required. Curing is accomplished by a short exposure to UV light, typically on the order of 30 second. Total thickness variation (TTV) on 300 mm diameter bumped wafers is <5 μm when using high-quality glass carriers with appropriate TTV and flatness. The bonding process occurs at room temperature and does not require added force. Very little thermal or mechanical stress is generated on the wafer.

The 3M WSS is used in production processes to create thin power device, LED, logic and DRAM wafers in diameters ranging from 100 to 300 mm. Typical post-thinning processes include silicon (Si) etch, chemical vapor deposition (CVD), physical vapor deposition (PVD), dielectric cure, and resist strip to support 3D wafer integration.

A key design feature that enables the low-force separation of the 3M adhesives is that they are designed to have high adhesion in shear between the adhesive and either the carrier or wafer surface, with controlled low adhesion normal to the substrate surface. This enables the bonded wafer to support downstream processes such as backgrinding and chemical–mechanical polishing (CMP), while allowing the adhesive to be removed with very low force applied normal to the wafer surface.

The debonding process is fast and simple. The first step in the process is to raster a standard IR marking laser across the glass carrier. The laser causes very fast local heating of the LTHC layer, which then decomposes. The LTHC layer is designed so that it does not re-weld after this decomposition. Once the laser process is complete, typically less than a minute, the glass carrier is lifted away from the stack with essentially no force.

The second step in the debonding process is to mechanically remove the WSS adhesive. As illustrated above, this is accomplished by laminating a de-taping tape over the surface of the adhesive and peeling it away from the wafer surface. High-angle peel normal to the wafer surface initiates low-force separation between the wafer and adhesive. The force required to remove the adhesive is very modest and the 3M authorized tool builders have incorporated the capability to adjust the down force and fine-tune the speed and position of the peel roller in order to optimize this process.

13.4
High-Temperature Material Solutions

The LC-5200 family of 3M WSS adhesives is designed for high-temperature applications. Each adhesive consists of the base LC-5200 formulation plus a specific amount of an adhesion control agent.

Isothermal thermogravimetric analysis (TGA) data at 250 °C show that the weight loss of the LC-5200 is about 1.5%, compared to 6% for the standard room-temperature WSS adhesive, LC-3200 (Figure 13.2).

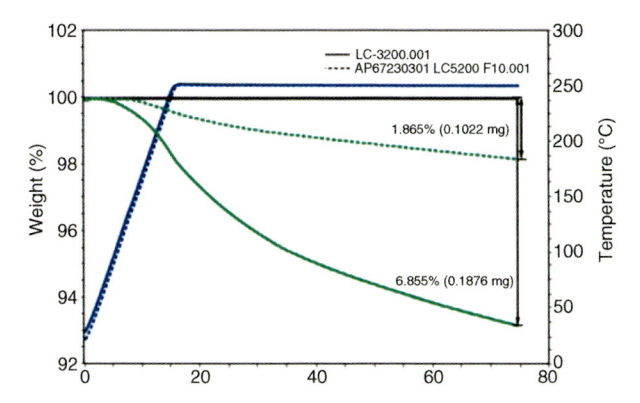

Figure 13.2 Typical TGA weight loss, 1 h at 250 °C; LC-3200, LC-5200 adhesives.

LC-5200 includes four commercially available formulations – each with a different level of adhesion control agent. The appropriate level of adhesion control agent depends on the details of the wafer materials, topography, and expected thermal history, but in general LC-5200-F1035 and LC-5200-F11 are typically suitable for inorganic dielectrics, while LC-5200-F12 and LC-5200-F14 are more often used for organic dielectrics. Many applications require bonding and processing of wafers with bumps or other significant topographies. Bumps are typically solder balls or solder-capped copper pillars of various sizes. LC-5200 has a viscosity around 1400 cps and is 100% solids, so it can be applied at an appropriate thickness for bumped wafers in a single spin coating step. Ultrathinning and processing of bumped wafers imposes additional constraints on the WSS adhesive. It must withstand high temperatures with no delamination and still have mechanical properties that allow it to be removed from around topography with low force so as to not damage the bumps – but still be strong enough to avoid tearing during the peel process. LC-5200 adhesives have a glass transition temperature in the range of 30–40 °C. Warming the adhesive to around 50 °C during the peel process reduces the modulus significantly and allows it to deform very easily and separate from features on the wafer surface.

Table 13.1 includes typical application examples with different wafer types and thermal histories. The wafers in the examples were bonded, thinned to 50 μm, run through the heat age prescribed, and then debonded. The peel force reported in the table is that which is required to remove a 25 mm wide strip from the wafer surface at an angle of 90° and a speed of 125 mm min^{-1}. A force of up to 2 N/25 mm is acceptable for automated debonding although lower forces are preferred.

The peel force will also vary as a function of bump density due to the contribution from mechanical interlocking between the adhesive and wafer topography.

Table 13.1 Typical application examples.

Sample	Dielectric	Features	Adhesive	Heat age	Peel force (N/25 mm)
1	SiN	Cu pillars	LC-5200-F11	1 h at 200 °C	1.7
2	SiN	Cu pillars	LC-5200-F12	1 h at 200 °C	0.8
3	Organic A	Cu pillars	LC-5200-F12	1 h at 200 °C	1.9
4	Organic A	Cu pillars	LC-5200-F14	1 h at 200 °C	1.0
5	Polyimide	Cu pillars	LC-5200-F12	2 h at 200 °C	0.6
6	Polyimide	SnAg solder balls	LC-5200-F14	2 h at 200 °C	1.0 (0.7 at 50 °C)
7	Polyimide	SnAg solder balls	LC-5200-F14	1 h at 200 °C	1.3 (0.5 at 50 °C)
8	Polyimide	Cu pillars	LC-5200-F11	1 h at 250 °C	1.3 (0.8 at 50 °C)
9	SiN	10 μm topography	LC-5200-F11	1 h at 200 °C plus reflow	1.0

13.5
Process Considerations

13.5.1
Wafer and Adhesive Delamination

The force required to separate the wafer and the WSS adhesive must be low enough to allow the adhesive to be mechanically removed but must also be high enough to prevent delamination during high-temperature processing. 3M offers high-temperature stable adhesives for these demanding applications, but if the correct adhesive is not matched to the application, the resulting low surface adhesion can cause delamination between the adhesive and wafer surface, as shown in Figure 13.3. This delamination is typically initiated by low-level volatiles generated during high-temperature process steps. The pressure generated from

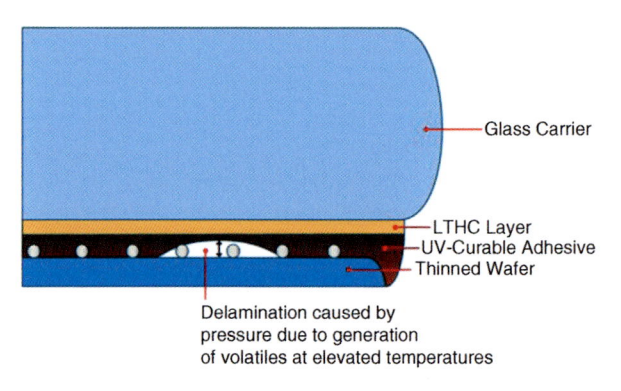

Figure 13.3 Delamination during high-temperature processing.

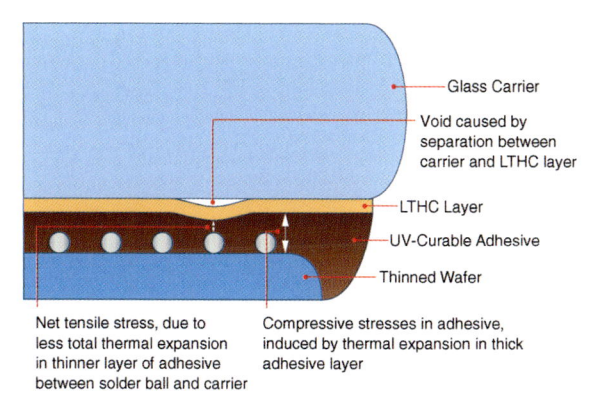

Figure 13.4 LTHC and carrier void.

Labels in figure:
- Glass Carrier
- Void caused by separation between carrier and LTHC layer
- LTHC Layer
- UV-Curable Adhesive
- Thinned Wafer
- Net tensile stress, due to less total thermal expansion in thinner layer of adhesive between solder ball and carrier
- Compressive stresses in adhesive, induced by thermal expansion in thick adhesive layer

these volatiles depends strongly on the time and temperature conditions to which the bonded stack is exposed. Appropriate selection of a WSS adhesive product with corresponding adhesion control agent level will prevent delamination while enabling effective bulk adhesive removal.

13.5.2
LTHC Glass Delamination

Stresses are also induced in the stack by differences in overall thermal expansion between areas with high and low feature density and locally in areas above and around individual feature geometries as shown in Figure 13.4. This typically manifests as a small spot of separation between the glass and the LTHC layer immediately above individual bumps, and is referred to as "spot" delamination. If too extensive, the spots can grow together and cause large areas of delamination. Prevention of this failure mode is easily achieved by having at least 40 μm of adhesive between the tops of the features and the glass to absorb the differential expansion.

13.6
Future Directions

13.6.1
Thermal Stability

3M is continuing to develop new adhesive and LTHC formulations and generate further improvements in thermal stability. The TGA data in Figure 13.5 compare the weight loss at 300 °C for LC-5200 versus an experimental formulation. The

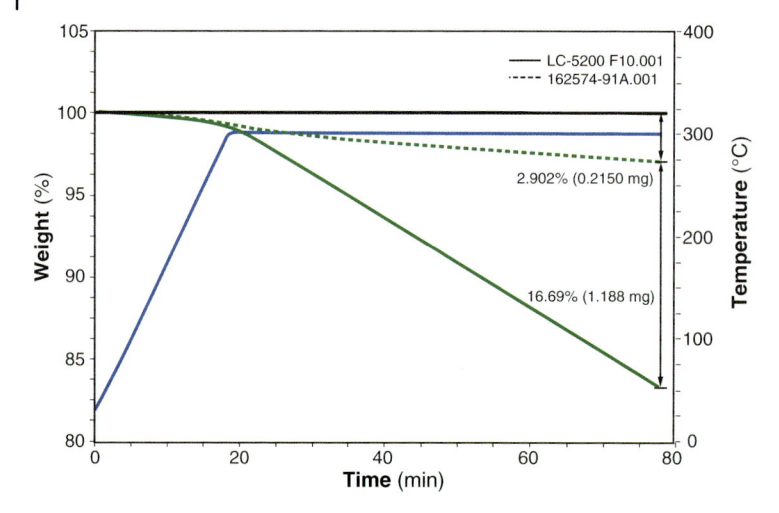

Figure 13.5 Typical TGA weight loss, 1 h at 300 °C; LC-5200 adhesive, experimental adhesive.

weight loss of both materials is roughly the same at 250 °C, but the new material is significantly better at 300 °C. Similar improvements have also been demonstrated with a new LTHC formulation. These materials will expand the range of applications available to the 3M WSS system.

13.6.2
Elimination of Adhesion Control Agents

These include the addition of a thin primer layer on the wafer surface that is easily removed after peeling the bulk layer WSS adhesive (Figure 13.6). This primer will be the same material for all wafer types, and will provide a common surface for adhering the bulk layer WSS adhesive, thus eliminating the need for variable adhesion control agent levels.

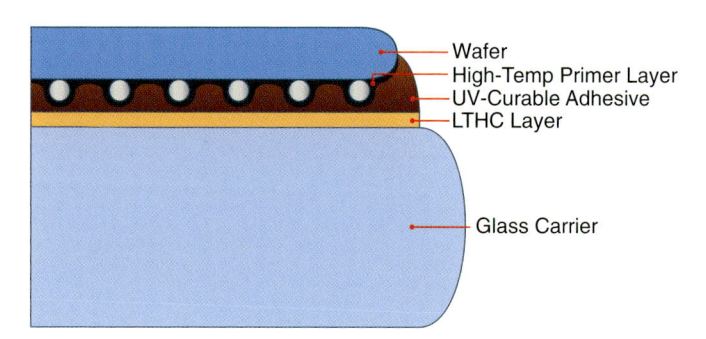

Figure 13.6 Second-generation WSS process and primer coat.

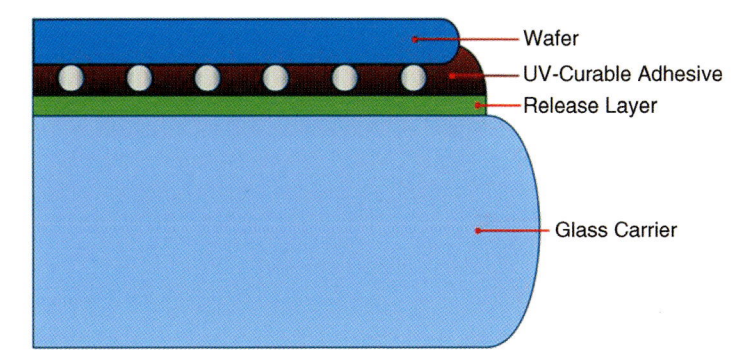

Figure 13.7 LTHC-free process.

13.6.3
Laser-Free Release Layer

A new release material that enables mechanical, laser-free removal of the glass carrier is also under development at 3M. This release layer replaces the LTHC layer and can be coated on the glass carrier in advance, or with *in situ* bonder tool processing (Figure 13.7). The release layer can, alternatively, be coated on a cured intermediate bulk WSS adhesive layer prior to bonding the glass carrier. Bonder tools are easily configured to enable a variety of process flows. Common to all of these processes, however, is the simplicity and performance of this release layer. Like the bulk layer adhesive, 3M's WSS mechanical release layer material is thermally stable, chemically resistant, and quick UV curing.

Together these solutions provide an *à la carte* approach to enable customer-defined process solutions for next-generation temporary bonding. The diagrams in Figure 13.1 represent a few of the most common customer process flows defining the future of temporary wafer handling and 3M WSS.

13.7
Summary

The 3M WSS is engineered to provide a simple, fast, and reliable solution for temporary wafer bonding.

The heart of the system is a family of 100% solids, UV-curable adhesives for temporary bonding of the wafer to a carrier, which is treated with a laser-activated release coating. These adhesives are solvent-free, bond at room temperature, require no drying step, and are compatible with post-thinning processes including Si etch, CVD, PVD, dielectric cure, and resist strip.

These 3M UV-curable adhesives are designed to securely attach and support the wafer to minimize stress during processing, after which a laser debonding process allows the wafer to be released from the carrier with virtually no force.

For high-temperature applications, the LC-5200 family of UV-curable adhesives is available in varying levels of adhesion, to meet specific material, topography, and thermal requirements. New 3M WSS adhesive and LTHC coating formulations, designed to offer improved high-temperature performance, are currently being evaluated for commercial release. In addition, 3M is proceeding with development of next-generation wafer support materials and processes – with the goal of offering even more practical and efficient ways to build the sophisticated 3D structures of the future.

Reference

1 Garrou, P., Ramm, P., and Koyanagi, M. (2014) Chapter 1: 3D IC integration since 2008, in *Handbook of 3D Integration* (eds P. Garrou, M. Koyanagi, and P. Ramm), Wiley-VCH Verlag GmbH, Weinheim.

14
Comparison of Temporary Bonding and Debonding Process Flows

Matthew Lueck

14.1
Introduction

The demands on temporary wafer bonding materials used for the fabrication of 3D integrated circuits (ICs) are rigorous. As compared to the waxes and tapes that are used for the thinning of wafers only, temporary wafer bonding systems must support the wafers through thinning plus be compatible with many fabrication processes (such as vacuum depositions, lithography, metal etches, solvent cleans, polymer cures) while maintaining adhesion and providing mechanical support to the wafer. Then the materials must quickly and cleanly debond without damage to the thin device layers. Because of these challenging requirements, a great deal of study must be done when selecting the temporary bond material(s) for a given 3D fabrication process. While many different systems have recently been developed, each material set has limitations and trade-offs that must be understood by the process design team. As a result, many companies and research institutions have undertaken to independently compare the performance of temporary wafer bonding systems through common fabrication steps in a 3D integration process. These studies are in addition to the work that has been done by the material suppliers and equipment vendors. See Chapters 10–13 by SUSS, EVG, TOK, and 3M for descriptions on their available processes. Chapter 9 discusses the general materials properties requirements for temporary bonding adhesives. Although carrierless designs for thin wafer handling have been reported [1], most thin wafer handling solutions are wafer support systems (WSSs) where the wafer to be thinned is temporarily bonded on a supporting wafer with an adhesive.

In general, research groups have evaluated the temporary bond systems based on the following requirements: total thickness variation (TTV) of the adhesive bondline, high temperature and vacuum compatibility, resistance to wet chemical processes, mechanical support of the thin wafer, debonding throughput, and debonding cleanliness.

Handbook of 3D Integration: 3D Process Technology, First Edition.
Edited by Philip Garrou, Mitsumasa Koyanagi, and Peter Ramm.
© 2014 Wiley-VCH Verlag GmbH & Co. KGaA. Published 2014 by Wiley-VCH Verlag GmbH & Co. KGaA.

14.2
Studies of Wafer Bonding and Thinning

The key requirements for a temporary wafer bonding system during the wafer bonding steps are throughput, minimization of thickness variation, and good adhesion to common surfaces. Table 14.1 contains a list of commercially available materials along with their method of bonding, carrier type, and range of coating thickness. The choice of material, from vendors such as Brewer Science (BSI), 3M, DuPont, TOK, and Thin Materials AG (T-MAT), will determine available adhesive thickness limits and subsequent process temperature limitations.

For a temporary wafer bond system, the method of carrier bonding and thickness of the adhesive will affect the achievable limits of TTV in the thinned wafer. For this reason, most common temporary wafer bond systems apply the adhesive by spinning it on, either to the device or to carrier wafer. In a study published in 2011, a group from ITRI compared temporary wafer bonding materials for use in a 3D integration process involving wafer thinning and backside bump interconnects [2]. For applications in which achieving the lowest possible TTV is important, this study found that a thin adhesive material (less than 50 µm) and thermocompression bonding should be used. Other studies have also reported low post-bond TTV for laser release [3], thermoplastic [4], and mechanical release temporary wafer bonding systems [5]. The mechanical release material from Thin Materials AG is bonded using a low-force, room-temperature bond.

14.3
Backside Processing

The ability to withstand common back-end-of-line (BEOL) semiconductor fabrication processes is what sets modern temporary wafer bonding systems apart from tapes and waxes that are used only for wafer thinning. Typical 3D integration process flows require that temporary wafer bonding materials are compatible with reactive ion etching (RIE), vacuum depositions, photolithography, solvent strips, acid-based metal etches, and electroplating; processes that are generally required to fabricate through-silicon via (TSV) interconnects in thinned wafers. These operations, and the order in which they are sequenced, impose certain requirements on the temporary bond material.

Regarding the compatibility of temporary wafer bond systems with various wet chemical steps, most research studies have not reported adhesion loss or undercut of any of the commercially available materials during the standard backside processes such as lithography, metal etches, electroplating, and even polymer dielectric cures [2–7]. However, vacuum depositions, such as plasma-enhanced chemical vapor deposition (PECVD), have been reported to compromise temporary bond systems. During PECVD depositions at 250 °C as part of the ITRI study, one temporary bond system exhibited dishing of the device wafers while the other system saw cracking of the device wafer [2]. During processing of a thermoplastic

Table 14.1 Some common temporary wafer bonding systems [2].

Material	Process temperature limitation (°C)	Available adhesive thickness (μm)	Process carrier	Equipment (spin coating/bonder/debonder)	Temporary bonding Device	Temporary bonding Carrier	Debonding Method	Debonding Temperature
T-MAT	~250	20–200	Si/glass	SUSS	Spin coat precursor w/plasma	Spin coat elastomer	(1). Mechanical release (lift-off) (2). Film remains on carrier	Room temperature
B/S (HT 10.10)	~220	<100	Si/glass	EVG SUSS	Spin coat adhesive	—	Thermal release (slide-off)	~180 °C
B/S (9001A) (ZoneBOND®)	~250	<120	Si/glass	EVG SUSS	Spin coat adhesive	Zone 2 treatment	(1). Lift-off then solvent release (2). Film remains on device	Room temperature
3M	~250	<125	Glass	Tazmo SUSS w/o debonder	Spin coat adhesive	Spin coat LTHC bond/UV cure	(1). YAG laser release (2). Film remains on device	Room temperature
DuPont	>350	2–20	Glass	SUSS	Spin coat adhesive	—	Excimer laser release	Room temperature
TOK	~250	<130	Perforated support plate	TOK	Spin coat adhesive	—	Solvent release	Room temperature

material, evolution of air pockets between the device and carrier wafers was seen during PECVD processing [3]. The combination of vacuum and high temperature during PECVD depositions can be particularly challenging for temporary bond systems, especially those containing uncured polymers or volatile compounds. Wafer pairs, even those exposed to similar temperatures near atmospheric pressure, exhibit defects only when under the higher vacuum conditions seen in the PECVD process [6]. These conditions are made worse by having non-CTE (coefficient of thermal expansion)-matched substrates or higher stress device layers [6,7]. In addition to the damage it can cause to the thin wafer, outgassing of these compounds is a contamination concern for the PECVD process chamber. Many temporary wafer bond systems allow for tailoring of the baking process or adhesive strength, which can be used to mitigate adhesion loss during high-temperature processing. Studies done on temporary wafer bonding for silicon interposer applications have shown successful PECVD processing at or below 280 °C without adhesion loss or blistering of the adhesives [5–8].

14.4
Debonding and Cleaning

For high-volume manufacturing in particular, the debonding step plays an important role in determining the viability of a given temporary wafer bond system. The debonding step must be done quickly, cleanly, and without imparting damage to the thinned wafer or its bump interconnects. Zoschke *et al.* discusses several scenarios for the optimal sequencing of temporary wafer bonding and debonding in a 3D fabrication process for different thin wafer device types and packaging assembly methods [4].

Fraunhoffer IZM at the All Silicon System Integration Dresden (ASSID) recently evaluated the debonding process of various temporary bond technologies; their assessment of four categories, done in late 2012, is shown in Figure 14.1 [8].

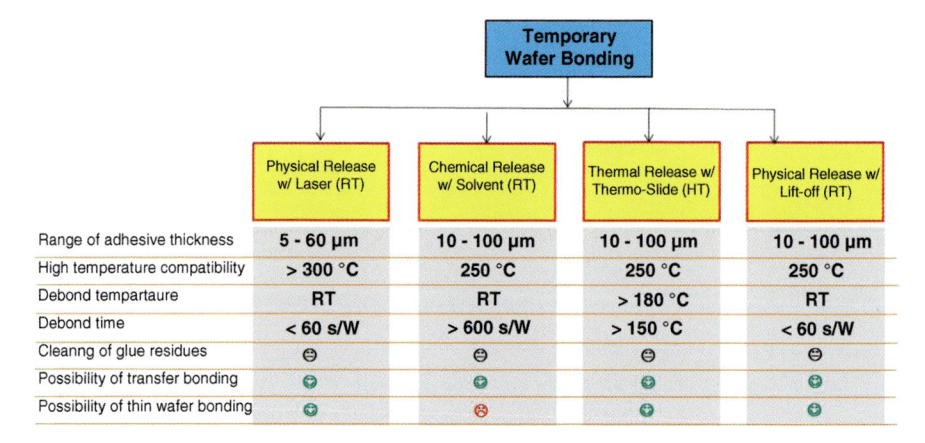

	Physical Release w/ Laser (RT)	Chemical Release w/ Solvent (RT)	Thermal Release w/ Thermo-Slide (HT)	Physical Release w/ Lift-off (RT)
Range of adhesive thickness	5 - 60 µm	10 - 100 µm	10 - 100 µm	10 - 100 µm
High temperature compatibility	> 300 °C	250 °C	250 °C	250 °C
Debond tempartaure	RT	RT	> 180 °C	RT
Debond time	< 60 s/W	> 600 s/W	> 150 °C	< 60 s/W
Cleanng of glue residues	☹	☹	☹	☹
Possibility of transfer bonding	☺	☺	☺	☺
Possibility of thin wafer bonding	☺	☹	☺	☺

Figure 14.1 Comparison of four categories of temporary wafer debonding methods [8].

Estimates are given of the wafer throughput expected for each debonding mechanism. They conclude that processes that rely on solvent release or thermal slide have lower throughput than those relying on room-temperature, dry debonding. And while studies have shown successful processes using slide debonding with large topography [6], many practitioners have come to the conclusion that room-temperature lift-off debonding is fundamentally less risky than thermal slide debonding and thus most material suppliers have already or are in the process of developing material sets that will release without solvent or heat. Some release layer-type temporary wafer bonding systems, in which the release is triggered either mechanically or by laser irradiation, have been covered in previous sections. A study has been done to show that these release materials are compatible with CMOS circuitry [9]. An RTI International study demonstrated mechanical release materials debonding cleanly from and without damage to solder bumps and Cu/Sn microbumps on the backside of thin silicon wafers [5]. A few newer room-temperature debond materials are briefly covered below.

Dow Corning has recently introduced a bilayer temporary bonding solution with a room-temperature debond [10]. This silicone-based system consists of an adhesive layer and a release layer that enables automated mechanical debonding on a flex frame at room temperature.

HD Microsystems has introduced HD3007, for temporary bonding of silicon wafers to carrier wafers by thermocompression [11]. Compared to other temporary adhesives, HD3007 reportedly offers the highest temperature budget for thin wafer backside processing. Debonding of the silicon and glass carrier wafer can be obtained by irradiation of the HD3007 bond layer through the glass wafer using a 248 nm excimer laser.

Dow Chemical has recently introduced a BCB-based temporary bonding solution XP-BCB. AP-3000 adhesion promoter is spun onto the carrier wafer while the device wafer is coated with XP-BCB with no adhesion promoter [12]. Debonding can be done mechanically at room temperature due to the lower adhesion of BCB to the device wafer than to the carrier wafer.

Acknowledgments

The author would like to gratefully acknowledge the contributions made by the 3D Microsystems Integration and Packaging team at RTI International. The author would also like to acknowledge the contributions made by Phil Garrou.

References

1 Bieck, F. *et al.* (2010) Carrierless design for handling and processing of ultrathin wafers. 60th IEEE Electronic Components and Technology Conference, p. 316.

2 Tsai, W.L., Chang, H.H., Chien, C.H., Lau, J.H., Fu, H.C., Chiang, C.W., Kuo, T.Y., Chen, Y.H., Lo, R., and Kao, M.J. (2011) How to select adhesive materials for

temporary bonding and de-bonding of 200 mm and 300 mm thin-wafer handling for 3D IC integration. Proceedings of the 61st Electronic Components and Technology Conference, Lake Buena Vista, FL, p. 989.

3 Shuangwu, M.H., Pang, D.L.W., Nathapong, S., and Marimuthu, P. (2008) Temporary bonding of wafer to carrier for 3D-wafer level packaging. Proceedings of the 10th Electronics Packaging and Technology Conference, pp. 405–411.

4 Zoschke, K., Wegner, M., Wilke, M., Jurgensen, N., Lopper, C., Kuna, I., Glaw, V., Roder, J., Wunsch, O., Wolf, M.J., Ehrmann, O., and Reichl, H. (2010) Evaluation of thin wafer processing using a temporary wafer handling system as key technology for 3D system integration. Proceedings of the 60th IEEE ECTC, May 2010, pp. 1385–1392.

5 Lueck, M., Garrou, P., Malta, D., Huffman, A., Butler, M., and Temple, D.S. (2012) Temporary wafer bonding for 2.5D and 3D integration applications. Presented at IMAPS Device Packaging Conference, Fountain Hills, AZ, April 2012.

6 Charbonnier, J., Cheramy, S., Henry, D., Astier, A., Brun, J., Sillon, N., Jouve, A., Fowler, S., Privett, M., Puligadda, R., Burggraf, J., and Pargfrieder, S. (2009) Integration of a temporary carrier in a TSV process flow. Proceedings of the 59th IEEE ECTC, May 2009, pp. 865–871.

7 Tamura, K., Nakada, K., Taneichi, N., Andry, P., Knickerbocker, J., and Rosenthal, C. (2010) Novel adhesive development for CMOS-compatible thin wafer handling. Proceedings of the 60th Electronic Components and Technology Conference (ECTC), pp. 1239–1244.

8 Wolf, J. (2012) 3D integration driving heterogeneous integration for system in packages. RTI 3D Architectures for Semiconductor Interconnect and Packaging, Redwood City, CA, December 2012.

9 Dang, B., Andry, P., Tsang, C., Maria, J., Polastre, R., Trzcinski, R., Prabhakar, A., and Knickerbocker, J. (2010) CMOS compatible thin wafer processing using temporary mechanical wafer, adhesive and laser release of thin chips/wafers for 3D integration. Proceedings of the 60th IEEE ECTC, May 2010, pp. 1393–1398.

10 Rosson, J., Meynen, H., Wang, S., John, R., Kim, S.W., Bourbina, M., and Fu, P.-F. (2012) Development of bi-layer, temporary bonding solution for 3D TSV, providing simple room temperature de-bond. SEMICON West 2012, SUSS MicroTec 3D Workshop.

11 Itabashi, T. and Zussman, M. (2010) High temperature resistant bonding solutions enabling thin wafer processing. Proceedings of the 60th IEEE ECTC, pp. 1877–1880.

12 Calvert, J. (2012) Dow temporary bonding adhesive development. SEMICON West 2012, SUSS MicroTec 3D Workshop.

15
Thinning, Via Reveal, and Backside Processing – Overview

Eric Beyne, Anne Jourdain, and Alain Phommahaxay

15.1
Introduction

The creation of through-silicon via (TSV) connections is a crucial element of 3D chip stacking. It is, however, not the only technology step. Perhaps even more challenging are the required precise wafer thinning and post-thinning wafer processing steps. Regardless of the TSV integration approach, via-middle or via-last, these process steps are, to a large extent, common. The approach is applicable to active device stacking as well as interposer-based ("2.5D") 3D stacking applications.

A schematic representation of the process flow, discussed in this chapter, is given in Figure 15.1. The active wafer or silicon interposer interconnect wafer with via-middle TSV, or without TSV in the case of via-last approaches, is bonded to a temporary carrier substrate for ease of handling after wafer thinning. This allows for the use of standard semiconductor equipment during post-thinning processing. The most critical aspect of this process flow is the choice of interfacial bonding material: it must combine a variety of properties and functions that are often contradictory. The bonding must be highly uniform in thickness across the wafer and physically, chemically, and thermally stable during the thinning and following backside processing steps. At the end of the process flow, the material must, however, be easily debondable from the thinned wafer, without leaving residues and risking wafer damage. A wide variety of approaches have and are being proposed and developed. The main directions for implementing what is referred to as a "thin wafer support system" are discussed in Section 13.3.

One of the first process steps that are generally performed on the wafers is the so-called edge trimming process. This is required to avoid problems with sharp edges of wafers after thinning or problems with wafer overlay in bonded stacks. The approaches to wafer edge trimming are discussed in Section 13.2.

The actual wafer thinning is performed on a stack consisting of a carrier substrate, a layer of temporary bonding glue, and the wafer to be thinned.

Handbook of 3D Integration: 3D Process Technology, First Edition.
Edited by Philip Garrou, Mitsumasa Koyanagi, and Peter Ramm.
© 2014 Wiley-VCH Verlag GmbH & Co. KGaA. Published 2014 by Wiley-VCH Verlag GmbH & Co. KGaA.

15 Thinning, Via Reveal, and Backside Processing – Overview

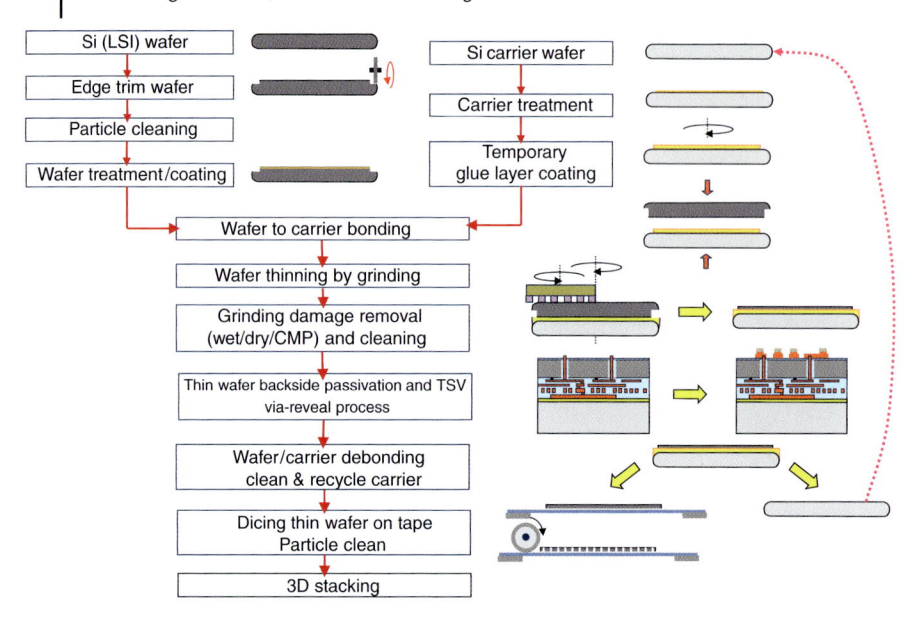

Figure 15.1 Generic process flow for wafer thinning and thin wafer backside processing by bonding to an intermediate temporary carrier.

Achieving a precise thickness control of the thinned wafer is critical for successful TSV-reveal or backside via-last TSV integration. This requires not only a high-precision grinding process, but also accurate thickness control of the carrier substrate and the temporary glue layer thickness. These aspects are discussed in Section 13.3.

After successful thinning of the Si device or interposer wafer, post-processing of the thin wafer-on-carrier can be performed. In the case of via-middle approaches, this step first consists of a "via-reveal" process, followed by backside passivation and TSV contact opening. When using the via-last approach, post-processing includes the TSV via etch, liner deposition, contact opening at the via bottom, barrier and seed deposition, and via Cu plating (partial or complete fill). After finishing the TSV-related backside processing, other processes may follow, such as interconnect redistribution layers and backside bumping. The wafer backside processes prior to wafer debonding are discussed in Section 13.4.

15.2
Wafer Edge Trimming

Si wafers are produced according to semi-standard specifications and exhibit a specifically rounded wafer edge shape. 3D integrations require wafer thinning to a thickness of 100 µm and below, leaving only part of the bevel area of the wafer edge intact. As this area has an angle to the wafer surface of less than 30°, this results in

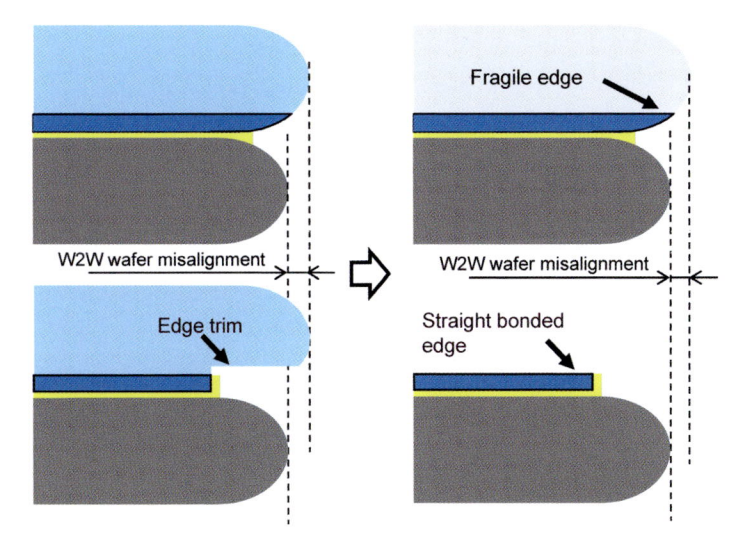

Figure 15.2 Schematic representation of the impact on wafer thinning on the edges of a bonded wafer pair and the need for wafer edge trimming prior to wafer thinning.

a razor sharp thin wafer edge. Such sharp edges are very fragile and may initiate wafer cracks that propagate easily in the thin wafer. An adequate solution to these problems is to modify the shape of the wafer edge by introducing a step profile at the wafer edge. This will result in a rectangular edge profile for the thinned wafer.

An appropriate method for realizing such a step profile is the use of a wafer dicer to trim the edge of the wafer (hence the term "wafer trimming"). This is achieved by turning the wafer on a rotary wafer stage below a stationary dicing blade with precision control of the dicing depth. An example is shown in Figures 15.2 and 15.3. After wafer trimming, particles generated by the trimming process must be removed prior to temporary wafer bonding.

An alternative to edge trimming prior to wafer bonding is to perform edge trimming after wafer bonding. This has the advantage that cleaner wafers are available for the delicate wafer bonding and tolerances in wafer-to-wafer alignment can be compensated in the post-bonding edge trimming. A disadvantage is that a deeper edge trimming is required (through the full wafer thickness) and depth

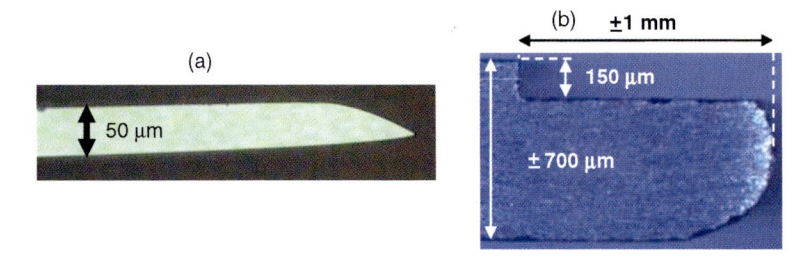

Figure 15.3 (a) Knife-sharp edge of a standard Si wafer thinned down to 50 μm. (b) Edge-trimmed wafer edge prior to bonding and thinning.

control needs to be more precise. It, however, also offers the possibility to remove some of the temporary bonding material from the wafer edge during edge trimming.

15.3
Thin Wafer Support Systems

Handling wafers with sub-100 µm thickness requires a support during handling, transport, and processing in a semiconductor process line. This thin wafer support system should be very stable during these operations but should also allow for easy wafer debonding after finishing the processes. This results in a complex set of requirements. These requirements can be listed as follows (Figure 15.4):

- Good thermal stability during post-processing: There is no outgassing or decomposition of the glue layer and no outgassing when processed in vacuum equipment, for example, Si etch or backside oxide or nitride deposition.
- Good chemical stability during post-processing: There is no attack by chemicals, particularly wet chemicals, that are used during the post-thinning processes, for example, cleaning, wet etch, and plating.
- Very good uniformity of the glue layer thickness: Any deviation in glue layer thickness across the wafer will result in thickness variations in the final thin wafer thickness.
- No voids or delamination after wafer bonding: Small defects will grow during the thinning and backside processing and may lead to catastrophic defects (thin wafer breakage).

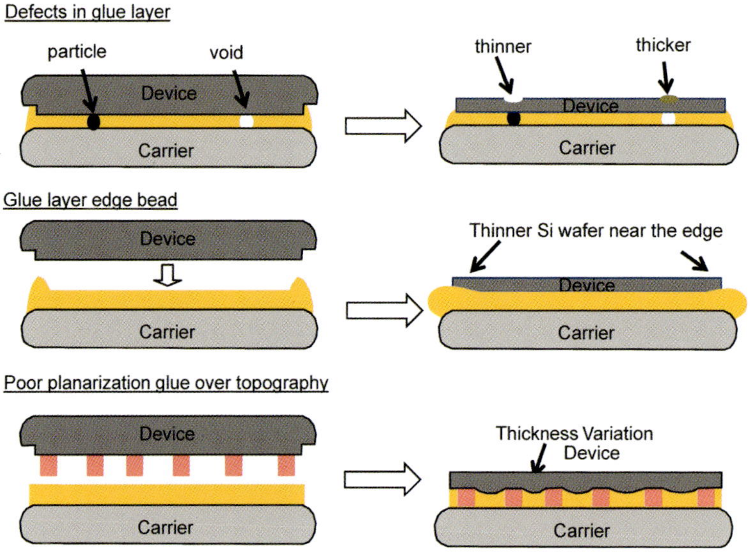

Figure 15.4 Typical defects during temporary wafer bonding.

- In many cases, solder materials are used on either the wafer front side or backside. This sets an upper limit for the temperatures that can be used for bonding or debonding the wafers.
- Despite the stability of the glue during bonding and processing, it should result in an easily wafer-level thin wafer debonding method.
- After debonding, any residues should be easily cleanable from the active wafer.
- The glue layer must also be removable from the carrier substrate to allow for reuse of the carrier substrate.

A wide variety of thin wafer support systems have been proposed. The different methods can be categorized based on the method of debonding:

- Laser ablation of polymeric glue material through a transparent carrier substrate.
- Thermal decomposition of a polymeric glue layer material.
- Thermoplastic glue materials with low viscosity at high temperature that allow for separation of the thin wafer from the carrier by "slide-off" action.
- Removal of the polymeric glue by dissolution.
- Removal of the carrier and glue layer from the thin wafer by "peel-off" action: adhesion layer with reduced resistance to peel-off forces but good adhesion with respect to shear.
- Glue-less methods:

  - *Electrostatic carrier solutions*: A charged carrier substrate supports the active wafer during the full thinning and backside processing.
  - *Membrane thinning*: Thinning a wafer without thinning the outer rim of the wafer results in a flat thinned wafer, such as "Tyco" grinding disc. (This does, however, require a protective organic coating on the wafer front side during backside processing that can be removed easily at the end of the process.)

All methods that involve high-temperature steps during debonding face strong challenges when dealing with temperature-sensitive wafers (e.g., DRAM) or when using solder-bumped wafers. Therefore, the room-temperature debondable methods are the favored solutions; in particular, the laser debonding and room-temperature peel debonding are the focus of the industry today.

An important aspect of the wafer carrier system is also the support or handle wafer used. Currently, either glass or silicon carrier substrates are used. Silicon has the advantage of being the standard substrate material in semiconductor processing. It is available with precise thickness control, exhibits perfect thermal matching to active Si wafers, has excellent thermal conductivity, and has a high strength. It is, however, not transparent and cannot be used in combination with laser-assisted debonding methods. Those methods use glass carrier substrates. The glass used needs to be thermally matched to silicon in the temperature range used during the processing. This requires adaptations of the semiconductor tools for handling the transparent carriers and handling the carrier wafer stacks.

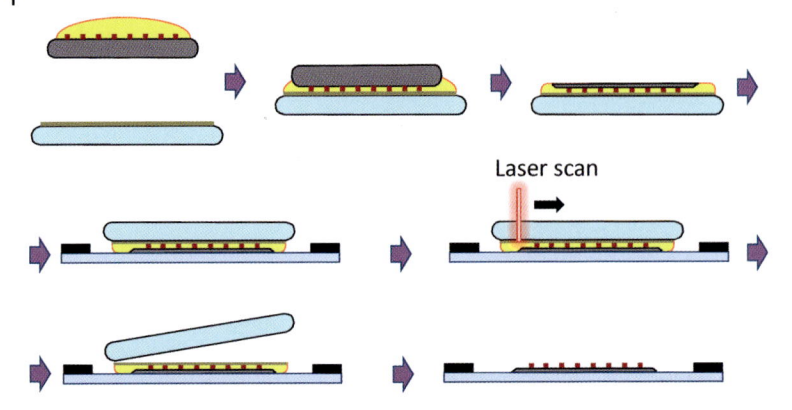

Figure 15.5 Schematic representation of the 3M laser debond temporary carrier process.

In the following sections, some of the most promising approaches are discussed in more detail.

15.3.1
Glass Carrier Support System with Laser Debonding Approach

In this approach, pioneered by 3M, a liquid adhesive is spin coated on the wafer, as illustrated in Figure 15.5. Subsequently, a glass carrier is vacuum bonded to this wafer. This carrier wafer has been coated before bonding with a special release layer. For the 3M process, this is a so-called light-to-heat conversion (LTHC) coating.

The use of a liquid bonding layers allows us to planarize features as big as 100 μm. After bonding, the glue material is cured with UV light, exploiting the transparency of the glass carrier wafer. Typically, a glass carrier larger than the standard Si wafer size is used. This avoids the need for a wafer trimming process; however, it does require handling of nonstandard wafers in subsequent processing steps.

After wafer thinning and backside wafer processing, the thin wafer can be released from the carrier. This process consists of first bonding the wafer pair to a dicing tape on a tape carrier. Next, a laser is used to debond the carrier from the thin wafer. The LTHC layer absorbs the energy from the laser light and debonds from the glass carrier. After scanning the glass carrier wafer, it can be easily removed from the wafer stack. In a next step, the adhesive layer is peeled off from the thin wafer on the tape carrier.

15.3.2
Thermoplastic Glue Thin Wafer Support System – Thermal Slide Debondable System

A typical process flow was pioneered by Brewer Science and is depicted in Figure 15.6. The process is based on the WaferBOND® HT10.10 thermoplastic

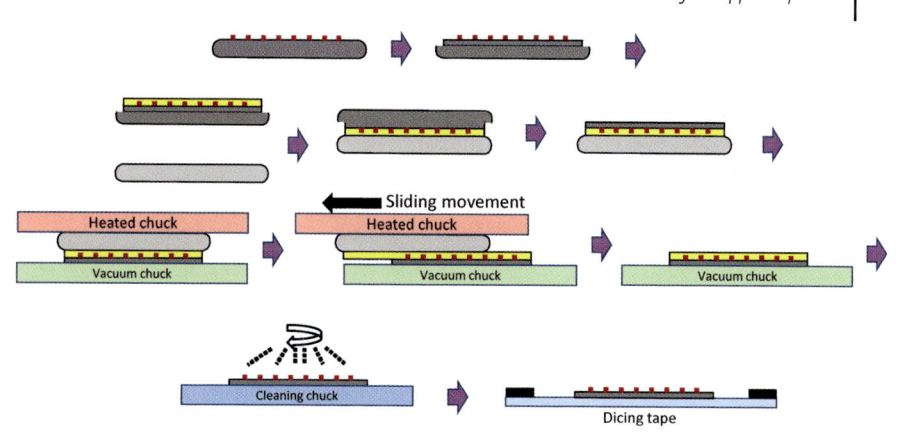

Figure 15.6 Schematic representation of the slide debond temporary carrier process as pioneered by Brewer Science.

high-temperature melt debonding material. After completion of the front-side processing, the wafer is edge trimmed and coated with the temporary bonding adhesive layer before being bonded to the carrier wafer (Si or glass). Upon completion of backside processing (including TSV reveal process, formation of redistribution layers, and microbumping), the thin wafer stack is ready for debonding. This process includes the following:

- The slide debonding sequence itself and thin wafer transfer to an alternative wafer support system (vacuum chuck or electrostatic chuck).
- Adhesive stripping from the thin wafer.
- Transfer to the dicing tape and release from the alternative wafer support system.

15.3.3 Room-Temperature, Peel-Debondable Thin Wafer Support Systems

This has become the main method for new developments. Wafer debonding is performed at room temperature, allowing for a high tool throughput, and is compatible with the presence of solder bumps on the thin wafers. It also does not require transparent or specially manufactured carrier substrates. A large variety of methods and approaches are being proposed and developed.

The method is based on achieving a good adhesion of the thin wafer to the carrier substrate with respect to shear forces, allowing for successful processing of the wafers, but resulting in a low resistance to peel forces. The methods can be roughly divided in two categories:

- *Zone-bond systems*: This is a method pioneered by Brewer Science and is illustrated in Figure 15.7. The adhesion to the wafer is modulated in two zones: low adhesion on

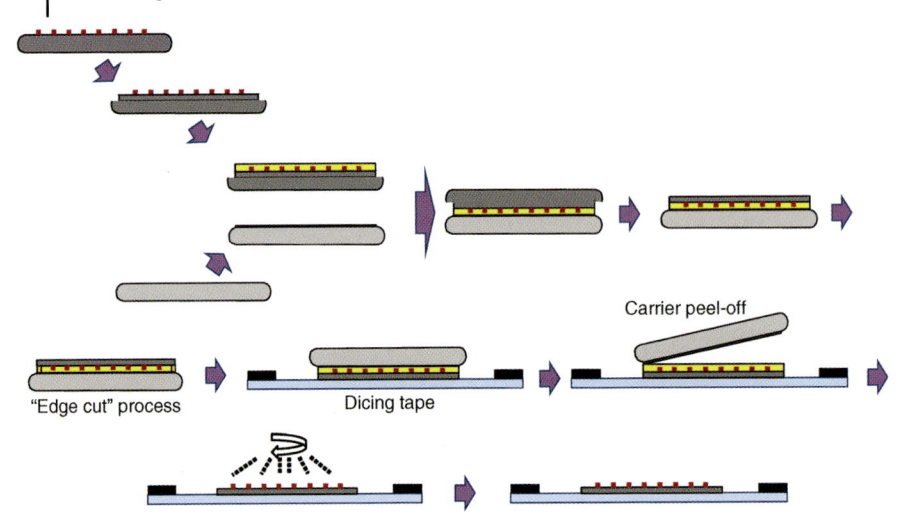

Figure 15.7 Schematic representation of a room-temperature peel-debondable temporary carrier process, in particular, the zone-bond process as pioneered by Brewer Science.

most of the wafer surface and high adhesion of the glue at the perimeter of the wafer (bonded zone). Before debonding, the adhesion strength of this perimeter zone is degraded to allow for peel debonding. This is typically performed by dissolving part of the perimeter glue layer.
- *Non-zone-bond systems*: In this method, the adhesion strength between the glue and the thin wafer and that between the glue and the active wafer are precisely controlled to be sufficient for wafer grinding and executing the post-grinding processing, but sufficiently low for allowing peel debonding. The absence of defined zones simplifies the process flow, but often multiple layer deposits are required to achieve the desired modulation of the adhesive properties. This method was pioneered by TMAT.

15.4
Wafer Thinning

Mechanical grinding is the most cost-effective method to reduce the thickness of a wafer. It is composed of a rough grinding step that removes a few microns of Si per second, and a fine grinding step (so-called polygrind) that removes only a few nanometers of Si per second, resulting in a somewhat polished surface. Mechanical wafer grinding does, however, introduce defects at the wafer surface, which may impact the mechanical properties of the wafer as well as the electrical characteristics of the integrated circuits. To avoid this degradation, a surface smoothening and stress-relief process is required. Therefore, understanding and minimizing the

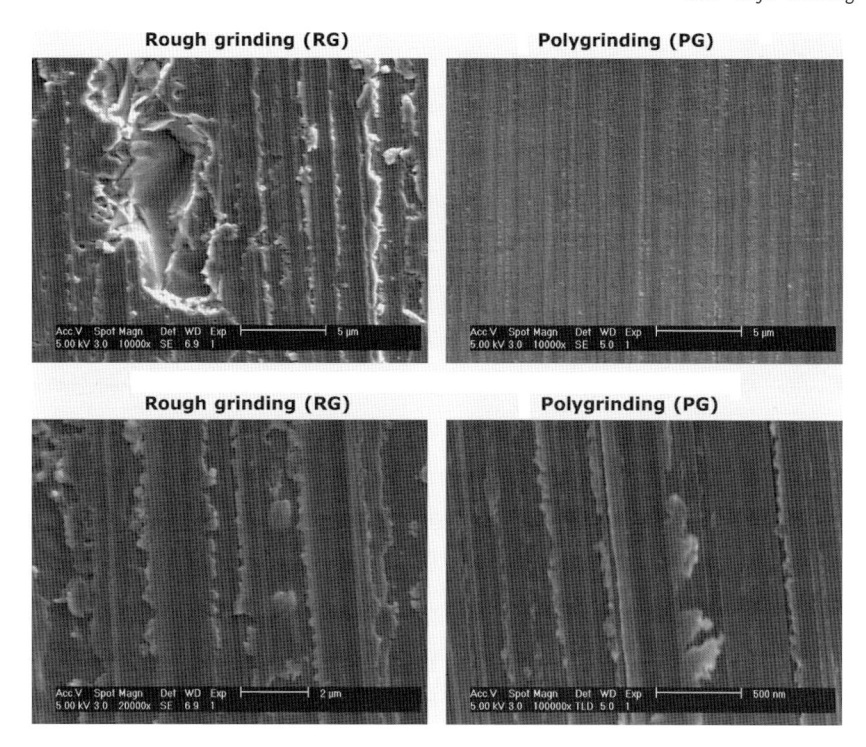

Figure 15.8 Brittle fracture on the surface of the Si wafer after grinding and picture of the subsurface damage.

process-induced damages during mechanical grinding is a key factor for the fabrication of ultrathin wafers.

In Figure 15.8, surface SEM pictures at different magnifications are given for rough and polygrind Si grinding. As can be observed, the grinding grooves on the polyground surfaces are shallower, denser, and more uniform. However, in both cases, some extrusions along the flanks of the grinding grooves are observed. On the rough ground surface, some regularly distributed dimples are visible. These are caused by brittle fracture of the silicon surface and may induce deep subsurface cracks as shown in Figure 15.9.

The nature of the surface after fine and rough grinding has been intensively studied by micro-Raman spectroscopy. During the grinding process, large compressive stress, shear stress, temperature, and rapid unloading lead to the formation of amorphous Si (a-Si) as a result of a sequential phase transformation of the silicon surface under the grinding conditions: Si-I → Si-II → a-Si. Two grinding mechanisms, namely *ductile* and *brittle* grinding, are identified, as shown in Figure 15.10. A ductile grinding mechanism leaves grooves on the surface, with extruded flanks (shallower damage layer). A brittle grinding mechanism causes

15 Thinning, Via Reveal, and Backside Processing – Overview

Brittle fracture

Subsurface crack

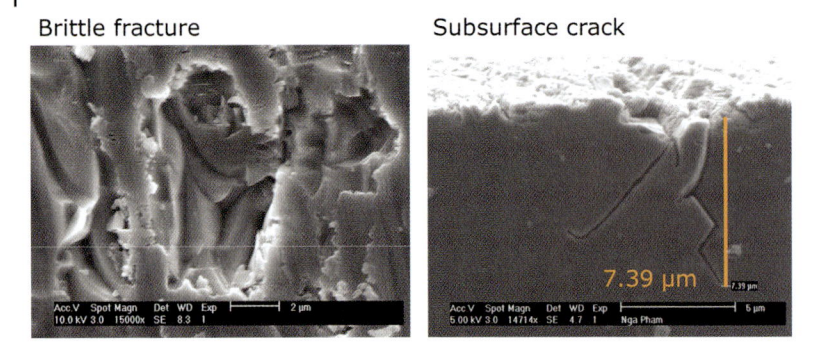

Figure 15.9 Brittle fracture on the surface of the Si wafer after grinding and picture of the subsurface damage.

fracture, with subsurface cracks (deep into the bulk material). The Si-II phase has a lower yield strength than Si-I allowing for the observed ductile deformation during grinding. In fine grinding, the ductile mechanism is dominant, while in rough grinding, a mixture of ductile and brittle mechanisms are observed.

When observed in detail, both rough and fine grinding show multilayer damaged surface structure (Figure 15.11):

- *Rough grinding*: a-Si (70 nm)/plastically deformed layer (3.5 μm)/stressed layer (20 μm).
- *Fine grinding*: a-Si (few nm)/plastically deformed layer (2 μm)/stressed layer (2 μm).

It is important to remove the plastically deformed Si layers in subsequent processing steps.

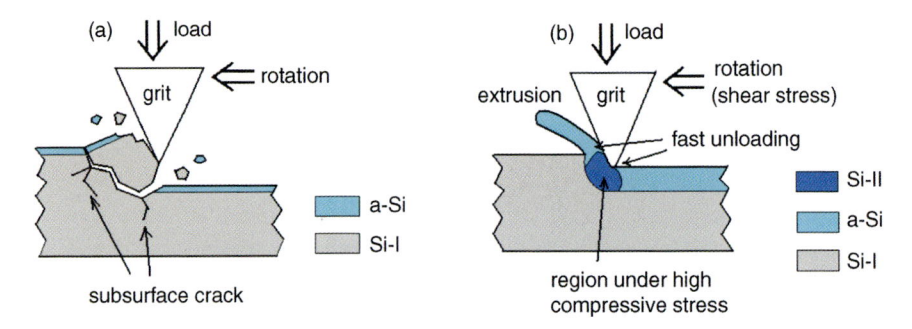

Figure 15.10 The brittle (a) and ductile (b) Si grinding mechanisms. Brittle grinding is caused by fracture of the Si-I subsurface structure. In ductile grinding, the Si-II phase is formed under grinding stress and Si is extruded from the surface, resulting an a-Si layer on the surface.

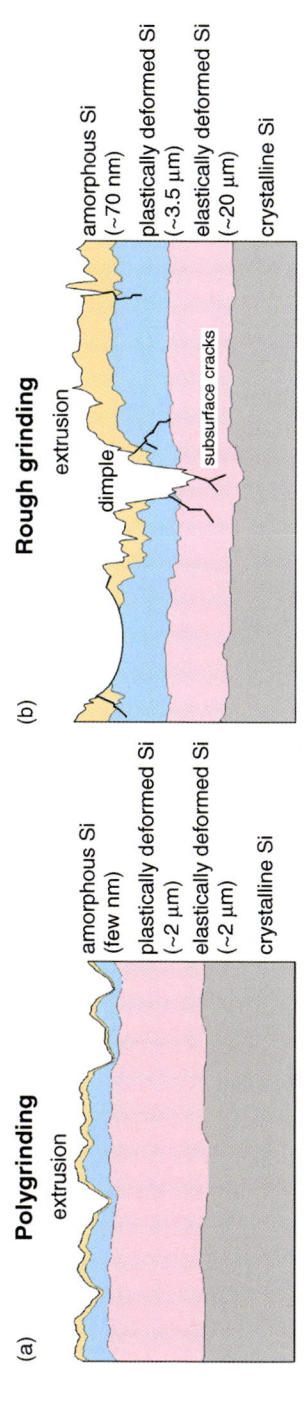

Figure 15.11 Structure of a ground Si surface, fine polygrinding (a) and rough grinding (b).

15.5
Thin Wafer Backside Processing

After wafer thinning on carrier, further processing of the wafer is required. This processing differs from via-middle and via-last process flows.

15.5.1
Via-Middle Thin Wafer Backside Processing: "Via-Reveal" Process

As the TSVs are already present in the thin wafer, the main purpose of the backside processing is to contact the buried TSVs from the wafer backside. Two approaches can be used for this "via-reveal" processing.

15.5.1.1 Mechanical Via Reveal

During wafer thinning, the Si grinding is continued until all Cu TSVs are reached. This implies grinding both Si and Cu at the final stages of the process. As mechanical grinding results in a damaged Si layer that needs to be removed, this process requires a final Cu/Si chemical–mechanical polish (CMP) step to remove the damaged Si layer and obtain flat Cu TSV. As a result of the mechanical grinding of the TSVs, Cu will come into contact with the Si wafer backside, resulting in some Cu contamination on the silicon wafer backside. This may be a risk for Si die with a thickness below 75 µm.

This approach is often used when thickness variations of the Cu TSVs are large (multiple µm of within-wafer variation) and/or the thin wafer thickness variation (total thickness variation (TTV)) is large (>4 µm TTV). For a given application, this comes down to realizing deeper TSV during via-middle processing (e.g., 60 µm deep instead of 50 µm) to cope with process tolerances and achieve a final 3D TSV target.

15.5.1.2 "Soft" Via Reveal

In this case, the mechanical grinding is stopped before the first Cu TSVs are reached. Some silicon will remain between the wafer backside and the tip of the TSVs, the so-called residual silicon thickness (RST). When the font-side TSV via etch depth is controlled within tight tolerances (below 1 µm within wafer variation) and the wafer thinning on carrier results in a thin wafer with a low total thickness variation, the variation of the RST can be very small, in the range of a few µm.

After grinding, the damaged silicon surface has to be removed. Wet or dry etching techniques can be used to remove the damaged backside Si layer whose thickness should not exceed the RST layer. An additional wet or dry etch, selective to the liner oxide, can then be used to reveal the TSV bottom tips (also referred to as TSV "nails"). With the proper control on RST variability, this can result in TSV nails that stick out a few µm from the wafer backside (e.g., 1–4 µm nail height variations), as illustrated in Figure 15.12. The Cu of the TSV remains enclosed in the oxide liner, strongly reducing the risk for backside Cu contamination. After this step, a backside passivation layer should be applied (typically oxide/nitride layer) to

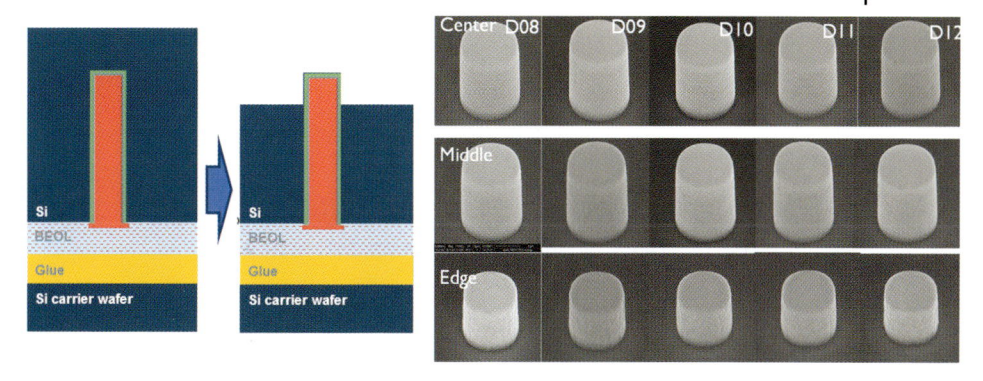

Figure 15.12 (a) "Soft" via reveal process: after wafer thinning, the damaged Si layer is removed and the TSV is exposed on the wafer backside, without etching the TSV liner oxide. In this example, (right) wet etching is used.

the wafer to avoid backside wafer contamination during later steps in the process. Depending on the temperature limitations of the temporary glue, this will require a chemical vapor deposition (CVD) at low temperature.

After this step, different routes can be used to open the contact to the TSV. A first method uses CMP polishing to remove the parts of the Cu nail that stick out from the wafer backside. This process, sometimes referred to as "knock-off" CMP, results in a flat backside surface with exposed TSV Cu, exposed liner ring, and the backside low-temperature CVD oxide or nitride layer. An example is shown in Figure 15.13.

A different approach uses a self-aligned process to open the oxide liner, without removing any of the TSV Cu. For this process, a planarizing photoresist is first deposited uniformly across the wafer backside, without photopatterning. A dry recess etch is then used to reduce the thickness of this layer. As the TSVs stand higher on the wafer backside and the resist easily planarizes over these relatively small features, the top of the TSVs is exposed first during this process. This allows subsequent etching of the backside passivation layer and the oxide liner layer from the tip of the TSV nails. After this process, the remaining resist is stripped from the wafer and the TSVs can be contacted for further processing. This process is illustrated in Figure 15.14.

Further processing can include the processing of backside redistribution layers and interconnect bump pads as illustrated in Figure 15.15. It is, however, also possible to directly use the Cu TSV as backside microbump for 3D stacking.

15.5.2
Via-Last Thin Wafer Backside Processing

As no TSVs are present in the Si, the backside or the thinned Si will reveal itself as a uniform Si surface after grinding and removal of the Si layer damaged by the grinding. A key challenge for via-last integration is the TTV of the thin wafer, as this will result in a required overetch of the via at the thinnest wafer locations. This may

15 Thinning, Via Reveal, and Backside Processing – Overview

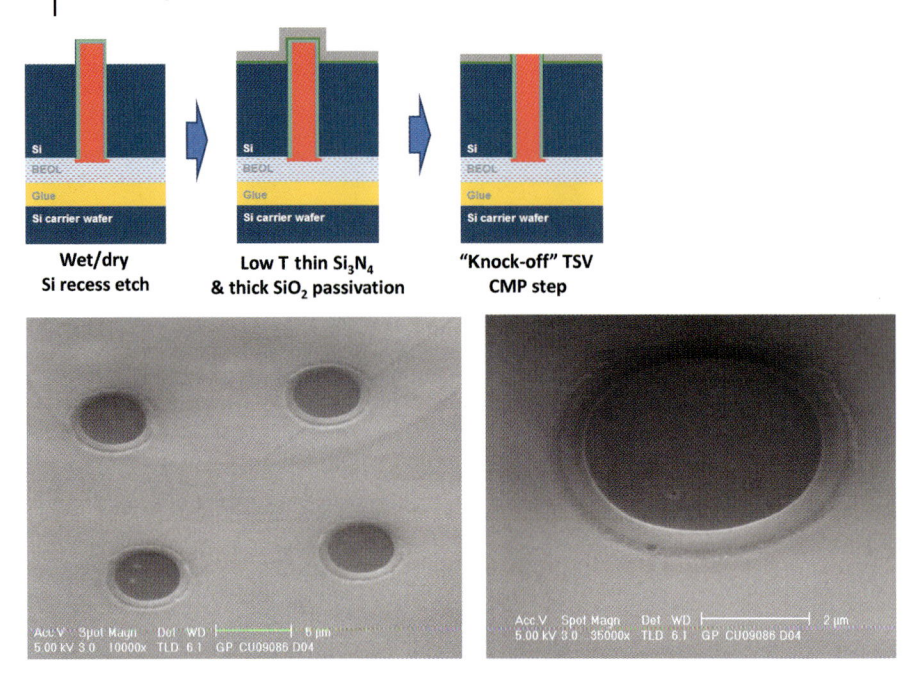

Wet/dry Si recess etch → **Low T thin Si$_3$N$_4$ & thick SiO$_2$ passivation** → **"Knock-off" TSV CMP step**

Figure 15.13 Backside CMP "knock-off" TSV reveal process.

result in so-called notching of the via at the bottom oxide/Si interface: Because of charge buildup on the oxide (STI and PMD layer on the active device side), a widening of the via at the bottom of the TSV hole may occur, resulting in a reentrant profile of the bottom via sidewalls.

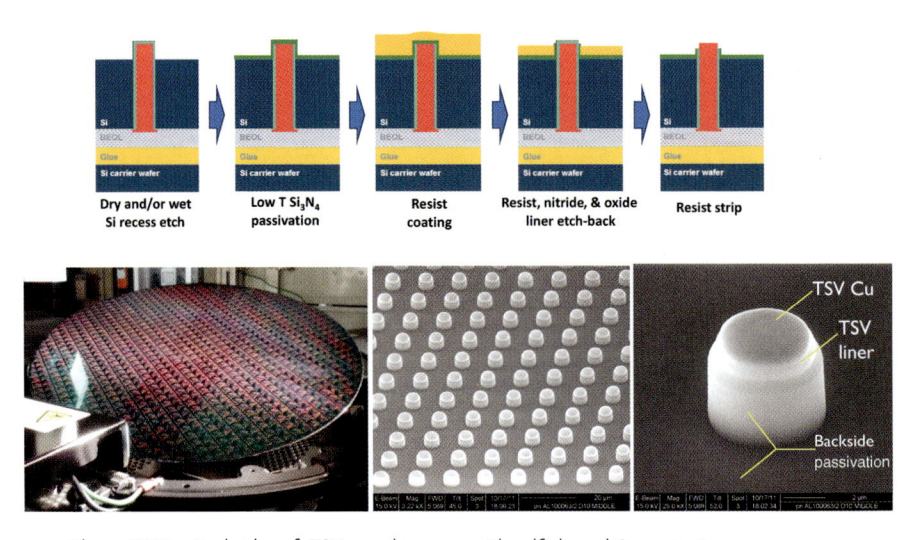

Dry and/or wet Si recess etch → **Low T Si$_3$N$_4$ passivation** → **Resist coating** → **Resist, nitride, & oxide liner etch-back** → **Resist strip**

Figure 15.14 Backside soft TSV reveal process with self-aligned Cu contact exposure.

Figure 15.15 Backside Cu RDL and Cu microbumps (25 μm diameter, 40 μm pitch) contacted to TSVs. (These are located below the topography visible in the redistribution layer (RDL) at the lower left side of the microbumps.)

As most of the temporary carrier solutions have a limited temperature budget, deposition of the TSV insulating liner, the adhesion and copper diffusion barrier, and the Cu seed layer all have to be performed at a reduced temperature, typically with a maximum temperature of around 200 °C. This typically limits the achievable aspect ratio of backside TSV processes.

A further key difference with via-middle integration is the creation of contact between the interconnect on the wafer front side and the TSV. This typically requires a selective anisotropic dry etch of the deposited liner and the STI/PMD insulating layers at the bottom of the TSV hole.

References

1 Yang, Y. et al. (2008) Process induced sub-surface damage in mechanically ground silicon wafers. *Semiconductor Science and Technology*, **23**, 075038.
2 Puligadda, R. et al. (2007) High-performance temporary adhesives for wafer bonding applications. *Materials Research Society Symposium Proceedings*, **970**.
3 McCutcheon J. and Bai, D. (2010) Advanced thin wafer support processes for temporary wafer bonding. IMAPS 2010 – 43rd International Symposium on Microelectronics, 2010, pp. 361–363.
4 Phommahaxay, A. et al. (2012) Ultrathin wafer handling in 3D stacked IC manufacturing combining a novel ZoneBOND™ temporary bonding process with room temperature peel debonding. IEEE-CPMT 3D Systems Integration Conference (3DIC), January 31–February 2, 2012, Osaka, Japan, pp. 1–4.
5 Jourdain, A. et al. (2012) Integration of the ZoneBOND™ temporary bonding material in backside processing for 3D applications. Proceedings of the IEEE-CPMT 4th Electronic System-Integration Technology

Conference (ESTC), Amsterdam, September 2012.
6 Jourdain, A. *et al.* (2013) Integration and manufacturing aspects of moving from WaferBOND® HT-10.10 to ZoneBOND® material in temporary wafer bonding and debonding for 3D applications. IEEE-CPMT 63rd Electronic Components and Technology Conference (ECTC), May 28–31, 2013, Las Vegas, NV.

16
Backside Thinning and Stress-Relief Techniques for Thin Silicon Wafers

Christof Landesberger, Christoph Paschke, Hans-Peter Spöhrle, and Karlheinz Bock

16.1
Introduction

This chapter summarizes the basic techniques for wafer thinning and explains their specific characteristics. It also represents the continuation of the chapter on wafer thinning in Ref. [1]. At that time, technical focus was on the process characteristics of wafer backside grinding and its influence on the fracture strength of thinned silicon. In this volume we will concentrate on state-of-the-art stress-relief techniques that are industrially applied in thin wafer manufacturing. Recently achieved research results on plasma etching of silicon wafers will be shown, which give new insights into the dependence of thinning techniques on the material properties of processed silicon wafers. Finally, possible process flows on wafer thinning and dicing will be briefly summarized and discussed.

16.2
Thin Semiconductor Devices

Let us have a quick look back in history. It was in 1970 that Mash *et al.* published their first ideas on ultrathin silicon devices [2]. They demonstrated the mechanical flexibility of 10 µm thin silicon and announced possible improvements of diode devices when reducing the thickness of the silicon substrate. Meanwhile, in the past 10 years, preparation and processing of very thin wafers have become a more and more important task in semiconductor industry. According to study works from Yole Développement, the ratio of thinned wafers to all shipped wafers will increase from 31% in 2010 to 55% in 2016 [3]. It is expected that until 2017, around 74% of silicon wafers will have to be thinned during product manufacture [4]. "Thin wafers" in this context means a thickness below 100 µm; silicon wafers below 40 µm in thickness are regarded as "ultrathin wafers."

There are different reasons for thinner devices: reduction of internal electrical resistance in the case of so-called vertical devices (e.g., power devices, IGBT, and others), increase of heat dissipation (high-frequency devices based on gallium

Table 16.1 Benefits of wafer thinning for different types of products.

Advantageous property	Applications
Reduce electrical resistance of the semiconductor substrate (= performance improvement)	Power devices (IGBT) Discrete semiconductors (diodes) Photodiodes Solar cells Light-emitting diodes (LED)
Improve heat dissipation	High-frequency devices based on GaAs
Save cost of semiconductor material	Solar cells
Enable very thin packages	RFID devices, laminated on or between foils Chip cards (logistics, identification, security, etc.) Thinner MEMS devices Flexible electronics Medical electronics and implantables
Enable 3D integration	Preparation of interposer substrates Preparation of TSV in thin device wafers
Increase functionality of electronic systems	Chip stacks for memory, logic, sensor devices Polytronic systems (hybrid integration of ultrathin silicon and organic electronics, displays, batteries, etc.)

arsenide (GaAs) wafers), and of course 3D integration technologies. In the last case, the reduced thickness enables or facilitates the preparation of through-silicon via (TSV) and interposer substrates. Furthermore, the reduction of the overall height of packaged semiconductor devices represent a very strong requirement for advanced wafer thinning and handling techniques. Table 16.1 summarizes the various advantages and applications of semiconductor devices that need to become thinner to increase product performance or to enable thin and flat packages.

16.3
Wafer Thinning Techniques

Silicon wafers are typically processed at a thickness of some 750 μm. In order to prepare wafers of a thickness below 100 μm, a large amount of silicon must be removed from the backside of fully processed device wafers. Industrial wafer thinning sequences generally use mechanical abrasion as the first step in wafer thinning because grinding processes allow for high throughput and high accuracy. As it was already explained by Kröninger [1], grinding processes induce micro defects in the silicon substrate. In order to restore a defect-free monocrystalline lattice, the distorted zone near the ground surface has to be removed. Various stress-relief techniques are available for this second step in wafer thinning:

ultrafine grinding, dry polishing, chemical–mechanical polishing (CMP), chemical–mechanical grinding (CMG), wet-chemical etching, and plasma dry etching. These techniques use very different types of removal mechanisms. In order to set up a cost-effective wafer thinning sequence, the relationship between material removal by stress relief, fracture strength of thinned silicon, and process efforts (i.e., cost) needs to be investigated. The following section will first explain the main process characteristics of stress-relief techniques. The very important aspect of mechanical strength will be discussed in subsequent sections.

16.3.1
Wafer Grinding

Wafer grinding is the most often used first step for wafer backside thinning. Wafer grinders use diamonds embedded in grinding wheels to abrade the silicon backside within a short time. During grinding, wafer and wheel rotate either in the same or the opposite direction. The grinding wheel consists of multimaterial blocks (ceramic, polymer, glass) with embedded diamonds of specified size. The blocks are placed along the outer rim of the wheel and the ring moves over the center of the wafer substrate, which is fixed on a rotating vacuum table. The geometry of the abrasive process is shown in Figure 16.1. Semiconductor grinders often use two types of spindle equipment: the first spindle (Z1) for rough grinding and the second spindle (Z2) for fine grinding. Industrial grinding processes for silicon wafers typically show removal rates of 100–600 $\mu m\, min^{-1}$ at spindle Z1 and 10–100 $\mu m\, min^{-1}$ at spindle Z2. Total thickness variation (TTV) after grinding is in the range of 1–2 μm.

Grinding wheels are generally specified by a mesh size, which represents the grit size of a mesh that was used to select diamonds of a specific diameter. Higher "mesh value" represents finer diamonds and results in smoother surfaces but lower material removal rates (micrometer per minute). Fine grinding performed at the second spindle is the first stress-relief process. In order to avoid handling of just rough-ground silicon wafers, which might be strongly bent, practically all grinding equipment keeps the wafer at the same vacuum chuck when transferring the wafer between the rough grinding and fine grinding stations.

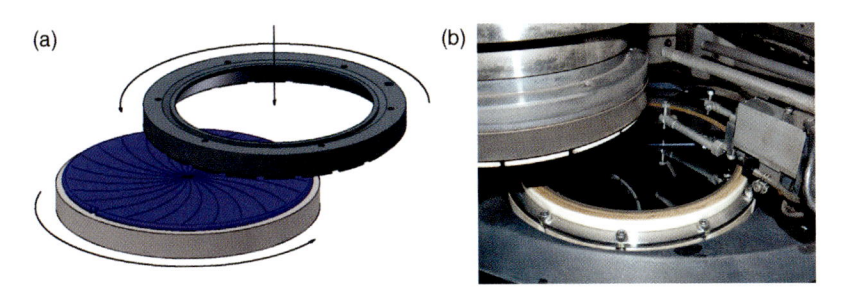

Figure 16.1 Geometry of rotating wafer (blue) and abrasive wheel during standard grinding process (a) and photograph (b) of grinding wheel (on spindle from top) and silicon wafer on chuck table (b).

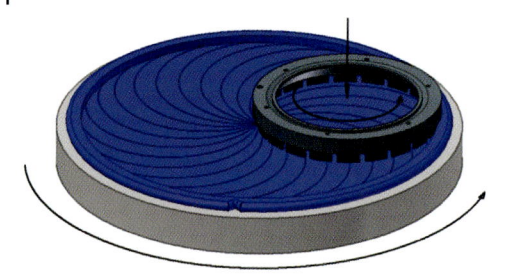

Figure 16.2 Schematic of the TAIKO grinding process. Only the center area of the wafer is abraded; a thicker stiffening ring at the outer rim of the silicon wafer is being kept.

Current trends in grinding technology are targeting for ultrafine-grinding wheels, which allow for further reduction of subsurface damage and increased breaking strength of thin silicon. Three-spindle wafer grinders are also offered from equipment suppliers.

Around 2008, the company DISCO introduced the "TAIKO" grinding process [5,6]. According to this concept, the grinding wheel has a smaller diameter than the wafer and does not move over the outer wafer rim (Figure 16.2). Thereby, a silicon stiffening ring (approximately 2 mm width) at the wafer edge is maintained. By such an approach, the thinned wafer area is kept flat and the substrate can be handled safely. An interesting aspect of the TAIKO concept is its capability to further process the front and rear sides of the wafer after grinding.

16.3.2
Wet-Chemical Spin Etching

Wet-chemical spin etching was first introduced by the Austrian company SEZ (now it is part of LAM). Wet-chemical spin etching probably was the first industrially applied stress-relief process for silicon wafers. It uses a mixture of nitric acid (HNO_3) and fluoric acid (HF) for wet-chemically etching of silicon. The principal mechanism is an oxidation of the silicon (Si) atoms at the surface and the subsequent removal of the oxide film by fluorine molecules. If such a process was done in an etching bath, the material removal would be strongly nonuniform due to the generation of nitrous fumes. Contrary to this, during spin etching, the medium flows over the surface of a rotating wafer and is spun off within a very short time. It was found that adding the heavyweight acid H_3PO_4 and, optionally, further wetting agents to the etchant results in an uniform etching behavior. Furthermore, spin etching equipment uses a moving arm for dispensing liquid etchant (Figure 16.3). The movement can be programmed in various steps. Thereby, it is possible to precisely control the material removal over the whole wafer surface.

Material removal of 20 μm can be done with an uniformity of better than 5%. This would result in a decrease of TTV by just 1 μm. Removal rate depends on HF concentration and the temperature of the etchant; typical values are in the range of

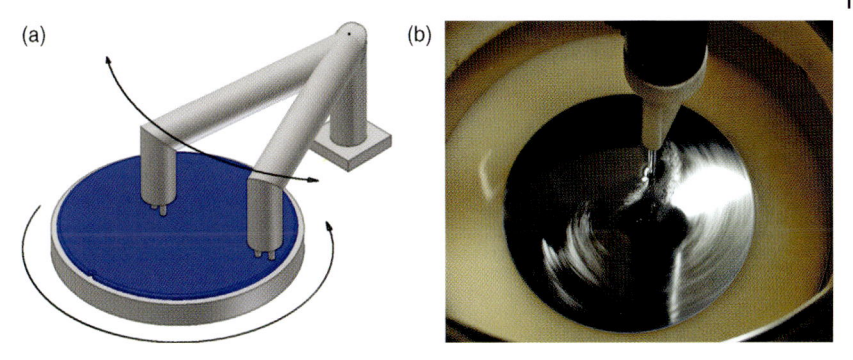

Figure 16.3 Schematic of the spin etching process (a) and photograph of the liquid flow over the rotating wafer (b).

10–40 μm per minute. Spin etching systems reuse the etchant internally for a certain period of cycles; the HF consume can be compensated. An overview on the influence of various process parameters is given in the literature [7]. At Fraunhofer EMFT, we found that removing of 200 μm of silicon by wet etching can be realized at a TTV of less than 2 μm by introducing a two-step process cycle. The etchant consumption to remove a 10 μm thick silicon layer is roughly 1 l for 10 wafers of 8 in. diameter. Damaged backside layers of ground silicon wafers can be removed effectively. The process gives a wafer with a mirror-like backside.

16.3.3
CMP Polishing

CMP is widely used as a standard process for final surface finish of bare silicon wafers as well as for planarization of silicon oxide layers. Stand-alone CMP machines use polishing plates of a large diameter. The wafer is held in a moving fixture and can be pressed onto the polishing pad by controlled forces (Figure 16.4). CMP delivers perfectly smooth surfaces without damaging the crystal lattice. Removal rate by CMP at silicon wafers is up to $1\,\mu m\,min^{-1}$. Abrasive material is a basic chemical etchant, so-called slurry, containing SiO_2 spheres with diameters typically in the range of 20–50 nm. These slurry particles need to be flushed away

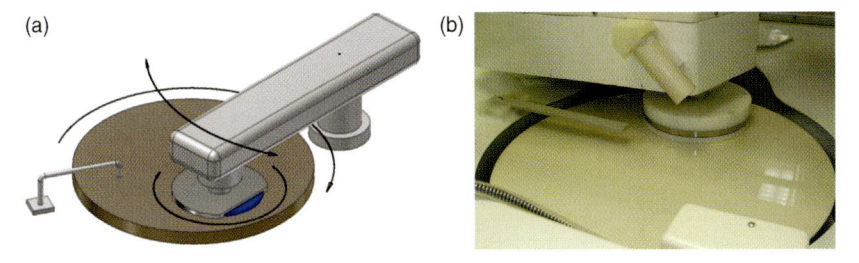

Figure 16.4 Schematic of a conventional CMP process (a) and a photograph of wafer processing (b). The wafer is held beneath the arm.

immediately after polishing. Today's equipment often uses integrated cleaning stations that are based on soft rotating brushes (scrubber). Generally, CMP processing is regarded a costly process due to handling of chemicals and required cleaning sequences.

According to our experiences, CMP generally results in a surface roughness of less than 0.5 nm and at a very low number of remaining particles after cleaning (below 10 per wafer). Even smoother surfaces (down to 0.2 nm) can be achieved by a two-step polish process. Thereby, CMP enables direct wafer bonding of polished silicon wafers. This represents a unique feature of the CMP process.

Grinding equipment companies also offer cluster tools for grinding and CMP polishing for an integrated process solution. Here, CMP can be regarded as a third spindle process within the grinding machine [8,9]. In such cases, the polishing plate shows diameters similar to grinding wheels and rotates over the backside of a ground wafer.

16.3.4
Plasma Dry Etching

Plasma etching of silicon using fluorine radicals is a well-known process since many years [10]. Today sulfur hexafluoride SF_6 is mainly used for dry etching of silicon. In a plasma chamber, this gas easily generates fluorine radicals, which then etch silicon and produce gaseous SiF_4 molecules. Figure 16.5 illustrates the processing situation.

Dry etching offers interesting advantages for wafer thinning. It does not induce mechanical stress to the wafer and so it allows for etching of microstructured wafers such as MEMS devices. However, high plasma power and high etching rates may cause high wafer temperatures. Therefore, application of plasma processes to thin wafers that are covered by a polymeric protective tape or adhesively bonded to a carrier requires the limitation of process temperature below 100 °C. At

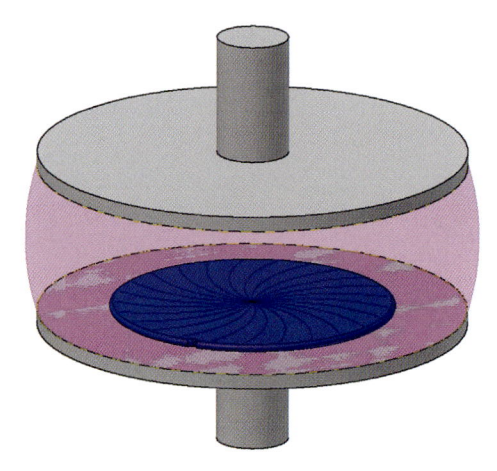

Figure 16.5 Schematic of plasma dry etching process.

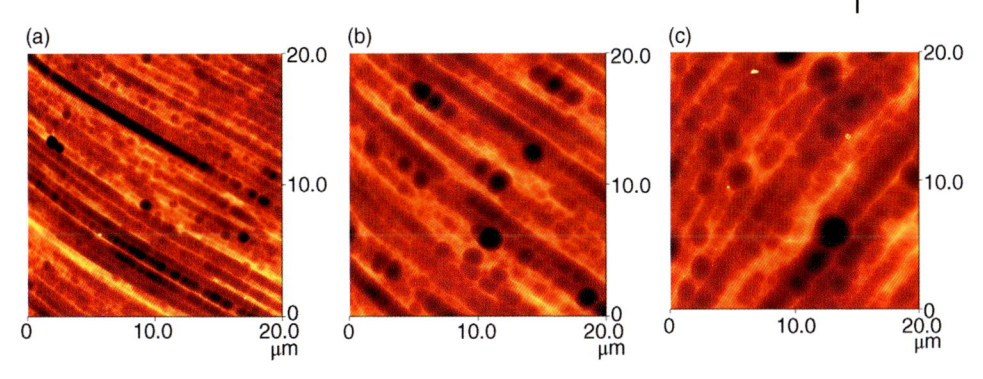

Figure 16.6 AFM pictures after dry etching performed at a fine-ground silicon wafer surface. Etch removal was 2 μm (a), 5 μm (b), and 10 μm (c). Initial grinding marks still can be identified.

Fraunhofer EMFT we used etching equipment from Surface Technology Systems (STS) and developed a uniform plasma etching process that allows for an etching rate of up to $2\,\mu m\,min^{-1}$ at moderate temperatures. The AFM pictures in Figure 16.6 show the resulting surface after dry-etch material removal of 2, 5, and 10 μm. The corresponding roughness values are 11, 9, and 8 nm. The initial roughness value after fine grinding has been 9 nm. Actually, the initial grinding marks still can be recognized. However, it was found that the fracture strength of dry-etched silicon was strongly increased after just 2 μm etch removal (see Figure 16.9). So we conclude that plasma etching removes surface material down to the nondistorted silicon lattice, but shows only poor capabilities to planarize and smooth the ground surface.

Jiun et al. reported that plasma dry etch was able to reduce the roughness of a ground silicon surface from 35 nm to approximately 10 nm. In comparison to our results, they started at a higher roughness value after grinding but finally reached a similar value after dry etching [11].

As with CMP, dry etch plasma chambers can also be an integral part of a grinding and the stress-relief process line.

16.3.5
Dry Polish

The "dry polish" process was introduced by DISCO [12]. The rotation geometry is similar to the grinding process (Figure 16.7). The basic idea was to develop a stress-relief process that leaves a smooth surface like with CMP, however, without the need to use wet-chemical slurries. Actually, dry polish is a real dry process without water or other fluids. An air-based ventilation system ensures that dust particles, which are generated during this polish step, are kept within the process chamber in a filter unit.

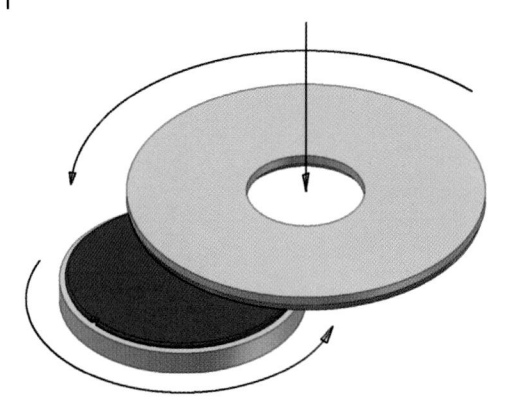

Figure 16.7 Schematic of a dry polish process.

16.3.6
Chemical–Mechanical Grinding (CMG)

CMG uses grinding wheels that contain chemically active abrasives. An example is ceria particles (CeO_2) that initiate chemical reactions at the silicon wafer surface resulting in an amorphous layer of the compound Ce–Si–O, which is finally mechanically removed. Guo *et al.* report a material removal rate in the range of $1\,\mu m\,min^{-1}$ and a surface roughness of 0.7 nm [13]. CMG can be run as a dry process (no liquid coolant) on standard grinding equipment.

16.4
Fracture Tests for Thin Silicon Wafers

Looking back, we find a method to calculate the breaking strength from a ball on a ring test geometry and an early investigation of the influence of ion implantation and annealing processes on the fracture strength of silicon published by Hu in 1982 [14].

Over the last decade, a lot of research work was carried out by various research groups to analyze the influence of wafer thinning techniques on the fracture strength of thinned silicon wafers. It was found that breaking tests are very sensitive to micro defects in the silicon material respectively at its surface or edges. The two main test methods are the ring-ball test and the three-point bending test. Figure 16.8 illustrates the basic experimental setup and bending situation. Both methods can be extended to tests where the mechanical stress is induced more uniformly over a larger area. These tests are called the "ring-on-ring" and four-point bending tests. An overview on these test methods and the related mathematical description can be found in the literature [15,16].

The main difference between these methods is the possible influence on the sidewall of the samples. With the three-point (or four-point) bending test, the

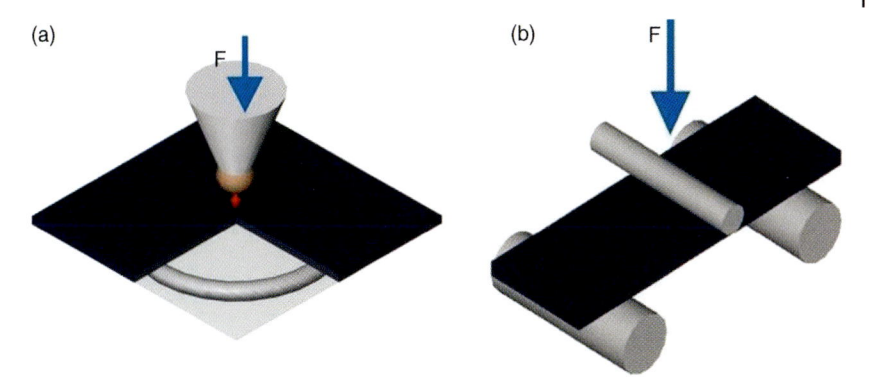

Figure 16.8 Experimental determination of fracture strength of thinned silicon samples: ring-ball test (a) and three-point bending test (b).

sidewalls of the sample undergo the same deflection as the bulk material. Therefore, this method is generally selected to analyze the influence of dicing technology to chip fracture strength.

With the ring-ball test, sidewalls of the samples are farthest away from the area of maximum deformation, which is located in the center, directly below the ball. Therefore, the ring-ball test is mainly sensitive to the backside treatment of the samples. In order to determine the effectiveness of a certain stress-relief technique, the amount of material removal should be varied and the resulting breaking force in the ring-ball test has to be measured.

An alternative method to the ring-ball test was proposed by Hawkins et al. [17] and Tsai and Lin [18]. They used a soft material on a baseplate and induced the pressing force via a ball onto the silicon sample, which was lying on the soft layer.

All fracture tests measure breaking force values. These values depend on sample geometry and thickness. In order to compare fracture test results from different experimental setups, it would be necessary to calculate the breaking stress from the breaking force. However, most often-used formulas are only valid as long as the sample deflection is lower than half of the sample thickness. For very thin silicon, this assumption is generally not fulfilled, and hence, the application of stress formulas is very questionable. Large deflections require FEM modeling of the bending geometry. A comprehensive and helpful discussion on this subject can be found in the literature [15].

Although the mathematical formulation of breaking tests is a complex task, the experiments are easy to conduct and they always offer the possibility to show relative changes in breaking behavior when different technological steps have been carried out during wafer processing. Therefore, the comparison of breaking forces (instead of breaking stress) is of the same relevance as long as the sample geometries are kept constant.

Further references to recent research work on fracture tests of thin silicon will be given and discussed in the following sections.

16.5
Comparison of Stress-Relief Techniques for Wafer Backside Thinning

Today, standard wafer-thinning techniques start with wafer backside grinding. This first step is generally called "rough grinding." Grinding-induced damage at the backside of semiconductor wafers need to be removed by further process steps, which are called "stress-relief processes." Table 16.2 summarizes the main techniques for such postprocessing after wafer rough grinding.

It is well known that fine grinding or ultrafine grinding using finer abrasive materials (e.g., smaller diamonds in the grinding wheel) result in increasingly smoother surfaces and, thereby, also in higher fracture strength values of thinned silicon materials [19,20].

Wet-chemical spin etching does not apply any force to the substrate under preparation. Nevertheless, it shows the highest material removal rate of all stress-relief processes, and it offers a unique possibility for removing an even larger amount of silicon within a short time and without damaging the silicon lattice. Spin etching smoothes the grinding marks to a large extent. However, it does not result in perfect planarization [21]. Spin etching requires handling and disposal of

Table 16.2 Stress-relief techniques for wafer thinning.

	Method of material removal	Abrasive medium	Removal rate temperature
Fine grinding	Mechanical abrasion	Fine sized diamonds embedded in grinding wheel, process uses DI water	10–100 $\mu m\,min^{-1}$ DI water cooling
Dry polish	Mechanical polishing	Solid-state abrasive surface, no water and no chemicals during operation	1 $\mu m\,min^{-1}$
CMG	Chemical etching and mechanical removal	Reactive abrasive, CeO_2 embedded in grinding wheel, dry process	Up to 1 $\mu m\,min^{-1}$
Spin etching	Wet-chemical etching	Liquid etchant: $HF + HNO_3 + H_3PO_4$, also other compositions	20–50 $\mu m\,min^{-1}$ $T < 40\,°C$
CMP	Chemical–mechanical polishing	"Slurry": SiO_2 particles in liquid medium, generally strongly basic	1–2 $\mu m\,min^{-1}$ $T < 40\,°C$
Plasma dry etching	Etching in a reactive plasma gas	Fluorine atoms or molecules (SF_6, CF_4, etc.), plasma chamber	1–10 $\mu m\,min^{-1}$ T depends on etching rate and plasma conditions ($T > 60\,°C$, potentially even $> 300\,°C$)

dangerous etchants (HF and others). Therefore, it has to be implemented in an adequate fabrication infrastructure.

Several publications showed optimum low roughness values after CMP polishing of silicon wafers. Also fracture strength of thin silicon samples after CMP generally results in top rankings. The drawbacks with CMP are rather high cost for equipment and processing media (pads, slurry, cleaning). Due to the low material removal rate, CMP can only be a short last step in wafer thinning.

Possible alternatives to CMP are dry polish and CMG. Zhou et al. report that CMG can produce a surface quality that is even superior to CMP [22]. Influences of dry-polish parameters on the fracture strength of thin silicon samples are reported in Ref. [23].

Maeda et al. showed that the depth of the noncrystalline layer after ultrafine grinding is approximately 50 nm, and it can be completely removed by CMP or dry polish [24].

With dry polish and CMG, only the backside of semiconductor wafers can be influenced. This means there is no chance for rounding of corners or an additional stress-relief effect for sidewalls of silicon dies. Such features need to be discussed when wafer thinning and dicing should be evaluated as a whole (also see following sections).

Plasma dry etching shows properties that appear to be quite different from other stress-relief techniques. Ring-ball breaking tests of dry-etched silicon samples showed that breaking strength is strongly improved already after removing a 2 μm layer from the backside of a fine-ground silicon wafer (Figure 16.9) [21].

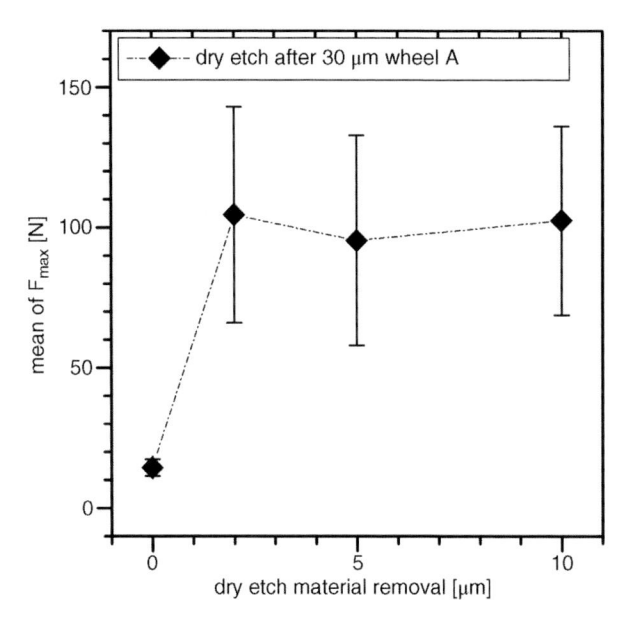

Figure 16.9 Breaking force after plasma dry etching for stress relief of ground silicon.

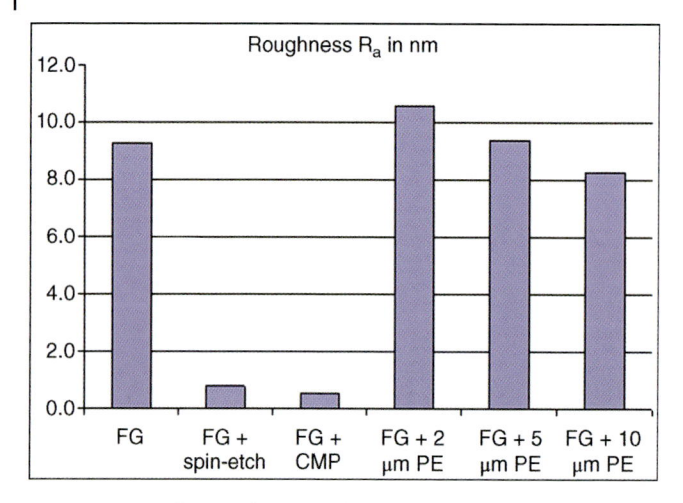

Figure 16.10 Surface roughness R_a after different stress-relief processes.

The surface roughness of a ground silicon wafer tends to become smoother after several minutes of dry etching. However, even after removing 10 µm of silicon, the roughness is not significantly better than the initial state after fine grinding as seen in Figure 16.10. This experimental result shows a unique property of plasma etching for stress relief: The breaking force is rapidly increasing after 2 µm etch removal, but it is not correlated with surface roughness as it is the case with other stress-relief techniques.

At Fraunhofer EMFT, we compared the influence of spin etching, CMP polishing, and plasma dry etching to the strength properties of fine-ground silicon wafers of a thickness of 200 µm [21]. The results of a ring-ball test series plotted in a Weibull diagram for the breaking force are shown in Figure 16.11. The y-axis represents the cumulated failure probability on a logarithmic scale. The zero value at the y-axis defines the line where 63% of silicon samples have been broken.

Figure 16.11 clearly shows the steady enhancement in fracture strength when just rough-ground wafers (blue line) are further treated by fine grinding (green lines, 10–40 µm material removal) and a stress-relief process. It is interesting to see that the very different processes - CMP, spin etching, and dry etching - result in the same maximum values of breaking strength. It is concluded that the damaged layers were removed in all cases and so the final breaking force is related to an undistorted silicon lattice. However, the required material removal depends on the stress-relief method. Dry etching and CMP polishing are most effective in eliminating stressed layers after grinding. Our experiences with these stress-relief techniques are summarized in Table 16.3

The Weibull plot in Figure 16.11 also shows a low slope for the breaking test series for the three stress-relief techniques. This means that the fracture test results show quite large scattering. Actually, this is a critical point because the mechanical reliability of thin silicon finally will be related to the low end of measured breaking forces. From our point of view, this large scattering is due to the separation

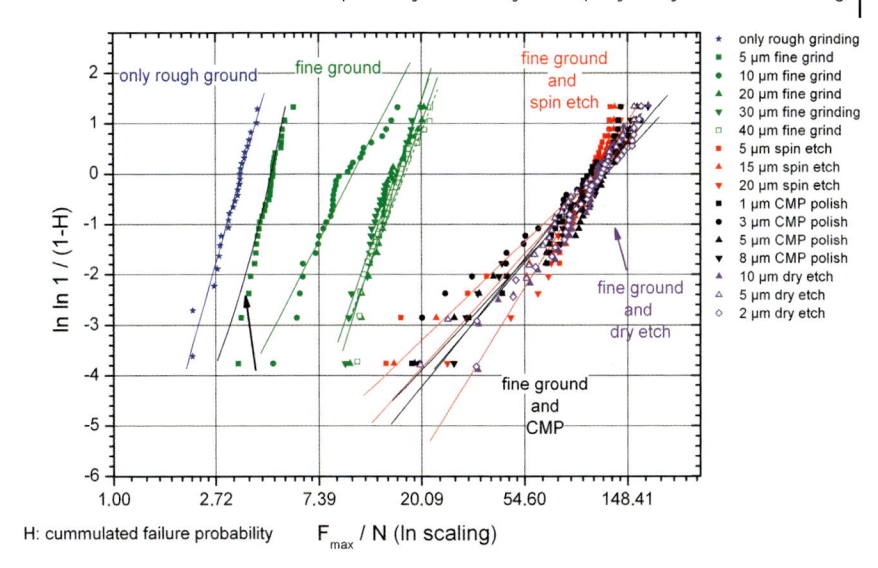

Figure 16.11 Comparison of resulting breaking force (measured by ring-ball test) after different grinding steps and subsequent stress-relief processes, spin etching, CMP, and plasma dry etching.

technique. As we used standard sawing for chip separation, there still is an influence from the sidewall of the test chips. This might rarely result in failures at lower breaking force. The very important point of dicing technology will be further discussed in the next section.

Table 16.3 Comparison of selected stress-relief techniques with respect to manufacture technology.

	CMP	Dry etching	Wet etching
Required material removal	1–3 μm	2 μm	5–15 μm
Process time	2–4 min	1–2 min	15–45 s
Surface properties	Perfectly smooth	Roughness of fine-ground wheel (Z2) is still present	Mirror like, not perfectly smooth
Fracture strength	Regarding the highest values of breaking force, there is no difference between these three stress-relief techniques		
Applicability	Well suited for subsequent wafer bonding; however, it requires cleaning of slurry residues	Take care of process temperature, enables stress relief at chip sidewalls as well, specific benefits for MEMS wafers	General stress relief, etch attack at structured silicon surfaces might be difficult to control

16.6
Process Flow for Wafer Thinning and Dicing

In the previous section, thinning techniques were compared in terms of their specific capability to remove grinding-induced damage at the backside of the whole wafer. However, in a final semiconductor product, high mechanical strength will be required for a single silicon die. According to standard wafer processing, dicing is done after thinning by means of rotating blades. However, this is again a mechanically abrasive process and can induce damage to the sidewall of chips. Therefore, for a target wafer thickness below 100 µm, and even more critically below 40 µm, dicing technology and backside stress-relief processes play an equally important role.

Figure 16.12 shows possible process flows for wafer thinning and dicing. The blue "process line" is related to very thin wafers. It also comprises the optional step for mounting of a rigid carrier substrate for secure handling and processing.

We did not find published research results on breaking tests that would include all dicing and thinning techniques that are supposed to show optimum die strength properties. The reason for this lack of a general comparison might be due to the fact that equipment and processes are still under development and no research group has access to all latest technologies on their own site. So, in the following paragraphs we will try to summarize the main statements on wafer thinning and dicing technologies from different publications. However, we cannot compare all different wafer-processing techniques against each other.

Two principal ways for thinning and dicing can be distinguished: "dicing before thinning" and "dicing after thinning." Two variations exist for the

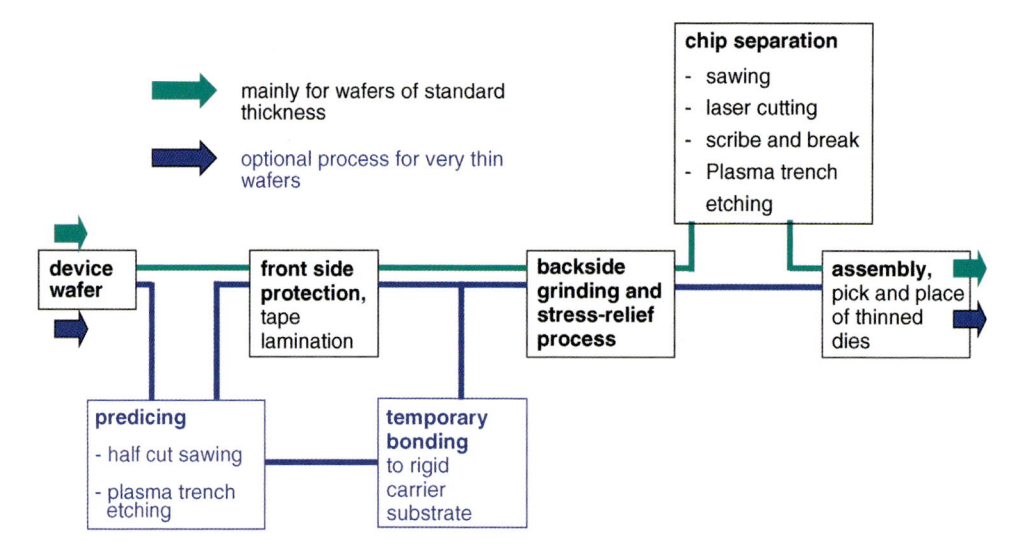

Figure 16.12 Process flows for wafer thinning and dicing for thick (green line) and thin (blue line) wafers.

"predicting first" approach: the so-called dicing before grinding (DBG) process (proposed by Toshiba) [25] and the dicing by thinning (DbyT) process (proposed by Fraunhofer) [26].

DBG is already widely used in the industry. It shows high throughput and low cost of consumables. However, as it is based on a conventional sawing process, front side and backside chipping as well as micro defects at the sidewall of chips are likely to occur. According to Takyu et al. backside chipping can produce micro defects up to 50 μm in depth and is thereby dominating the effect of die strength reduction [19]. Fracture strength of DBG chip samples can be increased by using plasma etching or CMP polishing as the last process step in backside thinning because these techniques allow for overetching of the chip sidewalls (either by slurry or etching gas). The positive effect using CMP for rounding of the corners at the chip backside was also described by Takyu et al. [19]. In another paper, this research group proposes to use DBG plus CMP for die thickness above 20 μm and DBG plus plasma for die thickness below 20 μm [27].

Contrary to DBG, the DbyT concept is based on the application of temporarily bonded rigid carrier substrates and focuses on plasma etching for die separation as well as for backside stress relief. It has already been shown that such concept allows for preparation of ultrathin and reliably bendable dies [28]. Introducing plasma dicing techniques eliminates front-side chipping and sidewall damage. Furthermore, backside chipping also cannot occur when opening of the front-side trenches is realized by a final plasma etch process at the rear side of wafers. Limitations of this concept are related to required changes in chip design and appropriate wafer support techniques during backside etching (see further comments below).

The "predicting first" approaches show the drawback that they are limited to semiconductor products that do not require further process steps after thinning. As it was shown in Table 16.1, applications such as power devices and 3D integration require processing concepts that allow for additional process steps at the backside of the thin wafer. In these cases, possibilities for "dicing after thinning" need to be considered. The main techniques of interest today are sawing, plasma dicing, and various types of laser cutting.

When studying published research results on these dicing techniques you will find one statement that is common understanding among the various authors: Damage at the sidewall of very thin silicon dies must be avoided or eliminated in order to end up with chips of high fracture strength. Many comparative studies have been conducted using three-point or four-point bending tests. Results on the variation of blade and laser dicing techniques can be found in the literature [19,29–31]. Additional investigations with plasma dicing were shown in Refs [30,32–35]. Plasma etching also can be used as an additional stress relief for the sidewalls of chips, for example, after laser cutting or DBG half-cut sawing. A strong increase in fracture strength after plasma stress-relief etching of the dicing lanes is generally observed [31,36].

Plasma dicing is based on lithographic definition of chip trenches. Consequently, on the one hand, it requires additional process steps in a wafer fab environment. On the other hand, it allows preparation of chips of any shape, for example, with rounded corners or hexagons. The chip outer dimensions are precisely defined within 1 or 2 µm. This would allow for using the chip corner as an alignment mark in die placement (e.g., in chip-to-wafer bonding applications) and it also enables self-assembly techniques of thin dies on base substrates. Plasma dicing shows practically damage-free sidewalls of thin silicon dies [33,37]. It might become the right solution for ultrathin wafers (below 40 µm) because in this case, the etching time is in the range of just a few minutes. During die separation, the wafer needs some kind of support or fixture. Panasonic proposes to use frame carriers as wafer and chip support during plasma dicing and reports on higher throughput for plasma dicing of small dies compared to blade dicing [35]. When polymeric tapes are used, the wafer temperature during etching must be limited to temperatures below 100 °C. Alternatively, tapes or reversible adhesives with higher temperature stability would help. The application of plasma dicing requires some adaptations in the design for the chip layout. For example, metal patterns are not allowed to cross the dicing lanes.

Laser dicing can be performed in two ways: material ablation or laser-induced cleaving/cracking. The latter was invented by Hamamatsu [38] and is also known as "stealth dicing." In this case the laser beam is focused into the center of the silicon substrate and creates distortions in the silicon lattice. Expanding the wafer afterwards results in die separation. By such method the saw-line width can be close to zero. Laser dicing has already been introduced to industrial manufacture. In the case of 10 µm thin wafers, it might become increasingly difficult to limit the focus spot size within the silicon substrate [29].

It should be mentioned that up to now most of research work on the analysis of wafer thinning and dicing techniques has been done on bare silicon wafers. However, a real device wafer will show different properties in terms of nonuniform substrate stress and complex multiple layers (e.g., low-k dielectrics) on the front and rear sides. This will definitely result in further challenges for dicing techniques and also for final assembly steps of very thin dies.

16.7
Summary and Outlook on 3D Integration

A variety of useful techniques for wafer thinning and dicing have been developed and established over the last years. Their specific influence on the quality of thin or ultrathin silicon dies is well understood. The main requirement for robust ultrathin dies is the absence of crystalline damage at both the rear side of chips and their sidewalls. Technical solutions and production equipment for wafer thinning and dicing for a final chip thickness of down to roughly 50 µm exist and are partly implemented in semiconductor industry. Aiming for even thinner wafers (below 40 µm) seems to be necessary in order to enable robust flexibility or ultrathin package solutions.

One of the main open questions concerns the introduction of cost-effective carrier technologies for thin wafer processing. Here, a final solution for high-volume manufacture cannot be clearly identified yet. We conclude that an overall manufacture concept for ultrathin wafers, including backside processing, die separation, and also assembly of ultrathin devices still represents an important task for upcoming technological developments.

In the last years the role of backside thinning and stress-relief technologies gained even more importance for 3D system integration, since these technologies determine the technological surface quality and the performance of the assembly technology processes like bonding, via-fill, or underfiller together with a significant part of the cost involved. The mechanical and electrical reliability of a 3D assembled system consisting of thin devices and thin Si interposers are determined to a large extent by the backside processes because backside technologies are used for their manufacturing. For a reliable cost-efficient 3D integration in particular, plasma processes offer an interesting advantage, since they can be applied for thinning, dicing, and via etching. All surface conditions on the thin chips, the interposer, and even in the TSV are comparable, leading to easier development of the different assembly technologies, since boundary conditions of the process development, the process windows, as well as the possible reliability issues and degradation mechanisms are depending on one plasma process-related surface condition and quality only. They complete themselves, therefore, to a possible modular approach for a 3D-oriented integration technology, where after dicing, thinning, and TSV, all additional assembly technologies like via fill, contact, and interconnect can be based on such a similar material and surface quality for ICs and interposers, leading to a more predictable reliable 3D integration volume manufacturing technology based on a similar possibly multipurpose equipment platform with also significantly improved cost-efficiency.

References

1 Kröninger, W. (2012) Fabrication, processing and singulation of thin wafers, in *Handbook of 3D Integration*, vol. **1** (eds P. Garrou, C. Bower, and P. Ramm), John Wiley & Sons, Inc., Hoboken, pp. 177–208.
2 Mash, D.H., Henshall, G.D., and Eales, B.A. (1970) Thin silicon device technology. *Journal of Physics D: Applied Physics*, **3**, 1199–1204.
3 Mounier, E. (November (2011)) Thin wafer handling, a market perspective. Presentation given at the Workshop "Be-flexible," Munich, Germany.
4 Mounier, E., Pizzagalli, A., and Rosina, M. (2012) Thin wafer and temporary bonding, equipment and materials market. Study from Yole Développement.
5 Yoshida, S. *et al.* (2010) Wafer grinding method. US patent 7,758,402 B2.
6 Klug, G. (2009) Advanced solutions for ultra-thin wafers and packaging. Proceedings of European Microelectronics and Packaging Conference (EMPC 2009).
7 Starzynski, J.S., Yellowaga, D., McFarland, J., Palmer, B., and Drews, S. (2007) Honeywell Silicon Polish Etchant I, Honeywell Materials White Paper.
8 http://www.accretech.jp/english/semicon/product/c/polish (accessed January 2013).
9 http://www.okamoto-sed.com/products.php (accessed January 2013).
10 Bhardwaj, J. and Ashraf, H. (1995) Advanced silicon etching using high-density plasmas. Proceedings of SPIE 2639, Micromachining and Microfabrication Process Technology, September 19, 1995, p. 224.

11 Jiun, H.H., Ahmad, I., Jalar, A., and Omar, G. (2006) Effect of wafer thinning methods towards fracture strength and topography of silicon die. *Microelectronics Reliability*, **46**, 836–845.

12 http://www.disco.co.jp/eg/products/polisher_etcher/8000_gp.html (accessed January 2013).

13 Guo, D.M., Tian, Y.B., Kang, R.K., Zhou, L., and Lei, M.K. (2009) Material removal mechanism of chemo-mechanical grinding (CMG) of Si wafer by using soft abrasive grinding wheels (SAGW) *Key Engineering Material*, **389–390**, 459–464.

14 Hu, S.M. (1982) Critical stress in silicon brittle fracture, and effect of ion implantation and other surface treatments. *Journal of Applied Physics*, **53** (5), 3576–3580.

15 Schoenfelder, S., Bagdahn, J., and Petzold, M. (2011) Mechanical characterisation and modelling of thin chips, in *Ultra-Thin Chip Technology and Applications* (ed. J. Burghartz), Springer.

16 Yeung, B. and Lee, T.-Y.T. (2003) An overview of experimental methodologies and their applications for die strength measurement. *IEEE Transactions on Components and Packaging Technologies*, **26** (2), 423–428.

17 Hawkins, G. *et al.* (1987)) Measurement of silicon strength as affected by wafer back processing. 25th Annual Reliability Physics Symposium, San Diego, CA, USA, April 1987.

18 Tsai, M.Y. and Lin, C.S. (2005) Determination of silicon die strength. Proceedings of the 55th Electronic Components and Technology Conference, 2005 (ECTC 2005), May 31–June 3, 2005.

19 Takyu, S., Kurosawa, T., Shimizu, N., and Harada, S. (2006) Novel wafer dicing and chip thinning technologies realizing high chip strength. Proceedings of the 56th Electronic Components and Technology Conference, 2006 (ECTC 2006), San Diego, CA, USA.

20 Vedantham, R. and Upadhyay, R. (2008) FAVS advances in Si thinning. Proceedings of the Forum "Be-Flexible 2008," Munich, Germany.

21 Landesberger, C., Paschke, C., and Bock, K. (2011) Influence of wafer grinding and etching techniques on the fracture strength of thin silicon substrates. *Advanced Materials Research*, **325**, 659.

22 Zhou, L., Eda, H., Shimizu, J., Kamiya, S., Iwase, H., Kimura, S., and Sato, H. (2006) Defect-free fabrication for single crystal silicon substrate by chemo-mechanical grinding. *CIRP Annals: Manufacturing Technology*, **55** (1), 313–316.

23 Sun, Wei, Zhu, W.H., Che, F.X., Wang, C.K., Sun, Anthony Y.S., and Tan, H.B. (2007) Ultra-thin die characterization for stack-die packaging. Proceedings of the 55th Electronic Components and Technology Conference, 2007 (ECTC 2007), pp. 1390–1396.

24 Maeda, N. *et al.* (2010) Development of ultra-thinning technology for logic and memory heterogeneous stack applications. 2010 Symposium on VLSI Technology: Digest of Technical Papers.

25 Sasaki *et al.* Method of dividing a wafer and method of manufacturing a semiconductor device. US Patent 5,888,883.

26 Feil, M. *et al.* Method of subdividing a wafer. US Patent 6,756,288 B1.

27 Takyu, S., Sagara, J., and Kurosawa, T. (2008) A study on chip thinning process for ultra thin memory devices. Proceedings of the 58th Electronics Components and Technology Conference (ECTC 2008), Lake Buena Vista, Fl, USA.

28 Landesberger, C., Klink, G., Schwinn, G., and Aschenbrenner, R. (2001) New dicing and thinning concept improves mechanical reliability of ultra thin silicon. International Symposium on Advanced Packaging Materials: Processes, Properties and Interfaces, 2001, Braselton, GA, USA, March 11–14, 2001.

29 Miyazaki, C., Shimamoto, H., Uematsu, T., and Abe, Y. (2009) Development of wafer thinning and dicing technology for thin wafer. IEEE International Conference on 3D System Integration, 2009 (3DIC 2009), San Francisco, CA, USA.

30 Chen, S., Kuo, T.-Y., Hu, H.-T., Lin, J.-R., and Yu, S.-P. (2005) The evaluation of wafer thinning and singulating processes to enhance chip strength. Proceedings of the 55th Electronic Components and Technology Conference, 2005 (ECTC 2005), pp. 1526–1530.

31 Li, J., Hwang, H., Ahn, E.-C., Chen, Q., Kim, P., Lee, T., Chung, M., and Chung, T. (2007) Laser dicing and subsequent die strength enhancement technologies for ultra-thin wafer. Proceedings of the 57th Electronic Components and Technology Conference, 2007 (ECTC 2007).

32 Lee, Wen Sheng, Khan, N., Kek, J., Chua, H. S., Tsutsumi, Y., Yew, L.C., Ho, Soon Wee, Eipa, M., Vempati, S., Kripesh, V., and Sundaram, V. (2009) Ultra thinning of wafer for embedded wafer packaging. 11th Electronics Packaging Technology Conference, 2009 (EPTC 2009).

33 Heinze, P., Amberger, M., and Chabert, T. (2008) Perfect chips: chip-side-wall stress relief boosts stability. *Future Fab International*, **25**, 111–117.

34 Schönfelder, S., Ebert, M., Landesberger, C., Bock, K., and Bagdahn, J. (2007) Investigations of the influence of dicing techniques on the strength properties of thin silicon. *Microelectronics Reliability*, **47**, 168–178.

35 Matsubara, N., Windemuth, R., Mitsuru, H., and Atsushi, H. (2012) Plasma dicing technology. Proceedings of the 4th Electronics Systems Integration Technology Conference, 2012 (ESTC 2012).

36 Schmitt, F. and Zernack, M. (2011) Thin wafer manufacturing and handling using low cost carriers, in *Ultra-Thin Chip Technology and Applications* (ed. J.N. Burghartz), Springer.

37 Landesberger, C., Scherbaum, S., and Bock, K. (2011) Ultra-thin wafer fabrication through dicing by thinning, in *Ultra-Thin Chip Technology and Applications* (ed. J.N. Burghartz), Springer.

38 Kumagai, M., Uchiyama, N., Ohmura, E., Sugiura, R., Atsumi, K., and Fukumitsu, K. (2007) Advanced dicing technology for semiconductor wafer: stealth dicing. *IEEE Transactions on Semiconductor Manufacturing*, **20** (3), 259–265.

17
Via Reveal and Backside Processing

Mitsumasa Koyanagi and Tetsu Tanaka

17.1
Introduction

Three-dimensional LSIs based on wafer or chip stacking using the through-silicon via (TSV) have been attracting considerable attention since we can now significantly reduce the wiring length, the wiring capacitance, and the chip size and increase the I/O data bandwidth, and hence dramatically improve the signal processing speed and decrease the power consumption [1–16]. However, there are still many issues that should be solved before starting volume production [17–19]. The wafer or chip stacking using TSVs and microbumps is the key for fabricating 3D LSIs. A wafer or chip has to be thinned from the backside before stacking. Therefore, the backside processing before stacking is very important for fabricating reliable 3D LSIs. So far various kinds of 3D integration technologies have been proposed. However, two kinds of technologies, namely, the via-middle (via-first) process and the back-via process, are considered to be promising as 3D technologies for mass production. In this chapter, the backside processing in the via-middle technology and the back-via technology are discussed.

17.2
Via Reveal and Backside Processing in Via-Middle Process

Three-dimensional LSI fabrication process flow using a via-middle process is shown in Figure 17.1. The deep Si trench for TSV is formed by using deep reactive ion etching (DRIE) after the transistor formation in the via-middle 3D technology, as shown in Figure 17.1 (step 1). Then the oxide liner is formed inside the trench by plasma-enhanced chemical vapor deposition (PECVD) or O_3-CVD and the thin barrier metal layer and copper (Cu) seed layer are deposited onto the oxide liner by physical vapor deposition (PVD), which is followed by the Cu electroplating, as shown in Figure 17.1 (step 2). The barrier metal layer, Cu seed layer, and electroplated Cu layer at the surface are removed by chemical–mechanical polishing (CMP) to form the TSV, as shown in Figure 17.1 (step 3). After the TSV

Handbook of 3D Integration: 3D Process Technology, First Edition.
Edited by Philip Garrou, Mitsumasa Koyanagi, and Peter Ramm.
© 2014 Wiley-VCH Verlag GmbH & Co. KGaA. Published 2014 by Wiley-VCH Verlag GmbH & Co. KGaA.

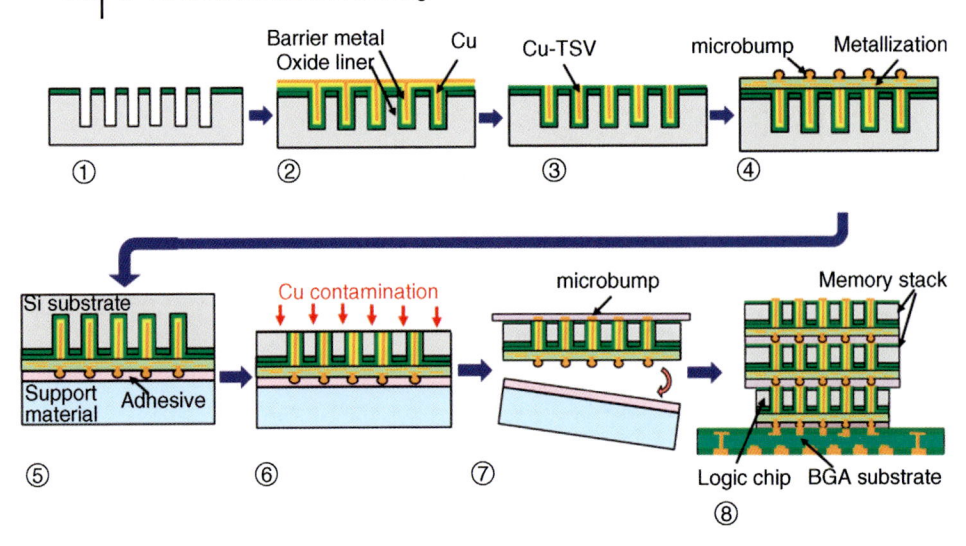

Figure 17.1 Three-dimensional integration process flow by via-middle technology. (1) Si deep RIE. (2) Oxide liner deposition, barrier metal deposition, and Cu electroplating. (3) Cu CMP. (4) BEOL/metal microbump formation. (5) Temporary bonding to support material. (6) Thinning of Si substrate; exposing Cu at TSV base by Si etching. (7) Formation of backside microbump; detaching of support material and dicing. (8) Multilayer stacking of logic/memory onto BGA substrate.

formation, multilevel metallization layers are formed at the back-end-of-line (BEOL) process and metal microbumps are formed onto them, as shown in Figure 17.1 (step 4). Such an LSI wafer with TSVs and metal microbumps is flipped and temporarily bonded to a support material, and thinned from the backside, as shown in Figure 17.1 (steps 5 and 6). Backside metal microbumps are formed on this thinned wafer, and then the thinned wafer is diced into chips after detaching the support material, as shown in Figure 17.1 (step 7). By repeating these sequences, we can easily obtain a 3D LSI, as shown in Figure 17.1 (step 8), where a stacked memory chip and a logic chip are stacked onto a ball grid array (BGA) substrate. This via-middle technology was widely employed in the early stage of 3D technology development. However, there were serious issues in this technology in that the Si substrate was easily contaminated by Cu impurities during the Si thinning (step 6). Both the backside of Si substrate and the base of Cu-TSV are simultaneously exposed in the Si thinning process shown in Figure 17.1 (step 6). As a result, the backside surface of the Si substrate is contaminated by Cu impurities during the CMP process for Si thinning, as shown in Figure 17.2. It is very difficult to completely eliminate Cu impurities from the Cu contaminated surface by cleaning once the Si surface is contaminated by Cu. These Cu impurities remained at the Si surface easily diffuse into Si substrate and generate the deep level at a mid-gap in the Si energy band at a relatively low temperature. It is well known that such a deep level acts as a generation–recombination center for

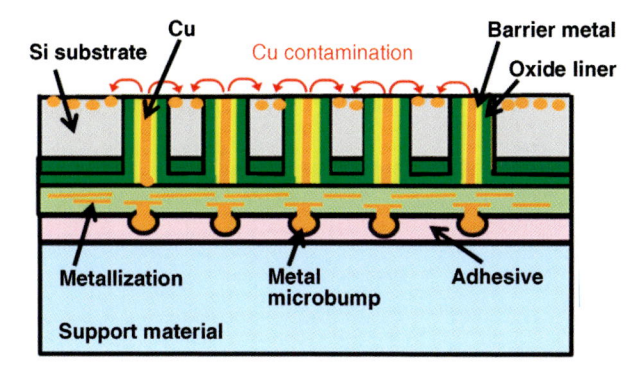

Figure 17.2 Cu contamination during backside processing in via-middle technology.

electrons and holes and degrades the carrier lifetime. We have evaluated the influences of Cu contamination on the carrier lifetime by a capacitance–time (C–t) method, the details of which are described in Chapter 24. The C–t plots after the Cu diffusion from the backside of Si substrate are shown in Figure 17.3 [20]. A relatively high step voltage is applied to the MOS capacitor in the C–t method. The capacitance value of the MOS capacitor suddenly deceases just after applying the step voltage, then gradually increases and finally reaches a final value C_f in the steady state of inversion condition at the Si surface. The time t_f to reach the final value is important in the C–t method, which is related to the carrier generation lifetime: the smaller the t_f, the shorter the carrier generation lifetime. The result of Figure 17.3 was obtained by intentionally diffusing Cu from the backside surface of the Si substrate at 200 °C. It is clear in the figure that the t_f rapidly decreases as the annealing time for Cu diffusion increases even at the low temperature of 200 °C. Thus, Cu atoms rapidly diffuse in the Si substrate and even in Si oxide.

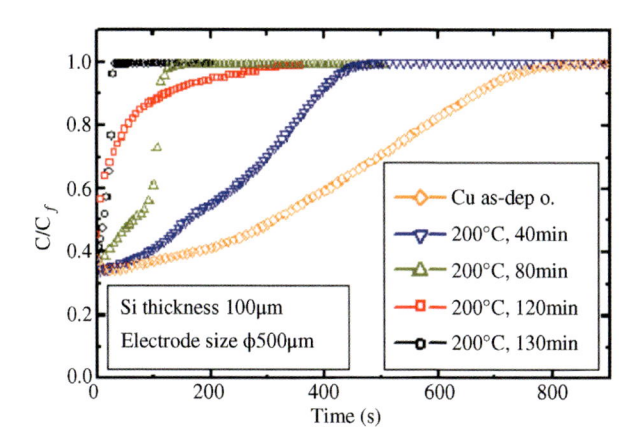

Figure 17.3 Degradation of C–t curves due to Cu contamination from the backside.

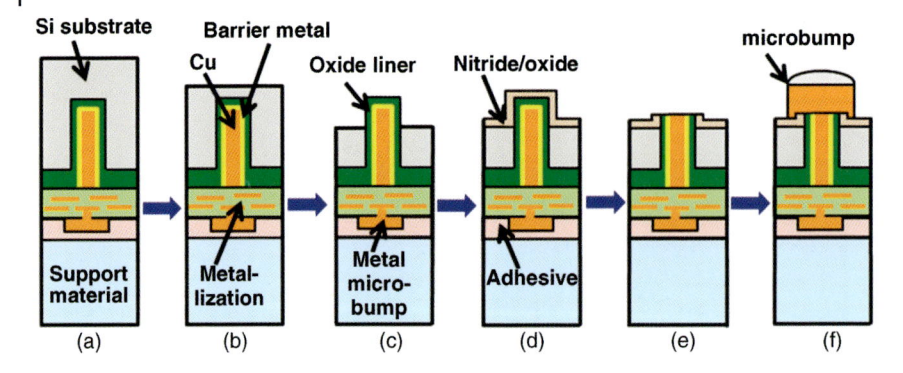

Figure 17.4 Via reveal process to avoid Cu contamination at the backside (CMP method). (a) Temporary bonding to support material. (b) Thinning of Si substrate by CMP. (c) Thinning of Si substrate by plasma etching. (d) Deposition of nitride/oxide film by PE-CVD. (e) Exposing Cu at TSV base by CMP. (f) Formation of backside microbump.

Therefore, it is very important to avoid Cu contamination during Si thinning in the 3D integration technology.

The via-middle process to avoid Cu contamination is shown in Figure 17.4 [21] where the Si substrate thinning by mechanical grinding and CMP is stopped just before exposing the base of the TSV, as shown in Figure 17.4b. Then the base of the TSV is exposed by plasma etching of the Si substrate, as shown in Figure 17.4c, which has a high etching selectivity to the oxide. After that, double layers of silicon nitride and oxide are deposited on the backside surface by PECVD at a relatively low temperature, and the top portion of the TSV base is removed by CMP, as shown in Figure 17.4d and e. Then Cu/Sn microbumps are formed on the exposed TSV base, as shown in Figure 17.4f. In this process, the backside surface of the Si substrate is protected by double layers of silicon nitride and oxide when Cu of TSV base is exposed by CMP. It is well known that the silicon nitride is an effective Cu diffusion barrier. Therefore, even if the backside surface is contaminated by Cu during CMP, Cu cannot easily diffuse through the silicon nitride layer to the Si substrate.

Another via-middle process to avoid the Cu contamination is shown in Figure 17.5 [22] where the photoresist etch-back method is used in removing the top portion of the TSV base instead of the CMP. After the TSV base is exposed by plasma etching of Si substrate and double layers of silicon nitride and oxide are deposited, the photoresist is spin-coated on the backside surface and then etched back to expose the top portion of TSV base, as shown in Figure 17.5c and d. Then double layers of silicon nitride and oxide on the top portion of the TSV base are removed by plasma etching or combination etching of wet etching and plasma etching, and finally Cu/Sn microbumps are formed on the exposed TSV base, as shown in Figure 17.5e and f. In this process, the backside surface of Si substrate is covered by the photoresist and double layers of silicon nitride and oxide. In addition, the backside surface is less contaminated by Cu because the Cu of the TSV base is exposed by plasma etching. Cu particles removed from the top part of the TSV base stuck on the

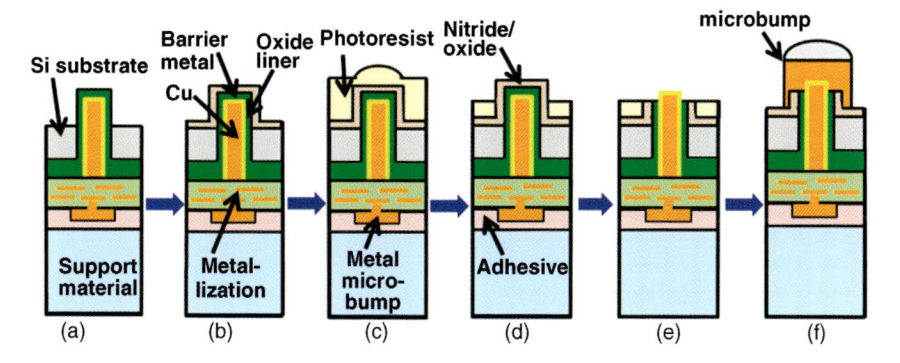

Figure 17.5 Via reveal process to avoid Cu contamination at the backside (resist lift-off method). (a) Thinning of Si substrate by plasma etching. (b) Deposition of nitride/oxide film by PE-CVD. (c) Coating of photoresist. (d) Photoresist etch-back. (e) Exposing barrier metal/Cu at TSV base by plasma and wet etching. (f) Formation of backside microbump.

backside surface when the CMP are used in exposing the Cu of the TSV base while fewer Cu particles remain on the backside and are lifted off during removing the photoresist in the photoresist etch-back method. A photograph of Cu-TSV after exposing the top of the TSV base by the photoresist etch-back method is shown in Figure 17.6. It is clearly seen in the figure that the Cu of TSV base is exposed.

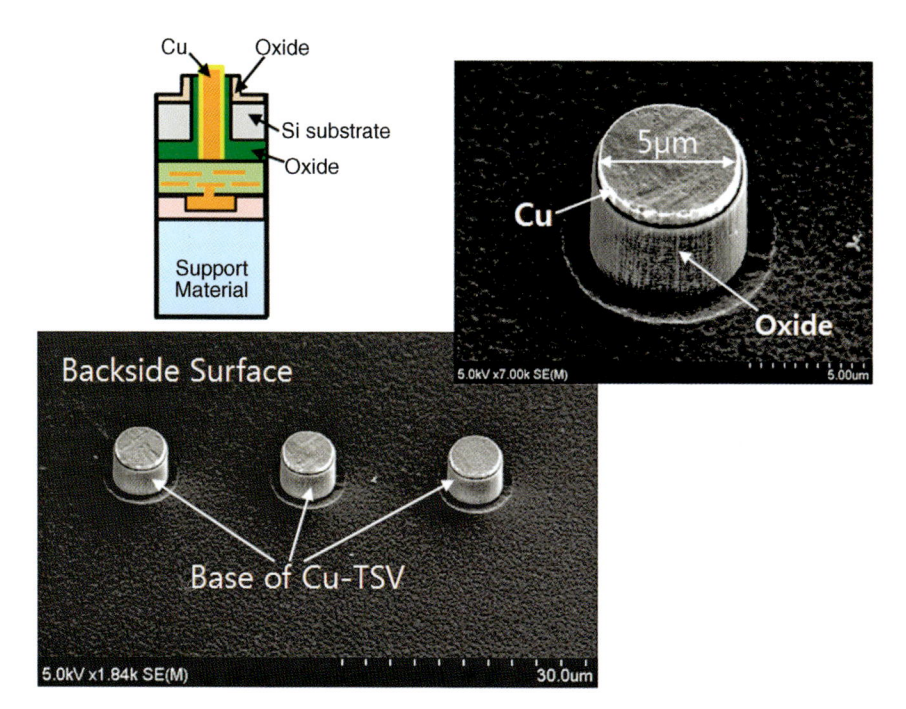

Figure 17.6 Micrographs of Cu-TSV base revealed by resist lift-off method.

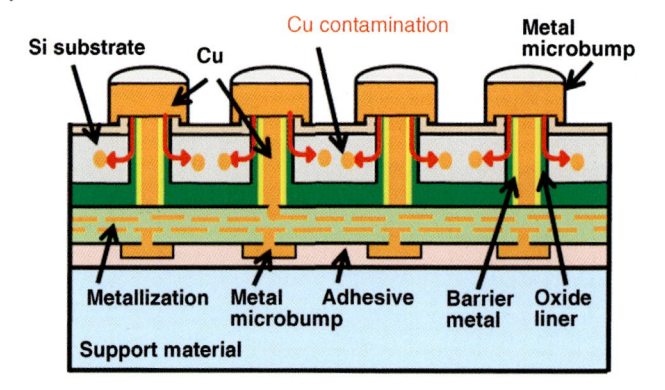

Figure 17.7 Cu diffusion along the oxide liner from the backside.

We can expect that Cu contamination at the backside surface is significantly reduced by employing the via-middle processes described in Figures 22.4 and 22.5. However, there is still a possibility that the Si substrate is contaminated by Cu impurities since Cu impurities can easily diffuse along the TSV oxide liner and then to the Si substrate, as shown in Figure 17.7. It is noted that the exposed oxide liner at the TSV base is highly contaminated by Cu during CMP in the via-middle process described in Figure 17.4. Furthermore, Cu microbumps contact with the TSV oxide liner in both processes, as shown in Figure 17.7. Therefore, Cu atoms might diffuse along the TSV oxide liner from Cu microbumps. To prevent such Cu diffusion, a thin metal barrier layer such as Ta, TaN, or TiN should be inserted underneath Cu microbumps.

17.3
Backside Processing in Back-Via Process

Three-dimensional LSI fabrication process flow using a back-via process is shown in Figure 17.8. TSVs are formed after completing the front-end-of-line (FEOL) and BEOL of CMOS process and thinning the Si substrate. First of all, a thick LSI wafer is glued to a support material and then thinned from the backside by mechanical grinding and CMP, as shown in Figure 17.8 (steps 1 and 2). Then a silicon nitride/oxide stacked layer is deposited by PECVD followed by the deep trench formation in the thinned Si substrate through the silicon nitride/oxide stacked layer by DRIE, and the oxide liner is formed within the Si trench, as shown in Figure 17.8 (step 3). After the bottom oxide of Si trench is removed by DRIE, the Si trench is filled with Cu by electroplating after a thin metal barrier layer and a Cu seed layer are deposited into the trench by PVD, and the electroplated Cu layer, Cu seed layer, and metal barrier layer on the backside surface are removed by CMP, as shown in Figure 17.8 (steps 4 and 5). Then metal microbumps are formed on the backside and the thinned wafer with TSVs and microbumps is diced into chips after

17.3 Backside Processing in Back-Via Process | 233

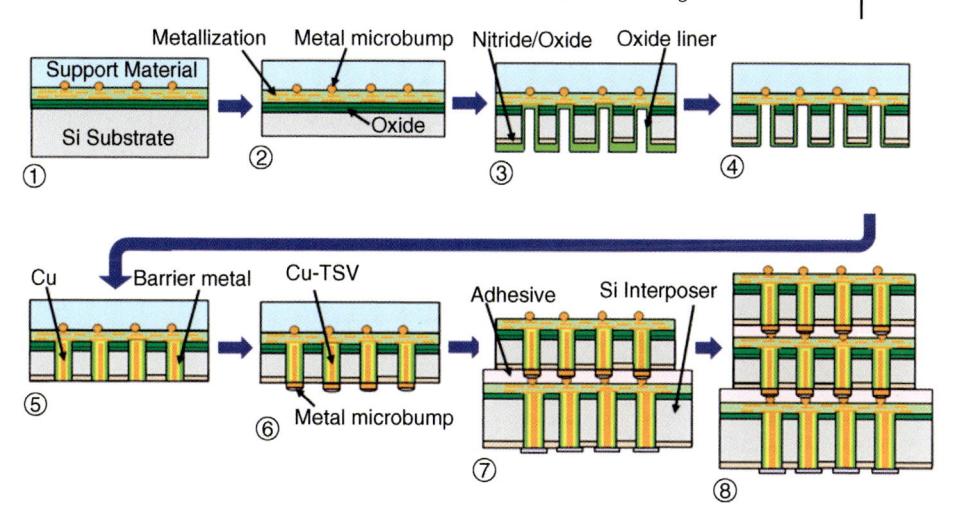

Figure 17.8 Three-dimensional integration process flow by back-via technology. (1) Temporary bonding to support material. (2) Thinning of Si substrate. (3). Nitride/oxide deposition; deep RIE; oxide deposition. (4) RIE of oxide at trench bottom. (5) Barrier metal deposition; Cu electroplating; Cu CMP. (6) Formation of backside microbump. (7) Detaching of support material and dicing; stacking onto Si interposer. (8) Multilayer stacking onto Si interposer.

detaching the support material, as shown in Figure 17.8 (steps 6 and 7). These chips are eventually stacked on a Si interposer to fabricate a 3D LSI, as shown in Figure 17.8 (step 8).

The processing temperature is limited below 300 °C in the back-via process because the backside processing is performed after the LSI wafer is glued to the support material using the adhesive, which has a low thermal tolerance. Therefore, usually, it is not easy to fabricate TSVs with a small diameter and high aspect ratio by the back-via process since it is difficult to conformally deposit the oxide liner into the Si trench with a small diameter and high aspect ratio in a low temperature. A cross-sectional scanning electron microscope (SEM) micrograph of Cu-TSV fabricated by the back-via process is shown in Figure 17.9, which was taken after the Si trench was filled by electroplating with Cu from the backside [23]. It is clearly seen that a narrow Si trench with the diameter of 5 µm and the depth of 30 µm is completely filled with Cu by electroplating since we succeeded in conformally depositing the oxide liner. It is also clear in the figure that Cu–TSV contacts with the first-level metallization (M1).

The Cu contamination is still a serious issue even in the back-via process technology. The backside surface and the exposed oxide liner at the top part of the TSV base are contaminated by Cu during Cu CMP described in Figure 17.8 (step 7). The Cu impurities on the backside surface might be protected by the backside silicon nitride, but Cu impurities on the exposed oxide liner easily diffuse along the oxide liner to the Si substrate. The process step to remove the bottom oxide of Si trench, as shown in Figure 17.8 (step 4) might be also a problem since

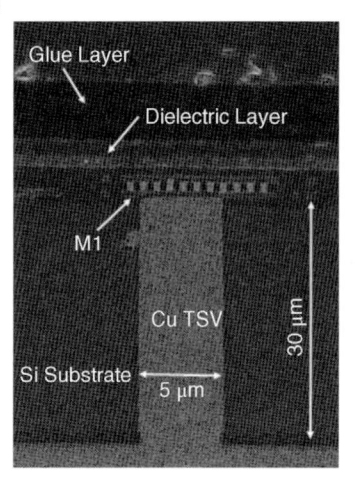

Figure 17.9 SEM cross-sectional view of Cu-TSV formed by back-via process.

the first level metallization material is slightly etched when the bottom oxide of Si trench is over-etched. The oxide liner on the Si trench sidewall is contaminated by Cu impurities, which are generated by overetching of the bottom oxide when Cu is used as the first level metallization material. These Cu impurities easily diffuse through the oxide liner to the Si substrate. To avoid such Cu contamination, a thin metal barrier layer should be inserted underneath the first level metallization and the overetching should be completely stopped by this thin metal barrier layer. Another possibility of Cu contamination in the back-via process is due to the Cu diffusion from Cu microbumps, as already described in Figure 17.7. The Cu diffusion from Cu microbumps is the problem in both the via-middle process and the back-via process.

17.4
Backside Processing and Impurity Gettering

To fabricate 3D LSIs, the LSI wafer or chip should be thinned from the backside. This means that devices located near the Si front surface come to be more easily affected by the backside condition of the thinned Si substrate in 3D LSIs. Specifically, the Cu contamination from the backside in the thinned Si substrate is the crucial issue in 3D LSIs [17]. The major part of such Cu contamination might be protected by the silicon nitride film deposited on the backside surface. However, this nitride film is usually deposited by a low-temperature PECVD, and hence, the ability to prevent the Cu diffusion is less compared with the Si_3N_4 film deposited by high-temperature low-pressure (LP)-CVD at around 650 °C. In addition, there might exist pinholes in such nitride film deposited at a low temperature. It is very difficult to completely remove these pinholes in volume production. Therefore the impurity gettering is very important in 3D LSI fabrication. It is well known that the intrinsic gettering (IG) is very effective at trapping and immobilizing Cu impurities in the Si substrate. Epiwafers with p-on-p$^+$ structure are also effective at preventing the Cu diffusion from the backside. However the IG and the p-on-p+ wafer might not be used in some applications. In such cases, another method, namely, the extrinsic gettering (EG) becomes important.

Crystal defects introduced into the backside surface act as the gettering sites for Cu impurities in the EG. Various kinds of EGs using the mechanical grinding, ion implantation, laser irradiation, and so forth have been proposed. It is challenging to determine how to employ such EGs in 3D LSIs. It is known that the mechanical grinding is very effective in introducing the damage and producing the crystal defect region (EG region) at the backside. However, the damage should be optimized in the EG. Four kinds of EG methods, UltraPoliGrind (UPG), dry polish (DP), Poligrind (PG), and #2000 mechanical grinding, are compared in Figure 17.10 where the width of the crystal defect region and the length of stress-field region from the backside surface measured by HRTEM are plotted [24]. The stress-field region means the region where the mechanical stress exists. As is obvious in the figure, #2000 mechanical grinding gives rise to the largest defect region width and stress-field region length, which means too much damage is introduced at the backside. We have confirmed by the C–t method that the carrier lifetime seriously degraded because of such large damages in the Si wafer treated by #2000 mechanical grinding, which enhance the Cu diffusion. Conversely, UPG and DP produce moderate defect region width and stress-field region length. As shown in Figure 28.6, we have confirmed that the carrier lifetime measured in the DP-treated wafer exhibited the smallest degradation even

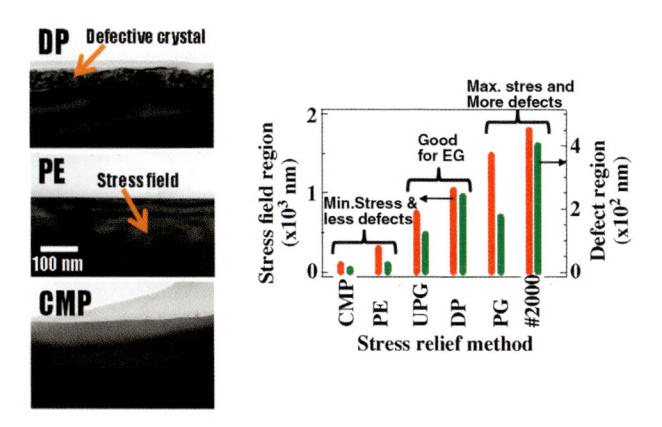

Figure 17.10 Defect region depth and stress-field region length after various backside treatments. (a) Stress field region in the HRTEM image for stress-relieved wafers. (b) Remnant stress and defect depth in various stress-relieved wafers.

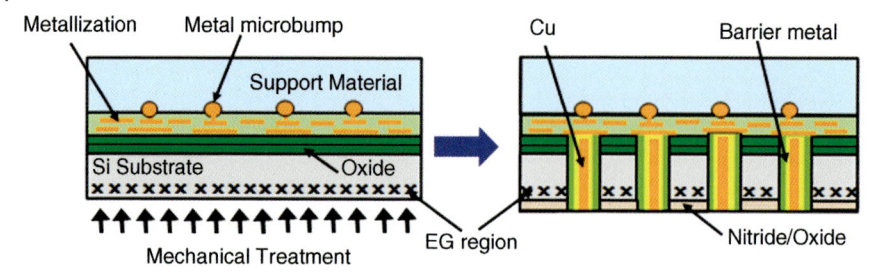

Figure 17.11 Formation of extrinsic gettering (EG) region at the backside in the back-via process.

after the intentional Cu diffusion. CMP and plasma etching (PE) give the smallest defect region width and stress-field region length, which means that CMP and PE hardly introduce the mechanical damages. However, the carrier lifetimes measured in wafers treated by both CMP and PE are seriously degraded after the intentional Cu diffusion.

From the results described above, we recommend introducing DP treatment after thinning the Si substrate to suppress the Cu diffusion in the back-via process, as shown in Figure 17.11. However, DP treatment cannot be used in the via-middle process because the protrusion of the TSV base occurs at the backside after the final thinning by PE, as shown in Figures 17.4c and 17.5c. Therefore, EG method using a phosphorus ion implantation, Ar ion implantation, or laser irradiation will be useful in the via-middle process, as shown in Figure 17.12. However, we have to be very careful when EG is used in the very thin Si substrate with the thickness of less than 50 μm because the Young modulus rapidly decreases as the Si thickness is reduced, as shown in Figure 17.13 [25]. This means that stacking of very thin Si dies (with less than 50 μm thickness) might cause the breakage of dies because the hardness significantly decreases in thin Si dies with a smaller Young modulus. It is noted that the PE-treated thin die also exhibits significant decrease of the Young modulus as the thickness decreases to less than 50 μm.

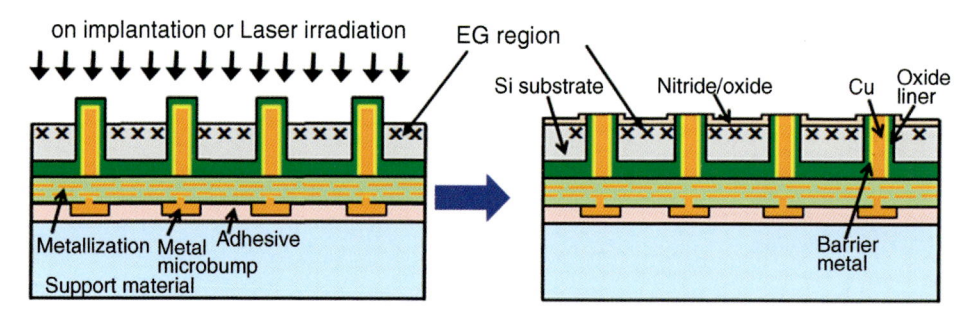

Figure 17.12 Formation of extrinsic gettering (EG) region at the backside in the via-middle process.

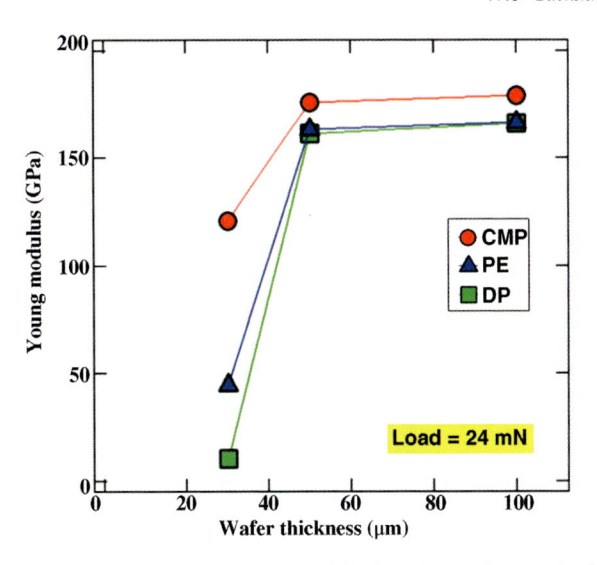

Figure 17.13 Change of Young modulus by reducing the Si wafer thickness.

17.5
Backside Processing for RDL Formation

A redistribution layer (RDL) formation is also included in the backside processing. An RDL is important because we can introduce the flexibility in connections between the upper die and lower die, which have different design rules for the wiring and microbumps. The Cu-RDL is formed on the backside by Cu electroplating after the TSV base Cu base is exposed in both the via-middle process and the back-via process. A thin metal barrier layer is inserted underneath the Cu-RDL. Cu/Sn microbumps are formed after the formation of Cu-RDL. Therefore, metal microbumps are allowed to form at the different positions from those of TSVs because we can electrically connect metal microbumps with TSVs by RDLs. A cross-sectional structure and a photomicrograph of the backside surface in the thinned Si die with Cu-TSVs, Cu-RDLs, and Cu/Sn microbumps is shown in Figure 17.14, which was fabricated by the back-via process [23]. An SEM cross-sectional view of a Cu/Sn microbump with the diameter of 15 μm is also shown in Figure 17.14. It is clearly observed in the figure that Cu/Sn microbumps are formed not on the TSV bases but at positions apart from TSVs, and microbumps are connected with TSVs by RDLs. Thus, the RDL allows us to introduce a flexible layout of TSVs and microbumps. We have fabricated a 3D stacked image sensor by the back-via process with RDLs. Figure 17.15 shows a cross-sectional photomicrograph of a fabricated 3D stacked image sensor where the image sensor chip (CIS), the analog circuit chip (CDS), and the ADC array chip (ADC), are stacked on the Si interposer [23]. The chip size is 5 mm × 5 mm and the chip thickness is 40 μm. The CIS chip and the CDS chip were fabricated by the 0.18 μm CMOS technology, and

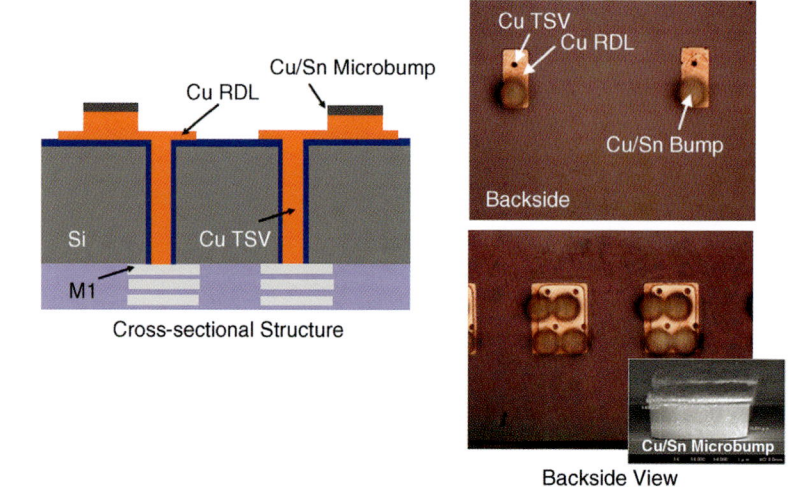

Figure 17.14 Cross-sectional structure and a photomicrograph of backside surface in the thinned Si die with Cu-TSVs, Cu-RDLs, and Cu/Sn microbumps.

the ADC chip was fabricated by the 90 nm CMOS technology. Cu-TSVs, Cu/Sn microbumps, and Cu-RDLs are formed in these chips by the back-via process. The position of the microbump is not always aligned to that of TSV because they are connected with TSVs by RDLs. Therefore, TSVs are not simultaneously observed at all three chips in the cross-sectional view of Figure 17.15. An SEM cross-sectional view of the ADC chip is also shown in Figure 17.15. It is clearly seen that TSVs with the diameter of 5 µm contact with the first-level Cu metallization and front-side Cu/Sn microbumps are formed on the multilevel metallization. We have confirmed that this 3D image sensor successfully operates.

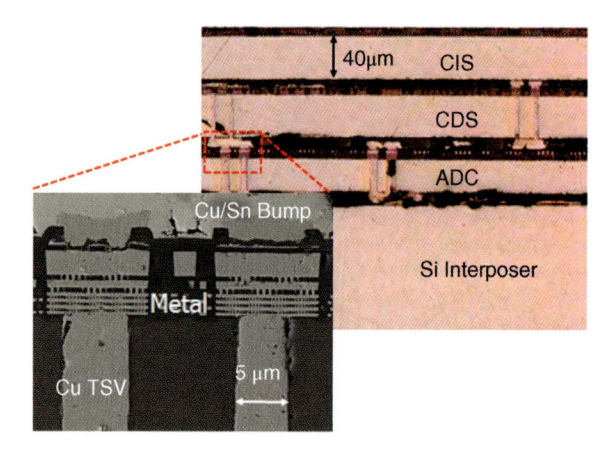

Figure 17.15 Cross-sectional photomicrograph of a fabricated 3D stacked image sensor.

References

1 Koyanagi, M., Kurino, H., Lee, K.W., Sakuma, K., Miyakawa, N., and Itani, H. (1998) Future system-on-silicon LSI chips. *IEEE Micro*, **18** (4), 17–22.

2 Matsumoto, T., Kudoh, Y., Tahara, M., Yu, K.-H., Miyakawa, N., Itani, H., Ichikizaki, T., Fujiwara, A., Tsukamoto, H., and Koyanagi, M. (1995) Three-dimensional integration technology based on wafer bonding technique using micro-bumps. Extended Abstract of the International Conference on Solid State Devices and Materials, pp. 1073–1074.

3 Kurino, H., Lee, K.W., Nakamura, T., Sakuma, K., Hashimoto, H., Park, K.T., Miyakawa, N., Shimazutsu, H., Kim, K.Y., Inamura, K., and Koyanagi, M. (1999) Intelligent image sensor chip with three dimensional structure. Technical Digest International Electron Devices Meeting, 1999 (IEDM 1999), pp. 879–882.

4 Lee, K.W., Nakamura, T., Ono, T., Yamada, Y., Mizukusa, T., Hashimoto, H., Park, K.T., Kurino, H., and Koyanagi, M. (2000) Three-dimensional shared memory fabricated using wafer stacking technology. Technical Digest International Electron Devices Meeting, 2000 (IEDM 2000), pp. 165–168.

5 Koyanagi, M., Nakagawa, Y., Lee, K.-W., Nakamura, T., Yamada, Y., Inamura, K., Park, K.-T., and Kurino, H. (2001) Neuromorphic vision chip fabricated using three-dimensional integration technology. Digest of Technical Papers, IEEE International Solid State Circuits Conference, 2001 (ISSCC 2001), pp. 270–271.

6 Ono, T., Mizukusa, T., Nakamura, T., Yamada, Y., Igarashi, Y., Morooka, T., Kurino, H., and Koyanagi, M. (2002) Three-dimensional processor system fabricated by wafer stacking technology. Proceedings of the International Symposium on Low-Power and High-Speed Chips (COOL Chips V), pp. 186–193.

7 Ramm, P., Bonfert, D., Gieser, H., Haufe, J., Iberl, F., Klumpp, A., Kux, A., and Wieland, R. (2001) Interchip via technology for vertical system integration. Proceedings of the IEEE International Interconnect Technology Conference (IITC), pp. 160–162.

8 Burns, J., Mcllrath, L., Keast, C., Lewis, C., Loomis, A., Warner, K., and Wyatt, P. (2001) Three-dimensional integrated circuits for low-power, high-bandwidth systems on a chip. Digest of Technical Papers, IEEE International Solid State Circuits Conference, 2001 (ISSCC 2001), pp. 268–269.

9 Lu, J.-Q., Jindal, A., Kwon, Y., McMahon, J.J., Rasco, M., Augur, R., Cale, T.S., and Gutmann, R.J. (2003) Evaluation procedures for wafer bonding and thinning of interconnect test structure for 3D ICs. Proceedings of the IEEE International Interconnect Technology Conference (IITC), pp. 74–76.

10 Koyanagi, M., Nakamura, T., Yamada, Y., Kikuchi, H., Fukushima, T., Tanaka, T., and Kurino, H. (2006) Three-dimensional integration technology based on wafer bonding with vertical buried interconnections. *IEEE Transactions on Electron Devices*, **53** (11), 2799–2808.

11 Koyanagi, M., Fukushima, T., and Tanaka, T. (2009) High-density through silicon vias for 3-D LSIs. *Proceedings of the IEEE*, **97** (1), 49–59.

12 Kang, U., Chung, H.-J., Heo, S., Ahn, S.-H., Lee, H., Cha, S.-H., Ahn, J., Kwon, D.M., Kim, J.H., Lee, J.-W., Joo, H.-S., Kim, W.-S., Kim, H.-K., Lee, E.-M., Kim, S.-R., Ma, K.-H., Jang, D.-H., Kim, N.-S., Choi, M.-S., Oh, S.-J., Lee, J.-B., Jung, T.-K., Yoo, J.-H., and Kim, C. (2009) 8 Gb 3D DDR3 DRAM using through-silicon-via technology. Digest of Technical Papers, IEEE International Solid State Circuits Conference, 2009 (ISSCC 2009), pp. 130–131.

13 Katti, G., Mercha, A., Van Olmen, J., Huyghebaert, C., Jourdain, A., Stucchi, M., Rakowski, M., Debusschere, I., Soussa, P., Dehaene, W., De Meyer, K., Travaly, Y., Beyne, E., Biesemans, S., and Swinnen, B. (2009) 3D stacked ICs using Cu TSVs and die to wafer hybrid collective bonding. Technical Digest International Electron Devices Meeting, 2009 (IEDM 2009), pp. 357–360.

14 Lin, J.C., Chiou, W.C., Yang, K.F., Chang, H.B., Lin, Y.C., Liao, E.B., Hung, J.P., Lin, Y.L., Tsai, P.H., Shih, Y.C., Wu, T.J., Wu, W.J., Tsai, F.W., Huang, Y.H., Wang, T.Y., Yu, C.L., Chang, C.H., Chen, M.F., Hou, S.Y., Tung, C.H., Jeng, S.P., and Yu, Doug C.H. (2010) High density 3D integration using CMOS foundry technologies for 28 nm node and beyond. Technical Digest International Electron Devices Meeting, 2010 (IEDM 2010), pp. 22–25.

15 Iyer, S.S. (2012) The evolution of dense embedded memory in high performance logic technologies. Technical Digest International Electron Devices Meeting, 2012 (IEDM 2012), pp. 781–784.

16 Sukegawa, S., Umebayashi, T., Nakajima, T., Kawanobe, H., Koseki, K., Hirota, I., Haruta, T., Kasai, M., Fukumoto, K., Wakano, T., Inoue, K., Takahashi, H., Nagano, T., Nitta, Y., Hirayama, T., and Fukushima, N. (2013) A 1/4-inch 8Mpixel back-illuminated stacked CMOS image sensor. Digest of Technical Papers, IEEE International Solid State Circuits Conference, 2013 (ISSCC 2013), pp. 484–485.

17 Murugesan, M., Bea, J.C., Kino, H., Ohara, Y., Kojima, T., Noriki, A., Lee, K.W., Kiyoyama, K., Fukushima, T., Nohira1, H., Hattori2, T., Ikenaga3, E., Tanaka, T., and Koyanagi, M. (2009) Impact of remnant stress/strain and metal contamination in 3D-LSIs with through-Si vias fabricated by wafer thinning and bonding. Technical Digest International Electron Devices Meeting, 2009 (IEDM 2009), pp. 361–364.

18 Mercha, A., Van der Plas, G., Moroz, V., De Wolf, I., Asimakopoulos5, P., Minas, N., Domae, S., Perry, D., Choi1, M., Redolfi, A., Okoro, C., Yang, Y., Van Olmen, J., Thangaraju, S., Sabuncuoglu Tezcan, D., Soussan, P., Cho, J.H., Yakovlev, A., Marchal, P., Travaly, Y., Beyne, E., Biesemans, S., and Swinnen, B. (2010) Comprehensive analysis of the impact of single and arrays of through silicon vias induced stress on high-k/metal gate CMOS performance. Technical Digest International Electron Devices Meeting, 2010 (IEDM 2010), pp. 26–29.

19 Farooq, M.G., G-Abe, T.L., Landers, W.F., Kothandaraman, C., Himmel, B.A., Andry, P.S., Tsang, C.K., Sprogis, E., Volant, R.P., Petrarca, K.S., Winstel, K.R., Safran, J.M., Sullivan, T.D., Chen, F., Shapiro, M.J., Hannon, R., Liptak, R., Berger, D., and Iyer, S.S. (2011) 3D copper TSV integration, testing and reliability. Technical Digest International Electron Devices Meeting, 2011 (IEDM 2011), pp. 143–146.

20 Bea, J., Lee, K., Fukushima, T., Tanaka, T., and Koyanagi, M. (2011) Evaluation of Cu contamination at backside surface of thinned wafer in 3-D integration by transient-capacitance measurement. *IEEE Electron Device Letters*, **32** (1), 66–68.

21 Ramaswami, S. (2011) Transitioning TSV technology to production. 3-D Architectures for Semiconductor Integration and Packaging Conference, December 12–14, 2011.

22 Beyne, E. (2011) Challenges in 3D system integration beyond TSV processing. 3-D Architectures for Semiconductor Integration and Packaging Conference, December 12–14, 2011.

23 Lee, K.-W., Ohara, Y., Kiyoyama, K., Konno, S., Sato, Y., Watanabe, S., Yabata, A., Kamada, T., Bea, J.-C., Hashimoto, H., Murugesan, M., Fukushima, T., Tanaka, T., and Koyanagi, M. (2012) Characterization of chip-level hetero-integration technology for high-speed, highly parallel 3D-stacked image processing system. Technical Digest International Electron Devices Meeting, 2012 (IEDM 2012), pp. 785–788.

24 Murugesan, M., Kino, H., Hashiguchi, A., Miyazaki, C., Shimamoto, H., Kobayashi, H., Fukushima, T., Tanaka, T., and Koyanagi, M. (2011) High density 3D LSI technology using W/Cu hybrid TSVs. Technical Digest International Electron Devices Meeting, 2011 (IEDM 2011), pp. 139–1142.

25 Murugesan, M., Kobayashi, H., Shimamoto, H., Yamada, F., Fukushima, T., Bea, J.C., Lee, K.W., Tanaka, T., and Koyanagi, M. (2012) Minimizing the local deformation induced around Cu-TSVs and CuSn/InAu-microbumps in high-density 3D-LSIs. Technical Digest International Electron Devices Meeting, 2012 (IEDM 2012), pp. 657–660.

18
Dicing, Grinding, and Polishing (Kiru Kezuru and Migaku)
Akihito Kawai

18.1
Introduction

For 3D IC with through-silicon via (TSV), because the Cu via-forming process decreases as wafers become thinner, wafer-thinning technology is important to overall cost. This section describes grinding, polishing, and dicing, and covers TSV topics regarding thinning and singulation.

18.2
Grinding and Polishing

18.2.1
Grinding General

18.2.1.1 Grinding Method
"Grinding," also known as back grinding (BG), is the process that grinds the backside of device wafers for assembly into packages. Figures 18.1 and 18.2 show two different methods, creep feed and in-feed. Creep feed is used for silicon wafers up to 6 in. and in-feed is a major process for silicon wafers of 8 in. or more.

Creep feed: The wafer on a nonrotating "chuck table," as shown in Figure 18.1, passes straight under a rotating grinding wheel, as shown in Figure 18.3. Usually the creep feed process is divided into three steps, so it needs a larger machine size. It should be noted that thickness control becomes more difficult after the grinding wheel is worn down.
In-feed: The wafer rotates on a chuck table, and the wafer is ground as a rotating grinding wheel is lowered.

For further discussion, we will assume that the grinding used is the in-feed method.

Handbook of 3D Integration: 3D Process Technology, First Edition.
Edited by Philip Garrou, Mitsumasa Koyanagi, and Peter Ramm.
© 2014 Wiley-VCH Verlag GmbH & Co. KGaA. Published 2014 by Wiley-VCH Verlag GmbH & Co. KGaA.

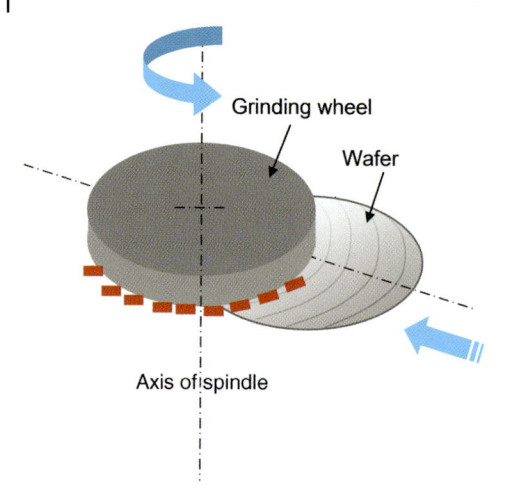

Figure 18.1 Creep feed.

18.2.1.2 Rough Grinding and Fine Grinding

In order to protect the device layer from directly touching the chuck table, prior to grinding, surface protection tape (called BG tape) is attached to the device side of the wafer. Then the wafer is turned over and secured to the chuck table with the wafer backside upward.

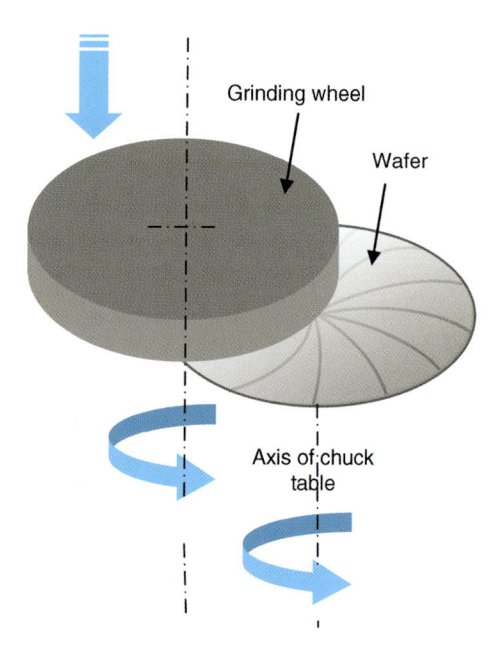

Figure 18.2 In-feed.

Figure 18.3 Grinding wheel.

Normally, silicon grinding is divided into two steps, rough grinding and fine grinding. Rough grinding removes most of the total grinding amount, and fine grinding removes the last 100–40 μm to reach the specified thickness. Rough grinding generally uses a grinding wheel with #320 diamond abrasive grains. It needs strong grinding force to remove a wafer backside layer such as oxides or nitrides. In order to improve productivity, the rough grinding wheel grinds the wafer at a higher process speed. Fine grinding commonly uses a grinding wheel with #2000 diamond grains, which is obviously finer than the rough grinding wheel. It removes the damage caused by the rough grinding at a lower process speed and makes a smoother surface on the backside of the wafer.

18.2.1.3 The Grinder Polisher

A typical fully automatic grinder polisher has three spindles and four chuck tables that are placed on a turntable, as shown in Figure 18.4. The Z1 spindle is for rough grinding, the Z2 spindle is for fine grinding, and the Z3 spindle is for polishing.

Focusing on the process point, the grinding wheel only covers less the half of the arc of the wafer as shown in Figure 18.5. In order to finish grinding at the specified thickness, an in-process contact gauge is used to measure the total thickness of the wafer, including the thickness of the wafer and that of the surface protection tape. Usually a contact gauge with two probes measures both the height of the wafer upper surface and that of the chuck table top surface (Figure 18.6).

Because grinding emits a large quantity of silicon particulate, cleaning of the wafers and the chuck tables by deionized water is required.

18.2.2 Thinning

Focusing on the thinnest wafers, currently wafer thinning up to 30 μm is already in mass production, and evaluations into producing wafers thinner than 30 μm has

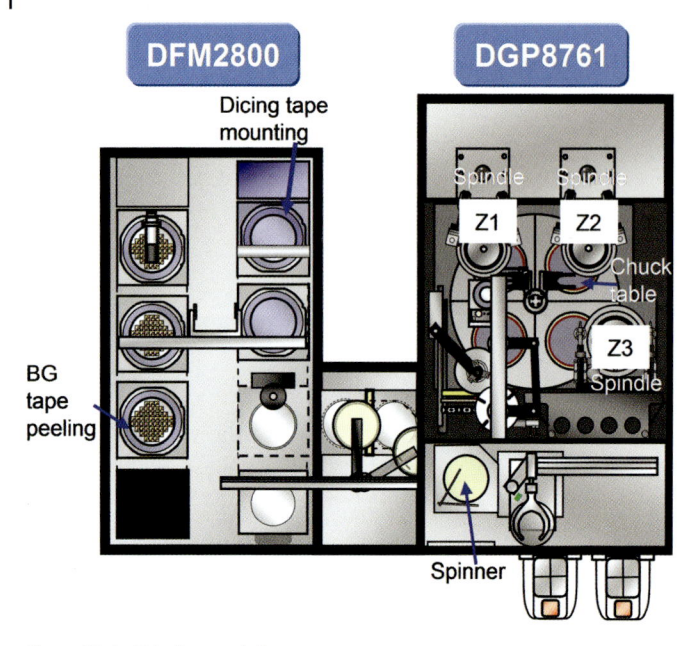

Figure 18.4 Grinding work flow.

also been initiated. A major issue in wafer thinning is wafer breakage. Reasons for breakage are as follows:

Low mechanical strength: As a wafer becomes thinner, the mechanical strength of the wafer decreases. Mechanical strength of the wafer is usually measured as die strength using the bending test.
Wafer warpage: A thick wafer has sufficient mechanical strength to resist the stress generated from the difference of tension between the front side and backside of the wafer. On the contrary, a thin wafer cannot resist the stress and becomes warped.

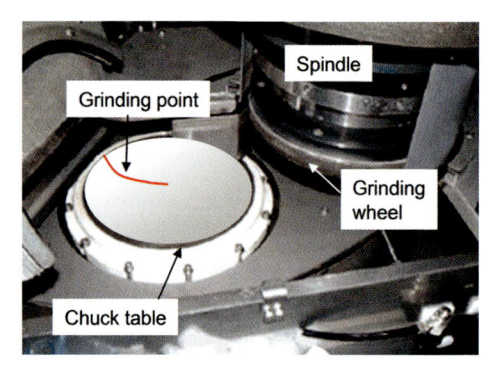

Figure 18.5 Grinding point.

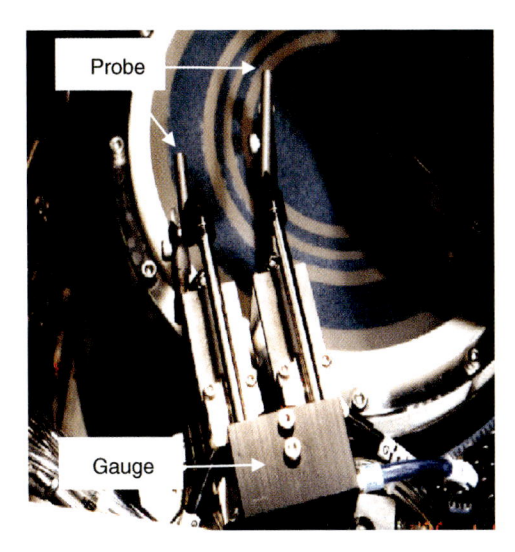

Figure 18.6 In-process gauge.

The thinner the wafer becomes, the larger the warpage. When transferring a thin wafer inside the machine, a warped wafer may contact the tables or transfer parts and this can result in wafer breakage.

Grinding damage: After grinding, subsurface damage such as micro cracking remains on the wafer backside. The depth of the micro cracks is a function of the grit size of the grinding wheel. A rough grinding wheel such as the #320 generates micro cracks deeper than 5 μm, while a fine grinding wheel such as the #2000 generates micro cracks of about 0.2 μm.

Edge chipping: Due to the influence of grinding water or an overaggressive feed speed, the wafer edge area flutters. This may generate edge chipping on the wafer outer circumference.

Another common cause for wafer breakage is silicon particulate that becomes trapped between the tape and wafer.

18.2.2.1 Stress Relief

When the required finished thickness of a ground wafer becomes thinner than 100 μm, the polishing process is introduced. In order to improve die strength and reduce wafer warpage, subsurface damage caused by fine grinding needs to be removed. This damage removal process is called stress relief.

Dry polishing: This is a dry process without using chemical compounds or water, so it is of low cost and environment-friendly. Dry polishing is mainly used for stress relief removing on the order of 2 μm of silicon from the wafer after fine grinding. It is a major process used in the memory device sector. For reference, the thickness of memory device such as DRAM is less than 100 μm.

Chemical–mechanical polishing (CMP): This is a wet process with slurry.
Plasma etching: This process etches silicon by fluoride gas plasma in a dry process.
Wet etching: This process etches silicon with a mixture of boric and nitric acids.

18.2.2.2 Die Attach Film

For stacking multiple dies inside a package, die attach film (DAF) is used instead of adhesive for die bonding. DAF is attached to the backside of the wafer after thinning, and then the silicon wafer and DAF together are blade diced into the specified die size. A uniform DAF amount can be attached to the backside of each die so bond-line thickness is well controlled at die bonding.

18.2.2.3 All-in-One System

In order to reduce the risk of wafer breakage during transferring thin wafers between BG tape peeling equipment and dicing tape mounting equipment, integration of the post-grinding processes is introduced. Grinding, polishing, BG tape peeling, and dicing tape mounting functions are all integrated into one system (see Figure 18.4).

18.2.2.4 Dicing Before Grinding

Dicing before grinding (DBG) reverses the usual process of dicing the wafer after grinding. In the DBG process, first the wafer is grooved. Then, die singulation naturally occurs when the wafer is thinned to the bottom of the groove. The DBG wafer is mounted on dicing tape and the BG tape is peeled off. In DBG, because the wafer is already diced when it is thinned, wafer handling is easier and there is a reduced risk of wafer breakage. Die singulation by back grinding reduces the incidence of backside chipping. In combination with the DAF separation by laser scribing, DAF is used in the DBG process. Currently, DBG is used in the memory device market as the process to produce thinned die (30–100 μm).

18.2.3
Grinding Topics for 3DIC Such as TSV Devices

18.2.3.1 Wafer Support System

Wafer support system (WSS) is the method of attaching support material such as silicon or glass to the front surface of a wafer (device surface). This is another solution to reduce the risk of wafer breakage during transferring and to reduce wafer warpage. TSV devices especially require safe handling for processing on multiple post-grinding equipment. So, commonly a silicon or glass substrate is bonded temporarily to the device surface with temporary bonding adhesives (Figure 18.7). The substrate is attached to the wafer during the several backside processes such as thinning, forming films, and etching. Finally, the substrate is removed and the wafer is mounted on dicing tape. Currently, various methods of removal have been introduced, such as thermal sliding or laser debonding. This is discussed in detail in other chapters of this book.

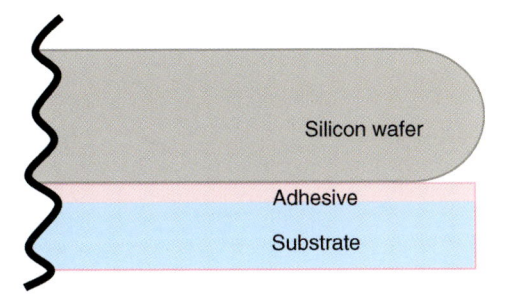

Figure 18.7 Bonded wafer (wafer with substrate).

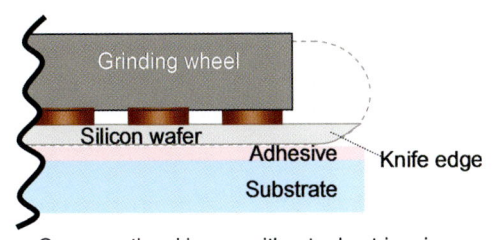

Cross-sectional image without edge trimming

Figure 18.8 Edge profiles without edge trimming.

18.2.3.2 Edge Trimming

Because the original wafer edge profile is beveled, the edge profile of the thinned wafer has a knife shape, as shown in Figure 18.8. Since the knife shape is prone to edge chipping, which is a factor in wafer breakage, a countermeasure called edge trimming is proposed. Edge trimming grooves the wafer edge with a dicing blade as shown in Figure 18.9. The straight wafer edge formed by trimming with a blade is effective to

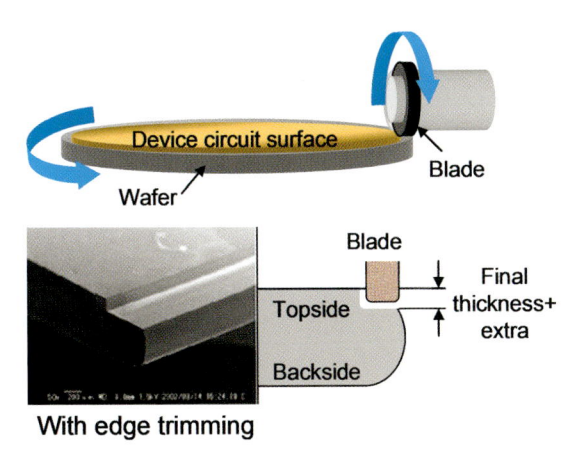

With edge trimming

Figure 18.9 Edge-trimming process diagram.

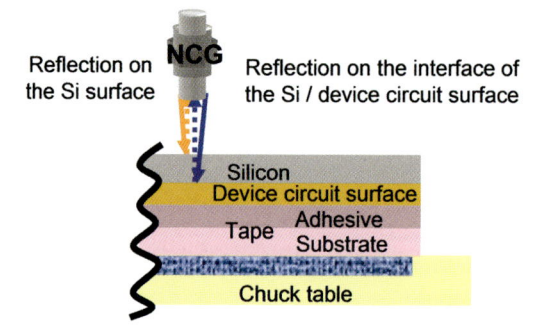

Figure 18.10 Non contact gauge.

prevent edge chipping and improve yield. For TSV devices, it prevents not only edge chipping but also edge contact, which is a factor in wafer breakage, caused from a bonding shift between the wafer and substrate.

18.2.3.3 Grinding to Improve Flatness

As described before, a contact gauge, the conventional wafer thickness measurement during grinding, measures total workpiece thickness. Wafer flatness after thinning is influenced by the thickness variation of materials other than the silicon, such as the back grinding tape, substrate, and temporary bonding adhesives. For TSV, if wafer thickness variation after thinning increases, required process time and cost also increase in the post-grinding processes, such as film forming, etching, and plating. In order to minimize the total process cost of ownership (CoO), wafer thickness control is important.

In order to remove the impact of thickness variation of materials other than the silicon, currently the Non contact gauge (NCG), shown in Figure 18.10, which measures only the bulk silicon thickness is effective. In pursuit of silicon total thickness variation (TTV), "Auto TTV" is introduced. This measures the wafer thickness during fully automated production using the NCG, and it adjusts the chuck table inclination automatically. This makes it possible to make wafer thickness uniform, as shown in Figure 18.11 and to maintain TTV at a constant level during continuous processing.

18.2.3.4 Higher Level of Cleanliness

Wafers with TSV devices require a high level of cleanliness after thinning, for loading to the front-end process equipment such as film-forming equipment and etching equipment. In order to prevent particle adhesion to the ground wafers with TSV devices, a wet process such as CMP is used, and the hydrophilic property of the wafer surface is maintained. By keeping the wafer in a wet condition during transfer to the cleaning process, the elimination of these particles becomes more

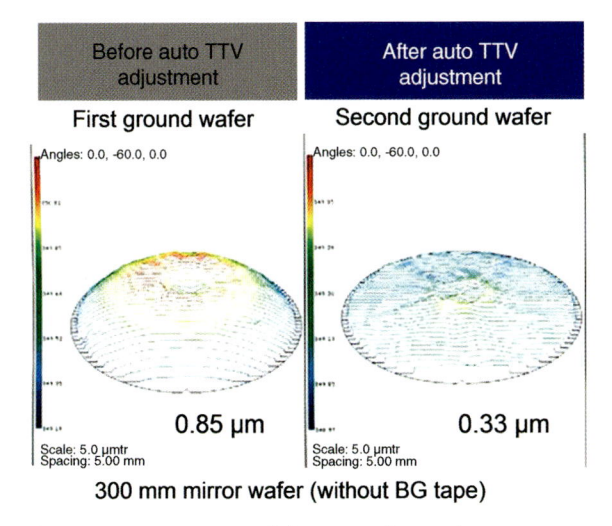

Figure 18.11 Comparison of silicon TTV after grinding.

achievable than when the wafer is dried after processing. To remove silicon particles, organic matter, and metal impurities, additional cleaning such as brush cleaning, chemical cleaning, and DHF cleaning are proposed at the spinner section. In addition, not only for thinning but also for edge trimming of wafers with TSV, in order to prevent wafer cross-contamination, a high level of cleanliness is required.

18.2.3.5 Via Reveal

When thinning wafers with TSV, such as via-middle, copper contamination is a big concern and therefore a countermeasure for copper contamination is required. After thinning the wafer up to the specified thickness, several micrometers above the copper vias, the silicon is removed by dry etching or wet etching to expose the passivation layer that covers the copper vias. A new passivation layer is formed on the whole backside of the wafer to cover the bulk silicon, and the copper vias are exposed by CMP or etching while the copper and bulk silicon are kept separated. In the case of a silicon interposer, which does not have transistors, the risk of electric migration is low, so copper vias can be exposed by polishing the silicon and copper vias simultaneously.

18.2.3.6 Planarization

Generally, the TSV wafer has bumps on the device surface, and it is necessary to make the temporary adhesive thicker (e.g., 60–100 μm) to cover the bumps. This causes unevenness of the temporary adhesive, and this unevenness is transferred to the wafer, so that the silicon wafer TTV tends to deteriorate. In addition to improving the uniformity when applying the temporary bonding adhesives, planarizing the adhesives with a diamond bit, as shown in Figure 18.12, is an

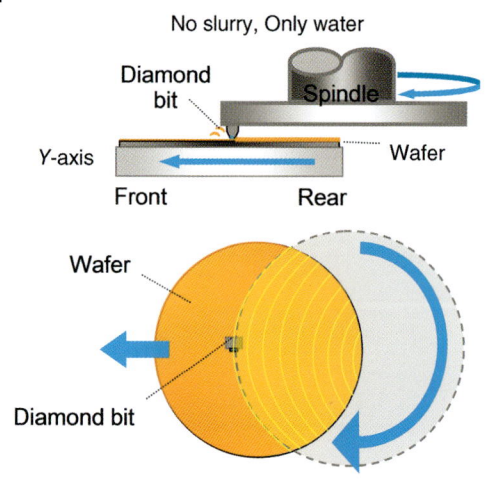

Figure 18.12 Planarization process point.

effective countermeasure for this issue. Planarization of adhesives before thinning may decrease the thickness variation after temporary bonding and improve the TTV for the finished silicon thickness.

18.3
Dicing

18.3.1
Blade Dicing General

18.3.1.1 Dicing Method

Dicing is a process to cut a wafer into individual die after thinning by grinding. The wafer is mounted on a tape frame using dicing tape, with the device side of the wafer upward. The device layer on the silicon wafer front surface is a compound material composed of insulating film, wiring materials, and test element group (TEG). The wafer is cut in straight lines, then rotated 90° and diced again. The separated dies are picked up from the dicing tape at the postprocess called die bonding.

In this section, general dicing methods such as blade dicing, laser grooving, and laser dicing are described. TSV topics on singulation are also discussed.

18.3.1.2 Blade Dicing Point

An ultrathin diamond blade is mounted on an air spindle and rotated at high speed to cut a silicon wafer. During blade dicing, the silicon wafer is impacted by the diamond grains on the blade, and finely broken silicon particles produced by the impacts are removed.

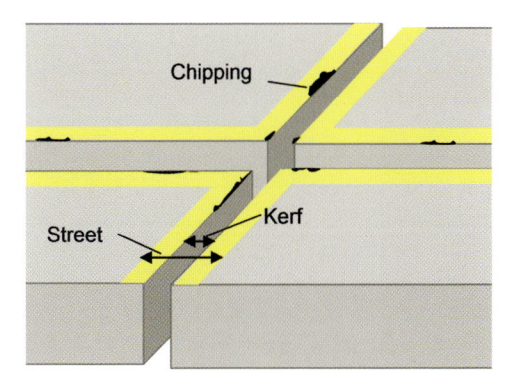

Figure 18.13 Basic items for dicing process.

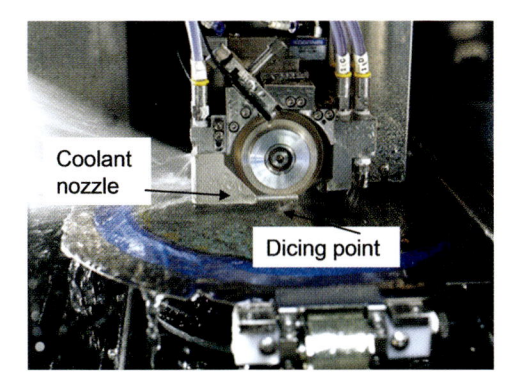

Figure 18.14 Blade dicing point.

The cut position prescribed on the device layer of the wafer is called the street. The actual amount of material that is removed during dicing is called the kerf, usually equal to the blade width (Figure 18.13).

The dicing point is made wet, as water, called wheel coolant, is supplied to the blade through the coolant nozzle (Figure 18.14). The cutting water from the coolant nozzle removes the heat generated at the dicing point and drains away silicon particles, during dicing.

18.3.1.3 Blade

Generally, the blades used for silicon dicing are hub blades manufactured by electroplating with a plating thickness of around 25–35 μm (Figure 18.15). Below are typical features of rough blades and fine blades:

Rough blade composed of rough diamond grit: Because the cutting amount per single rotation of the blade is larger, a strong cutting force and faster feed speed are available. Clogging caused from cutting ductile material such as the TEG and viscous material such as DAF is smaller. Cutting quality, including chipping, deteriorates due to the larger breakage intrinsic to larger diamond grains.

Figure 18.15 Hub blade.

Fine blade composed of finer diamond grit: Cutting quality improves due to the smaller breakage by smaller diamond grains per unit time. Because clogging caused from cutting viscous and ductile material is larger, feed speed needs to be slower.

18.3.1.4 Optimization of Process Control

It is important to balance productivity with cutting quality in terms of optimization of the process results. The combination of faster feed speed and blade with rough diamond grit is effective in increasing productivity. The combination of slower feed speed and blade with fine diamond grit is preferred to improve cutting quality.

To optimize the processed condition of the device wafer, the materials of the device layer must be considered. Cutting an insulating layer harder than silicon is prone to surface chipping. Cutting the wiring materials, such as TEG, softer than silicon, causes blade loading.

18.3.1.5 Dicer

Typically, the fully automatic dicer has two spindles and one chuck table. In order to cut a wafer precisely with high positioning accuracy, alignment is necessary before dicing, as shown in Figure 18.16. In alignment, a unique alignment target is registered to the dicer beforehand. The dicer recognizes the distance between the target and cuts street and then it adjusts the cut direction (θ-axis alignment) by using the target.

18.3.1.6 Dual Dicing Applications

The introduction of dual spindles increases the variety of blade dicing applications (Figure 18.17).

Dual cut: The same type of blade attached to the two spindles can cut two lines simultaneously. In order to improve productivity and reduce costs, dual cut made its first appearance in the mass production of 8 in. wafers.

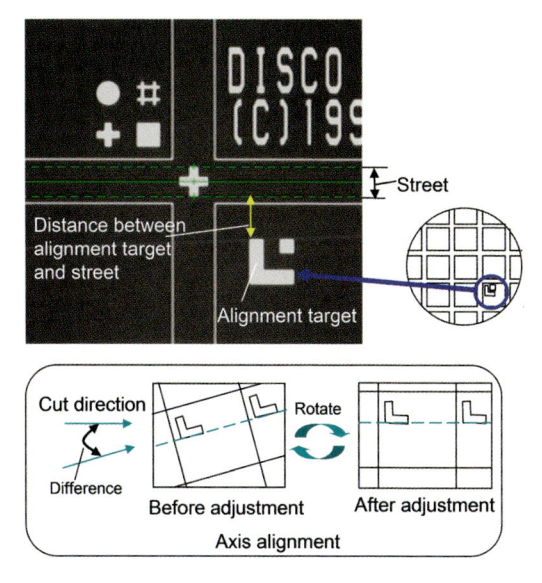

Figure 18.16 Alignment process.

Step cut: A different type of blade attached to each spindle can cut the same line twice. A typical application is when the first step grooves the device layer with a wider blade to remove the TEG, then a second step cuts through the silicon with a narrower blade. The blade type and cutting conditions used can be optimized for each step, one for cutting the device layer and the other for cutting the silicon. Step cut may improve cutting quality, such as front-side and backside chipping, compared to the results achieved when cutting with single blade. Step cut is widely used for high-value-added devices.

18.3.2
Thin Wafer Dicing

As packages become thinner and smaller, the requirements for silicon dicing increase and become more difficult. As described before, the thickness of the thinnest wafer is around 50 μm or less. In order to stack die inside a package, DAF

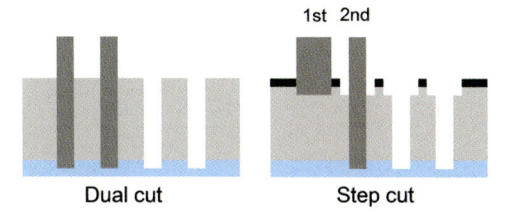

Figure 18.17 Dual dicing applications.

is attached to the backside of the wafer. In the following, typical issues for dicing thin wafers are described:

Die strength: As a die becomes thinner, the mechanical strength of the die decreases. Thin die are not rigid and therefore larger chipping, both as absolute chipping size and as a ratio of the die thickness, tends to occur, and the required quality may not be achieved. Therefore, cutting quality needs to be improved.

Blade loading when cutting TEG and DAF: Step cut is a suitable method for cutting thin wafers, with the second step optimizing cutting conditions for cutting silicon and DAF. As the wafer becomes thinner, the ratio of DAF thickness per total thickness of workpieces increases and the level of difficulty of blade dicing increases.

An example of process conditions for thin wafers, such as 50 μm, attached to DAF follows. In order to reduce surface chipping and prevent the blade from loading, a finer diamond grit blade and slower feed speed are used for cutting the device layer. To keep the blade in good condition, a rough diamond grit blade, which has strong cutting force and greater blade wear, is used for cutting the DAF. Blade dicing has enhanced capability by optimizing process conditions to meet an acceptable cutting quality.

18.3.3
Low-k Dicing

Mechanical strength of the low-k oxide, the interlayer insulator having a low-dielectric constant, decreases as the dielectric constant decreases, and delamination (peeling of the low-k film) may occur from dicing with a conventional blade even at a slower feed speed. As a countermeasure for this issue, laser grooving is effective. First, the wiring layer, including the low-k film, is removed by grooving with laser ablation. Then the rest of the silicon is cut by blade dicing. This method avoids the low-k film delamination, and it also improves productivity by using a fast laser grooving feed speed and increasing the feed speed of blade dicing.

18.3.4
Other Laser Dicing

Laser dicing was developed to be used for challenging materials, based on the increasing difficulties of blade dicing. Two types of laser were introduced, ablation and stealth dicing (SD). In the future, if different size dies are aligned on a single wafer, the laser will be more capable than blade dicing. Noncontinuous processing is realized by both ablation and SD.

18.3.4.1 Ablation
In ablation, a short-pulse laser is focused on the wafer surface. When sufficient energy is absorbed by the irradiated materials, it causes the materials to vaporize

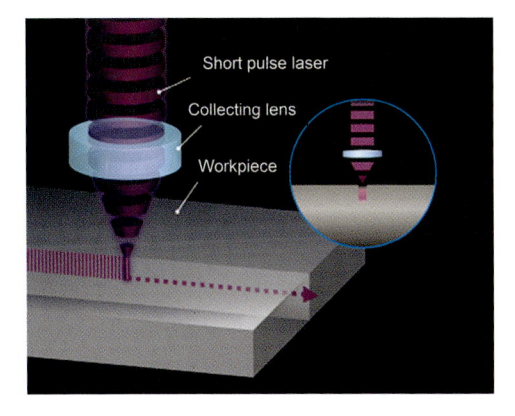

Figure 18.18 Process image of ablation.

(Figure 18.18). Ablation can process a wafer without adding mechanical force and with less thermal distortion than conventional ablation.

18.3.4.2 Laser Full Cut Application

For dicing thin wafers, laser ablation can not only decrease chipping on the die but also improve feed speed. Although conventional laser ablation is likely to have the issue of low die strength, this can be solved by the optimization of energy distribution at the process point by the uniquely developed (beam shaping system (BSS) and a newly developed laser head. Because of these improvements, the die strength from singulation by laser ablation is equal to, even better than, that of die processed by blade dicing (Figures 18.19 and 18.20).

Low-k grooving
Thickness 50 μm

Ablation full cut
Thickness 50 μm

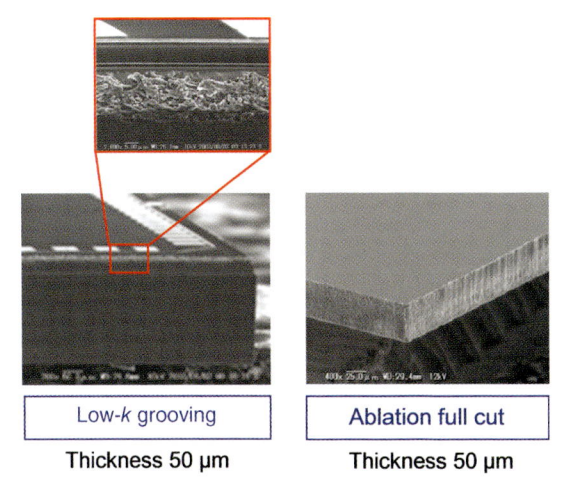

Figure 18.19 Processing example of laser ablation.

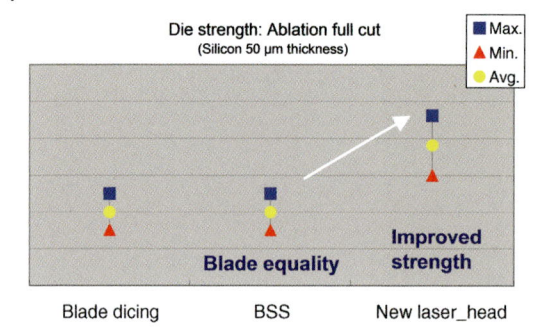

Figure 18.20 Die strength comparison between blade dicing and ablation full cut.

18.3.4.3 Stealth Dicing (SD)

Stealth dicing is the process that forms a modified layer by focusing the laser inside of the wafer, and dividing the wafer into die by tape expansion, as shown in Figure 18.21. Because the SD process only affects material subsequently inside of the wafer, without contacting the wafer front side or backside, it generates no front-side or backside chipping (Figure 18.22). For dicing thin wafers under 100 μm, SD achieves higher die strength compared to blade dicing (Figure 18.23).

It is a perfect dry process requiring no water. SD is suitable for MEMS wafers, which prohibit using cutting water due to the lower mechanical strength of the wafer surface structure such as with a membrane.

SD makes the kerf width as narrow as possible, almost zero, so street reduction can be realized. The number of die produced from a single wafer can be increased, especially for wafer with small die. Also, SD improves feed speed. This leads to greater improvement in productivity compared to conventional blade dicing.

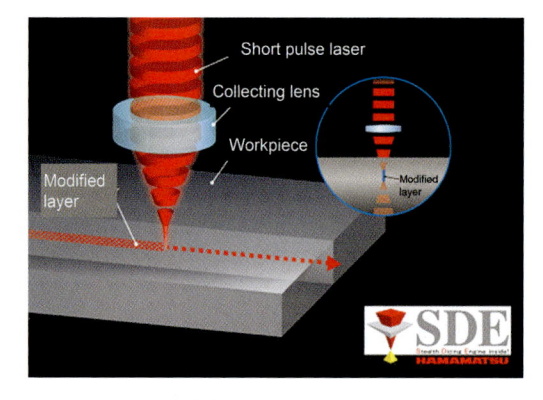

Figure 18.21 Stealth dicing process.

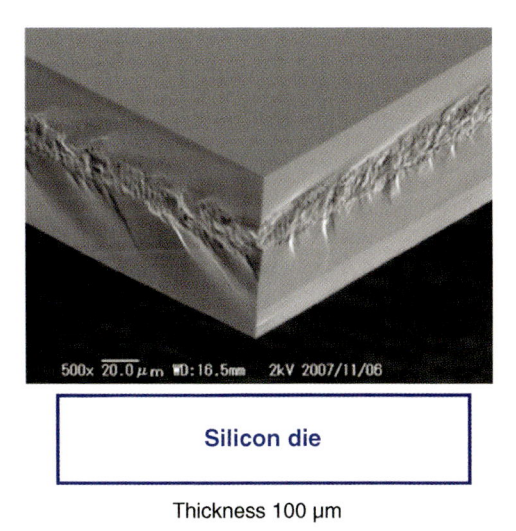

Figure 18.22 Stealth dicing processing example.

18.3.5
Dicing Topics for 3D-IC Such as TSV

Three types of cutting can be considered, such as cutting a single wafer into die, cutting a wafer with stacked die, and dicing completely encapsulated packages. It is necessary to determine the suitable singulation method for pickup at die bonding.

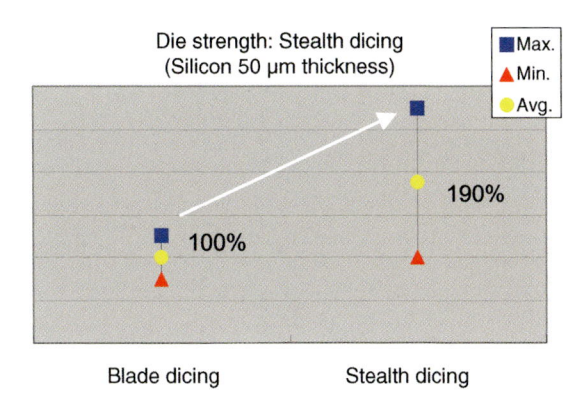

Figure 18.23 Die strength comparison between blade dicing and stealth dicing.

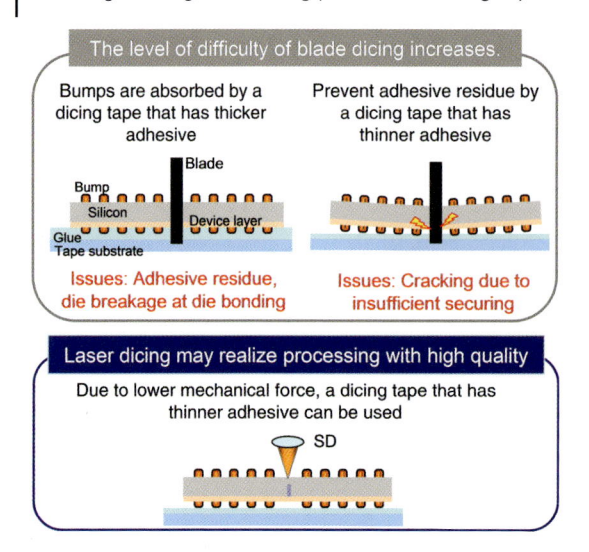

Figure 18.24 Issue of bumps on the wafer backside.

18.3.5.1 Cutting of Chip on Chip (CoC) and Chip on Wafer (CoW)

The major issues in cutting a single TSV wafer, such as CoC or CoW, into die are die strength and bumps on the wafer backside.

Die strength: Multiple vias (TSV) are formed vertically inside single die, so die strength of die with TSV decreases compared to that of normal die.

Bumps on wafer backside: A typical wafer with TSV is thin with bumps or pillars on both sides of the wafer. Bumps on the wafer backside are a new challenge to both blade dicing and SD, due to issues with securing the wafer to dicing tape (Figure 18.24).

The methods that can be considered for cutting CoC and CoW are blade dicing and SD.

Blade Dicing Because an insufficient securing strength may generate cracking on a wafer, stable securing of the wafer is required to maintain the cutting quality in blade dicing. In order to secure the wafer stably during dicing, bumps on the wafer backside need to be absorbed fully by a dicing tape that has thicker adhesive. This type of dicing tape introduces other new issues. It may leave adhesive residue on the die backside and it needs a greater pickup force at die bonding. For die bonding, a higher pickup force can be a reason for die breakage, so a lower pickup force is preferred. In order to use blade dicing for singulation of wafers with bumps on the backside, development of a dicing tape suitable for both blade dicing and die bonding is necessary.

Stealth Dicing (SD) SD improves die strength of thin wafers because it is a chip-free process as described before. Also, SD has lower mechanical force and can be proposed as a method for cutting with less damage. This means that it is unnecessary to absorb bumps fully into the tape. It is easier to secure a wafer with bumps on the backside for SD than for blade dicing.

Note: In cases where no bumps occur on the wafer outer circumference, the use of usual SD tape expansion may not be able to expand the outer part and the countermeasure may be required.

A de facto standard of temporary bonding process has not been determined and is under investigation at each manufacturer of TSV. So the process flow changes depending on the type of temporary bonding adhesives and removal methods in use.

Dicing tape uses a removal method conducted at room temperature. In this case, SD comes before removal. SD processes from the wafer backside, the wafer is mounted on dicing tape, and finally the carrier wafer is removed. This is the simplest process flow and easiest process to irradiate by SD from the wafer backside.

If thermal removal is conducted, the removal comes before the SD. Because there is no heat-resistant dicing tape that is sufficient for use in thermal removal, tape mounting comes after removal. In case where the TEG is placed on the street, no SD irradiation is transmitted and the SD must be applied from the wafer backside. The application of SD irradiation from the dicing tape side and the applicable dicing tape are under development (Figure 18.25).

18.3.5.2 Singulation of CoW and Wafer on Wafer (WoW)

A major issue for singulation of CoW and WoW (see Figure 18.26) is the dicing of compound materials. Multiple stacked wafers or mold-encapsulated packages consist of silicon and filling materials among the bumps. The filling materials

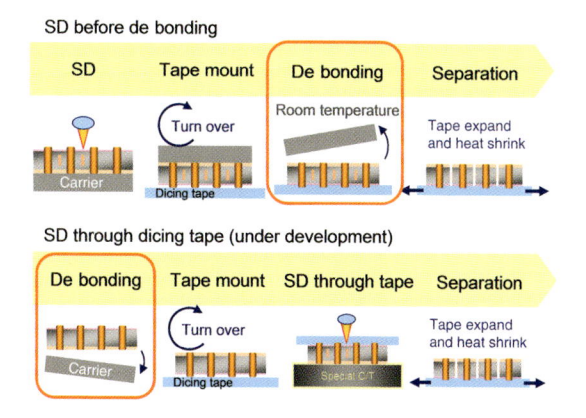

Figure 18.25 Example of the SD process flow.

Figure 18.26 WoW dicing example. (This work is carried out by the WoW Allaiance at the University of Tokyo.)

include underfill, nonconductive film (NCF) and nonconductive paste (NCP). Their typical components are usually resin and filler.

Blade dicing can be used for dicing these materials. Different from the conventional blade used for dicing silicon, with an increased percentage of the resin inside the chip, a blade having the feature of less loading is required for blade dicing both silicon and resin.

The other issue regarding singulation of CoW and WoW is dicing of wafers with low-k layers. Each individual wafer may require grooving by laser or full cut by laser ablation. These are still under development.

18.4
Summary

DISCO offers *Kiru*, *Kezuru*, and *Migaku* (dicing, grinding, and polishing) technologies and solutions to be the total solution provider in the field of 3D IC.

Further Reading

Geng, H. (2005) *Semiconductor Manufacturing Handbook*, McGRAW-HILL, pp. 20.1–20.22.

Hirose, M. (2007) Thinning wafer technology for 3D packages. Technical Digest of the International 3D System Integration Conference (3D-SIC 2007), pp. 22.1–22.12.

Kim, Y.S., Maeda, N., Kitada, H., Fujimoto, K., Kodama, S., Kawai, A., Arai, K., Suzuki, K., Nakamura, T., and Ohba, T. (2013) Advanced wafer thinning technology and feasibility test for 3D integration. *Microelectronic Engineering*, **107**, 65–71.

Kobayashi, Y. (2008) Electronic Journal Technical Symposium No.188, pp. 99–114.

Ohba, T. (2011) Through-silicon-via technology for stacked thin Si device wafers. Proceedings of IEEE IWJT, pp. 61–66.

19
Overview of Bonding and Assembly for 3D Integration

James J.-Q. Lu, Dingyou Zhang, and Peter Ramm

19.1
Introduction

Three-dimensional integrated circuit (3DIC) technology is quite different than traditional assembly technology. The major form factor difference is that 3DIC integrates multiple functional IC chips vertically, instead of packing a single chip or multiple chips into a 2D package. This in turn enables heterogeneous integration of materials, processes and functional IC chips [1–6]. Moreover, 3D integration has more choices in terms of 3D process flows: *chip-on-chip* (CoC), *chip-on-wafer* (CoW), and *wafer-on-wafer* (WoW), as shown in Figure 19.1.

Three-dimensional integration consists of four unique processes: through-silicon via (TSV) formation, alignment, bonding, and thinning, as discussed in the rest of this book. Bonding is a key enabling technology for any 3D integration; it glues multiple functional IC chips vertically and provides some combinations of mechanical, thermal, electrical, and optical connections between chips and wafers permanently, or offers temporary mechanical support for wafer thinning and thinned wafer handling. Bonding is closely related to alignment and thinning, while it can also be one of the assembly processes, such as microbump (or microsolder) bonding between chips, CoW self-assembly, or CoW bonding based on template alignment.

Three-dimensional integration requires bonding of the stack layers to one another and of the 3D stack to the substrate [1–5]. The choice of bonding technology depends on composition of the chips and their required process conditions, such as temperature limitations. The large spectrum of corresponding application-specific 3D integration technologies can be reasonably classified in the following main categories:

- Stacking of packaged devices (or substrates) using the CoC approach
- Stacking of bare chips (with or without TSVs) using the CoC or CoW approaches
- Stacking of wafers with wafer bonding with TSVs using the WoW approach

Handbook of 3D Integration: 3D Process Technology, First Edition.
Edited by Philip Garrou, Mitsumasa Koyanagi, and Peter Ramm.
© 2014 Wiley-VCH Verlag GmbH & Co. KGaA. Published 2014 by Wiley-VCH Verlag GmbH & Co. KGaA.

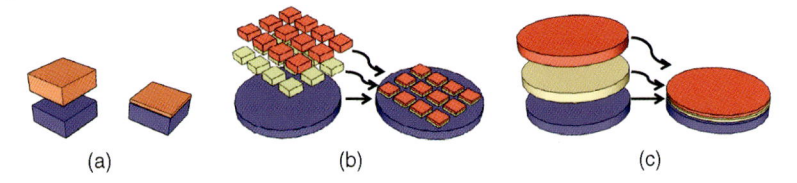

Figure 19.1 Major 3D integration technology platforms: chip-on-chip (CoC), chip-on-wafer (CoW), and wafer-on-wafer (WoW) vertical integration.

This chapter provides an overview of the bonding and assembly technologies for 3D integration; more details of specific bonding and assembly technologies can be found in the references and other sections in this chapter.

19.2
Direct, Indirect, and Hybrid Bonding

Bonding attaches two substrate materials either directly or indirectly, as shown in Figure 19.2. *Direct bonding* attaches two similar or dissimilar materials without an intermediate material. *Indirect bonding* attaches two materials using an intermediate material. In some cases, *hybrid bonding* is desired to form a bonding interface, where two or more materials coexist on one or both sides of the bonding surfaces. Various materials have been investigated for different bonding processes; some examples are given herein.

Examples of direct bonding include Si–Si hydrophobic bonding (with identical or different properties, e.g., crystal orientation and impurity) [8–11], Si–SiO$_2$ bonding [10], plasma-enhanced chemical vapor deposition (PECVD) SiO$_2$–SiO$_2$ bonding [12,13], Cu–Cu bonding (or metal bonding in general) [14–16], benzocyclobutene (BCB) to BCB bonding (or adhesive bonding in general) [7,17–21].

Examples of indirect bonding include Si–Si hydrophilic bonding, where a thin oxide layer is formed at the bonding interface [11]; BCB and a variety of polymers or polyimides to bond two other materials, such as Si chips or wafers [17–21]; metal layer (such a thin Ti film) to bond two Si wafers [22,23]; and intermetallic compounds (IMCs) to bond two chips or wafers [24–32].

Hybrid bonding is mainly used to form electrical connections between two materials, with nonmetal bonding to provide additional thermal/mechanical support [1,7,33–37]. The metal patterns are often formed within an insulator matrix

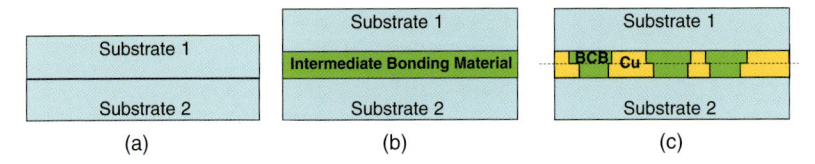

Figure 19.2 Major 3D bonding approaches: direct bonding, indirect bonding using an intermediate material, and hybrid bonding (with Cu–BCB patterns as an example) [7].

on each side of the bonding surface, such as copper pads within an oxide matrix (Cu–oxide) [33–36] or a BCB matrix (Cu–BCB) [37], microbumps within an underfill [38]. Other polymers may be used as the insulator matrix material.

Different bonding definitions are used in the literature, depending on how the system or the bonding process is considered. For example, BCB–BCB bonding may be referred to as indirect bonding if the bonding system is considered to consist of two wafers, with BCB to function as an intermediate bonding material. Similarly, SiO_2–SiO_2 bonding and Cu–Cu bonding may also refer to indirect bonding for bonding two functional silicon wafers.

Regardless of the terminology, all bonding approaches provide some combinations of mechanical, thermal, and electrical connections between functional chips or wafers.

19.3
Requirements for Bonding Process and Materials

Because front-end-of-line (FEOL) and back-end-of-line (BEOL) processes are completed or partially completed before 3D integration, the following bonding conditions are highly desired for any 3D approaches:

- Compatible with the BEOL IC process, that is, with a bonding temperature at or lower than 400 °C, with bonding materials that are relatively thin and of low stress
- High thermal and mechanical stability of the bonding interface over the ranges of BEOL and packaging processing conditions
- No outgassing, or with outgassing channels, during bonding to avoid void formation
- Seamless bonding interface with high bond strength to prevent delamination

To satisfy the last condition, strong chemical bonds must be formed between the bonding surfaces over the entire chip–wafer pair during the bonding process. This further requires (i) atomic-scale flatness and/or (ii) diffusion of massive atoms across the bonding interface during bonding or postbond anneal. Figure 19.3 shows the impacts of bonding surfaces. Atomic-scale flatness (i.e., in Angstrom scale) is particularly important for rigid bonding surfaces (e.g., for Si–Si or SiO_2–SiO_2 bonding) [10,11], while not very important for soft bonding surfaces (e.g., for Cu–Cu or BCB–BCB bonding because Cu atoms diffuse across the bonding interface during grain regrowth at a relative low bonding temperature of less than 400 °C [23] and BCB reflows and is cross-linked at higher than 180 °C) [17]. Furthermore, the rigid bonding surface must be extremely clean, that is, free of particulates and defects [10], while soft bonding surfaces (e.g., for Cu–Cu or BCB–BCB bonding) can tolerate some level of particulates and defects.

Figure 19.4 shows a variety of direct and indirect bonding examples and materials, while Figure 19.5 shows some hybrid bonding examples and their materials. Note that hydrophobic and hydrophilic silicon bonding starts from prebonding at room temperature, followed by a postbond anneal at high temperatures (∼1100 °C) [8],

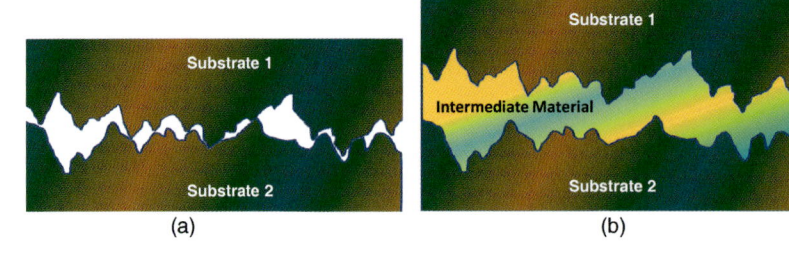

Figure 19.3 Impacts of bonding surfaces. (a) Surface roughness can prevent atomic-scale bonding between two rigid bonding surfaces, while soft bonding surfaces can be still bonded by surface deformation and interface diffusion. (b) An intermediate material (e.g., BCB) that is wet to the bonding surface can facilitate the bonding between two rigid rough surfaces.

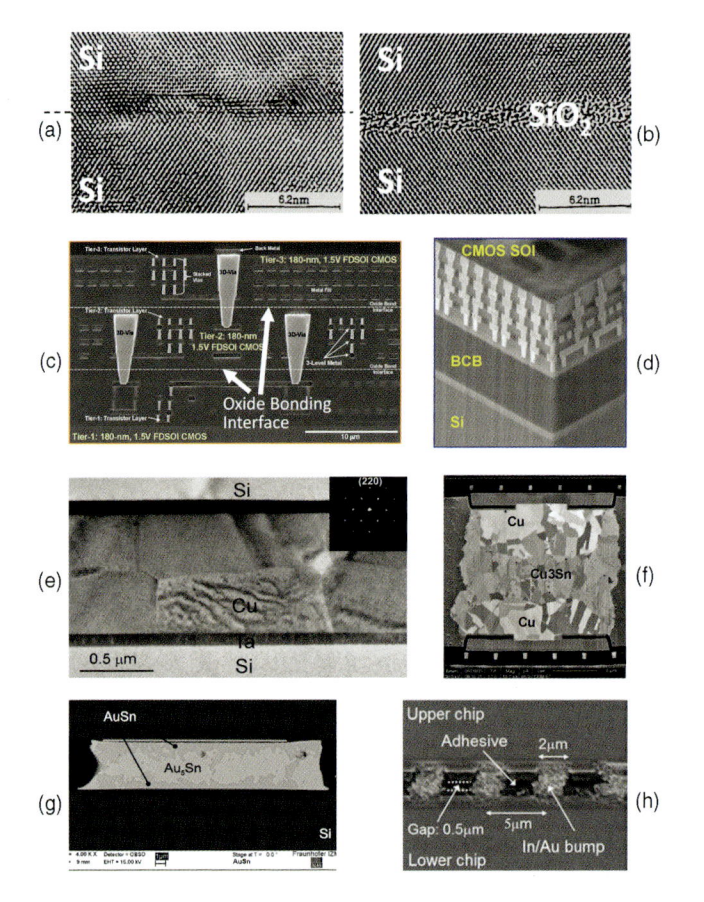

Figure 19.4 Direct and indirect bonding examples and materials. High-temperature bonding: hydrophobically (a) and hydrophilically (b) bonded silicon wafers after heating to 1100 °C [8]. Low-temperature ($\leq$400 °C) bonding: (c) PECVD SiO_2–SiO_2 bonding [12]; (d) BCB–BCB bonding [7]; (e) Cu–Cu bonding [14]; (f) Cu–Sn IMC bonding [26]; (g) Au–Sn eutectic bonding [39]; and (h) Au–In eutectic bonding [40].

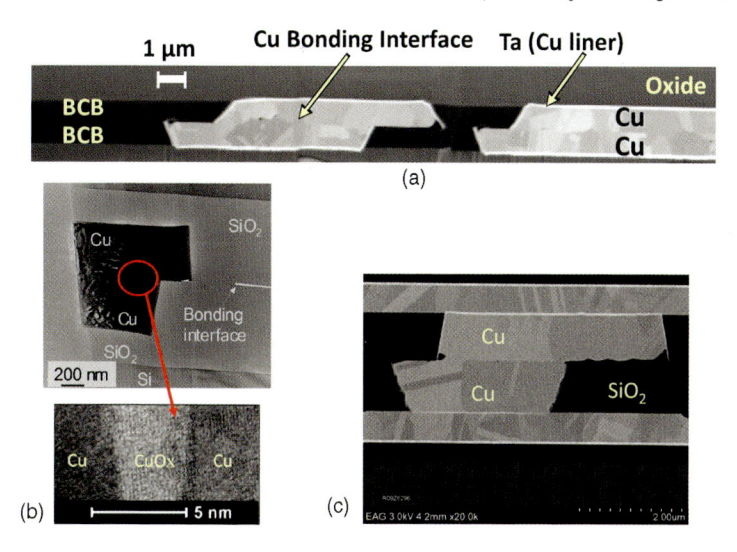

Figure 19.5 Hybrid bonding examples and materials. (a) Cu–BCB hybrid bonding [37], where BCB–BCB is bonded at lower temperature (~250 °C), followed by raising the temperature above 350 °C for Cu–Cu bonding. (b and c) Cu–PECVD oxide hybrid bonding [33,35], where prebonding is done at room temperature and the permanent bond is formed during the postbond anneal at above 350 °C. An ~4 nm thick CuO_x crystalline layer (b) is present before the postbonding anneal [33].

while low temperature (~400 °C) hydrophobic bonding is possible [41]. Besides the high-temperature Si-based bonding, other bonding process and materials discussed in this chapter are compatible with the CMOS BEOL IC process. Some important bonding aspects are discussed below; more details can be found from the corresponding references and other chapters of this book series.

Cu–Cu bonding, BCB–BCB bonding, and PECVD oxide bonding can satisfy all the requirements listed [7,42]. In particular, copper diffusion across the bonding interface with Cu grain regrowth, during Cu–Cu bonding at temperatures higher than 350 °C, leads to high bonding strength. Further thermal processes enhance the bonding strength with more Cu interdiffusion at the bonding interface [7,15,23,42]. BCB reflows at ~180 °C and forms BCB cross-link networks rapidly at 250 °C without outgassing, leading to very strong bonds. More thermal processes up to 400 °C can further complete the BCB cross-link networks, producing higher bond strength [7,17,20,42]. PECVD oxide bonding does require better surface flatness and cleanness compared to Cu–Cu bonding and BCB–BCB bonding, but not as stringently as for Si–Si and thermal oxide bonding [7,42]. PECVD oxide with Si–OH bonds at the bonding surface is less dense compared to thermal oxide; during thermal anneal at ~275 °C, Si–OH bonds break and bonding surface atoms rearrange to form strong Si–O–Si bonds (and release a small amount of H_2O molecules, which can easily diffuse in oxide) [7,12,13,42].

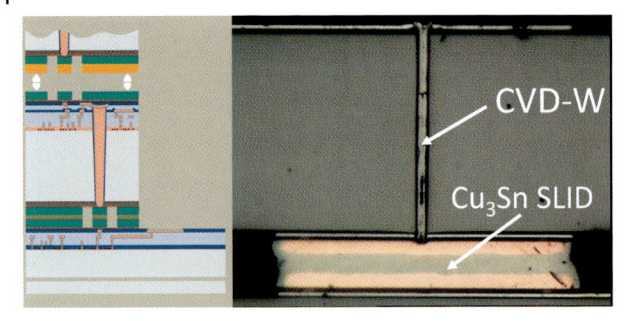

Figure 19.6 3D TSV technology based on SLID bonding process [25].

Solid–liquid interdiffusion (SLID) bonding [24–26] (or IMC bonding, for example, with Cu–Sn system) and eutectic bonding (e.g., with Au–In and other solder systems) can be conducted in air atmosphere [24–32]; this is advantageous for CoC and CoW bonding. For the copper–tin SLID system, the substrates to be bonded are deposited with Cu–Sn layer structures. At typical wafer bonding temperatures of 240–270 °C, the deposited tin is completely transformed into the Cu_3Sn intermetallic compound (Figure 19.6). This ε-phase is thermodynamically stable with a melting point above 600 °C. Using appropriate film thicknesses, the tin is consumed and the solidification is completed within a few minutes, leaving only unconsumed copper on both sides. The key advantage of the SLID technology is that both the mechanical and the electrical interconnects are formed simultaneously - so the 3D integration process is completed and no postprocessing is required. Three-dimensional integration can be executed subsequently for a next device layer to be stacked. This key ability of modularity is a consequence of the fact that the IMC melting temperature is considerably higher than the wafer bonding temperature; this is quite different from eutectic bonding, where the metal system remelts by reheating above the eutectic temperature. However, the materials for SLID or eutectic bonding melt during the bonding process. It is very difficult to control the lateral size of the bonding pad and the vertical height of the bonding bump (or microsolder) with appropriate bonding force (or pressure); thus, large bond pads (i.e., small interchip interconnection density) and thick bonding materials are often needed. After bonding, the IMCs or solders are usually brittle; this is a potential reliability concern [27].

Besides the desired conditions listed above, depending on the applications, additional requirements for the bonding materials may be needed for some combinations of mechanical, thermal, electrical, and optical connections between functional chips or wafers permanently, or for temporary mechanical support for wafer thinning and thinned wafer handling.

Temporary wafer bonding requires more stringent temperature limits for bonding and debonding, while not very stringent limits for other requirements [43–46]. However, it is extremely important to select the materials and processes for damage-free debonding because of the poor mechanical properties of thinned wafers [43].

19.4
Bonding Quality Characterization

A variety of methods have been developed to inspect and characterize the bonding quality, including structural, mechanical, thermal, and electrical properties of the bonding interface. Accordingly, key quantities related to the bonding quality are void/defect, bonding strength, electric resistance, and reliability [2]. All the structural, mechanical, thermal, and electrical properties may affect bonding reliability.

It is highly desirable to have a void-free and defect-free bonding interface because voids/defects directly affect the bonding quality and reliability. However, it is inevitable that voids and/or defects are formed at the bonding interface, particularly during the stage of bonding process development. The first goal for bonding process development is therefore to reduce the size and density of voids and/or defects, which depend on a number of factors, such as surface roughness (see Figure 19.3), outgassing at the bonding interface, particulates, or metal grain regrowth. Major effective methods to inspect the voids/defects are (i) nondestructive methods with limited resolution and materials, for example, infrared (IR) imaging and scanning acoustic microscopy (SAM), and (ii) destructive methods with high resolution, for example, focused ion beam (FIB) scanning electron microscopy (SEM) and transmission electron microscopy (TEM). Figure 19.7a–d shows TEM images of void and/or defect evolution at the Cu–Cu bonding interface with temperatures; Figure 19.4a, b, and e, and Figure 19.5b also show TEM images of other bonding interfaces, while the rest of images in Figures 19.4 and 19.5 show either FIB/SEM or SEM images of the bonding interfaces. Figure 19.7e shows an SAM image of a 150 mm prime Si wafer pair bonded hydrophobically at 400 °C for fabrication of a bidirectional insulated gate bipolar transistor (BD-IGBT); a void-free bonding interface is obtained for groove grid pitches of 3.95 and 7.9 mm [41]. A similar IR image was obtained for the same wafer pair. SAM and IR imaging methods were also used for bonding interface inspection of BCB [47] and other bonding processes [10,47,48].

Bond strength is not only an indication of the structural and mechanical bonding quality but also affects the thermal, mechanical, and electrical properties, as well as the overall bonding reliability. Permanent bonding requires a high bond strength (with a critical energy of $\sim$1 J m^{-2} or higher) to avoid delamination of the bonding interface. Temporary bond strength is not well quantified, but it has to be high enough to hold wafers for wafer thinning and polishing, while in some cases be low enough for debonding [43]. Major bond strength characterization methods include (i) razor blade test, (ii) crack opening method (or double cantilever beam test), and (iii) four-point bending technique. The razor blade test as the first-pass test for bonding is a simple, but effective method to quickly check the bond strength by inserting a sharp razor blade at the edge of a bonding interface. If the bonded wafer pair cannot be separated easily or is shattered to pieces, the bonding passes the initial test. The crack opening method is to insert a razor blade of a specific thickness between the bonded wafer pair, thus to quantitatively determine

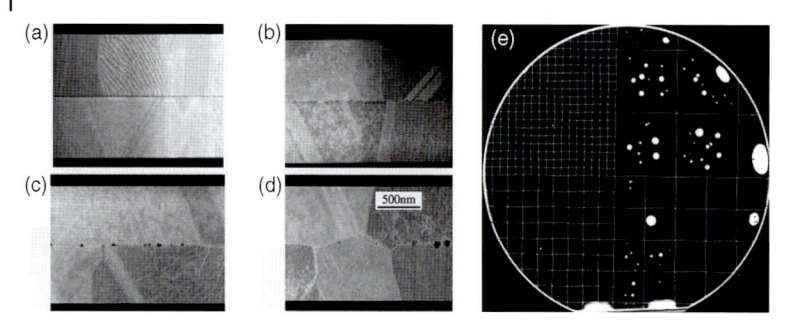

Figure 19.7 TEM cross section of Cu–Cu nonthermocompression bonding. (a) As bonded at room temperature and after a 30 min. postbonding anneal at (b) 200 °C, (c) 300 °C, and (d) 400 °C [49]. (e) Ultrasonic image of a 6 in. Si wafer pair bonded hydrophobically at 400 °C with variable outgassing grooves (i.e., groove grid pitches of 3.95, 7.9, 15.8, and 30.6 mm), showing void-free bonding interface for groove grid pitches of 3.95 and 7.9 mm [41].

the energy of the bonded surfaces (i.e., surface energy) by recording the time-dependent crack length of the bond connection split-up [50,51]. Measurement inaccuracy increases with a high surface fracture toughness or with brittle bonding materials (e.g., IMCs). The four-point bending technique is based on a theory of fracture mechanics in which delamination is modeled as a crack propagating along the interface between two bonded materials [19,20]. A beam specimen of bonded wafers or chips is prepared by forming a precrack on one of the bonded materials. With the beam specimen mounted in a special fixture, a load cell measures the applied load and an actuator measures the displacement. The bonding strength (i.e., the critical adhesion energy) can be determined from the load versus displacement curve [19]. Figure 19.8 shows examples of bond strength

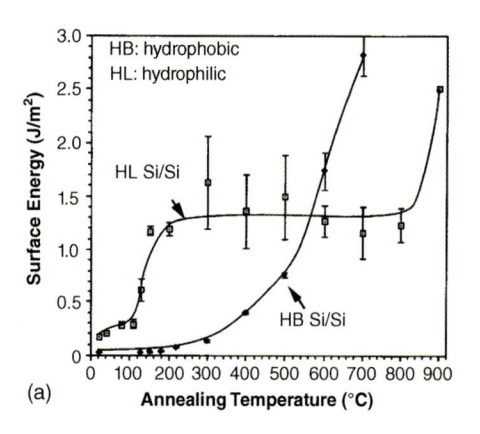

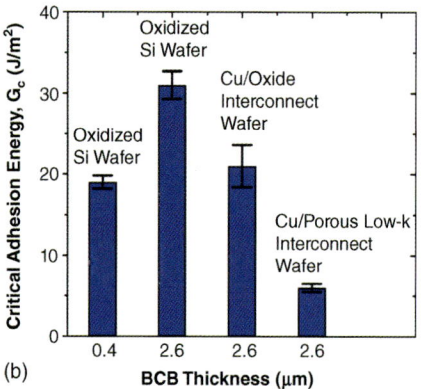

Figure 19.8 (a) Bonding energy of bonded hydrophilic and hydrophobic Si wafers measured by the crack opening method [9]. (b) Critical adhesion energy measured by a four-point bending technique for wafers with oxide, Cu–oxide, and Cu–low-k interconnects, bonded to an oxidized Si wafer using BCB [52].

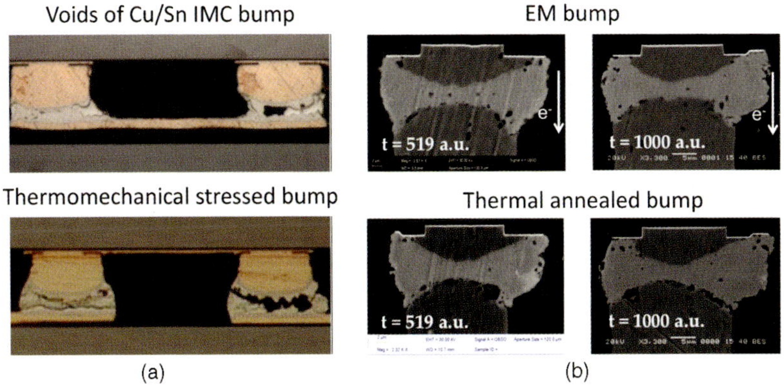

Figure 19.9 Examples of bonding reliability tests. (a) Cross section of Cu–Sn IMC bonding: void due to irregular bump versus failed bump after the thermomechanical stressing test of thermal cycle (−40 to 125 °C) [30]. (b) Electromigration (EM) stressing at 180 °C versus thermal annealing at 170 °C for Cu–Sn–Cu structure [54].

characterization for direct Si bonding using the crack-opening method [9] and for BCB bonding using a four-point bending technique [18,52].

Bonding reliability is increasingly important for 3D ICs with volume production in progress. A variety of traditional reliability tests are often combined for testing 3D bonding reliability, including the sheer test, mechanical shock test, thermal shock and cycling test, electromigration (EM) test, and other electrical tests [1–4,18,21,23,26–32,48,50–55]. In particular, the reliability tests have recently focused on bonding using IMCs [26–32,53–55], because IMC bonding is likely to be the choice for volume production of the wide I/O DRAM [42,53–57]. Figure 19.9 shows a couple of examples; much more research work is in progress (see also the corresponding chapters in this book). Relatively fewer reliability tests are reported for other bonding systems (e.g., Si–Si, Si–SiO$_2$, SiO$_2$–SiO$_2$, Cu–Cu and BCB–BCB), partially because they are quite reliable once they are well bonded [10,15,42,52] and partially because the semiconductor industry has not yet been focusing on these systems for volume production of 3DICs.

19.5
Discussion of Specific Bonding and Assembly Technologies

Three-dimensional integration has been recognized as a necessary approach to extend Moore's law and for more-than-Moore applications. Industrial development has started with a CoC 3D process flow [3], with the TSV as the core element, gearing up for wide I/O 3D DRAMs [56,57] and for giant field programmable gate arrays (FPGAs) with a TSV Si interposer [58], while image sensors using TSVs have been in the market for a few years [3].

Several 3D processing flows have been demonstrated with a number of materials investigated. A few major bonding and assembly technologies are briefly discussed as follows; more research and development efforts are in progress.

Temporary bonding–debonding: One of the key processing concerns in the CoC and CoW 3D platforms is how to handle the wafers for wafer thinning, TSV reveal, and other postthinning processes on the fragile thin wafers. Therefore, a number of organizations and companies have been developing their temporary bonding and debonding technologies and materials, including solvent-assisted debonding, tape debonding, slide-off debonding, and ZoneBOND^{TM} with thermoplastics, laser-degradable, or chemically dissolvable materials [43–46, 59]. Figure 19.10 shows an example of schematic process flow based on the ZoneBOND process. More research work is still in progress to qualify the temporary bonding–debonding process and equipment for volume production of 3D ICs – see Chapter 10 of this book for full details on these temporary bonding and debonding processes.

CoC bonding: CoC bonding enables stacking known-good-dies (KGDs) with different chip sizes, materials, and technologies to realize 3D systems. However, it suffers from low throughput because of chip level processes. To improve the

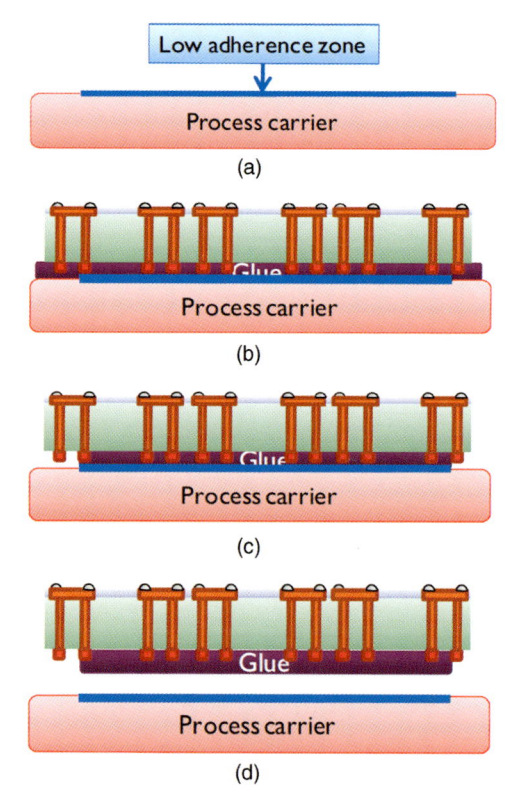

Figure 19.10 Schematic ZoneBOND temporary bonding–debonding process flow for thin wafer handling [44].

19.5 Discussion of Specific Bonding and Assembly Technologies

2 Wide-I/O DRAM dies

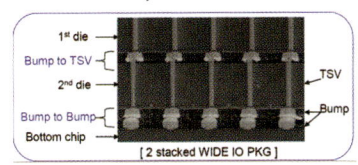

[2 stacked WIDE IO PKG]

- Microbumps: 20×17μm², 50μm pitch
- Fabricated TSVs (D=7.5μm):
  R=0.22–0.24Ω, C=47.4fF

Figure 19.11 Microbump bonding for a 1.2 V 12.8 GB s^{-1} 2 Gb mobile wide-I/O DRAM with 4 × 128 I/Os using TSV-based stacking [56].

throughput, the bonding process has to be conducted in air atmosphere. Microsolder and microbump technologies are developed, particularly SLID bonding (or intermetallic bonding, e.g., using the Cu–Sn system) and eutectic bonding (e.g., using Au–Sn, Au–In, and other solders). Figure 19.11 shows an example of microbump bonding for a 12.8 Gb s^{-1} 2 Gb mobile wide I/O DRAM with 4 × 128 I/Os using TSV-based stacking [56]. Figure 19.12 shows an example of In–Au bonding using a water droplet-assisted chip self-assembly technique [60].

CoW bonding: CoW bonding has all the advantages of CoC while using wafer-level processes to improve throughput and reliability. However, the key challenges associated with the present CoW platform are (i) the alignment accuracy limit with current pick-and-place processes, (ii) constraints in thermal budget and bonding conditions, and (iii) the difficulty to form more than two IC strata. The water droplet-assisted chip self-assembly technique with Au–In bonding can provide a submicrometer CoW alignment accuracy [60], where chip and wafer surfaces are selectively treated to be hydrophilic and hydrophobic, so that water droplets can be used to self-align the chips (Figure 19.12). Another CoW 3D platform using an alignment template on a host wafer can overcome all the three limits and can

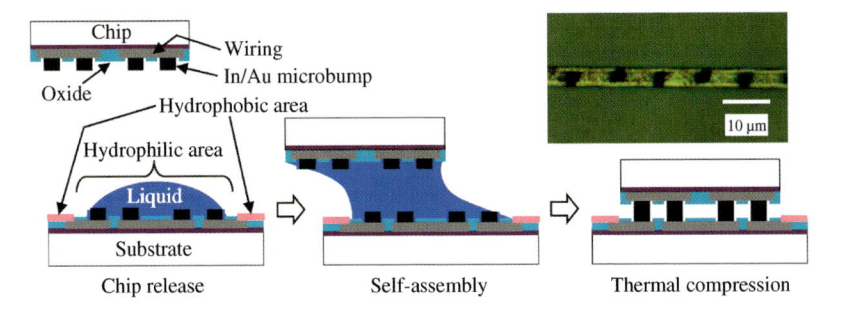

Figure 19.12 A schematic process flow for self-assembly of chips to substrates using liquid surface tension. The top-right cross-sectional image shows In/Au microbumps after self-assembly and the following thermal compression [59].

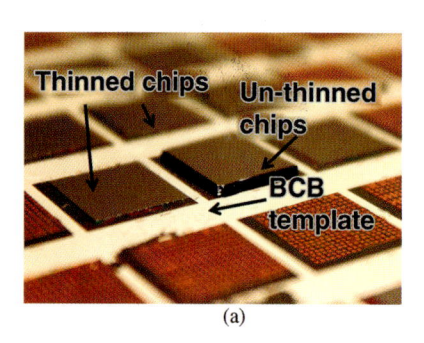

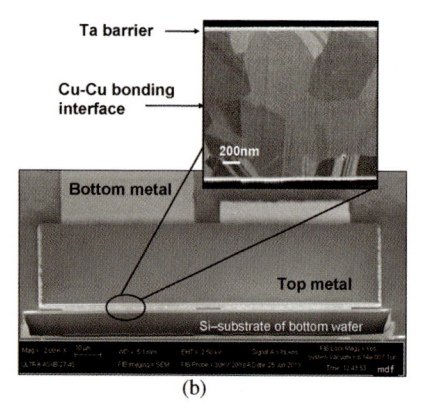

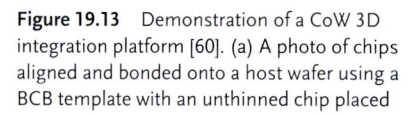

Figure 19.13 Demonstration of a CoW 3D integration platform [60]. (a) A photo of chips aligned and bonded onto a host wafer using a BCB template with an unthinned chip placed for comparison. (b) FIB-SEM images of the aligned and bonded Cu–Cu daisy-chain structure, demonstrating good alignment accuracy ($\sigma < 2\,\mu m$) and Cu–Cu bonding.

enable submicrometer alignment accuracy, easy and controllable bonding, and high-throughput 3D integration fabrication [60]. Moreover, the temporary bonding–debonding process can be eliminated, that is, full-thickness chips can be used and the thinning process can be carried out after all chips are bonded on the host wafer. Figure 19.13 shows a feasibility demonstration of a CoW 3D integration platform, where almost any bonding materials can be used and, except for the rough pick-place step, all the 3D integration processes can be conducted at wafer level [61].

Bonding for the WoW 3D platform: The WoW platform can use monolithic full wafer-level processes without the need of temporary bonding–debonding process, leading to high throughput and reliability as well as potential high yield. However, potential challenges of the WoW platform could be the high entry cost of full fab integration, the yield loss due to the lack of KGDs, and inflexibility in design and fabrication aspects, particularly in cases when the sizes or the substrate materials of different functional chips cannot be the same. Stringent wafer bonding conditions as listed in Section 19.3 are often required; the choice of the corresponding TSV technology for fabrication of a 3D integrated product has significant consequences for the wafer bonding technology to be applied. Fortunately, wafer bonding technology has been long investigated. Several wafer bonding technologies are developed and characterized [1–5]. Figure 19.14 shows the most commonly applied techniques as oxide fusion bonding, adhesive bonding, and metal bonding for 3D TSV integration.

Since the CoC platform is widely used with wire bonding or microbump bonding, 3D IC technology development and applications will continue advancing in the coming years in an evolutionary fashion, instead of revolutionary fashion, from CoC to CoW and WoW platforms; except for some applications, such as 3D stacking of pure memory. With the TSV technology getting mature for low-

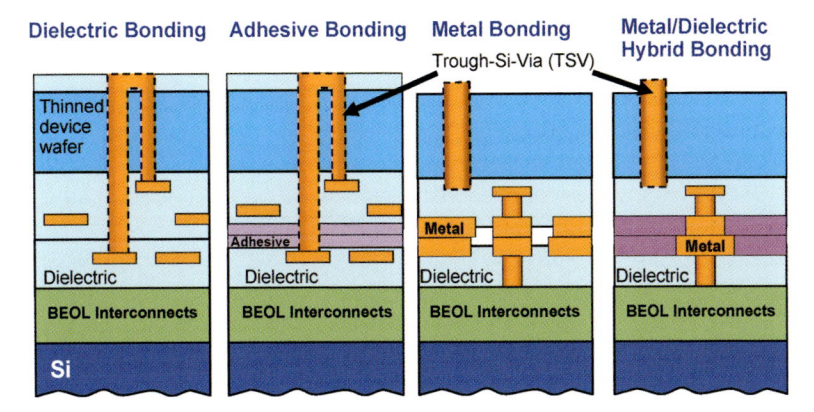

Figure 19.14 Schematic of four typical wafer bonding schemes for WoW 3D integration with TSVs (not in scale) [1–5].

hanging-fruit products, 3D processing materials, technologies, and equipment will be further developed for TSV, alignment, bonding, and thinning processes toward full CoW and WoW 3D integration once the infrastructure is in place. We believe that the research and development in materials and processes will provide a good set of solutions for processing integration, thermal–mechanical stress, yield and test, as well as cost reduction, thus to realize 3D integration of future smart systems for computing, information technology, mobile and mixed signal applications, biomedical devices, power, and smart lighting applications [62].

19.6
Summary and Conclusions

Three-dimensional integration can vertically stack and interconnect heterogeneous materials, technologies and functions. A number of technology concepts have been introduced and several sets of full 3D processes have been demonstrated. Reliable wafer bonding is an enabling technology for 3D integrated system fabrication.

Various bonding approaches and materials have been investigated:

- Permanent bonds to provide some combinations of mechanical, thermal, electrical, and optical connections between functional chips or wafers.
- Temporary mechanical support for wafer thinning and thinned wafer handling (see Chapter 10).

With the application-specific constraints of bonding conditions (e.g., temperature, ambience, integrity, and throughput), three types of permanent bonding (i.e., direct, indirect, and hybrid bonding) are investigated with various electrically conducting and insulating materials. Examples of electrically conducting bonding materials include Si, Cu (or other metals), IMC or SLID (e.g., Cu–Sn), eutectic metals (Au–Sn, Au–In, and other solders). These materials provide vertical electrical connections and mechanical and thermal connections. Examples for electrically insulating bonding materials include SiO_2, polymers (e.g., BCB), polyimides, and underfill. They provide a robust mechanical attachment between functional chips, while insulating the chips electrically and thermally. A few combinations of electrically conducting and insulating materials are used for hybrid bonding approaches to utilize the benefits of both types of bonding materials. Temporary bonding approaches, materials, and equipment were also developed; more work is in progress to qualify the process and equipment for 3D volume production. Various bonding characterization methods are used to quantify the void/defect, bonding strength, electrical resistance, and reliability, with the goal to provide reliable bonds between 3D chips.

Three-dimensional IC technology development and applications will continue advancing in the coming years in an evolutionary fashion from CoC to CoW and WoW platforms, with preferred bonding approaches and materials:

- *CoC bonding:* polymers, IMCs, or solders – in air atmosphere
- *CoW bonding:* microbumps (Cu–Sn or Au–In), metals, or hybrid metal/dielectric – in air atmosphere or vacuum
- *WoW bonding:* all possible bonding materials and approaches

Three-dimensional technology provides a viable path to a future paradigm combining semiconductor technology with more-than-Moore applications, thus future smart systems can easily interact with people and their environment.

References

1 Lu, J.-Q. (2009) 3D hyper-integration and packaging technologies for micro-nano-systems. *Proceedings of the IEEE*, **97** (1), 18–30.
2 Ramm, P., Lu, J.J.-Q., and Taklo, M. (eds) (2012) *Handbook of Wafer Bonding*, Wiley-VCH Verlag GmbH, Weinheim.
3 Garrou, P., Lu, J.-Q., and Ramm, P. (2012) Three-dimensional integration, in *Handbook of Wafer Bonding* (eds P. Ramm, J.J.-Q. Lu, and M. Taklo), Wiley-VCH Verlag GmbH, Weinheim, pp. 301–328.
4 Garrou, P., Bower, C., and Ramm, P. (eds) (2008) *Handbook of 3D Integration: Technology and Applications of 3D Integrated Circuits*, Wiley-VCH Verlag GmbH, Weinheim.
5 Tan, C.S., Gutmann, R.J., and Reif, L.R. (eds) (2008) *Wafer Level 3-D ICs Process Technology*, Springer.
6 International Technology Roadmap for Semiconductors (ITRS) (2011) http://public.itrs.net/.
7 Lu, J.-Q., McMahon, J.J., and Gutmann, R.J. (2008) 3D integration using adhesive, metal, and metal/adhesive as wafer bonding interfaces. *MRS Proceedings*, **1112**, 69.
8 Gösele, U., Stenzel, H., Reiche, M., Martini, T., Steinkirchner, H., and Tong, Q.-Y. (1996) History and future of semiconductor wafer bonding. *Solid State Phenomena*, **47–48**, 33–44.
9 Tong, Q.-Y., Cha, G., Gafiteanu, R., and Gösele, U. (1994) Low temperature wafer direct bonding. *IEEE Journal of Microelectromechanical Systems*, **3** (1), 29–35.
10 Tong, Q.-Y. and Gösele, U. (1999) *Semiconductor Wafer Bonding: Science and Technology*, Wiley-Interscience.
11 Plößl, A. and Kräuter, G. (March 1999) Wafer direct bonding: tailoring adhesion

between brittle materials. *Materials Science and Engineering*, **25** (1–2), 1–88.
12. Burns, J.A. *et al.* (2006) A wafer-scale 3-D circuit integration technology. *IEEE Transactions on Electron Devices*, **53** (10), 2507–2516.
13. Guarini, K.W. *et al.* (2002) Electrical integrity of state-of-the-art 0.13μm SOI CMOS devices and circuits transferred for three-dimensional (3D) integrated circuit (IC) fabrication. International Electron Devices Meeting (IEDM 2002), December 8–11, 2002, pp. 943–945.
14. Chen, K.N., Fan, A., Tan, C.S., Reif, R., and Wen, C.Y. (2002) Microstructure evolution and abnormal grain growth during copper wafer bonding. *Applied Physics Letters*, **81** (20), 3774–3776.
15. Chen, K.N. and Tan, C.S. (2011) Thermocompression Cu–Cu bonding of blanket and patterned wafers, in *Handbook of Wafer Bonding* (eds P. Ramm, J.J.-Q. Lu, and M. Taklo), Wiley-VCH Verlag GmbH, Weinheim, pp. 161–180.
16. Patti, R.S. (2006) Three-dimensional integrated circuits and the future of system-on-chip designs. *Proceedings of the IEEE*, **94** (6), 1214–1224.
17. Niklaus, F., Stemme, G., Lu, J.-Q., and Gutmann, R.J. (2006) Adhesive wafer bonding. *Journal of Applied Physics*, **99** (3), 031101.
18. Lu, J.-Q., Jindal, A., Kwon, Y., McMahon, J.J., Rasco, M., Augur, R., Cale, T.S., and Gutmann, R.J. (2003) Evaluation procedures for wafer bonding and thinning of interconnect test structures for 3D ICs. IEEE International Interconnect Technology Conference, pp. 74–76.
19. Kwon, Y., Seok, J., Lu, J.Q., Cale, T.S., and Gutmann, R.J. (2006) Critical adhesion energy of benzocyclobutene (BCB)-bonded wafers. *Journal of the Electrochemical Society*, **153** (4), G347–G352.
20. Kwon, Y., Seok, J., Lu, J.-Q., Cale, T.S., and Gutmann, R.J. (2005) Thermal cycling effects on critical adhesion energy and residual stress in benzocyclobutene (BCB)-bonded wafers. *Journal of the Electrochemical Society*, **152** (4), G286–G294.
21. Kwon, Y., Jindal, A., Augura, R., Seok, J., Cale, T.S., Gutmann, R.J., and Lu, J.-Q. (2008) Evaluations of bonding and thinning integrity of stacked wafers for three dimensional (3D) integration. *Journal of the Electrochemical Society*, **155** (5), H280–H286.
22. Yu, J., Wang, Y., Lu, J.-Q., and Gutmann, R.J. (2006) Low-temperature silicon wafer bonding based on Ti/Si solid-state amorphization. *Applied Physics Letters*, **89**, 092104.
23. Ko, C.-T. and Chen, K.-N. (2012) Low temperature bonding technology for 3D integration. *Microelectronics Reliability*, **52** (2), 302–311.
24. Ramm, P., Klumpp, A., Merkel, R., Weber, J., and Wieland, R. (2004) Vertical system integration by using inter-chip vias and solid–liquid interdiffusion bonding. *Japanese Journal of Applied Physics*, **43** (7A), L829–L830.
25. Ramm, P., Klumpp, A., Weber, J., Lietaer, N., Taklo, M., De Raedt, W., Fritzsch, T., and Couderc, P. (2010) 3D integration technology: status and application development. Proceedings of the ESSCIRC/ESSDERC, Seville, Spain, pp. 9–16.
26. Huebner, H., Penka, S., Barchmann, B., Eigner, M., Gruber, W., Nobis, M., Janka, S., Kristen, G., and Schneegans, M. (2006) Microcontacts with Sub-30μm pitch for 3D chip-on-chip integration. *Microelectronic Engineering*, **83** (11), 2155–2162.
27. Sakuma, K., Sueoka, K., Kohara, S., Matsumoto, K., Noma, H., Aoki, T., Oyama, Y., Nishiwaki, H., Andry, P.S., Tsang, C.K., Knickerbocker, J.U., and Orii, Y. (2010) IMC bonding for 3D interconnection. Proceedings of the 60th Electronic Components and Technology Conference, 2010 (ECTC 2010), pp. 864–871.
28. Huffman, A., Lueck, M., Bower, C., and Temple, D. (2007) Effects of assembly process parameters on the structure and thermal stability of Sn-capped Cu bump bonds. Proceedings of the 57th Electronic Components and Technology Conference, 2007 (ECTC 2007), pp. 1589–1596.
29. Sakuma, K., Andry, P.S., Dang, B., Maria, J., Tsang, C.K., Patel, C., Wright, S.L., Webb, B., Sprogis, E., Kang, S.K., Polastre, R., Horton, R., and Knickerbocker, J.U. (2007) 3D chip stacking technology with low-volume lead-free interconnections.

Proceedings of the 57th Electronic Components and Technology Conference, 2007 (ECTC 2007), pp. 627–632.

30 Khan, N., Wee, D.H.S., Chiew, O.S., Sharmani, C., Lim, L.S., Li, H.Y., and Vasarala, S. (2010) Three Chips Stacking with Low Volume Solder Using Single Re-Flow Process. Proceedings of the 60th Electronic Components and Technology Conference, 2010 (ECTC 2010), pp. 884–888.

31 Huang, S.-Y., Zhan, C.-J., Chung, S.-C., Fan, C.-W., Chen, S.-M., Chang, T.-C., and Chen, T.-H. (2011) Bonding and reliability assessment of 30μm pitch solder micro bump interconnection with various UBM structure for 3D chip stacking. 6th International Microsystems, Packaging, Assembly and Circuits Technology Conference (IMPACT 2011), Taipei, Taiwan, pp. 454–457.

32 Reed, J.D., Lueck, M., Gregory, C., Huffman, C.A., Lannon, J.M., Jr., and Temple, D. (2010) Reliability and ultra-low temperature bonding of high density large area arrays with Cu/Sn–Cu interconnects for 3D integration. 2010 International Interconnect Technology Conference, Burlingame, CA, USA, pp. 1–3.

33 Di Cioccio, L. (2011) Cu/SiO$_2$ hybrid bonding, in *Handbook of Wafer Bonding* (eds P. Ramm, J.J.-Q. Lu, and M. Taklo), Wiley-VCH Verlag GmbH, Weinheim, pp. 237–259.

34 Gueguen, P., Ventosa, C., Cioccio, L.D., Moriceau, H., Grossi, F., Rivoire, M., Leduc, P., and Clavelier, L. (2010) Physics of direct bonding: applications to 3D heterogeneous or monolithic integration. *Microelectronic Engineering*, **87** (3), 477–484.

35 Enquist, P., Fountain, G., Petteway, C., Hollingsworth, A., and Grady, H. (2009) Low cost of ownership scalable copper direct bond interconnect 3D IC technology for three dimensional integrated circuit applications. IEEE International Conference on 3D System Integration (3DIC 2009), pp. 1–6.

36 Enquist, P. (2011) Metal/silicon oxide hybrid bonding, in *Handbook of Wafer Bonding* (eds P. Ramm, J.J.-Q. Lu, and M. Taklo), Wiley-VCH Verlag GmbH, Weinheim, pp. 261–278.

37 Lu, J.-Q., McMahon, J.J., and Gutmann, R.J. (2012) Hybrid metal/polymer wafer bonding platform, in *Handbook of Wafer Bonding* (eds P. Ramm, J.J.-Q. Lu, and M. Taklo), Wiley-VCH Verlag GmbH, Weinheim, pp. 215–236.

38 Nimura, M., Sakuma, K., Ogino, H., Enomoto, T., Shigetou, A., Shoji, S., and Mizuno, J. (2012) Hybrid solder-adhesive bonding using simple resin planarization technique for 3D LSI. 3rd IEEE International Workshop on Low Temperature Bonding for 3D Integration (LTB-3D), Tokyo, Japan, p. 81.

39 Oppermann, H. and Hutter, M. (2012) Au/Sn solder, in *Handbook of Wafer Bonding* (eds P. Ramm, J.J.-Q. Lu, and M. Taklo), Wiley-VCH Verlag GmbH, Weinheim, pp. 119–138.

40 Koyanagi, M. and Motoyoshi, M. (2011) Eutectic Au–In bonding, in *Handbook of Wafer Bonding* (eds P. Ramm, J.J.-Q. Lu, and M. Taklo), Wiley-VCH Verlag GmbH, Weinheim, pp. 139–159.

41 Picard, J. (2012) Thinned wafer low temperature silicon direct bonding for power device fabrication. M.S. Thesis. Rensselaer Polytechnic Institute, Troy, NY.

42 Lu, J.-Q. (2012) Advances in materials and processes for 3D-TSV integration. *ECS Transactions*, **45** (6), 119–129.

43 Puligadda, R. (2011) Temporary bonding for enabling three-dimensional integration and packaging, in *Handbook of Wafer Bonding* (eds P. Ramm, J.J.-Q. Lu, and M. Taklo), Wiley-VCH Verlag GmbH, Weinheim, pp. 329–345.

44 Phommahaxay, A. *et al.* (2012) Ultrathin wafer handling in 3D stacked IC manufacturing combining a novel ZoneBOND™ temporary bonding process with room temperature peel debonding. IEEE International 3D Systems Integration Conference (3DIC), 2012, Osaka, Japan.

45 Kawano, M., Komatsu, T., Fehkührer, A., Schachinger, M., Wiesbauer, H., Burggraf, J., Burgstaller, D., Matthias, T., Wimplinger, M., and Lindner, P. (2012) Temporary-bonding and LowTemp® debonding technology for various applications. 3rd IEEE International Workshop on Low Temperature Bonding for 3D Integration, Tokyo, Japan, p. 149.

46 Chen, M.-C., Hsieh, F., and Hu, D.-C. (2012) A new carrier structure for TSV thin wafer handling. IEEE 62nd Electronic Components and Technology Conference (ECTC 2012), San Diego, CA, USA, pp. 561–563.

47 Zschech, E. and Diebold, A. (2011) Metrology and failure analysis for 3D IC integration. *AIP Conference Proceedings*, **1395**, 233–239.

48 Brand, S., Czurratis, P., Hoffrogge, P., Temple, D., Malta, D., Reed, J., and Petzold, M. (Oct. (2011)) Extending acoustic microscopy for comprehensive failure analysis applications. *Journal of Materials Science: Materials in Electronics*, **22** (10), 1580–1593.

49 Moriceau, H., Rieutord, F., Fournel, F., Di Cioccio, L., Moulet, C., Libralesso, L., Gueguen, P., Taibi, R., and Deguet, C. (2012) Low temperature direct bonding: an attractive technique for heterostructures build-up. *Microelectronics Reliability*, **52** (2), 331–341.

50 Gosele, U. and Tong, Q.-Y. (1998) Semiconductor wafer bonding. *Annual Review of Materials Science*, **28**, 215–241.

51 Martini, T., Steinkirchner, J., and Gösele, U. (1997) The crack opening method in silicon wafer bonding: how useful is it? *Journal of the Electrochemical Society*, **144** (1), 354–357.

52 Lu, J.-Q., Cale, T.S., and Gutmann, R.J. (2008) *Handbook of 3D Integration: Technology and Applications of 3D Integrated Circuits*, vol. 1 (eds P. Garrou, C. Bower, and P. Ramm), Polymer Adhesive Bonding Technology, pp. 249–259.

53 Smith, K., Hanaway, P., Jolley, M., Gleason, R., Strid, E., Daenen, T., Dupas, L., Knuts, B., Marinissen, E.J., and Dievel, M.V. (2011) Evaluation of TSV and micro-bump probing for wide I/O testing. IEEE International Test Conference (ITC), 2011, Anaheim, CA, USA, pp. 1–10.

54 Wang, Y., Chae, S.-H., Dunne, R., Takahashi, Y., Mawatari, K., Steinmann, P., Bonifield, T., Jiang, T., Im, J., and Ho, P.S. (2012) Effect of Intermetallic Formation on Electromigration Reliability of TSV-Microbump Joints in 3D Interconnect. IEEE 62nd Electronic Components and Technology Conference (ECTC), 2012, San Diego, CA, pp. 319–325.

55 Lee, C.-K. *et al.* (2012) Wafer bumping, assembly, and reliability assessment of μbumps with 5 μm Pads on 10 μm Pitch for 3D IC Integration. IEEE 62nd Electronic Components and Technology Conference (ECTC), San Diego, CA, USA, pp. 636–640.

56 Kim, J.-S., Oh, C.S., Lee, H., Lee, D., Hwang, H.-R., Hwang, S., Na, B. *et al.* (2011) A 1.2 V 12.8 GB/s 2 Gb mobile Wide-I/O DRAM with 4×128 I/Os using TSV-based stacking. IEEE International Solid-State Circuits Conference Digest of Technical Papers (ISSCC), pp. 496–498.

57 Zhan, C.-J., Tzeng, P.-J., Lau, J.H., Dai, M.-J., Chien, H.-C., Lee, C.-K., Wu, S.-T. *et al.* (2012) Assembly process and reliability assessment of TSV/RDL/IPD interposer with multi-chip-stacking for 3D IC integration SiP. IEEE 62nd Electronic Components and Technology Conference (ECTC), San Diego, CA, USA, pp. 548–554.

58 Banijamali, B., Ramalingam, S., Nagarajan, K., and Chaware, R. (May 2011) Advanced reliability study of TSV interposers and interconnects for the 28nm technology FPGA. IEEE 61st Electronic Components and Technology Conference (ECTC), pp. 285–290.

59 Charbonnier, J., Cheramy, S., Henry, D., Astier, A., Brun, J., Sillon, N., Jouve, A., Fowler, S., Privett, M., Puligadda, R., Burggraf, J., and Pargfrieder, S. (2009) Integration of a temporary carrier in a TSV process flow. IEEE 59th Electronic Components and Technology Conference, San Diego, CA, USA, pp. 865–871.

60 Fukushima, T., Ohara, Y., Murugesan, M., Bea, J.-C., Lee, K.-W., Tanaka, T., and Koyanagi, M. (2011) Self-assembly technologies with high-precision chip alignment and fine-pitch microbump bonding for advanced die-to-wafer 3D integration. IEEE 61st Electronic Components and Technology Conference (ECTC), pp. 2050–2055.

61 Chen, Q., Zhang, D., Wang, Z., Liu, L., and Lu, J.-Q. (2011) Chip-to-wafer (C2W) 3D integration with well-controlled template alignment and wafer-level bonding. IEEE 61st Electronic Components and Technology Conference (ECTC), pp. 1–6.

62 Lu, J.-Q. (2012) 3D hyper-integration: past, present and future. *Future Fab International*, **41**, 81–87.

20
Bonding and Assembly at TSMC

Douglas C.H. Yu

20.1
Introduction

Three-dimensional (3D) integration using wafer-scale through-silicon via (TSV) technology has achieved an important milestone by moving from the development stage into production at TSMC in 2012. Homogeneous integration was first implemented by dividing a large die into several smaller ones and integrating them on an Si interposer with significantly improved die yield and reduced cost. Heterogeneous integration offers the most efficient means of packaging dissimilar devices, different technology nodes, and functionalities into a single package with the shortest possible inter-die interconnect length. 3D IC is an attractive and low-cost substitute for a large multifunctional system-on-chip (SoC) die, which can be partitioned into a few smaller dies, each manufactured with its specific function and the best technology node and integrated together by 3D IC technology. The significant reduction in power consumption, improved performance, and smaller form factor has been widely discussed and will be reported in the other chapters.

As discussed in this book and other published literature, there are a wide variety of 3D integration approaches using TSV for vertical die stacking. It is important to understand the limitations and advantages of each process alternative before determining the most viable and attractive process options. TSMC is committed to delivering the most efficient and cost-competitive 3D solution in the market. The selection of the most viable technology needs to fulfill not only the most advanced and cost-competitive technology available in the market, but also the vision that the technology is sustainable for broad applications and extendable to future product roadmaps. It is, thus, very important to select a process flow not only from the technology point of view but also for its cost, yield, and risk considerations, along with the extendibility into future technology nodes and fulfilling the need for a wide spectrum of future applications.

Chip-on-wafer-on-substrate (CoWoS™) is one of the 3D TSV technologies that attached the chips first on a wafer through a chip-on-wafer (CoW) process. The CoW is then diced and attached to the substrate (CoW-on-substrate) to form the final CoWoS assembly. TSMC has successfully developed, optimized, qualified the CoWoS process, and moved into production. All the key enablers of the CoWoS technology, such as the TSV, microbumping, wafer handling, bonding and debonding, fine-pitch redistribution layers (RDLs) fabricated on both sides of the wafer, have been thoroughly investigated and demonstrated to meet stringent yield and reliability requirements. It provides the end-to-end solution to the customers that offers the full advantages of 3D chip stacking and system integration with four major differentiators when compared to the alternative approaches, such as chip-on-substrate (CoS) or chip-on-chip (CoC):

1) *High yield*: Excellent microbump joining yield has been demonstrated.
2) *Easy handling*: Large die size and asymmetrical chip dimensions can be handled easily.
3) *Risk mitigation*: Risk is minimized for thin wafer handling and transportation.
4) Mature wafer processes are done in a foundry environment for high volume, throughput, and high yield.

In particular, wafer and chip warpage control and joining have been major challenges for the CoC process, but were easily solved with the CoW approach for 3D chip stacking. CoWoS also minimizes the warpage introduced by thermal expansion mismatch between silicon and organic substrates. When the key technology challenges are resolved, CoWoS has proven to be capable of achieving a stable and consistently high yield process. Test vehicles have been qualified and the technology is in volume production [1].

20.2
Process Flow

CoWoS process basically has two major assembly steps. First, individually diced top dies are flip joined on the bottom die, which is still in the wafer form or could be in the form of a reconstructed wafer. During stacking, the wafer provides the mechanical support. The joining between the top dies and the wafer is through microbumps, which are at much finer pitch than the much coarser C4 bumps on the organic substrate. CoW-first offers a much broader process window, which is demonstrated by the excellent stacking yield and warpage control because both sides of the joint are silicon with matched coefficients of thermal expansion (CTEs). This advantage also enables CoW-first capable of stacking large die with finer microbump pitch, which is critically important to extend to the future generation of products as microbump pitch continues to scale down. After CoW stacking, the next step is dicing the wafer and performing CoW strata stacking on the organic substrate (Figure 20.1c).

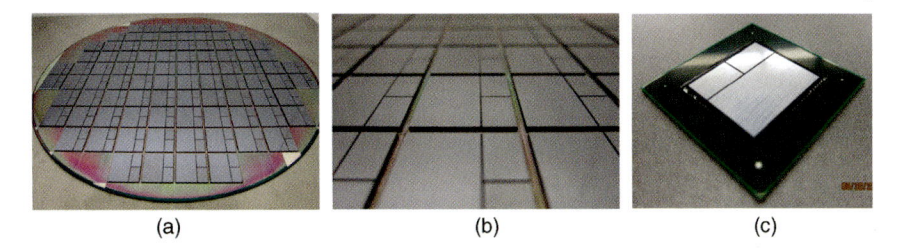

Figure 20.1 CoWoS process example showing two major stacking steps. First, CoW is performed on the wafer-level as seen in parts (a) and (b). And finally (CoW)oS to complete the process is demonstrated in part (c) [1]. (a) CoW each with three chips attached and repeated throughout the bottom wafer. (b) Enlarged image shows three-chip integration. Notice the saw street gap is larger than the chip-to-chip gap. (c) CoWoS^{TM} after completion of final on substrate stacking. Note that in this case, the top dies have not been protected by molding compound. Reprinted with permission from IEEE. Courtesy of TSMC.

20.3 Chip-on-Wafer Stacking

Depending on whether the designated stacking is face-to-face (F2F) or face-to-back (F2B), there are at least two different approaches to perform CoW stacking. If it is F2B, the approach is to form C4 bumps on the front side (the side with active devices or high-density RDL) of the wafer and then protect the C4 bumps with wafer carrier and temporary bonding. The wafer is ground from the backside to reveal TSV, followed by backside RDL and microbump formation. Finally, CoW stacking is performed using microbump joining [2]. Figure 20.2 shows the process flow of such an approach.

Alternatively for F2F, CoW stacking is performed using traditional wafer-form pick-and-place flip-chip tools with the microbump on the front side of the wafer. Although the joining is through microbumps instead of C4 bumps, there is no fundamental difference in production flow. The wafer can go through the solder reflow process to complete the joining, depending on if traditional solder reflow or thermal compressive bonding (TCB) is used.

After joining all the individual dies on the wafer, capillary underfill is applied. In an alternative approach, where nonflow underfill (NUF) or wafer-level underfill (WLUF) is used, it is to be applied before die placement prior to TCB. After underfill application, the CoW wafer front side will be protected with carrier and proceed for wafer thinning and backside (bottom wafer backside) processing. Typical TSV wafer thinning, TSV revealing, backside RDL (if any) processing, under-bump-metallization (UBM), and C4 solder bumping are then processed on the backside of the wafer to complete the wafer-level processes. The CoW wafer is then diced. Isolated CoW chips are ready for stacking on substrate. As the CoW stacking occurs when the bottom die is still in thick wafer form, and when the CoW strata wafer is thinned from backside, the challenging issue of thin wafer handling can be minimized or fully avoided. CoW also enables poststacking processing or

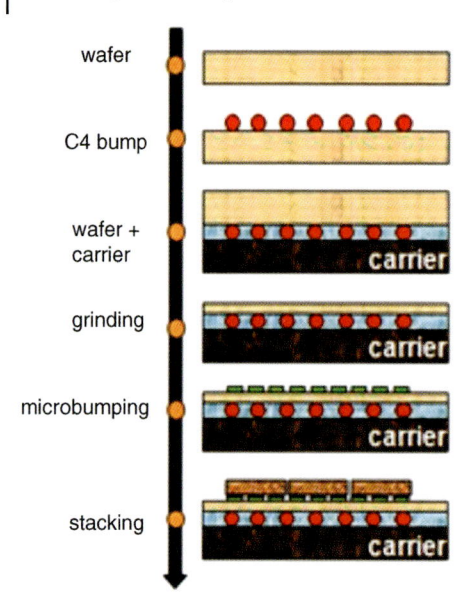

Figure 20.2 CoW process flow with C4 bumps on the front side (the side with the devices) of the wafer [2]. Reprint with permission from IEEE. Courtesy of TSMC.

testing to be performed in the wafer form, which greatly improves the overlay alignment, resolution tolerance, process throughput, and yield control of the poststacking process steps. Figure 20.3 shows a post-CoW test result. Little or no impact from CoW assembly on ring oscillator performance was compared before

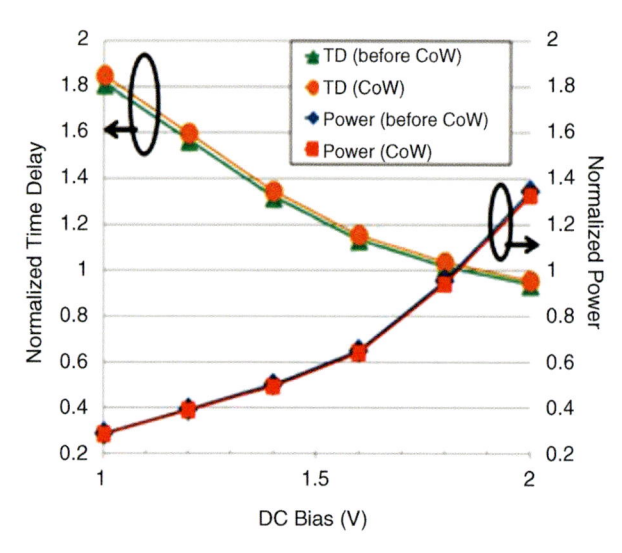

Figure 20.3 Comparison of ring oscillator's normalized power and time delay before and after CoW stacking process. Both the power and time delay have comparable performance before and after CoW stacking process [2]. Reprinted with permission from IEEE. Courtesy of TSMC.

and after the joining. This example also demonstrated the possibility of highly sensitive wafer level functional testing to be done immediately after the CoW joining.

20.4
CoW-on-Substrate (CoWoS) Stacking

There is no fundamental difference between CoW on substrate stacking and traditional flip-chip packaging processes. C4 bumps are used for the joining, chip placement on substrate with the same alignment accuracy and throughputs, same reflow, same underfill capillary process and curing, same functional testing and sorting. This, in fact, is one of the reasons why CoWoS is attractive and production friendly. Nearly all of the flip-chip process learning and experiences can be applied directly to (CoW)oS processes.

20.5
CoWoS Versus CoCoS

Alternative stacking sequences such as chip-on-chip-on-substrate (CoCoS), either CoC-first or CoS-first, are suffering a variety of different challenges [3]. For CoC-first, thin wafer and thin die handling and postjoining testing are major issues. All chips to be assembled need to be tested, thinned, backside processed, diced, and sorted onto proper containers or carriers individually. The bottom die then needs to be placed onto a proper supporting or protective stage or carrier before a top die can be bonded on top of it. This is not a trivial process step, as the bottom die has C4 bumps on the backside and microbumps on the front side. Handling a double-side-bumped large-dimensional chip for bonding is a new technological challenge for the industry. Coplanarity control becomes a nightmare if there is no proper supporting mechanism, not to mention if the CoC bonding is using thermal compression bonding.

For CoS-first, a large-dimension thinned bottom chip joined to organic substrate inevitably results in large warpage of the assembled strata, as seen in Figure 20.4. Subsequent top die joining to the CoS strata becomes extremely difficult due to the large warpage of the CoS bottom strata coupled with the fact that Co(CoS) joining is using fine-pitch microbumps with strict coplanarity and lateral dimension alignment requirements.

Alternative CoCoS assembly flow uses TCB and NUF or WLUF. Warpage control becomes less of a problem by using TCB. But TCB with a double-side-bumped chip is not a straightforward process even by using a specially designed handling jig, bonding head, or bump structure. Both TCB and NUF or WLUF are very challenging for achieving high yield in high-volume mass production. And the industry as a whole has little experience in the long-term reliability of such assembly. The industry needs to learn from both high-volume production and

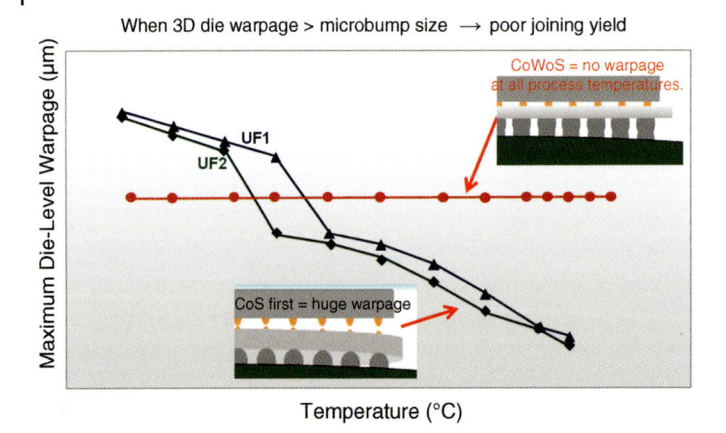

Figure 20.4 Three-dimensional assembly yield is tightly correlated to warpage control during assembly. CoS-first introduces large warpage both at soldering temperature and at reliability stressing, which, in turn, results in low assembly yield and poor mechanical reliability, such as temperature cycle. Courtesy of TSMC.

long-term field application in order to accumulate sufficient data and experience to optimize the technologies. Many of the above-mentioned issues associated with CoCoS are either currently without solution, or may impact the joining yield, or require extensive effort, cost, and time to resolve.

20.6
Testing and Known Good Stacks (KGS)

Another often mentioned 3D issue is testing and known good dies (KGD) or known good stacks (KGS). Testing is a cost-adding step in the production flow. Proper testing can help to remove the bad components, improves the final product yield, and thus reduces the overall production cost. However, testing also increases the product-handling steps and helps to increase the risk of producing bad parts, particularly when the dies are thin with stacked strata. Testing is often required when products are handed over from one supply chain step to the next. In 3D stacking, this becomes a complicated issue.

The true advantage of 3D TSV integration, compared to traditional packaging technology-based 3D integration such as package-on-package (PoP), CoC, or multichip module (MCM), is to integrate different or the same type of chips with the shortest interconnect length, along with using fine pitch and high-density silicon-based wafer interconnects instead of low-density organic substrate-based technology. Obviously, such 3D integration needs to be based on wafer processes. When different chips are stacked onto the bottom wafer (either an interposer or a large active chip serves as the bottom chip), CoW, the stack can be tested at the wafer form (wafer probing, known as the CP test) for KGS before the wafer is sawn into individual strata. Such testing is not possible if the stacking is starting from

CoS, unless special design-for-testing (DfT) is implemented through the organic substrate. The effort for implementing such DfT bears extra cost to the chip and substrate design and, in some cases, could be far too complicated to implement in organic substrate. CoWoS, thus, is a more cost-effective 3D assembly flow in terms of KGS and substrate design.

20.7
Future Perspectives

CoWoS processes using extremely large Si interposers (larger than 1 in. × 1 in. square) as the bottom die have been demonstrated with high yield using various test vehicles and prototype chips. Mechanical and electrical reliability tests have passed the stringent industrial component-level reliability requirement.

CoWoS stacking scheme offers a wide application window for large die, fine pitch, and complex heterogeneous 3D system integration. The fundamental process flow is robust and adaptable to different joining technologies (reflow, TCB, or Cu–Cu bonding), different underfill materials (capillary, nonflow, wafer-level), and different molding processes (wafer or chip). Such adaptability is crucial and allows the 3D system integration to evolve according to rapid change in market, products, and application needs.

References

1 Chiou, W.C., Yang, K.F., Yeh, J.L., Wang, S. H., Liou, Y.H., Wu, T.J., Lin, J.C., Huang, C. L., Lu, S.W., Hsieh, C.C., Teng, H.A., Chiu, C.C., Chang, H.B., Wei, T.S., Lin, Y.C., Chen, Y.H., Tu, H.J., Ko, H.D., Yu, T.H., Hung, J.P., Tsai, P.H., Yeh, D.C., Wu, W.C., Su, A.J., Chiu, S.L., Hou, S.Y., Shih, D.Y., Chen, Kim, H., Jeng, S.P., and Yu, C.H. (2012) Symposium on VLSI Technology (VLSIT), 2012, pp. 107–108.

2 Lo, T., Chen, M.F., Jan, S.B., Tsai, W.C., Tseng, Y.C., Lin, C.S., Chiu, T.J., Lu, W.S., Teng, H.A., Chen, S.M., Hou, S.Y., Jeng, S.P., and Yu, C.H. (2012) IEEE International Electron Devices Meeting (IEDM), 2012, pp. 793–795.

3 Wakiyama, S., Ozaki, H., Nabe, Y., Kume, T., Ezaki, T., and Ogawa, T. (2007) Novel low-temperature CoC interconnection technology for multichip LSI (MCL). Proceedings of the 57th Electronic Components and Technology Conference, 2007 (ECTC 2007), pp. 610–615.

21
TSV Packaging Development at STATS ChipPAC

Rajendra D. Pendse

21.1
Introduction

The advent of 3D through-silicon via (TSV) technology may be viewed as a convergence of the independent trajectories of system-on-chip (SoC) at the Si level and system in a package (SiP) at the package level. As such, it represents a superset of the SoC and SiP integration approaches and presents a paradigm change for both the Si and packaging technologies as we know them today.

Table 21.1 outlines STATS ChipPAC's view of the variants of TSV technology, the affected products and applications, the changes in package architecture that they bring about, the associated input/output (I/O) densities emanating from the Si device and the estimated timing of introduction for each variant. The application of TSV technology for "partitioning and reintegration" of Si is a natural application for large devices with die sizes approaching the limits for lithography tools (steppers): It may be viewed as providing another dimension for the continuation of the proverbial Moore's law for die sizes that would otherwise exceed the limits of the reticle sizes used in steppers. This comprises the first variant of the technology as shown in Table 21.1 and has already been embraced by the leading foundries and customers for field programmable gate array (FPGA) products. The second variant is targeted at mobile platforms and represents a "true 3D" implementation; the primary driver is the increase in band width between the application processor and memory with concomitant decrease in parasitics, which allows the attainment of higher performance in products such as smartphones and tablets without an increase in power dissipation. This implementation of TSV technology is under active development at STATS ChipPAC in joint work with key application processor customers and will be described in detail later. The third variant is an adaptation of the first and second variants but targeted more toward traditional computing, that is, a GPU or CPU communicating with memory wherein performance and power requirements necessitate this approach. The final (fourth) variant may be viewed as

Handbook of 3D Integration: 3D Process Technology, First Edition.
Edited by Philip Garrou, Mitsumasa Koyanagi, and Peter Ramm.
© 2014 Wiley-VCH Verlag GmbH & Co. KGaA. Published 2014 by Wiley-VCH Verlag GmbH & Co. KGaA.

Table 21.1 Applications, markets, and introduction timing for different variants of TSV technology.

	Integration scheme	Application	Previous configuration	Future configuration with TSV	Drivers	I/O density	Approximate timing
1	Partitioning and reintegration of monolithic Si (2.5D)	FPGA; large ASIC; GPU			Cost reduction; Si node/Moore's law extension	<1 μm L/S; ~10^3–10^4 mm^{-2}	2011–2012
2	Integration of disparate functions (for mobile) (3D)	AP + Wide I/O memory			Higher bandwidth at low power	10 μm L/S; 10^1–10^2 mm^{-2}	2014
3	Integration of disparate functions (for computing) (2.5D)	GPU/CPU + memory			High power/performance through minimization of parasitics	1–5 μm L/S; 10^2–10^3 mm^{-2}	2014–2015
4	Integration of disparate functions (for computing) (3D)	CPU + HBM memory			Higher performance through removal of memory bottleneck	<1 μm L/S; >10^4 mm^{-2}	~2015–2016

the "Holy Grail" of TSV technology as it is targeted directly at speeding the CPU–memory interface in high-end computing systems by essentially relieving the so-called Von Neumann bottleneck; however, this version requires specialized Si design, structure, and unique cooling solutions, hence its deployment is likely to occur further out in time. The primary development focus at STATS ChipPAC has been in the second variant in keeping with the company's leadership in the packaging of mobile platforms with parallel development activity in the third variant.

21.2 Development of the 3DTSV Solution for Mobile Platforms

One of the major challenges in the implementation of TSV technology has been the partitioning and handoff points between the Si fabrication (foundry) and package manufacturing (outsourced semiconductor assembly and test (OSAT)) operations (package manufacturing houses are henceforth generally referred to as OSATs). Figure 21.1 shows the model that is followed currently by STATS ChipPAC and other OSAT houses.

STATS ChipPAC has developed both the so-called mid-end and back-end processing capabilities [1–6]. Mid-end processes require completely new know-how and entail fablike manufacturing operations and significantly higher capital intensity (dollar investment per unit output), whereas the back-end processes are extensions of existing, more traditional packaging capabilities.

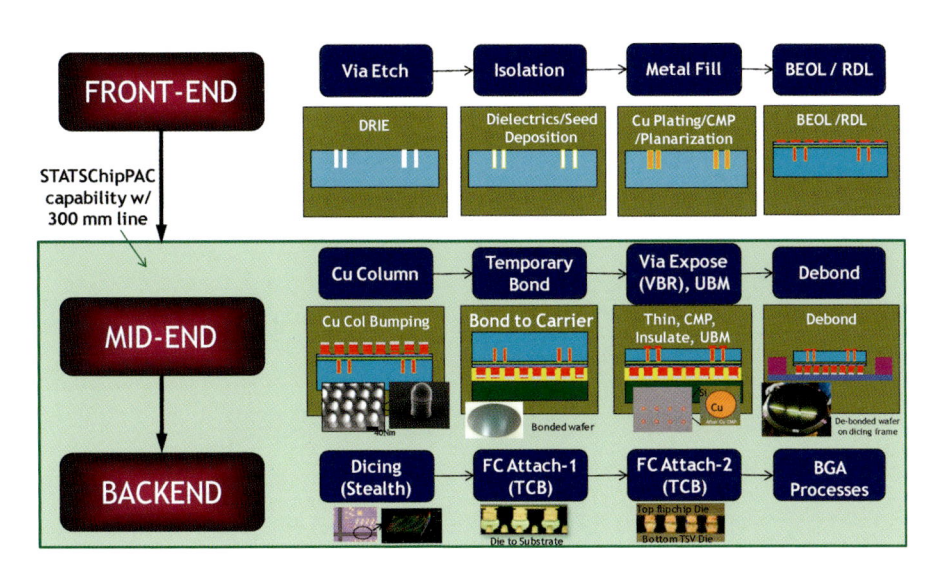

Figure 21.1 Overall 3D TSV process flow and capabilities put in place by STATS ChipPAC.

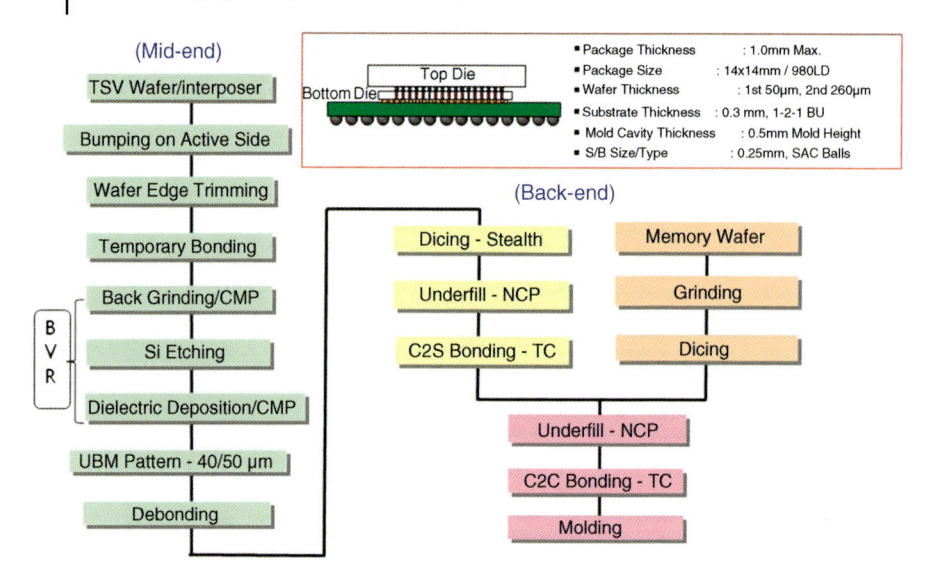

Figure 21.2 Process flow and package test vehicle used for 3D TSV development.

Figure 21.2 shows the typical package structure and process flow that was developed using a 28 nm test vehicle targeted toward smartphone and tablet product configurations. STATS ChipPAC developed each of the unit processes depicted in Figure 21.2, that is, the processing begins with an incoming full-thickness device wafer from an Si foundry with the TSVs already embedded within the wafer structure but not yet revealed. STATS ChipPAC is currently tooled for a manufacturing capacity in the range of 5000 wafers per month. Key features of the individual unit processes are described below.

Shown in Figure 21.3 are features of the microbumping unit process, which are an extension of the plated bump process already in high-volume manufacturing (HVM) for flip-chip products with a demonstrated capability of 40 μm pitch for Cu

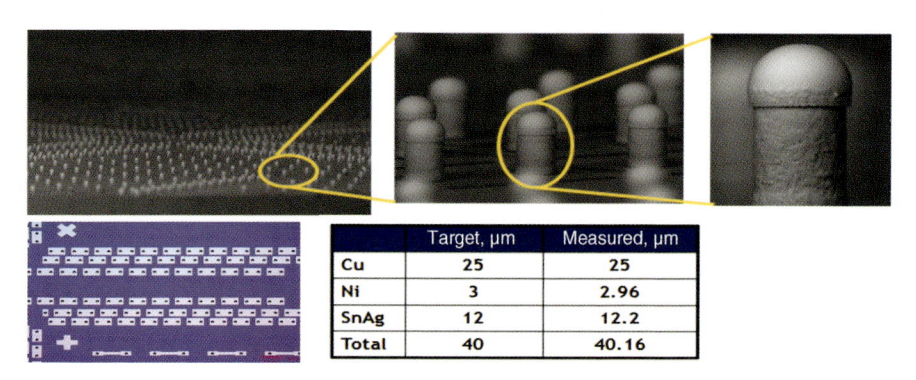

	Target, μm	Measured, μm
Cu	25	25
Ni	3	2.96
SnAg	12	12.2
Total	40	40.16

Figure 21.3 Plated microbumping process (Cu column bumps).

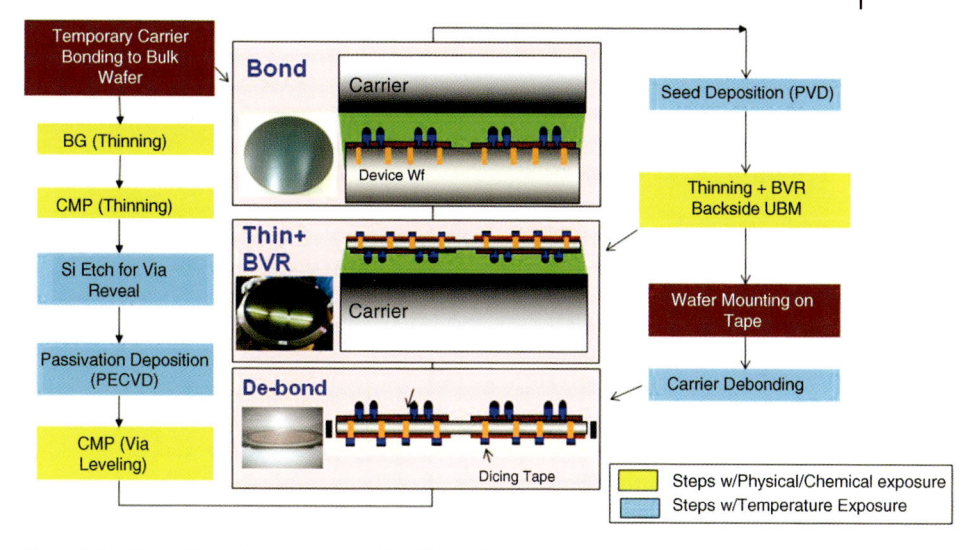

21.2 Development of the 3DTSV Solution for Mobile Platforms

Figure 21.4 Illustration of temporary bonding/debonding processing.

columns with "caps" composed of Sn/Ag solder. The associated process for formation of landing pads for the face-to-back bonding of the top die also use plating but typically have an Ni/Au metallurgy.

A combination of mechanical grinding, chemical–mechanical polishing (CMP), and wet etching is used for backside via reveal (BVR) following a temporary bonding of the wafer (Figures 21.4 and 21.5). An UV-based temporary bonding process with laser-assisted release was selected in view of the manufacturing ease, throughput, and process stability relative to other alternate options. IR microscopy is used as the metrology tool for monitoring and managing the thicknesses of the adhesive layer. The BVR process is followed by passivation of the exposed via "nails" using plasma-enhanced chemical vapor deposition (PECVD) and final fine grinding with CMP to form the finished via structure. The wafer is now debonded with simultaneous transfer to the dicing tape.

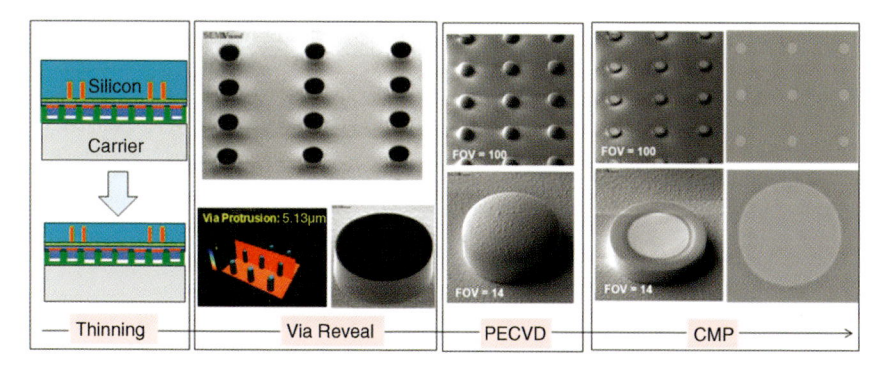

Figure 21.5 Illustration of thinning, via reveal, PECVD, CMP processing.

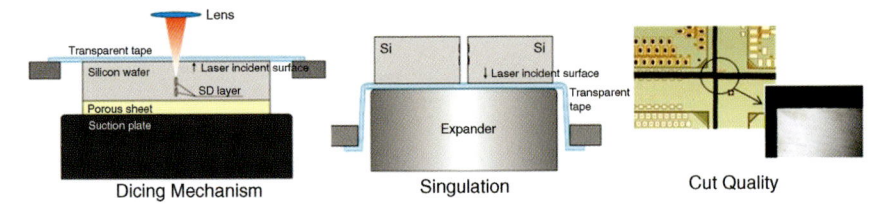

Figure 21.6 Stealth dicing process.

The back end processing begins with wafer dicing. We have developed "stealth dicing" (Figure 21.6) that lends itself well to the dicing of thin wafers below about 100 μm nominal thickness and allows the use of narrow scribe street widths in the range of 50 μm or less.

This is followed by the microbonding of the bottom die to the package substrate and the subsequent microbonding of the top die in a face-to-back configuration to the bottom die. Both of the bonding steps typically use the thermocompression bonding technique with nonconductive paste (NCP) or nonconductive film (NCF) (Figure 21.7). The die-to-substrate bonding is accomplished typically at 80 μm staggered pitch whereas the die-to-die bonding is at sub-50 μm pitch (40 μm typical). In some applications, the die-to-substrate bonding is performed using mass reflow process with subsequent capillary underfilling as opposed to thermocompression bonding with NCP and NCF. For some applications, a molding step is performed using film-assisted molding (FAM) technology to leave the back side of the top die exposed for subsequent attachment of an optional heat spreader.

The final package structure shown in Figure 21.2 was subjected to a full suite of reliability testing per standard JEDEC test procedures to demonstrate a robust structure and manufacturing process (Figure 21.8).

STATS ChipPAC also developed solutions for the third variant show in Table 21.1, commonly referred to as "2.5D TSV Integration," wherein a separate passive interposer with prefabricated TSVs is used to integrate a GPU or CPU

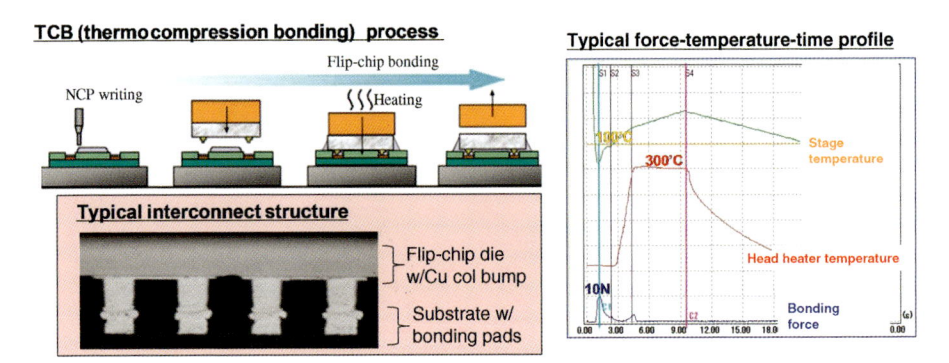

Figure 21.7 Thermocompression bonding process for fine-pitch interconnection.

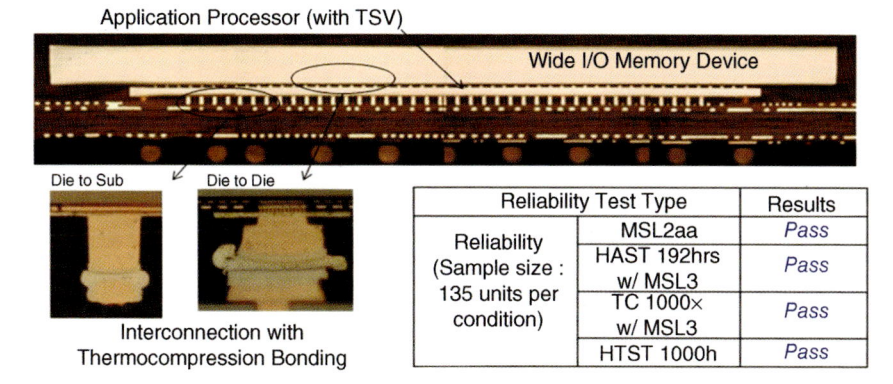

Figure 21.8 Final package structure and reliability test results.

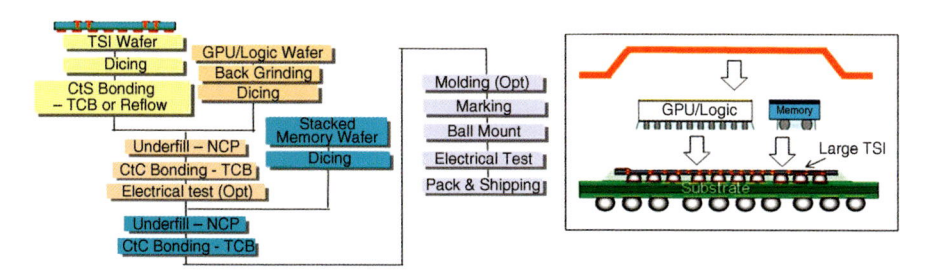

Reliability Test Type		Results
Reliability (Sample size: 135 units per condition)	MSL2aa	Pass
	HAST 192hrs w/ MSL3	Pass
	TC 1000× w/ MSL3	Pass
	HTST 1000h	Pass

Figure 21.9 Process flow for 2.5D integration with large thin through-Si interposer (large TSI).

device with a memory device. The process flow for this structure is slightly different (Figure 21.9) but leverages the technology developed for the 3D TSV process described above with the possible exception of the handling of large thin through-silicon interposers (TSIs).

21.3
Alternative Approaches and Future Developments

Development of infrastructure and handoff points between the Si foundry and OSAT organizations has been a key area of focus in recent months. Also, the refinement of process tools and individual unit processes for yield and throughput improvement is in progress as the industry braces for a possible start of production in the 2014 time frame particularly for the second and third variants in Table 21.1.

The 3D TSV approach entails the fabrication of TSVs directly in the live Si wafer – this calls for more advanced fabrication techniques, design tools, and standard cell development at the Si level. Some industry experts have therefore suggested that the 2.5D approach with a suitable technique and supply source for the interposer is likely to emerge as a stopgap solution. In STATS ChipPAC's

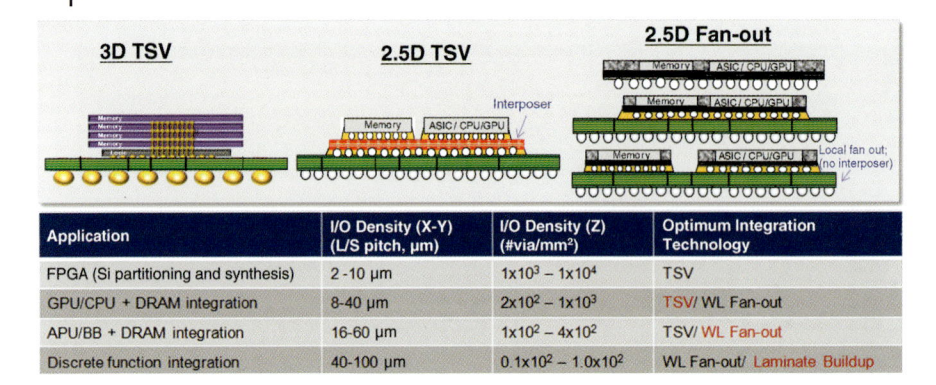

Application	I/O Density (X-Y) (L/S pitch, μm)	I/O Density (Z) (#via/mm²)	Optimum Integration Technology
FPGA (Si partitioning and synthesis)	2-10 μm	$1 \times 10^3 - 1 \times 10^4$	TSV
GPU/CPU + DRAM integration	8-40 μm	$2 \times 10^2 - 1 \times 10^3$	TSV/ WL Fan-out
APU/BB + DRAM integration	16-60 μm	$1 \times 10^2 - 4 \times 10^2$	TSV/ WL Fan-out
Discrete function integration	40-100 μm	$0.1 \times 10^2 - 1.0 \times 10^2$	WL Fan-out/ Laminate Buildup

Figure 21.10 Alternate approaches for 2.5D integration using FOWLP/eWLB.

assessment, there are multiple alternatives for the interposer concept that go beyond the Si TSV approach, yet fulfill the necessary density requirements, as illustrated in Figure 21.10. Notably, fan-out wafer level packaging (FOWLP) or eWLB (eWLB is a trade name popularized by STATS ChipPAC) technology offers an appealing approach to accomplish a 2.5D structure as it provides design rules within the sweet spot of applications represented by the second and third variants in Table 21.1, and is particularly suitable for mobile platform packaging (smartphones and tablet computers). FOWLP- and eWLB-based solutions are currently in active development with leading IC vendors in the smartphones and tablet space and production solutions are expected in the 2013 to 2014 time frame.

References

1 Pendse, R. (2011) 3D packaging evolution from an outsourced semiconductor assembly and test (OSAT) perspective. The ConFab Conference, PennWell, Las Vegas, NV.
2 Pendse, R. (2011) The packaging application space for 2.5D/3D and recent developments at OSATs. 3-D Architectures for Semiconductor Integration and Packaging Conference, RTI International, Burlingame, CA, USA.
3 Yoon, S.W., Kang, K.T., Choi, W.K., Lee, H.T., Yong, C.B., and Marimuthu, P. (2012) 3D TSV micro Cu pillar chip-to-substrate/chip/ assembly/packaging. International Wafer Level Packaging Conference, SMTA, Santa Clara, CA, USA.
4 Yoon, S.W., Na, D.J., Choi, W.K., Yong, C.B., Kim, Y.C., and Marimuthu, P. (2012) TSV MEOL (mid-end-of-line) and its assembly/ packaging technology for 3D/2.5D solutions. ICEP-IMAPS All Asia Conference (IAAC), ICEP, Tokyo, Japan.
5 Yoon, S.W., Na, D.J., Choi, W.K., Kang, K.T., Yong, C.B., Kim, Y.C., and Marimuthu, P. (2011) 2.5D/3D TSV processes development and assembly/packaging technology. IEEE 13th Electronics Packaging Technology Conference, 2011 (EPTC 2011), Singapore.
6 Yoon, S.W., Hsiao, Y.K., Yong, C.B., Choi, W.K., Kim, Y.C., Kang, G.T., and Marimuthu, P. (2011) 3D TSV middle-end processes and assembly/packaging technology. 18th European Microelectronics and Packaging Conference, 2011 (EMPS 2011), Brighton, England.

22
Cu–SiO₂ Hybrid Bonding

Léa Di Cioccio, S. Moreau, Loïc Sanchez, Floriane Baudin, Pierric Gueguen, Sebastien Mermoz, Yann Beilliard, and Rachid Taibi

22.1
Introduction

Bonding of metal surfaces is extensively used for MEMS sealing, power devices, heat dissipation, or 3D interconnections. For these applications, techniques such as thermocompression, with or without eutectic alloys or adhesives layers, bumps with low-temperature solders or direct bonding are implemented techniques [1–4].

Moreover, for more-Moore and more-than-Moore applications, low-temperature bonding and metal bonding are becoming the main drivers of the latest developments. As copper is the main metal used for CMOS interconnects, a high-density Cu interconnection between layer structures, is expected to be important for future three-dimensional integration of discrete electronic devices fabricated on the basis of different technology/design concepts.

To address that, copper–oxide surface direct bonding presents many attractive advantages: compatibility with front-end or back-end requirements for the sequential approach, low-temperature process, through-interconnect layer compatibility, high accuracy of alignment during bonding, and, the last but not least as the surfaces are bonded at room temperature, no underfill is needed and this technique allows tough a very high interconnect density.

Direct wafer bonding refers to a process by which two mirror-polished wafers are put into contact and held together at room temperature by adhesive forces, without any additional intermediate materials [5]. Direct wafer bonding can be achieved in a clean room at room temperature with ambient air, without any pressure and without any adhesive materials. However, high bond strength and a void-free bonding interface are required in order to provide reliable and high-performance devices.

Copper–oxide surface direct bonding is a technology that enables 3D metal interconnects at the same time as a bond between two piled-up 2D processed strata. Otherwise, with an insulating layer at the bonding interface, two through-silicon vias (TSVs) of different depths are needed to connect the upper and the lower layer. A single TSV technology is feasible because of localized electrical conductive

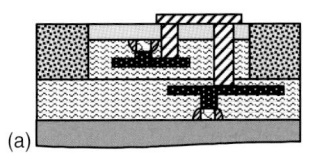

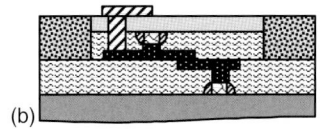

Figure 22.1 Die-to-wafer integration scheme with oxide bonding. This bonding technology needs two through-silicon vias: die-to-wafer integration scheme (a) with patterned Cu bonding (b).

pads at the bonding interface obtained with copper–oxide surface direct bonding (Figure 22.1).

Direct copper experiments have led to 8 μm pitch interconnections of thin Cu electrodes at around 350 °C using oxide bonding technologies for a local Cu thermocompression bonding [6]. The diffusion of Cu atoms across a thick adsorbate (mainly oxide) layer was obtained after long-time annealing in order to obtain sufficient bonding strength and electrical conductivity. Also demonstrated were 6 μm pitch bumpless Cu electrodes fabricated with the applied damascene process. The surfaces were bonded at room temperature under vacuum after an Ar sputtering of the surfaces in order to remove the native copper oxide layer. The use of high vacuum conditions increased the bonding process complexity, such as difficulties in sample handling and alignment procedure, as well as low manufacturing throughput and high cost. More recently, this process was modified allowing the bonding at atmospheric pressure in an O_2 atmosphere after the same surface preparation under vacuum [7].

The best process has to be easily integrated into the dual damascene back-end-of-line (BEOL) process, using standard BEOL process steps. So we aimed to develop a process based on the surface preparation using chemical–mechanical polishing (CMP), with a bonding at room temperature in ambient air without added bonding pressure during the bonding. The bonding annealing should be in the range allowed by the BEOL requirements, that is, under 400 °C [8–16]. This process is described in the following sections.

22.2
Blanket Cu–SiO$_2$ Direct Bonding Principle

22.2.1
Chemical–Mechanical Polishing Parameters

Damascene copper CMP is widely used in production lines as an efficient back-end process for copper removal down to the oxide. Standard damascene CMP process is the most used process to planarize metal levels in microelectronics. However, to avoid leakage, the process usually induces a dishing of the copper pads on patterned surfaces. This surface topology leads to important inconvenience: copper

Figure 22.2 Damascene CMP process with dishing controlled on copper pads.

pads dishing and oxide erosion. They are critical issues if one wants a direct copper contact between two face-to-face copper pads. At CEA-LETI, an optimized CMP-based process was implemented on a patterned surface to reach a high surface planarization level (Figure 22.2). The CMP optimizations (pad-slurry matching) target to minimize the copper dishing and oxide erosion with respect to the layout, and a standard wet cleaning is applied to remove the slurry residues and the remaining particles.

Wafers with 5 μm copper pads and a copper density of 20% were designed to achieve direct Cu–SiO$_2$ bonding. After CMP-based preparation, a dishing of less than 10 nm was obtained on the pads all over the wafer.

After cleaning, the wafers were introduced separately in an EVG SmartView™ alignment tool. Alignment was optically achieved because of specific alignment marks. The bonding wave is then initiated with a tip at the center of the wafers. Bonding occurs at room temperature, ambient air, and atmospheric pressure.

In Figure 22.3, one can see that using this process, the sealing of the different interfaces is perfect.

Simulation of copper pad deformation with respect to temperature, pad size, and height was done with ANSYS software, considering the elastic properties of silicon, silicon dioxide, and elastoplastic properties of copper. It can be seen in Figure 22.4 that the vertical displacement of the Cu pads varies with the pad geometry. Also the results show that a small dishing of copper pad will be overcome during the

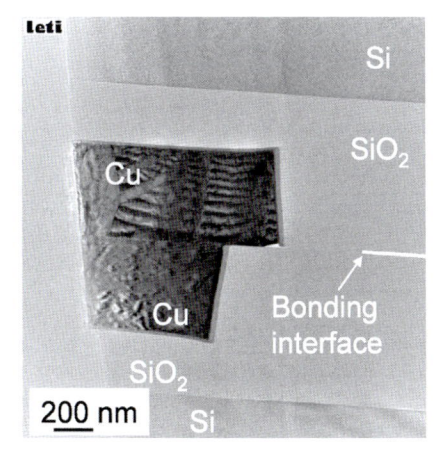

Figure 22.3 TEM image of bonded copper pad – the sealing of the different interfaces is perfect.

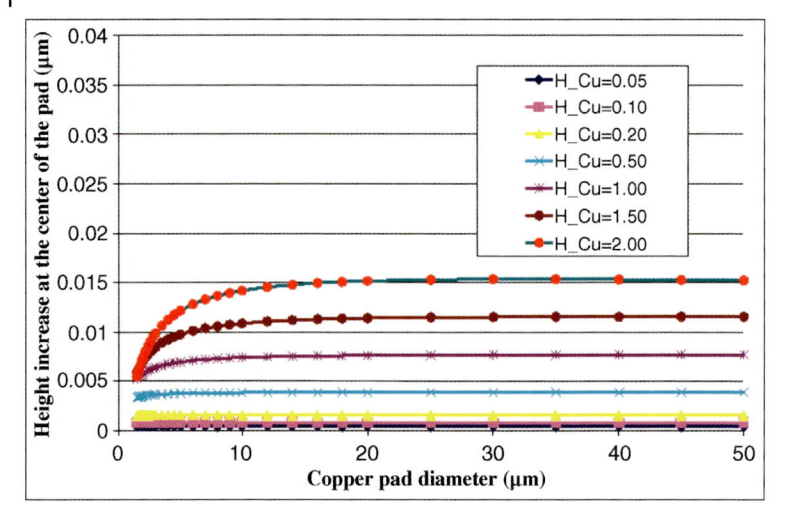

Figure 22.4 Simulation of the vertical displacement of copper pads at 200 °C, as a function of the pad geometry. H is the height of the copper line in micrometers.

postbond annealing even at low temperature, such as 200 °C. It is also important to point out that the monitoring of the CMP step and the knowledge of the layout needed for a good bonding is mandatory.

Very good bonding and high bonding toughness can then be obtained with respect to this bonding behavior. On Figure 22.5, acoustic observations confirm the high quality of the bonding. Bonding toughness was recorded using the double cantilever technique [17]; it ranged from 1 J m^{-2} at 200 °C up to 6.6 J m^{-2} at 400 °C, confirming that the mechanism is identical to intermetallic bonding described earlier [18]. With a 200 °C postbonding anneal, it was already possible to grind down to 5 µm the top silicon of the bonded pair Figure 22.5.

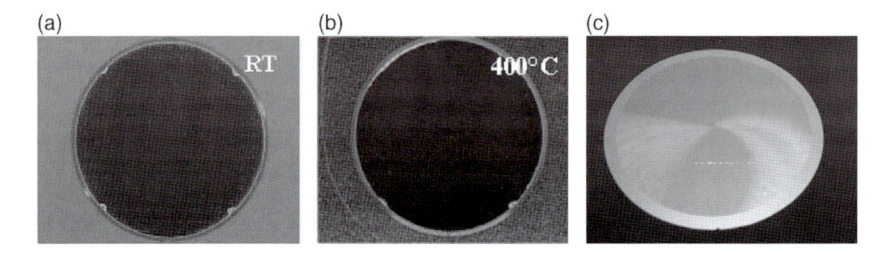

Figure 22.5 (a) Acoustic image of bonded patterned wafers (hybrid oxide-metal) at room temperature. (b) Same wafers after a 400 °C anneal. (c) Picture of a bonded pair ground down to 5 µm after a 200 °C postbonding anneal.

22.3
Aligned Bonding

22.3.1
Wafer-to-Wafer Bonding

Right after CMP, the patterned wafers are bonded at atmospheric pressure in an EVG SmartView alignment tool. The used alignments marks (for optical recognition) are standard crosses in a box with verniers allowing measurement accuracy at 0.2 µm or less depending on specific mark design. Alignment measurements are done with an infrared microscope (Figure 22.6).

Since alignment marks are present in every die of the wafer, it was possible to create a mapping of the misalignment of the bonded pair. No misalignment above 1.2 µm was recorded and the mean value was 0.6 µm with $\sigma = 0.4$ µm. These results confirm that since the process is done at room temperature, no huge misalignment is induced during the bonding, besides the bond is initiated by a pin in the center of the wafers.

22.3.2
Die-to-Wafer Bonding in Pick-and-Place Equipment

Chip-to-wafer (C2W) direct bonding was performed using the alignment capability of SET FC300 equipment and two sets of alignment patterns defined on chip and host wafer surfaces. To limit the particle contamination during processing, specific developments were realized on the FC300 equipment. A local microenvironment was created in the equipment to protect the wafer surface during the bonding of the chip (Figure 22.7). The bonding process is almost the same as that for wafer-to-wafer bonding. An extra cleaning is added after chip dicing.

On 300 mm wafers, the alignment accuracy was measured after bonding by reading the vernier patterns through the chip and the wafer with an infrared

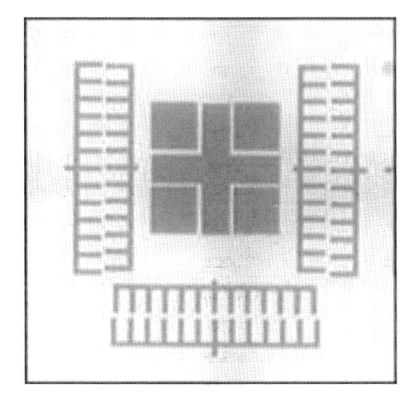

Figure 22.6 Alignment marks after bonding of patterned wafers.

Figure 22.7 Illustration of the microenvironment realized by the SET Company on the FC300 equipment to limit the particulate contamination on the wafer during the chip-to-wafer bonding.

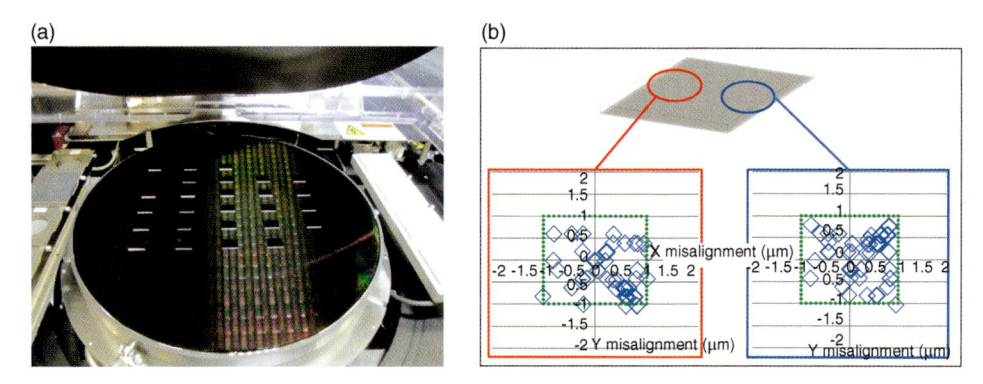

Figure 22.8 (a) Bonded chips on a 300 mm wafer. (b) Chip-to-wafer misalignment in both x- and y-directions measured on the left and right alignment patterns of 50 chips. Vernier scale resolution is 200 nm.

microscope (Figure 22.8). The measured misalignment remains lower than 1 μm in both the *x*- and *y*-axes on left and right alignment verniers [19,20].

22.3.3
Die-to-Wafer by the Self-Assembly Technique

The self-assembly process is based on the use of small waterdrops to align and bond dies to a wafer substrate. The obtained chip alignment results from waterdrop surface tension minimization, and the hybridization is performed because of direct bonding after the evaporation of the waterdrop as presented in Figure 22.9.

The drop containment sort was analyzed and its impact on the self-alignment process yield for dummy silicon oxidized chip was checked. As a result, we showed that with an adequate drop containment sort, that is, that the edges of the chip pad are hydrophobic (contact angle around 110°), a submicronic alignment with a 90% process yield can be obtained [21].

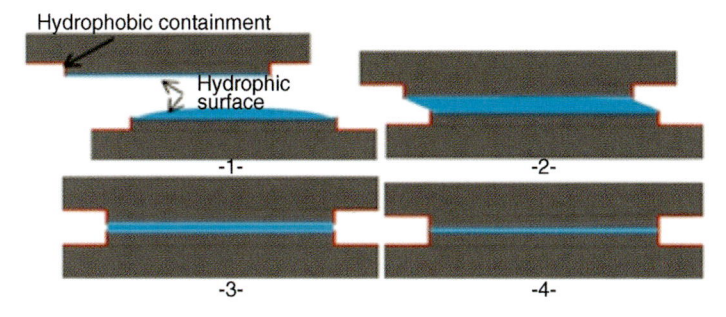

Figure 22.9 The different steps of the self-assembly process. (1) Inaccurate prealignment with pick-and-place tool (around 200 μm). (2) Drop spreading and chip is realigned with submicrometer accuracy (around 400 nm). (3) Drop evaporation. (4) Direct bonding hybridization.

In line with these results, the self-assembly process was transferred to chips with Cu-patterned structures. Hydrophobic preparation and integration were developed to maximize the wetting contrast between the bonding site and the rest of the wafer, taking care of the presence of Cu interconnections on both the chip and the wafer surfaces. First, the etch steps needed to create the hybridization pads were adapted to the patterned Cu–oxide surface. Copper is firstly etched with a wet chemical bath, typically based on a mixture of sulfuric acid (H_2SO_4) and hydrogen peroxide (H_2O_2). Then, the silicon oxide and the bulk silicon are etched by plasma. In a third step, a few nanometers of hydrophobic polymer are deposed at the wafer level. By using an adapted resist-stripping method, the lift-off effect occurs and the polymer film is selectively removed from the bonding areas. After the resist stripping, bonding areas are hydrophilized to reach a contact angle lower than 5° without degrading the hydrophobic containment (contact angle around 110°). Figure 22.10 presented C2 W structures that were obtained with such an integration process for patterned Cu surfaces.

Scanning electron microscope (SEM) observation of the bonding interface was performed (Figure 22.11) for the self-assembled structures after a 300 °C annealing. The bonding interface is not observable implying a perfect closure after the

Figure 22.10 Observation of the obtained self-aligned C2 W structures (patterned Cu surfaces).

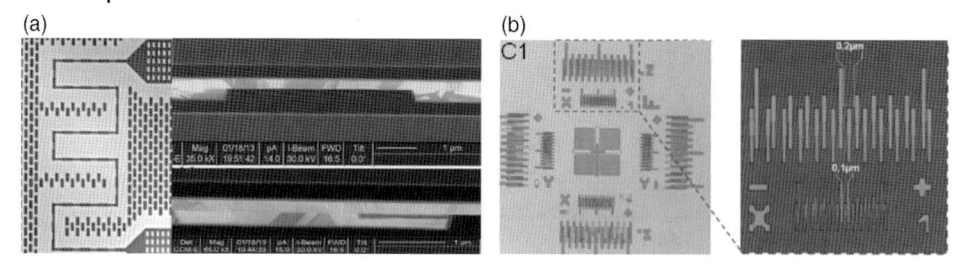

Figure 22.11 (a) Infrared observation of daisy chain after bonding. (b) SEM observation of the Cu–Cu contact (left part) after 300 °C annealing, and vernier at 200 nm is obtained (right part).

bonding and annealing sequence. Alignment accuracy was measured on specific vernier patterns – a misalignment as low as 200 nm was measured

22.4
Blanket Metal Direct Bonding Principle

Different stack of metals with or without diffusion barrier layers were tested for direct metal bonding Cu, W, and Ti. As deposited, the metal layers present a surface roughness that is too high to allow the direct bonding. A CMP step is therefore added prior to bonding to recover a surface roughness of less than 0.4 nm RMS. It has been verified that each surface exhibits a hydrophilic behavior with a contact angle of less than 5° for at least a few hours. After CMP, the metal surfaces include a thin oxide layer that has been measured by X-ray reflectometry (XRR) or ellipsometry and confirmed by X-ray photoelectron spectroscopy (XPS). The thickness of metal oxides is summarized in Table 22.1. This means that at room temperature, the bonding is obtained because of the metal oxide.

Bonding tests, following the specific surface treatments including planarization, cleaning, and symmetric metal–metal bonding (e.g., Cu–Cu, Ti–Ti, W–W), were performed at room temperature without applying any external force (similar to silicon direct bonding). No bonding defect could be seen from room temperature up to 400 °C (Figure 22.12).

The bonding strength of the different metal–metal bonding was evaluated after annealing at different temperatures with the double cantilever technique, Figure 22.13 [21]. It is worth mentioning that the Cu–Cu or Ti–Ti bonding appears to be much stronger than conventional SiO_2–SiO_2 bonding, while the W–W bonding leads to weaker bonding energy. However, this technique was developed for brittle samples and the energy dissipation generated by plasticity in the metal layer (ductility effect) is not taken into account and might significantly affect the

Table 22.1 Metal oxide layer thickness before bonding.

	Copper	Tungsten	Titanium
Oxygen-enriched layer (nm)	3	1.5	3.5

22.4 Blanket Metal Direct Bonding Principle

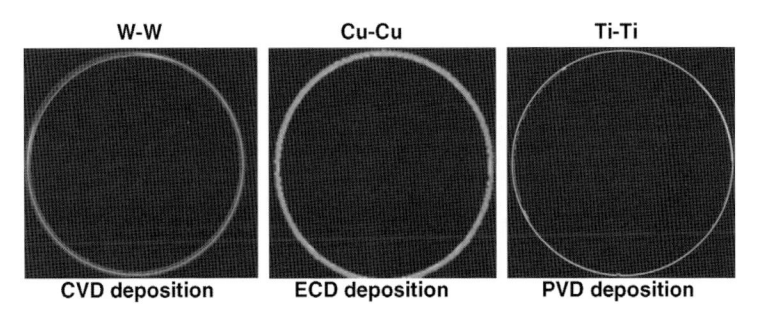

Figure 22.12 Scanning acoustic microscopy of the different metal bonding after a 400 °C annealing.

measure. Furthermore this ductility effect is different for each metal with respect to its nature and deposition parameter processes. However, the strength of these bondings will allow postprocessing such as thinning, TSV processing, etching, and so on. The variation of the bonding strength with temperature could be explained by the metal interdiffusion kinetics through the bonding interface and the native oxide layer instability. The oxide dissolution is also helped by plasticity mechanisms beginning at 0.3 of the melting temperature of the considered metal, Figure 22.14. It has been previously reported [18] that a high amount of metal bonds are recovered at the interface above 200 °C for copper and 700 °C for tungsten. For titanium, the getter effect allows a perfect metal bonding at 300 °C in that range of thickness [22].

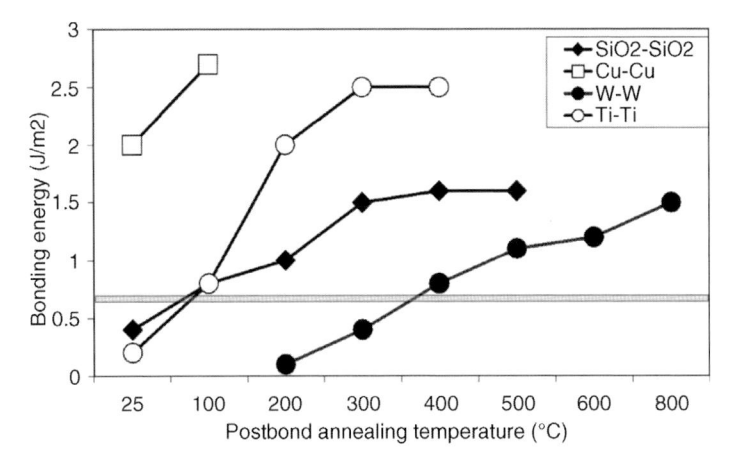

Figure 22.13 Bonding energies of the different metal bonding for different temperatures.

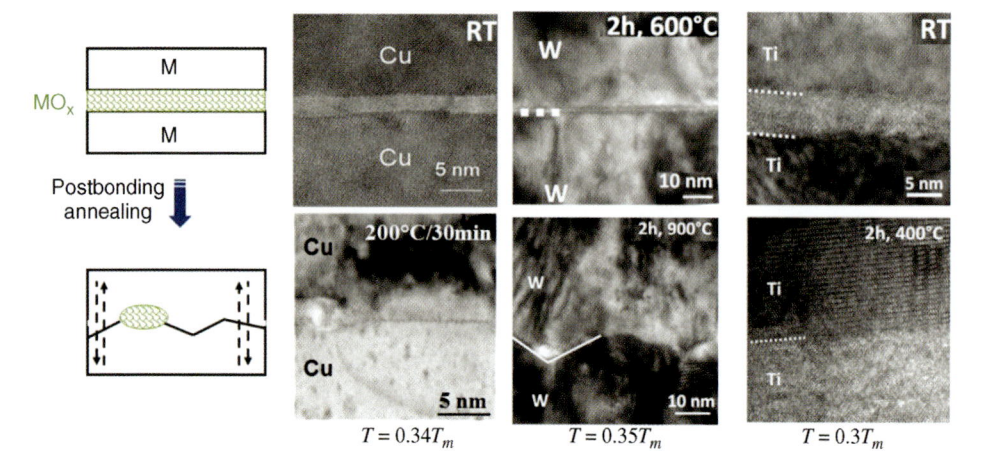

Figure 22.14 TEM observation of the metal bonding interface. At room temperature, each bonding interface presents an oxide layer with a thickness (twice that of each stack before bonding) in accordance with the measure of Table 22.1. At 0.3 Tm, the oxide layer is unstable and is mainly dissolved. Some oxide nodules of a few nanometers are visible at triple-grain boundaries.

22.5
Electrical Characterization

22.5.1
Wafer-to-Wafer and Die-to-Wafer Copper-Bonding Electrical Characterization

NIST (stand-alone and bonded) and daisy chains were patterned on wafers with a copper line of 500 nm. After bonding and annealing at 200 °C done in the way it was described earlier, the upper wafers were thinned down to 50 µm. Then via etching and deposition of a rerouting layer was done. Finally, benzocyclobutene (BCB) and under-bump-metallization (UBM) passivation are achieved. Figure 22.15 shows a schematic description of the full integration used for this study.

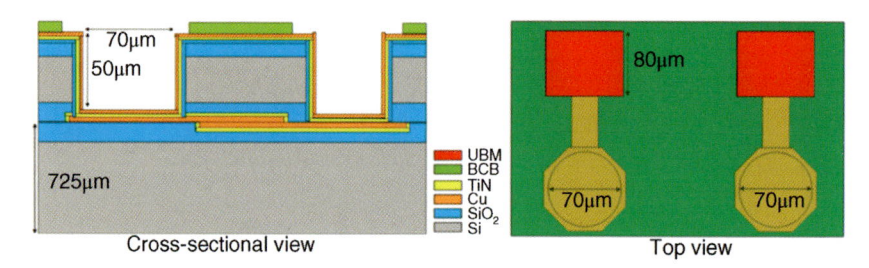

Figure 22.15 Schematic diagram of the full integration used for the tested dice. The top wafer of the bonded pair is thinned down to 50 µm followed by the vias etching and the deposition of the rerouting layer. Finally, BCB and UBM passivation are done to allow probing and packaging.

22.5 Electrical Characterization | 305

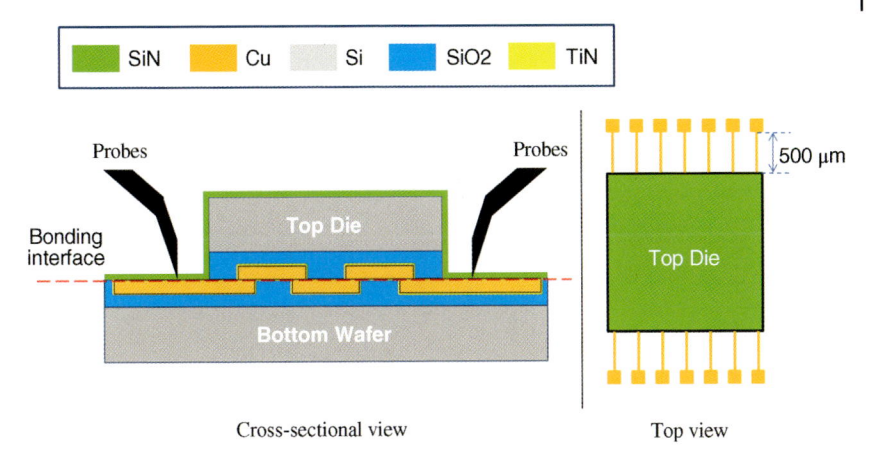

Figure 22.16 Schematics diagrams of a bonded die with deported electrical characterization probes (left) in contact with the deported pads 500 μm away from the top die (right).

For D2W bonding, in order to characterize the bonding interface without any further steps of lithography and etching, the bottom wafer was designed to integrate deported contact pads. An SiN encapsulation was required to prevent oxidation of the copper lines. Figure 22.16 shows the final integration and the location of the test probes.

Stand-alone NIST were used to extract a reference value for copper resistivity (Table 22.2). Copper-line resistivity was measured with four-point probes electrical characterization on NIST structures:

$$\rho = (R \times S)/L, \tag{22.1}$$

where, R (ohm) is the measured copper line resistance, L is the copper line length ($L = 640$ μm), and S is its section ($S = 3 \times 0.5$ μm^2).

A copper resistivity of 2.0×10^{-2} Ω μm has been extracted.

In Table 22.3, we present a comparison between wafer-to-wafer (W2W) and die-to-wafer (D2W) experimental resistance along with theoretical resistance calculated because of the extracted resistivity. As already shown by Taibi et al. [23,24], W2W structures are in very good accordance with the theoretical value. Furthermore, the bonding interface offers a very low resistance since the results between NIST a and NIST c are identical, showing that the current passes through the entire bonded line. The same behavior is observed on the D2W structures. Indeed, the electrical resistances are very close to both the theoretical resistance and the W2W structures, showing a good bonding quality and low resistance interface.

The various tested daisy chains are listed in Table 22.3. DC5 is the daisy chain with interconnection density of 1.5×10^6 cm^{-2}. Table 22.4 presents both theoretical and experimental electrical measurements of the five DC. The global resistance represents the measured resistance on the whole structure, while the resistance per node corresponds to the unitary connection resistance. We can see that all the global resistances associated to the D2W bonding are slightly below the W2W

Table 22.2 Resistance of bonded and stand-alone NIST (for different bonded schemes).

NIST structures	NIST a	NIST b	NIST c	NIST d
Contact area (μm^2)	640 × 3	340 × 3	640 × 3	/
Theoretical resistance (Ω) Wafer-to-wafer	4.27	6.26	4.27	8.54
Experimental resistance (Ω) Die-to-wafer	4.39	6.34	4.31	8.59
Experimental resistance (Ω)	4.34	6.37	4.47	8.62

Table 22.3 Description of the various daisy chains used for the reliability tests.

Daisy chain name	Number of connection	Contact area (μm^2)	Lower linewidth (μm)	Upper linewidth (μm)	Pitch
DC1	10 136	9	3	3	17 µm on x; 85 µm on y
DC2	4872	25	5	5	25 µm on x; 66 µm on y
DC3	10 772	9	5	3	10 µm on x; 30 µm on y
DC4	149 342	9	3	3	7 µm on x; 19 µm on y
DC5	29 422	9	3	3	7 µm on x; 7 µm on y

Table 22.4 Daisy chains electrical results.

Daisy chains	DC1	DC2	DC3	DC4	DC5
Number of connection	10 136	4872	10 772	14 934	29 422
Contact area	$3 \times 3\,\mu m^2$	$5 \times 5\,\mu m^2$	$3 \times 3\,\mu m^2$	$3 \times 3\,\mu m^2$	$3 \times 3\,\mu m^2$
Wafer-to-wafer					
Global resistance (Ω)	801	310	681	1180	2340
Resistance per node (mΩ)	79	63.6	63.2	79	79.5
Die-to-wafer					
Global resistance (Ω)	793	304	661	1148	2297
Resistance per node (mΩ)	78.2	62.4	61.3	79	79.5

resistances, which implies a lower resistance per node. These results confirm the very good bonding quality of our D2W process as well as the absence of impact on the global electrical performances compare to the W2W process.

22.5.2
Reliability

W2W bonding underwent reliability tests. The focus was on the daisy chain behavior.

22.5.3
Thermal Cycling

Daisy chains of 200 °C were subjected to a thousand thermal cycles from −65 to +150 °C. No degradation of resistance was recorded from all the different

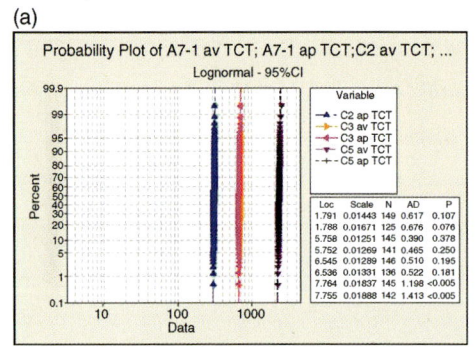

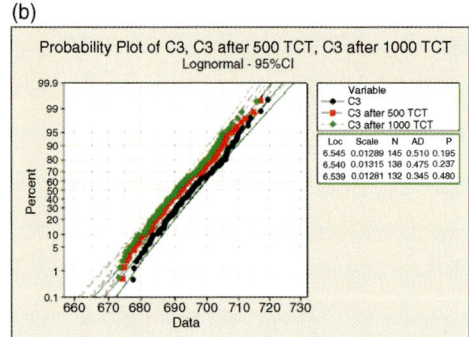

Figure 22.17 (a) Thermal cycling over 500 cycles (−65 °C/150 °C) on 200 °C bonded daisy-chain structures). No degradation was recorded from previous (av) and after thermal cycles (ap) on electrical resistance. (b) Cumulative resistance on daisy chain subjected to 500 and 1000 cycles (green dot line). No degradation was recorded from previous (av) and after thermal cycles (ap).

structures tested: from 1 to 30 K connections, from 3 to 15 μm width and for symmetrical or asymmetrical (DC3) daisy chains. No delamination at any interface was recorded even for DC3. The bonded interface is though stable enough at 200 °C to pass these tests (see Figure 22.17).

22.5.4
Stress Voiding (SIV) Test on 200 °C Postbonding Annealed Samples

SIV tests were carried out at a thermal stress of 175 °C for 500 h on various daisy chains. Resistance measurements were done before and after storage, and are represented in Figure 22.18. Of the tested dies, 99% had a shift resistance of less than 3%. A resistance decrease for dies located at the edges of the 200 mm wafer

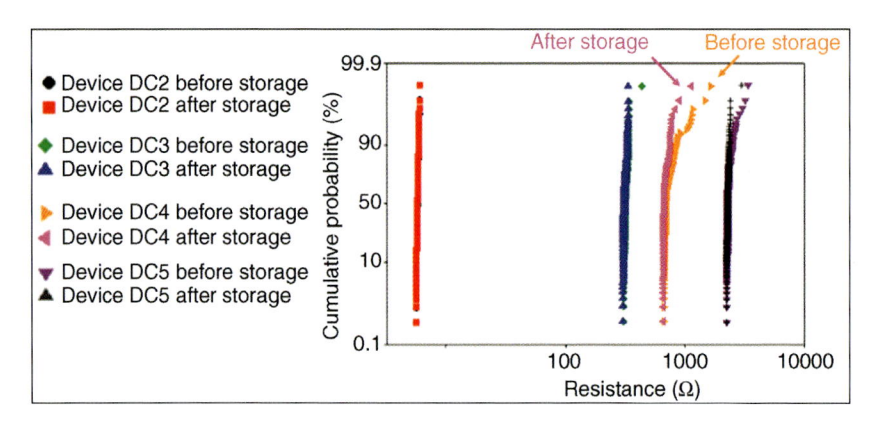

Figure 22.18 Cumulative probability of daisy chain resistance before and after baking at 175 °C for 500 h. No shift was observed after storage. One can see the decrease of resistance for the tested dice initially with a high resistance.

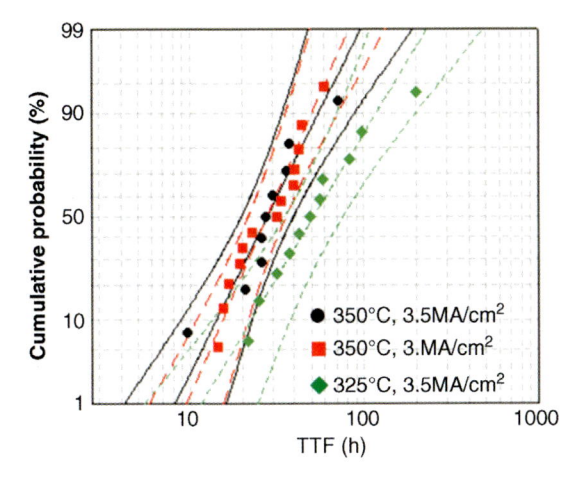

Figure 22.19 EM tests on daisy chain for two temperatures and two current densities. Only one mechanism is visible.

was observed. These dies were bonded with a higher dishing of the copper pads and an improvement of the resistance with storage time is explained by a closure of the interface with long annealing, even at low temperatures. Another important result to note is that even with long anneal, no short circuit occurred. So copper diffusion, into the oxide, if any, do not degrade the structure even with a pitch as low as 7 µm.

22.5.5
Package-Level Electromigration Test

Electromigration (EM) tests are achieved on packaged daisy chains, the ones that have been subjected to a 200 °C postbonding anneal. Each device undergoes tests with two temperatures (325 and 350 °C) and two current stress conditions (3 and 3.5 MA cm^{-2}). The time to failure (TTF) is obtained for a 10% resistance variation. Only one degradation mechanism was observed, the difference in slope 0.5 and 0.6 for the different tests is not significant as the number of tested die is high. For the 350 °C tests, one has to note that the current density is as shown in Figure 22.19.

The Cu–TiN interface appears as the dominant pathway for EM phenomena and not the bonded interface as would be expected. Voids are nucleated at the Cu–TiN interface. They grow until they cut the copper lines and short circuit the device.

Voids appear on the top or bottom lines independent of bonding or stress test conditions. Associated with these defects, one can see copper extrusion at the cathode side of the line (Figure 22.20). The extrusion happens at the oxide-bonding interface, like a mode I solicitation [23].

In fact, bonded daisy chains do not react as standard 3D chains with lines and via. In the latter case, the EM phenomena is confined in the lines as the TiN barrier of the via is more resistive, in the former case, direct bonded daisy chains react as

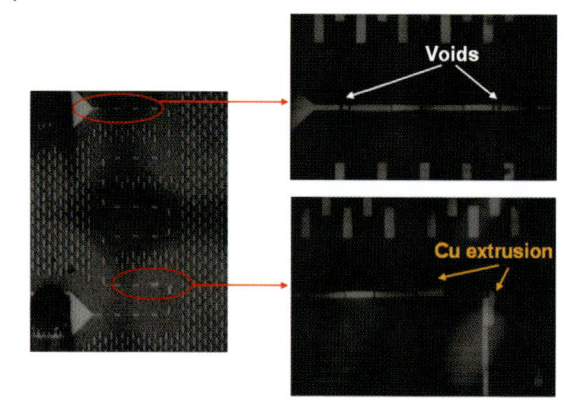

Figure 22.20 Infrared microscopy picture (through BCB, oxide, and silicon stack) of a daisy chain after EM test, showing voids and copper extrusion at each side of the copper line.

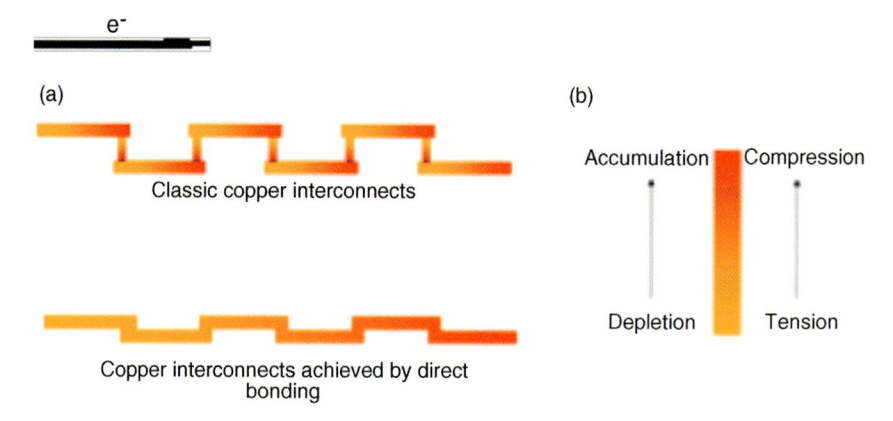

Figure 22.21 Stress and matter concentration in (a) classic 3D interconnect and (b) copper interconnect that achieved direct bonding.

single copper lines, pointing out that the numerous bonded interfaces have no impact on the device (Figure 22.21).

22.6
Conclusions

Cu–SiO$_2$ bonding has been demonstrated on W2W bonding as well as D2W bonding. In both cases, high bonding quality and alignment were presented with scanning acoustic microscopy and infrared analyses. A very low contact resistance was achieved with multiple contact areas, and the contact resistivity was lowered down to $\rho_c = 22.5$ mΩ μm^2 with 29 422 connections daisy chains. Reliability tests have shown the quality of the technology. The results show that EM reliability on bonded devices is not assumed to be dominated by the bonding interface. No failures were observed for structures under thermal stress below 200 °C and thermal cycling.

Acknowledgments

This work is supported by SOITEC and ST microelectronics.

References

1 Chen, K.N., Fan, A., Tan, C.S., and Reif, R. (2004) Contact resistance measurement of bonded copper interconnects for three-dimensional integration technology. *IEEE Electron Device Letters*, **25** (1), 10–12.
2 Labie, R., Ruythooren, W., Baert, K. *et al.* (2008) Resistance to electromigration of purely intermetallic micro-bump interconnections for 3D-device stacking. International Interconnect Technology Conference, 2008 (IITC 2008), p. 19.
3 McMahon, J.J., Lu, J.-Q., and Gutmann, R.J. (2005) Wafer bonding of damascene-patterned metal/adhesive redistribution layers for via-first three-dimensional (3D) interconnect. Proceedings of the 55th Electronic Components and Technology Conference, 2005 (ECTC 2005), pp. 331–336.
4 Temple, D. (2008) Bonding for 3-D integration of heterogeneous technologies and materials. *ECS Transaction*, **16**, p. 3–13.
5 Ventosa, C., Rieutord, F., and Libralesso, L. (2008) Hydrophilic low temperature direct wafer bonding. *Journal of Applied Physics*, **104**, p. 123524.
6 Enquist, P., Fountain, G., Petteway, C. *et al.* (2009) Low cost of ownership scalable copper direct bond interconnect 3D IC technology for three dimensional integrated circuit applications. IEEE International Conference on 3D System Integration, 2009 (3DIC 2009).
7 Shigetou, A. and Suga, T. (2009) Modified diffusion bond process for chemical mechanical polishing (CMP)-Cu at 150°C in ambient air. Proceedings of the 55th Electronic Components and Technology Conference, 2009 (ECTC 2009).
8 Di Cioccio, L., Gueguen, P., Grossi, F. *et al.* (2008) 3D Technologies at CEA-Leti Minatec. IMAPS Conference on Device Packaging.
9 Gueguen, P., Di Cioccio, L., and Rivoire, M. (2008) Copper direct bonding for 3D integration. International Interconnect Technology Conference, 2008 (IITC 2008), pp. 61–63.
10 Leduc, P., Assous, M., Di Cioccio, L. *et al.* (2008) Enabling technologies for 3D chip stacking. International Symposium on VLSI. Technology, Systems and Applications, 2008 (VLSI-TSA 2008), pp. 76–78.
11 Sillon, N., Astier, A., Boutry, H. *et al.* (2008) Enabling technologies for 3D integration: from packaging miniaturization to advanced stacked ICs. IEEE International Electron Devices Meeting, 2008 (IEDM 2008), pp. 1–4.
12 Gueguen, P., Di Cioccio, L., Gergaud, P. *et al.* (2009) Copper direct bonding characterization and its interests for 3D integration. *Journal of the Electrochemical Society*, **156** (10), H772–H776.
13 Gueguen, P., Di Cioccio, L., Gonchond, J.P. *et al.* (2008) 3D vertical interconnects by copper direct bonding. *Materials Research Society Symposium Proceedings*, **1112**, 81.
14 Gueguen, P., Ventosa, C., Di Cioccio, L. *et al.* (2010) Physics of direct bonding; applications to 3D heterogeneous or monolithic integration. *Journal of Microelectronics Engineering*, **87**, 477–484.

15 Di Cioccio, L., Gueguen, P., Signamarcheix, T. et al. (2009) Enabling 3D interconnects with metal direct bonding. International Interconnect Technology Conference, 2009 (IITC 2009), pp. 152–154.

16 Di Cioccio, L., Gueguen, P., Taibi, R. et al. (2009) An innovative die to wafer 3D integration scheme: die to wafer oxide or copper direct bonding with planarised oxide inter-die filling. IEEE International Conference on 3D System Integration, 2009 (3DIC 2009).

17 Maszara, W.P. et al. (1988) Bonding of silicon-wafers for silicon-on-insulator. *Journal of Applied Physics*, **64**, 4943–4950.

18 Di Cioccio, L. et al. (2011) An overview of patterned metal/dielectric surface bonding: mechanism, alignment and characterization. *Journal of the Electrochemical Society*, **158**, 81–86.

19 Augendre, E. et al. (2011) Chip-to-wafer technologies for high density 3D integration. MINaPAD.

20 Sanchez, L., Bally, L., Montmayeul, B. et al. (2012) Chip to wafer direct bonding technologies for high density 3D integration. IEEE 62nd Electronic Components and Technology Conference, 2012 (ECTC 2012), pp. 1960–1964.

21 Mermoz, S., Sanchez, L., Di Cioccio, L. et al. (2012) Impact of containment and deposition method on sub-micron chip-to-wafer self-assembly yield. IEEE International 3D Systems IntegrationConference, 2011 (3DIC 2012).

22 Baudin, F., Di Cioccio, L., Delaye, V., Chevalier, N., Dechamp, J., Moriceau, H., Martinez, E., and Bréchet, Y. Direct bonding of titanium layers on silicon. *Microsystem Technologies*, **19**, 647–653.

23 Taïbi, R. et al. (2010) Full characterization of Cu/Cu direct bonding for 3D integration. Proceedings of the 55th Electronic Components and Technology Conference, 2010 (ECTC 2010), pp. 219–225.

24 Taibi, R. et al. (2011) Investigation of stress induced voiding and electromigration phenomena on direct copper bonding interconnects for 3D integration. IEEE International Electron Devices Meeting, 2011 (IEDM 2011), pp. 6.5.1–6.5.4.

23
Bump Interconnect for 2.5D and 3D Integration
Alan Huffman

The use of bump interconnects in electronic packaging is now ubiquitous. As 3D integration technologies were first being envisioned, bump interconnects were always seen as one of the primary methods for interconnecting devices. Bump interconnects provide short connection lengths, low parasitics, and high mechanical stability – all of which are needed for 3D systems. More recently, the term 2.5D has been used to describe the use of silicon and glass as substrate materials that provide a platform for integrating devices. 2.5D technologies offer the ability to integrate devices with fine-line interconnects and the vertical interconnections can be placed in the substrate, simplifying processing and reducing the risk to ICs. As 2.5D/3D technology has been evolving, bump interconnects have continued to advance as well, with higher interconnect densities and ever smaller dimensions. The widespread acceptance of 2.5D and 3D integration is ushering in the next paradigm change in electronic packaging, and bump interconnects play an important role in the evolution of this technology. This chapter will take an in-depth look at various bump technologies, their evolution, and their application to 2.5D and 3D technology.

23.1
History

Bump interconnect was first developed at IBM in the early 1960s as a way to enable face-to-face electrical connection between two devices. IBM's original implementation of bump interconnects was solder-coated copper balls placed between the IC and the substrate [1]. IBM further developed the bump interconnect in the following years, moving toward a bump consisting wholly of high-lead solder (i.e., 95% Pb, 5% Sn), which was dubbed C4 for controlled collapse chip connection. The ductile solder provided a more reliable interconnection between the chip and substrate and served to limit the transfer of TCE-induced stress to the chip. The C4 bumps were deposited by evaporation and thus the process was rather costly and primarily limited to high-end applications such as mainframe computers.

Throughout the decade of the 1990s, several groups worldwide were working to further develop bumping processes to bring cost down and reduce bump dimensions. Electroplated deposition of high lead and eutectic Sn/Pb solders using Ti/Cu/Ni under-bump-metallization (UBM) was developed at MCNC [2,3] and then brought to high-volume manufacturing by Unitive Electronics, which was later acquired by Amkor in 2002. Bumps fabricated by stencil deposition of solder paste with Al/Ni-V/Cu UBM were being developed by Flip Chip Technologies (a joint venture of Delco and Kulicke & Soffa), now Flip Chip International [4]. Once major OEMs, such as Intel [5] and Motorola [6,7], revealed their use of bump interconnects in portable consumer products and desktop computers, wafer bumping was quickly adopted by other companies and high-volume capacity was established over the first half of the 2000s at companies such as Amkor, ASE, SPIL, and STATS ChipPAC, among others. As CMOS technologies continued to develop, the need for even higher bump density led to the development of Cu pillar bumps to further reduce interconnect pitch and increase the number of connections on the chip. Originally developed by Advanpack and implemented by Fujitsu and Intel [8–10], Cu pillar bump technology was eventually licensed to many outsourced semiconductor assembly and tests (OSATs) that offered high-volume access for leading edge applications. Today, bump interconnect technology in all of its different forms has become a mainstream packaging technology and is implemented in a wide array of applications and products.

The use of bumps to interconnect devices in vertically stacked configurations was envisioned from the very beginning of 3D integration development [11]. There are three primary bump structures used in 2.5D and 3D integration applications today: (i) the C4 solder bump, (ii) the Cu pillar bump, and (iii) the Cu bump. In general, the selection bump type to use in a particular application is dependent on the pitch and dimension of the bumps required and the compatibility of the bump with the assembly process for building the 2.5/3D structure. The chart in Figure 23.1 from Sematech [12] illustrates these three types of bumps (with the Cu pillar shown as two variations that are based on assembly process) and the general ranges of bump pitch for which each type is typically used.

Bonding Method	C4 FC (Controlled Collapse Chip Connect)	C2 FC (Chip Connect)	TC/LR (Local Reflow) FC	TC FC
Schematic Diagram	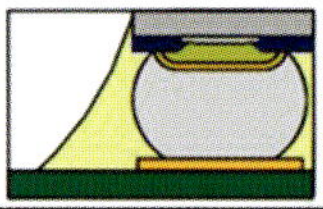	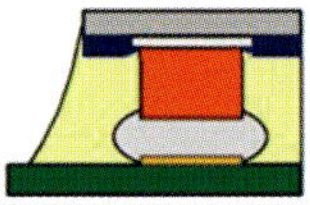	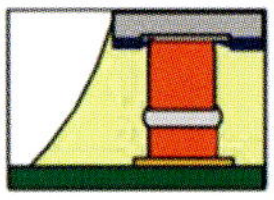	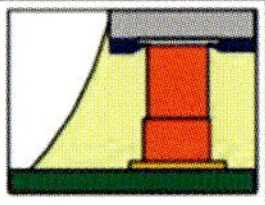
Major Bump Pitch Range at Application	> 130 μm	140 μm ~ 60 μm	80 μm ~ 20 μm	< 30 μm

Figure 23.1 Bump interconnect variants for 2.5D and 3D.

23.2
C4 Solder Bumps

The implementation of C4 solder bumps in 2.5D and 3D integration is limited to larger interconnect pitches and certain applications, but is important in enabling certain structures and applications. The fabrication of C4 solder bumps for 2.5D and 3D is done similarly to process flows for typical wafer-level packaging applications. Solder is deposited on the surface of the device over I/O pads that have been prepared with an UBM, which provides the mechanical and electrical interface between the solder and the IC. The implementation of C4 bumps can be limited if the bump size or pitch required is too small. While C4 bumps can be reliably fabricated down to 25 μm diameter on 50 μm pitch [13], bumps smaller than 50–75 μm are normally fabricated using other types of bumps (i.e., copper pillar). However, larger bump sizes can be used for 2.5 and 3D applications, most notably for Si interposer (2.5D) applications where they are used to provide the interconnects between the interposer and an organic substrate. As a notable example, C4 bumps are used in the silicon interposer present in the Virtex-7 FPGA device from Xilinx shown in Figure 23.2. In this device, 100 μm C4 bumps are used to connect the Si interposer to an organic laminate substrate where larger bump sizes are compatible with the I/O density requirement and provide the necessary mechanical strength to tolerate the stresses generated by the coefficient of thermal expansion (CTE) mismatch between the interposer and substrate [14].

Solder bumps can also be used for vertical stacking applications where the interconnect density is compatible with the fabrication and assembly capabilities of C4 technology. For example, multiple device layers with solder bumps on one side of the device and solder bump bond pads on the other can be stacked simply through placement and gang reflow. IBM demonstrated such a process using 25 μm solder bumps on thinned Si chips, as shown in Figure 23.3 [15]. The UBM of most C4 processes is reliable enough to withstand multiple reflows without degrading [16], which allows for the bumps in devices placed first to stand up to additional reflows required to attach subsequent devices. This structure can only be extended so far, as each additional level also adds more weight that must be borne

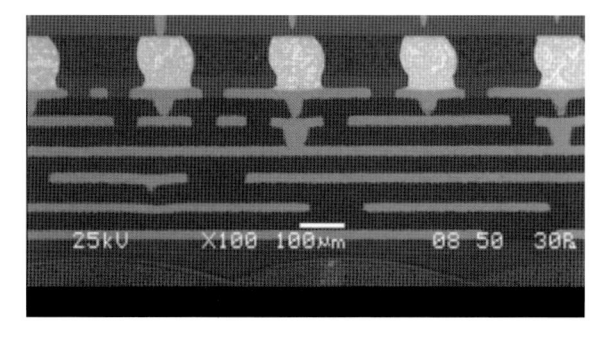

Figure 23.2 A view of 100 μm C4 bumps connecting Si interposer (top) to organic substrate.

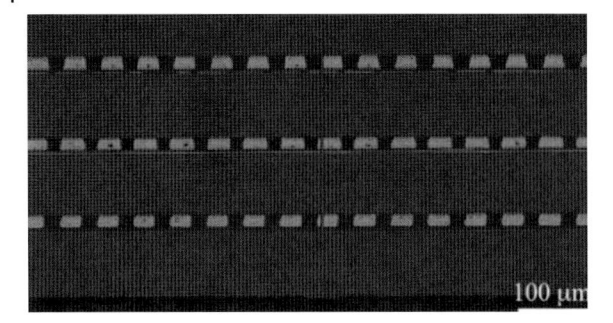

Figure 23.3 Stacked Si chips with 25 µm C4 bumps.

by the bumps in lower levels when they liquefy in reflow processing. Eventually, the weight of the devices will overcome the buoyant force of the molten solder bumps, causing the lower level bumps to collapse completely when they are reflowed and thus destroying the bump interconnects in that level.

23.3
Copper Pillar Bumps

Copper pillar bumps are a more recent development in bump interconnects. Developed by APS, Cu pillar bumps were originally envisioned as replacements for solder bumps to provide finer pitch capability with high reliability when assembled to organic substrates [17]. Cu pillar bumps are typically formed in a process similar to electroplated solder bumps; a thick layer of Cu is plated prior to the solder and later reflowed to form a rounded cap on the top of the Cu post, as shown in Figure 23.4. This structure allows for the amount of solder used in the bump to be greatly reduced compared to C4 bumps of similar dimensions, but the Cu post

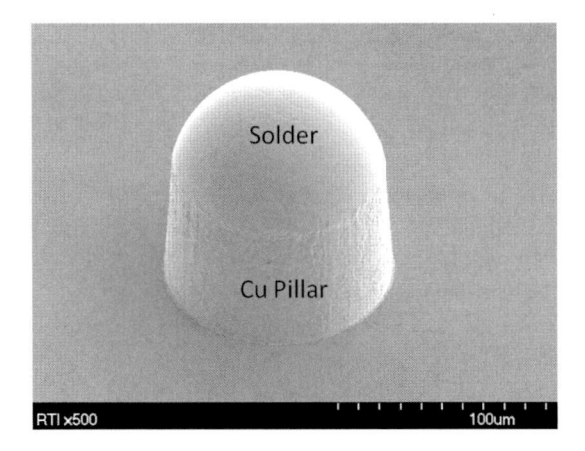

Figure 23.4 Cu pillar bump (source: RTI International)

provides much of the standoff height to the bump. Because of this structure, Cu pillar bumps do not collapse when reflowed, so their maximum diameter after reflow does not increase as solder bumps do. This means that they can be fabricated with less space between adjacent bumps without increasing the risk of bridging, so higher interconnect densities can be realized without decreasing bump diameter. Cu pillars can be structured to provide taller standoff height than solder bumps of the same diameter, which can improve underfilling and reliability.

There are two primary variations of Cu pillar bumps, based on the amount of solder that is deposited on the top of the Cu post which in turn determines the assembly process for the bump array. Cu pillar bumps with a substantial amount of solder in the cap (~10 µms or more) can be referred to as flip-chip (FC) Cu pillar bumps. FC Cu pillar bumps can be assembled similarly to C4 solder bumps using conventional placement and gang reflow processes. FC Cu pillar bumps tend to be larger diameters and pitches but can still be extended down to 50–60 µm pitches successfully. As the diameter of the Cu pillar bump reaches about 40 µm, thermocompression assembly processes become necessary to achieve good assembly yield and can be referred to as TC Cu pillar bumps. The table in Figure 23.5 from Amkor illustrates how the assembly method moves from reflow to thermocompression at 40–50 µm pitch [18].

The use of Cu pillar bumps has particularly enabled 2.5D and 3D technology in recent years. Because high interconnect density is needed for many 2.5 and 3D applications, fine-pitch Cu pillar bumps are an option that provides good reliability. As one of the most widely reported applications of Cu pillar bumps for 2.5D applications, Xilinx used 45 µm pitch Cu pillar bumps fabricated by Amkor to attach four individual field programmable gate array (FPGA) chips to a Si interposer [19]. Figure 23.6 shows the Cu pillar bumps bonding an FPGA device to the Si interposer.

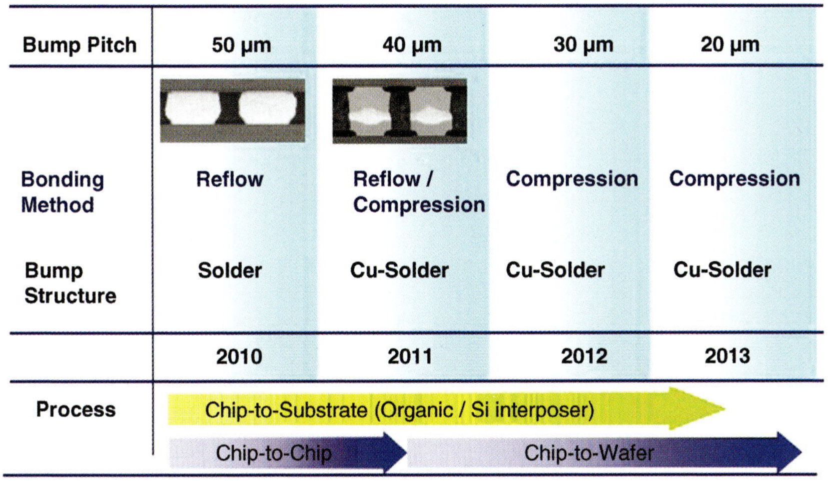

Bump Pitch	50 µm	40 µm	30 µm	20 µm
Bonding Method	Reflow	Reflow / Compression	Compression	Compression
Bump Structure	Solder	Cu-Solder	Cu-Solder	Cu-Solder
	2010	2011	2012	2013
Process	Chip-to-Substrate (Organic / Si interposer)			
	Chip-to-Chip		Chip-to-Wafer	

Figure 23.5 Cu pillar bump pitch and assembly methodology.

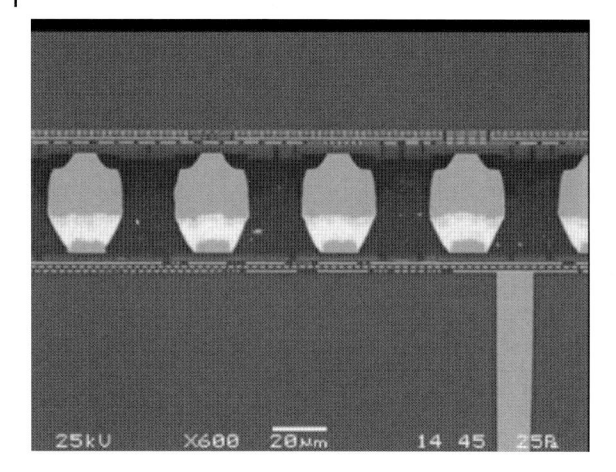

Figure 23.6 A view of 45 μm pitch Cu pillar bumps bonding an FPGA (top) to an Si interposer.

Thermocompression bonding is typically used for assembling Cu pillar bumps with pitches less than 50 μm. When the standoff height of the bumps is small, assembly yield can be impacted by warpage of the organic substrate during reflow, making conventional gang reflow processes inconsistent. In addition, it is preferable to use an underfill applied prior to assembly instead of capillary underfill after bonding due to the difficultly of achieving uniform underfill spreading when the bump density is high and the chip-to-substrate gap is small. In such cases, thermocompression is used to achieve high bonding yields with thicker solder layers [20].

When the solder layer thickness of the Cu pillar bump is reduced to only a few micrometers, a point is reached where there is no longer enough solder for typical reflow processes to be used for assembly, and thermocompression bonding is needed. In this case, the solder layer is completely converted into intermetallic phases as reacts with the Cu. Such bonding is called solid–liquid interdiffusion (SLID) bonding and was described by Bernstein in [21] for microelectronic applications. When Cu pillar bumps with thin layers of solder or Sn are thermocompression-bonded to Cu pads or other Cu pillars with thin solder layers, a layer of Cu–Sn intermetallic is formed at the bonding interface. Such structures have been studied extensively for 2.5D and 3D applications by numerous groups and have been demonstrated at bump pitches ≤ 10 μm [22–26]. SLID bonding with this type of Cu pillar bump is typically done in the 250–300°C range, but the resulting Cu–Sn intermetallic layers melt at temperatures far exceeding the bonding temperatures (415°C for Cu_6Sn_5 and 670°C for Cu_3Sn). This means that bonds formed by SLID bonding do not melt in subsequent bonding processes, making it a good bump interconnect choice for chip-stacking applications where the bumps will see multiple thermocompression cycles. Figure 23.7 shows a SLID-bonded Cu pillar bump that used Sn as the solder layer [25]. After the bonding, the Sn has been fully converted to Cu–Sn intermetallic phases.

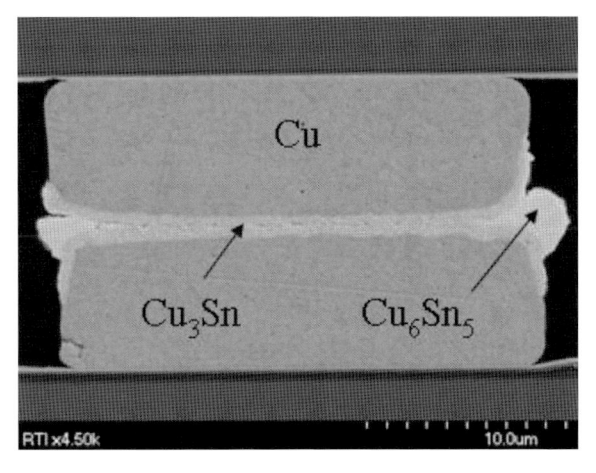

Figure 23.7 Cu pillar bumps bonded by SLID using Sn.

23.4
Cu Bumps

Bumps consisting solely of Cu are less commonly used but nonetheless have been demonstrated in 2.5D and 3D applications. Cu bumps are generally formed in an electroplating process similar to Cu pillar bumps and solder bumps. Without a solder layer to melt and form the bond, thermocompression bonding is the only option for joining Cu bumps and relies on Cu–Cu diffusion across the bonding interface. Temperatures on the order of 400°C or higher for long periods of time (∼1 h) can be required to drive the diffusion of Cu atoms across the bonding interface [27], but some devices are not tolerant of such high processing temperatures. High bonding pressures are also needed to ensure that the Cu bumps are in good physical contact with each other during the bonding process. It is also generally necessary to planarize the Cu bumps in some way, so the bonding yield is not affected by bumps that are slightly higher or lower than average. Finally, preparation of the Cu bump surface to remove oxides prior to bonding is critical to allow diffusion of Cu atoms to occur during the bonding operation. Several different options have been demonstrated for this, including wet etching [28], ion beam etching [29], and exposure to oxide-reducing atmospheres [30]. The practical application of many Cu bump bonding methodologies is difficult to extend to volume assembly processes using standard industry equipment, so the use of Cu bumps is still rather restricted.

Lower temperature bonding of Cu bumps has been demonstrated by various groups. RTI International (Figure 23.8) [31] and IMEC [32] have both published results on bonding of Cu bumps using thermocompression processes at 300°C. Fujitsu has demonstrated a novel Cu bump process in which the bumps are planarized using a mechanical cutting process, as shown in Figure 23.9. This

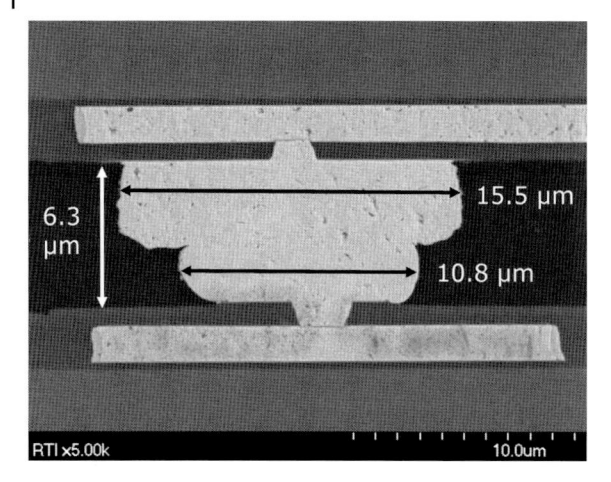

Figure 23.8 Cu bumps bonded at 300 °C.

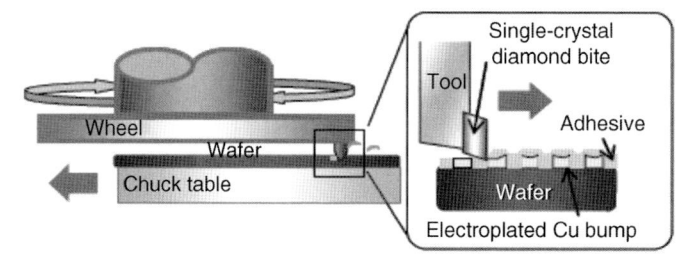

Figure 23.9 Fujitsu precision Cu bump-cutting process.

procedure creates a damage layer in the surface of the bumps and this proves to be advantageous in the bonding process, allowing the bonding temperature to be reduced as far as 200°C to achieve bonding [33].

23.5
Electromigration

As bump sizes continue to decrease for 2.5D and 3D systems, electromigration issues become a more important consideration in the choice of bump interconnect. Electromigration is caused when high current flow through the bump physically moves atoms from one area of the interconnect to another. This causes degradation in the bump's electrical and mechanical properties over time, ultimately leading to failure of the interconnect. Figure 23.10 shows a case of typical electromigration damage in a eutectic Sn/Pb bump, where cracking due to void accumulation has occurred at the cathode (top) side of the bump [34]. Over time, as the cross-sectional area of the bump is reduced by voiding and cracking, electromigration

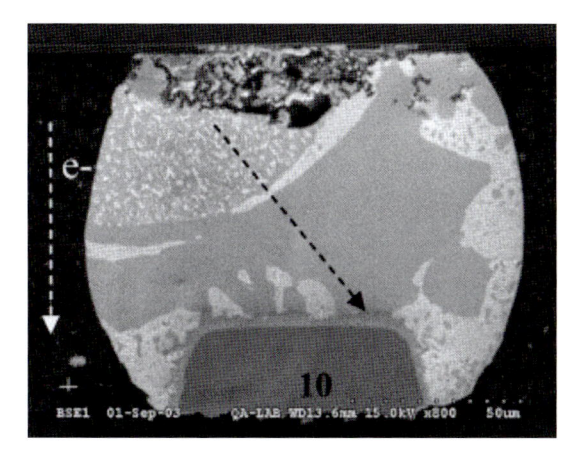

Figure 23.10 Electromigration damage in SnPb solder bump.

damage tends to worsen more quickly as the decreased conductor cross section causes increased resistance and resistive heating which contribute to the rate of damage.

Lead-free solders and Cu pillar bumps can provide significantly improved electromigration performance. Darveaux presented electromigration data [35] that compared high Pb, eutectic SnPb, SnAg, and Cu pillar bumps of the same dimensions (90 μm UBM diameter, 150 μm bump pitch) and showed significantly longer lifetimes for the Cu pillar bump. In a related study, Syed et al. tested the electromigration performance of 25 μm diameter Cu pillar bumps with SnAg solder cap, which showed lifetimes in excess of 5500 h at current densities greater than 35 000 A cm^{-2} ([36], Figure 23.11). Cross-sectional analysis of a sample pulled

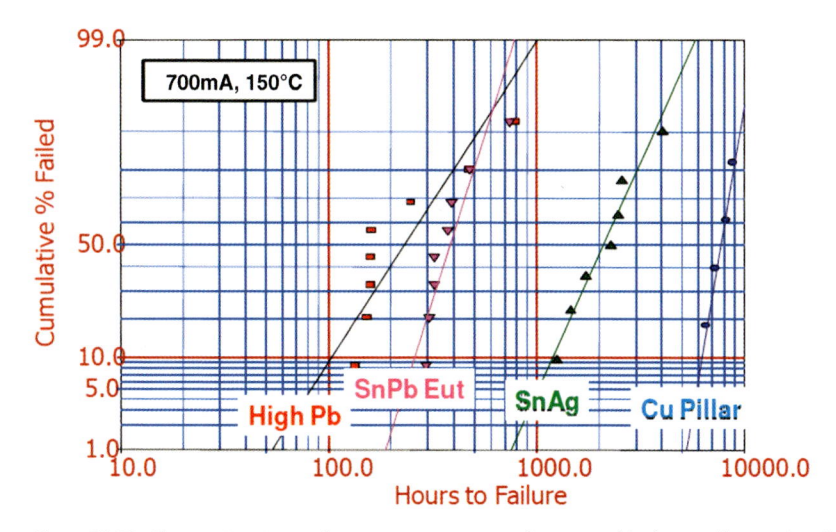

Figure 23.11 Electromigration performance comparison of various solder bump alloys to Cu pillar.

from testing after 2775 h of testing showed that the solder in both current stressed and unstressed bumps had been fully converted to Cu–Sn intermetallic phases.

References

1 Davis, E. et al. (1964) Solid logic technology: versatile high performance electronics. *IBM Systems Journal*, **8**, 102.
2 Yung, E. and Turlik, I. (1991) Electroplated solder joints for flip chip applications. *IEEE Transactions on Components, Hybrids and Manufacturing Technology*, **14**, 549.
3 Mis, J., Rinne, G., Deane, P., and Adema, G. (1996) Flip chip production experience: design, process, reliability and cost considerations. Proceedings of the International Symposium on Hybrid Microelectronics (ISHM), Minneapolis, p. 20.
4 Elenius, P. (1997) Flex on cap – solder paste bumping. Proceedings of IEEE Electronic Components and Technology Conference, San Jose, p. 248.
5 Shukla, R., Murali, V., and Bhansali, A. (1999) Flip chip CPU package technology at Intel: a technology and manufacturing overview. Proceedings of the 49th Electronic Components and Technology Conference, 1999, pp. 945–949.
6 De Haven, K. and Deitz, J. (1994) C4: an enabling technology. Proceedings of the 44th Electronic Components and Technology Conference, p. 1.
7 Brooks, R. et al. (1993) Direct chip attach: a viable chip mounting alternative. Proceedings of the International Conference and Exhibition on Multichip Modules, Denver, p. 595.
8 Wang, T., Tung, F., and Foo, L. (2001) Studies on a novel flip-chip interconnect structure: pillar bump. Proceedings of the 51st Electronic Components Technology Conference, p. 945.
9 Kawahara, T. (2000) SuperCSP™. *IEEE Transactions on Advanced Packaging*, **23** (2), 215–219.
10 Chipworks (2006). Intel D920 Presler and T2300 Yonah Copper Pillar Bump Technology Package Analyses, April 3.
11 Takahashi, S., Hayashi, Y., Kunio, T., and Endo, N. (1992) Characteristics of thin-film devices for a stack-type MCM. Proceedings of the 1992 IEEE Multi-Chip Module Conference, pp. 159–162.
12 Arkalgud, S. (2012) Scaling 2.5D/3D: the next R&D challenge. 2nd Annual IEEE Global Interposer Technology Workshop, Atlanta, GA, November 2012.
13 Huffman, A., LaBennett, R., Bonafede, S., and Statler, C. (2003) Fine-pitch wafer bumping and assembly for high density detector systems. *IEEE Nuclear Science Symposium Conference Record*, **5**, 3522–3526.
14 Banijamali, B., Ramalingam, S., Nagarajan, K., and Chaware, R. (2011) Reliability study of TSV interposers and interconnects for the 28nm technology FPGA. Proceedings of the 61st Electronic Components Technology Conference, pp. 285–290.
15 Maria, J., Dang, B., Wright, S.L., Tsang, C. K., Andry, P., Polastre, R., Liu, Y., Wiggins, L., and Knickerbocker, J.U. (2011) 3D chip stacking with 50 μm pitch lead-free micro-C4 interconnections. Proceedings of the 61st Electronic Components Technology Conference, pp. 268–273.
16 Li, L., Jang, J.W., and Allmen, B. (2001) Shear property and microstructure evaluation of Pb-free solder bumps under room temperature and multiple reflow/high temperature aging. Proceedings of the International Symposium on Advanced Packaging Materials Processes, Properties, and Interfaces, pp. 347–353.
17 Wang, T., Tung, F., and Foo, L. (2001) Studies on a novel flip-chip interconnect structure: pillar bump. Proceedings of the 51st Electronic Components Technology Conference, p. 945.
18 Yoo, M. (2012) 3D IC technology: OSAT perspective. SEMICON Taiwan SiP Global Summit 2012, Taipei, Taiwan.
19 Chaware, R., Nagarajan, K., and Ramalingam, S. (2012) Assembly and reliability challenges in 3D integration of 28nm FPGA die on a large high density 65nm passive interposer. Proceedings of the

62nd Electronic Components Technology Conference, pp. 279–283.
20. Patterson, D. (2012) 2.5/3D packaging enablement through copper pillar technology. *Chip Scale Review*, **16** (3), 20–26.
21. Bernstein, L. (1966) Semi-conductor joining by the solid–liquid inter-diffusion (SLID) process. *Journal of the Electrochemical Society*, **113** (12), 1282–1288.
22. Tomita, Y. et al. (2000) Copper bump bonding with electroless metal cap on 3 dimensional stacked structure. Proceedings of the 3rd Electronics Packaging Technology Conference, Sheraton Towers, Singapore, pp. 286–291.
23. Tanida, K. et al. (2003) Ultra-high-density 3D chip stacking technology. Proceedings of the 43rd Electronic Components and Technology Conference, New Orleans, LA, pp. 1084–1089.
24. Klumpp, A. et al. (2003) Chip-to-wafer stacking technology for 3D system integration. Proceedings of the 43rd Electronic Components and Technology Conference, New Orleans, LA, pp. 1080–1083.
25. Huffman, A., Lueck, M., Bower, C., and Temple, D. (2007) Effects of assembly process parameters on the structure and thermal stability of Sn-capped Cu bump bonds. Proceedings of the 57th Electronic Components and Technology Conference, Reno, NV, pp. 1589–1596.
26. Reed, J., Lueck, M., Gregory, C., Huffman, A., Lannon, J., and Temple, D. (2010) Low temperature bonding of high density large area array interconnects for 3D integration. Proceedings of the 4th International Symposium on Microelectronics, 2010 (IMAPS 2010), Raleigh, NC, USA, pp. 28–35.
27. Tan, C.S. (2009) Cu-to-Cu thermo-compression bonding and its recent progress. International Workshop on 3D Design and Nanoelectronic System Technologies, Atlanta, GA.
28. Lannon, J., Gregory, C., Huffman, A., Lueck, M., and Temple, D. (2009) High density Cu–Cu interconnect bonding for 3D integration. Proceedings of the 59th Electronic Components and Technology Conference, San Diego, CA, pp. 355–359.
29. Kim, T.H., Howlader, M.M.R., Itoh, T., and Suga, T. (2003) Room temperature Cu–Cu direct bonding using surface activated bonding method. *Journal of Vacuum Science and Technology A*, **21** (2), 449–453.
30. Farrens, S. (2008) Wafer and die bonding technologies for 3D integration. MRS Symposium: Materials and Technologies for 3-D Integration, December 2008, Boston, MA, vol. **1112**, pp. 55–65.
31. Lannon, J., Gregory, C., Huffman, A., Lueck, M., and Temple, D. (2009) High density Cu–Cu interconnect bonding for 3D integration. Proceedings of the 59th Electronic Components and Technology Conference, pp. 355–359.
32. Swinnen, B., Ruythooren, W., DeMoor, P., Bogaerts, L., Carbonell, L., DeMunck, K., Eyckens, B., Stoukatch, S., Sabuncuoglu Tezcan, D., Tokei, Z., Vaes, J., Van Aelst, J., and Beyne, E. (2006) 3D integration by Cu–Cu thermo-compression bonding of extremely thinned bulk-Si die containing 10 µm pitch through-Si vias. Proceedings of the International Electron Devices Meeting, pp. 1–4.
33. Sakai, T. et al. (2011) Cu–Cu thermocompression bonding using ultra precision cutting of Cu bumps for 3D-SIC. Proceedings of the IMAPS International Conference and Exhibition on Device Packaging, Fountain Hills, AZ, USA, pp. 146–149.
34. Chen, K.-M., Wu, J.D., and Chiang, K.-N. (2006) Effects of pre-bump probing and bumping processes on eutectic solder bump electromigration. *Microelectronics Reliability*, **46** (12), 2104–2111.
35. Darveaux, R. (2012) Escalating challenges in developing complex solutions for next generation package and interconnect technologies. Proceedings of the IMAPS International Conference and Exhibition on Device Packaging, Fountain Hills, AZ, USA.
36. Syed, A., Dhandapani, K., Moody, R., Nicholls, L., and Kelly, M. (2011) Cu pillar and µ-bump electromigration reliability and comparison with high Pb, SnPb, and SnAg bumps. Proceedings of the 61st ECTC, Lake Buena Vista, FL, USA, May 31–June 3, pp. 332–339.

24
Self-Assembly Based 3D and Heterointegration

Takafumi Fukushima and Jicheol Bea

24.1
Introduction

Chip-to-wafer 3D integration approaches can improve production yield in 3DIC fabrication because known good dies (KGDs) can be selectively used and wafer-level processing can be applied after pick-and-place chip assembly. Therefore, chip-to-wafer 3D integration is thought to be a promising candidate for 3D chip-stacking methods [1,2]. However, conventional chip-to-wafer 3D integration has a serious trade-off between assembly throughput and alignment accuracies of KGDs on wafers, as shown in Figure 24.1. To overcome the problem, multichip self-assembly as a massively parallel high-precision alignment technique with bonding has been proposed using the surface tension of liquid [3]. A large number of KGDs can be simultaneously, instantly, and precisely aligned to wafers by multichip self-assembly. In this section, self-assembly based 3D integration technologies are introduced. In particular, key parameters affecting alignment accuracies obtained by self-assembly of KGDs are descried. In addition, this section deals with how to interconnect the self-assembled KGDs to wafers or the other KGDs to be stacked.

24.2
Self-Assembly Process

Multichip self-assembly requires liquid droplets as a hosting fluid in addition to wafers having high and low wettability regions to the liquid. The self-assembly processes are quite simple. Figure 24.2 shows snapshots from a short movie of self-assembly of a chip observed with two high-speed cameras [4]. First, hydrophilic regions of chip self-assembly areas and hydrophobic regions of the surrounding areas or vice versa are formed on host wafers. For example, hydrophilic assembly areas are formed with SiO_2 on the host wafers by standard photolithography, whereas the surrounding hydrophobic areas are formed with Si [5,6] or fluorocarbon [7,8]. Self-assembled monolayer (SAM) techniques are often employed for forming the hydrophobic regions with silane coupling agents [9–11].

Handbook of 3D Integration: 3D Process Technology, First Edition.
Edited by Philip Garrou, Mitsumasa Koyanagi, and Peter Ramm.
© 2014 Wiley-VCH Verlag GmbH & Co. KGaA. Published 2014 by Wiley-VCH Verlag GmbH & Co. KGaA.

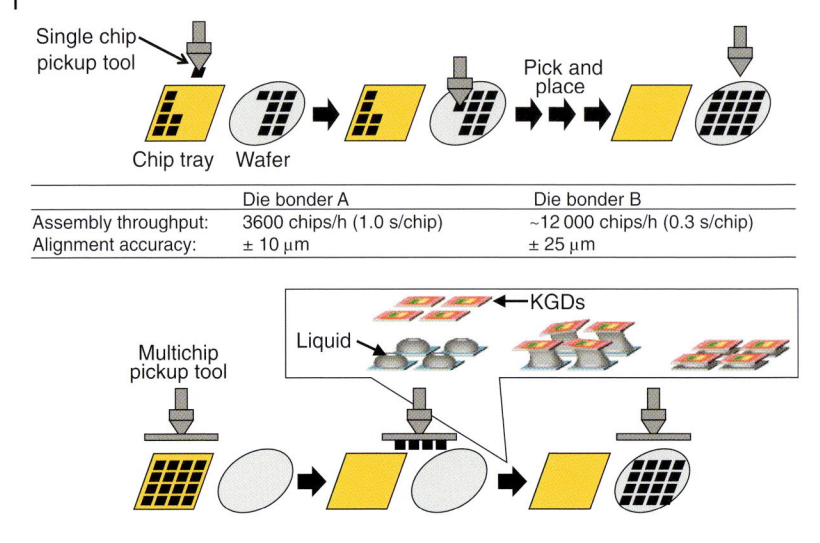

	Die bonder A	Die bonder B
Assembly throughput:	3600 chips/h (1.0 s/chip)	~12 000 chips/h (0.3 s/chip)
Alignment accuracy:	± 10 μm	± 25 μm

Figure 24.1 Comparison between traditional pick-and-place chip assembly and surface-tension-driven multichip self-assembly techniques for chip-to-wafer 3D integration. (a) Traditional pick-and-place assembly in one-by-one processing. (b) Self-assembly using liquid surface tension in batch processing.

In Figure 24.2, water droplets are supplied to a hydrophilic assembly area by dispensing. Dip coating [12] and steam [13] or mist [14] have been reported to simultaneously supply liquid droplets to the host wafers. After the chip is roughly prealigned to the corresponding assembly area, they are released to the surface of water droplets. Immediately after that, the chip is precisely aligned to the assembly

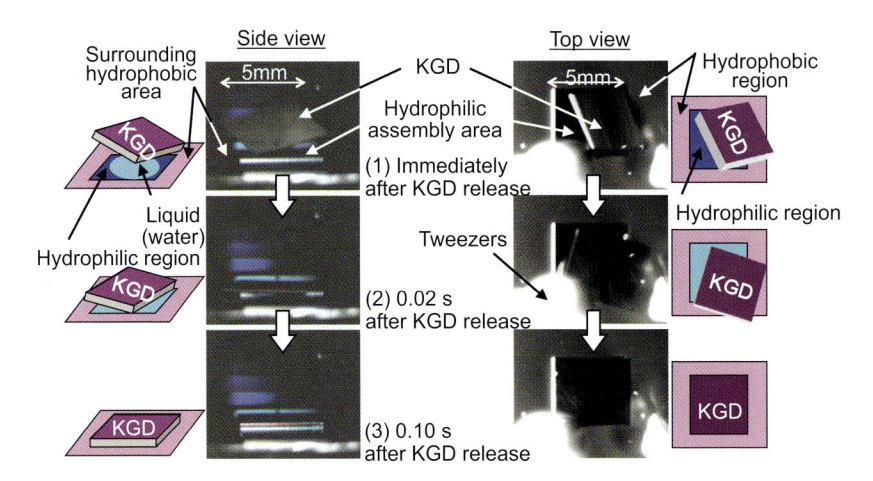

Figure 24.2 Snapshots from a short movie of chip self-assembly to a host wafer with hydrophilic assembly area and the surrounding hydrophobic areas.

area for a short time. Pure water [7,15–17], diluted fluoric acid [5,6], water-soluble adhesives [17,18], and water-soluble flux [19] as hosting fluids have been employed for self-assembly with hydrophilic assembly areas. The diluted fluoric acid can assist load-free room-temperature oxide-to-oxide direct chip bonding after self-alignment. Conversely, hydrocarbon [10,11], adhesives such as acrylic resin [9] and epoxy resin [20], and molten solder [21,22] have been applied to hosting fluids in self-assembly with hydrophobic assembly areas. As seen in Figure 24.2, the chip can be precisely aligned to a hydrophilic assembly area formed on a host wafer within 0.1 s.

24.3
Key Parameters of Self-Assembly on Alignment Accuracies

This section discusses what parameters depend on alignment accuracies of chip self-assembly. The driving force acting on KGDs self-assembling to wafers in x-direction is surface tension and it can be expressed by the following equation:

$$F_x = -\frac{\partial E}{\partial x} = -\mu \frac{\partial A}{\partial x},$$

where E and A indicate surface free energy and area of a chip, respectively. As is clearly seen in this equation, the driving force F_x is proportional to liquid surface tension μ, and thus it can be said that the first key parameter is surface tension of liquid as a hosting fluid used in self-assembly processes. Water is well known to be a liquid having the highest surface tension at room temperature except for toxic mercury. The value is 71.99 mN m^{-1} (= mJ m^{-2}) at 25 °C. As shown in Figure 24.3a, liquid with higher surface tension indicates higher alignment accuracies. Actually, water has been previously employed for self-assembly experiments [7,15–17]. Molten solders have higher surface tension than water at elevated temperature. Therefore, traditional self-alignment techniques in electronic packaging use their high surface tension to slightly shift KGDs toward accurate

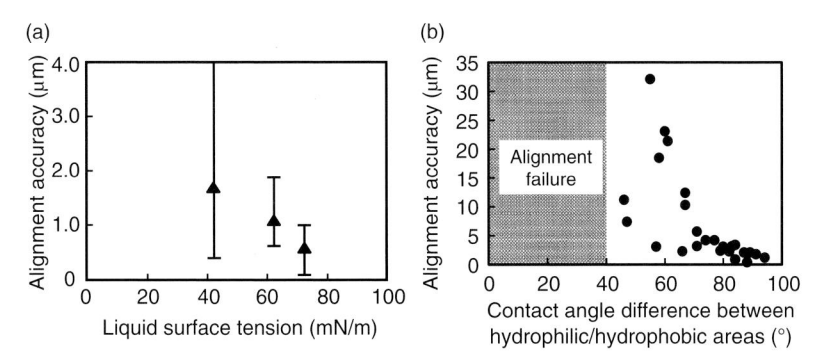

Figure 24.3 Dependence of alignment accuracies in self-assembly on liquid surface tension (a) and wetting contrast between hydrophilic assembly area and the surrounding hydrophobic area (b).

position after the KGDs with solder bumps are mechanically mounted by pick-and-place flip-chip bonding [23,24].

The second key parameter is contact-angle differences, called wettability contrast, between hydrophilic assembly areas and the surrounding hydrophobic areas or vice versa. If the contact-angle differences are low, liquid droplets can easily spread outside the assembly areas to give low alignment accuracies. In contrast, if the differences are high, liquid droplets can be confined to assembly areas to give high alignment accuracies. Figure 24.3b shows relationship between contact angle differences and alignment accuracies. When the differences of water contact angles between assembly areas and the surrounding areas is more than 90°, the alignment accuracies is found to be within 1–2 μm [4]. Zhou and coworkers reported a very high hydrophobic surrounding area indicating nearly 180° in water contact angle by using black silicon surfaces to tightly confine hosting fluids, although their receptor sites (assembly areas) shows 69° in water contact angle [25].

Impact of the other parameters such as wafer tilt [6] and liquid volume [26,27] on self-assembly accuracies has been studied so far. Sato *et al.* describe alignment accuracies obtained by self-assembly depends on initial misalignment that means chip positions in x, y, z, and θ directions before chip release [15]. Zhou and coworkers presented that higher alignment accuracies can be obtained by the forced wetting to press KGDs into the liquid surface. Consequently, liquid droplets can spuriously increase the wettability in assembly areas and perfectly wet all over the areas to give high alignment accuracy, although the assembly areas do not have high hydrophilicity for hosting fluids [25].

Chip size accuracies are another important parameter to precisely align KGDs to wafers. Fukushima *et al.* presented self-assembly with chips that are fabricated by plasma dicing before grinding technique to give high precision chip size accuracy ±1 μm, by which high alignment accuracies of approximately 0.4 μm are obtained [6]. Standard saw dicing with ceramic blades can provide high chip size accuracies ±1–2 μm, which can have a high potential to achieve submicron alignment accuracies. However, chip sizes are not well controlled in current semiconductor industries. F. Sun *et al.* reported pad-assisted self-assembly and fully – or partially – wetted self-assembly techniques with chips with/without hydrophobic surrounding areas, as shown in Figure 24.4. With an initial placement quality of less than half a pad pitch, this technique achieves submicrometer alignment accuracy [26].

24.4
How to Interconnect Self-Assembled Chips to Chips or Wafers

Self-assembled KGDs have to be interconnected to the others chips stacked above and below the KGDs or target wafers, such as Si interposers and LSI wafers. Self-assembly-based 3D integration approaches are simply divided into two categories. One is self-assembly of KGDs in a facedown bonding manner, where the KGDs have metal microbumps for flip-chip bonding and are directly self-assembled to the interposer or LSI wafers without temporal carrier or support wafers. The other is

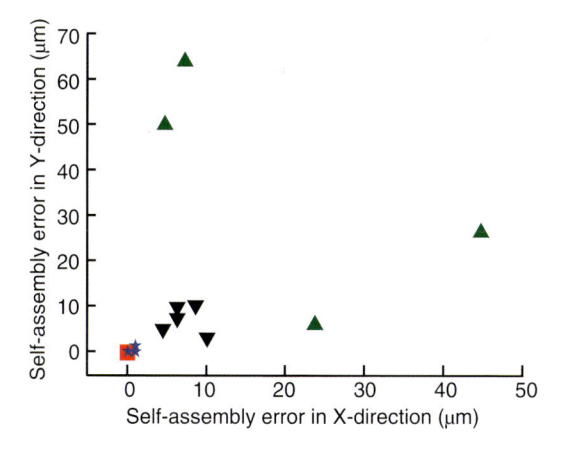

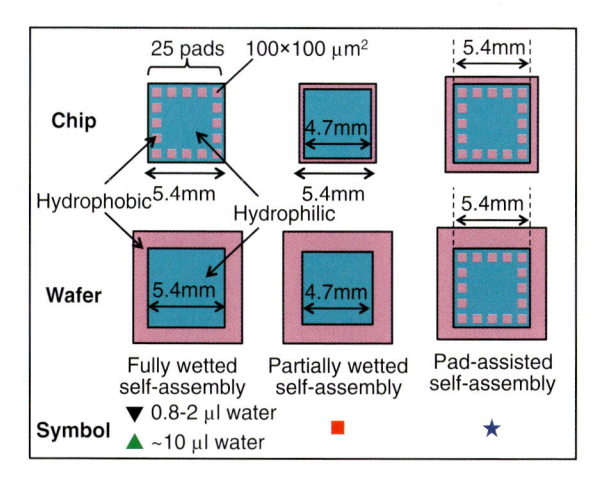

Figure 24.4 Impact of alignment accuracies in self-assembly on liquid volume and design of hydrophilic/hydrophobic areas on chips and wafers [26].

self-assembly of KGDs in a faceup bonding manner, with which Fukushima et al. proposed reconfigured wafer-to-wafer 3D integration as a new 3D integration using self-assembly [5]. In this section, flip-chip-to-wafer 3D integration and reconfigured wafer-to-wafer 3D integration are introduced.

24.4.1
Flip-Chip-to-Wafer 3D Integration

Self-assembly of KGDs having metal microbump electrodes to wafers has been demonstrated by using water surface tension [4,7,8,11,28]. The representative facedown self-assembly process, such as flip-chip bonding, is shown in Figure 24.5a. Liquid droplets are put onto the assembly areas formed on the corresponding wafers on which the surrounding areas are rendered hydrophobic, which is a similar

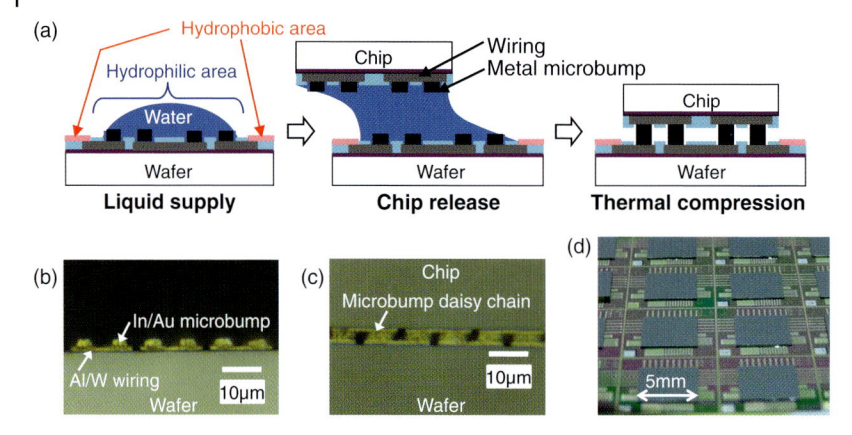

24 Self-Assembly Based 3D and Heterointegration

Figure 24.5 A self-assembly process for direct multichip-to-wafer 3D integration (a), cross-sectional images of microbumps formed on a wafer before self-assembly (b) and after self-assembly and the subsequent thermal compression (c), and a perspective view of an array of self-assembled chips on the wafer (d).

procedure to self-assembly of KGDs without microbumps. Immediately after KGDs are released to the droplets, the KGDs are self-assembled to the wafers, followed by liquid evaporation. The subsequent thermal compression results in 3D microbump interconnects.

Chapuis et al. carried out surface tension-driven self-assembly of MEMS chips using water droplets [8]. The MEMS chips with $200 \times 200\,\mu m^2$ solder bumps (Bi 44.7%, Pb 22.6%, In 19.1%, Sn 8.3%, and Cd 5.3%) having a low melting temperature of 47 °C were successfully self-assembled to the wafer with Cr/Au pads with alignment accuracies of ~20–100 µm. Mastrangeli et al. (IMEC) presented their self-assembly research for 3D integration [11]. They employed SAM technology to realize their chip-to-wafer 3D integration using self-assembly. They demonstrated flip-chip self-assembly of chips with Ti/Au pads to a wafer with 120 µm pitch/diameter bumps in water using hexadecane as a hosting fluid.

Fukushima et al. presented fine-pitch flip-chip self-assembly with chips having 3 µm thick In/Au microbump arrays with a bump pitch of 10 or 20 µm [4]. Figure 24.5b and c shows cross-sectional views of microbumps formed on wiring before and after self-assembly and the subsequent thermal compression. As seen from Figure 24.5c, the chips are structurally connected to the wafers. Figure 24.5d shows a photograph of the array of the self-assembled chips having the microbumps. These chips are obtained by standard saw dicing to give high chip size accuracies ±2 µm. The chip alignment accuracy resulting from the flip-chip self-assembly is high and found to be 0.8 µm and 0.2 µm in x and y directions. The 2500 In/Au microbump daisy chain is achieved by the flip-chip self-assembly and the resulting daisy chain shows ohmic contact. The resistance is sufficiently low enough to employ 3D LSI fabrication based on chip-to-wafer bonding and comparable to that obtained from traditional pick-and-place chip assembly.

24.4.2
Reconfigured-Wafer-to-Wafer 3D Integration

Reconfigured wafer-to-wafer 3D integration is an advanced multichip-to-wafer 3D integration methodology using faceup self-assembly with temporary bonding to carrier wafers and multichip transfer from the carrier wafers to the other target wafers. Fukushima *et al.* call the carrier wafers having self-assembled KGDs "reconfigured wafers." Figures 24.6a and b show a concept of reconfigured wafer-to-wafer 3D integration and its process flow [5,17,28–30]. First, various kinds of first-layer KGDs with microbumps are sorted from various device wafer after testing. Then, the KGDs are self-assembled on the carrier wafer and temporarily bonded to the wafer in a facedown bonding manner. After that, the carrier wafers are aligned and bonded to the corresponding target interposer or LSI wafers with the same design of microbumps to the self-assembled KGDs. Consequently, microbumps-to-microbumps bonding is formed between the KGDs and the target wafers, and then, the KGDs are detached from the carrier wafers and transferred to the target wafers. The following processes are underfilling, multichip thinning including backside grinding and chemical–mechanical polishing (CMP), and through-silicon via (TSV)/microbump formation. Finally, these KGDs are vertically stacked in layers by repeating the multichip transfer processes.

Another example of chip-to-wafer 3D integration approaches using faceup chip bonding is reported by LETI. Berthier *et al.* (LETI) combined self-assembly using

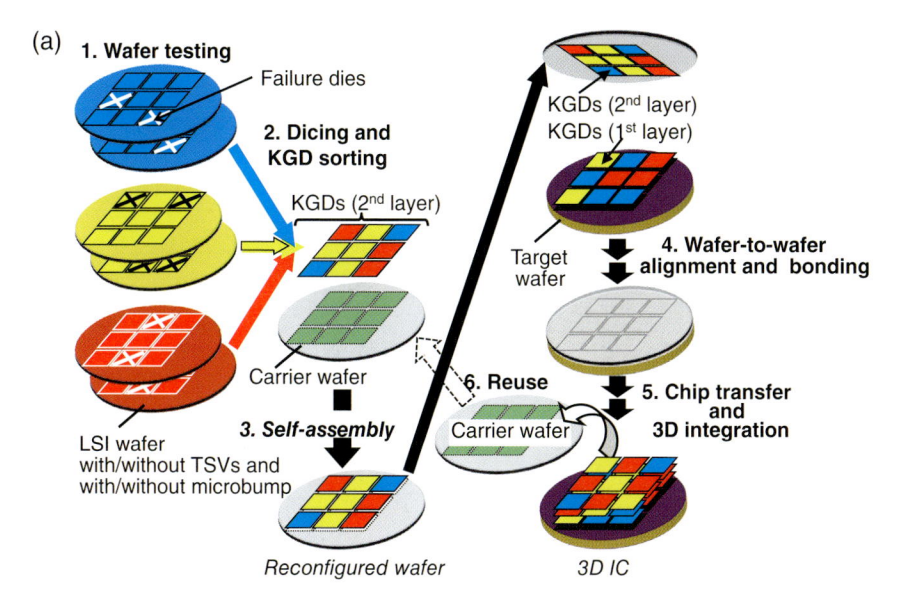

Figure 24.6 Concept of reconfigured wafer-to-wafer 3D integration (a) and the process flow including photos of self-assembled chips with metal microbumps on the top surface and the self-assembled chips transferred to a target wafer (b).

(b)

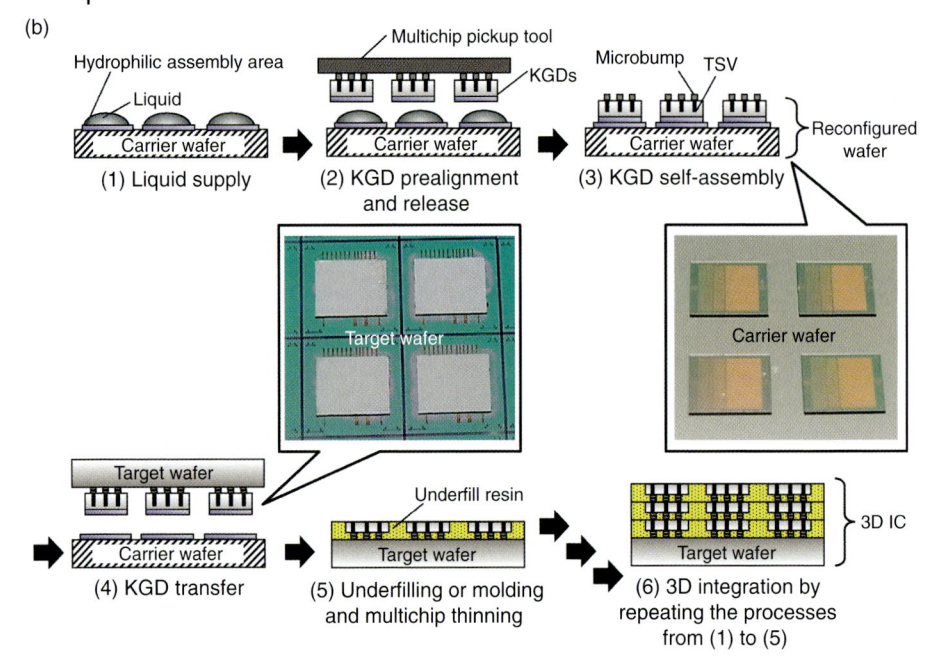

Figure content:

(1) Liquid supply — Hydrophilic assembly area, Liquid, Carrier wafer
(2) KGD prealignment and release — Multichip pickup tool, KGDs, Carrier wafer
(3) KGD self-assembly — Microbump, TSV, Carrier wafer, Reconfigured wafer
(4) KGD transfer — Target wafer, Carrier wafer
(5) Underfilling or molding and multichip thinning — Underfill resin, Target wafer
(6) 3D integration by repeating the processes from (1) to (5) — Target wafer, 3D IC

Figure 24.6 (Continued)

water surface tension with direct oxide–oxide bonding for 3D integration [16]. In their chip self-assembly experiments, SAM technology was employed to render assembly areas hydrophilic and 50 μm deep trench structures were formed around the hydrophilic areas to tightly confine chips to the assembly areas.

References

1 Klumpp, A., Merkel, R., Ramm, P., Weber, J., and Wieland, R. (2004) Vertical system integration by using inter-chip vias and solid–liquid interdiffusion bonding. *Japanese Journal of Applied Physics*, **43**, L829.

2 Fukushima, T., Yamada, Y., Kikuchi, H., and Koyanagi, M. (2006) New three-dimensional integration technology using chip-to-wafer bonding to achieve ultimate super-chip integration. *Japanese Journal of Applied Physics*, **45**, 3030.

3 Fukushima, T., Yamada, Y., Kikuchi, H., and Koyanagi, M. (2005) New three-dimensional integration technology using self-assembly technique. Technical Digest: International Electron Devices Meeting (IEDM), pp. 359–362.

4 Fukushima, T., Iwata, E., Ohara, Y., Murugesan, M., Bea, J.-C., Lee, K.-W., Tanaka, T., and Koyanagi, M. (2012) Multichip-to-wafer 3D integration technology using chip self-assembly with excimer lamp irradiation. *IEEE Transactions on Electron Devices*, **59**, 2956.

5 Fukushima, T., Yamada, Y., Kikuchi, H., Konno, T., Liang, J., Ali, A.M., Sasaki, K., Inamura, K., Tanaka, T., and Koyanagi, M. (2007) New three-dimensional integration technology based on reconfigured wafer-on-wafer bonding technique. Technical Digest: International Electron Devices Meeting (IEDM), pp. 985–988.

6 Fukushima, T., Iwata, E., Konno, T., Bea, J.C., Lee, K.W., Tanaka, T., and Koyanagi, M.

(2010) Surface tension-driven chip self-assembly with load-free hydrogen fluoride-assisted direct bonding at room temperature for three-dimensional integrated circuits. *Applied Physics Letters*, **96**, 154105.

7 Fukushima, T., Konno, T., Kiyoyama, K., Murugesan, M., Sato, K., Jeong, W.-C., Ohara, Y., Noriki, A., Kanno, S., Kaiho, Y., Kino, H., Makita, K., Kobayashi, R., Yin, C.-K., Inamura, K., Lee, K.-W., Bea, J.-C., Tanaka, T., and Koyanagi, M. (2008) New heterogeneous multichip module integration technology using self-assembly method. Technical Digest: International Electron Devices Meeting (IEDM), pp. 499–502.

8 Chapuis, Y.A., Debray, A., Jalabert, L., and Fujita, H. (2009) Alternative approach in 3D MEMS-IC integration using fluidic self-assembly techniques. *Journal of Micromechanics and Microengineering*, **19**, 105002.

9 Srinivasan, U., Liepmann, D., and Howe, R.T. (2001) Microstructure to substrate self-assembly using capillary forces. *Journal of Microelectromechanical Systems*, **10**, 17.

10 Xiong, X., Hanein, Y., Fang, J., Wang, Y., Wang, W., Schwartz, D.T., and Böhringer, K.F. (2003) Controlled multibatch self-assembly of microdevices. *Journal of Microelectromechanical Systems*, **12**, 117.

11 Mastrangeli, M., Ruythooren, W., Celis, J.-P., and Hoof, C.V. (2011) Challenges for capillary self-assembly of microsystems. *IEEE Transactions on Electronics Packaging Manufacturing*, **1**, 133.

12 Mastrangeli, M., Ruythooren, W., Van Hoof, C., and Celis, J.-P. (2009) Conformal dip-coating of patterned surfaces for capillary die-to-substrate self-assembly. *Journal of Micromechanics and Microengineering*, **19**, 045015.

13 Fang, J. and Böhringer, K.F. (2006) Wafer-level packaging based on uniquely orienting self-assembly (the DUO-SPASS processes). *Journal of Microelectromechanical Systems*, **15**, 531.

14 Chang, B., Sariola, V., Jääskeläinen, M., and Zhou, Q. (2011) Self-alignment in the stacking of microchips with mist-induced water droplets. *Journal of Micromechanics and Microengineering*, **21**, 015016.

15 Sato, K., Ito, K., Hata, S., and Shimokohbe, A. (2003) Self-alignment of microparts using liquid surface tension: behavior of micropart and alignment characteristics. *Precision Engineering*, **27**, 42.

16 Berthier, J., Brakke, K., Grossi, F., Sanchez, L., and Di Cioccio, L. (2010) Self-alignment of silicon chips on wafers: a capillary approach. *Journal of Applied Physics*, **108**, 054905.

17 Fukushima, T., Iwata, E., Ohara, Y., Murugesan, M., Bea, J.-C., Lee, K.-W., Tanaka, T., and Koyanagi, M. (2011) A multichip self-assembly technology for advanced die-to-wafer 3D integration to precisely align known good dies in batch processing. *IEEE Transactions on Electronics Packaging Manufacturing*, **1**, 1873.

18 Sato, K., Lee, K., Nishimura, M., and Okutsu, K. (2007) Self-alignment and bonding of microparts using adhesive droplets. *IEEE Transactions on Electronics Packaging Manufacturing*, **8**, 75.

19 Ito, Y., Fukushima, T., Lee, K.-W., Choki, K., Tanaka, T., and Koyanagi, M. (2013) Reductant-assisted self-assembly with Cu/Sn microbump for three-dimensional heterogeneous integration. *Japanese Journal of Applied Physics*, **52**, 04CB09.

20 Kim, J.-M., Yasuda, K., and Fujimoto, K. (2005) Resin self-alignment processes for self-assembly systems. *Journal of Electronic Packaging*, **127**, 18.

21 Chung, J., Zheng, W., Hatch, T.J., and Jacobs, H.O. (2006) Programmable reconfigurable self-assembly: parallel heterogeneous integration of chip-scale components on planar and nonplanar surfaces. *Journal of Microelectromechanical Systems*, **15**, 457.

22 Morris, C.J., Stauth, S.A., and Parviz, B.A. (2005) Self-assembly for microscale and nanoscale packaging: steps toward self-packaging. *IEEE Transactions on Advanced Packaging*, **28**, 600.

23 Wale, M.J. and Edge, C. (1990) Self-aligned flip-chip assembly of photonic devices with electrical and optical connections. *IEEE Transactions on Components, Hybrids, and Manufacturing Technology*, **13**, 780.

24 Hayashi, T. (1992) An innovative bonding technique for optical chips using solder bumps that eliminate chip positioning

adjustments. *IEEE Transactions on Components, Hybrids, and Manufacturing Technology*, **15**, 225.

25 Chang, B., Shah, A., Routa, I., Lipsanen, H., and Zhou, Q. (2012) Surface-tension driven self-assembly of microchips on hydrophobic receptor sites with water using forced wetting. *Applied Physics Letters*, **101**, 114105.

26 Sun, F., Leblebici, Y., and Brunschwiler, T. (2011) Surface-tension-driven multichip self-alignment techniques for heterogeneous 3D integration. Proceedings of the IEEE Electronic Components and Technology Conference (ECTC), pp. 1153–1158.

27 Lee, K.-W., Kanno, S., Kiyoyama, K., Fukushima, T., Tanaka, T., and Koyanagi, M. (2010) A cavity chip interconnection technology for thick MEMS chip integration in MEMS-LSI multichip module. *Journal of Microelectromechanical Systems*, **19**, 1284.

28 Fukushima, T., Iwata, E., Ohara, Y., Noriki, A., Inamura, K., Lee, K.-W., Bea, J.-C., Tanaka, T., and Koyanagi, M. (2009) Three-dimensional integration technology based on reconfigured wafer-to-wafer and multichip-to-wafer stacking using self-assembly method. Technical Digest: International Electron Devices Meeting (IEDM), pp. 349–352.

29 Fukushima, T., Hashiguchi, H., Bea, J., Ohara, Y., Murugesan, M., Lee, K.-W., Tanaka, T., and Koyanagi, M. (2012) New chip-to-wafer 3D integration technology using hybrid self-assembly and electrostatic temporary bonding. Technical Digest: International Electron Devices Meeting (IEDM), pp. 789–792.

30 Fukushima, T., Bea, J.-C., Kino, H., Nagai, C., Murugesan, M., Hashiguchi, H., Lee, K.-W., Tanaka, T., and Koyanagi, M. Reconfigured-wafer-to-wafer 3D integration using parallel self-assembly of chips with Cu-SnAg μbumps and a non-conductive film. IEEE Transactions on Electron Devices, Accepted.

25
High-Accuracy Self-Alignment of Thin Silicon Dies on Plasma-Programmed Surfaces

Christof Landesberger, Mitsuru Hiroshima, Josef Weber, and Karlheinz Bock

25.1
Introduction

Highly accurate placement of semiconductor devices is a strong requirement for manufacture technologies for 3D system integration. The introduction of self-assembly techniques is supposed to both increase die placement accuracy and simplify technical construction of die bonding equipment. During the past years, several research groups proposed and analyzed process schemes for self-alignment of microelectronic devices. An informative review on the various techniques is given in Ref. [1]. Specific results on self-alignment based on surface tension forces in liquid environment are presented in other literature [2–5].

Generally, concepts for fluidic self-assembly require substrate surfaces that show selective wetting behavior for specific liquids. This chapter reports on a surface programming process that results in an opposite wetting behavior on metal and polymer surfaces after fluorine plasma treatment. The main objectives of this research work were the preparation of thin silicon chips of different size, the development of a fluidic self-alignment process, and the determination of the resulting alignment accuracy.

25.2
Principle of Fluidic Self-Alignment Process for Thin Dies

Thin silicon chips with a thickness of 50 μm are of lightweight and are able to swim on droplets of fluids such as water. If the liquid shows a specific wetting behavior with respect to the underlying substrate, then droplet and chip can move and self-align on the substrate. Such a concept requires a surface programming step for the substrate that defines areas of selective wettability. The principal process flow is shown in Figure 25.1.

At first metal patterns are prepared on a polymeric surface, in this case lithographically defined aluminum pads on a polyimide layer on a silicon wafer substrate. Aluminum pads represent the target area for the final chip position. The

Handbook of 3D Integration: 3D Process Technology, First Edition.
Edited by Philip Garrou, Mitsumasa Koyanagi, and Peter Ramm.
© 2014 Wiley-VCH Verlag GmbH & Co. KGaA. Published 2014 by Wiley-VCH Verlag GmbH & Co. KGaA.

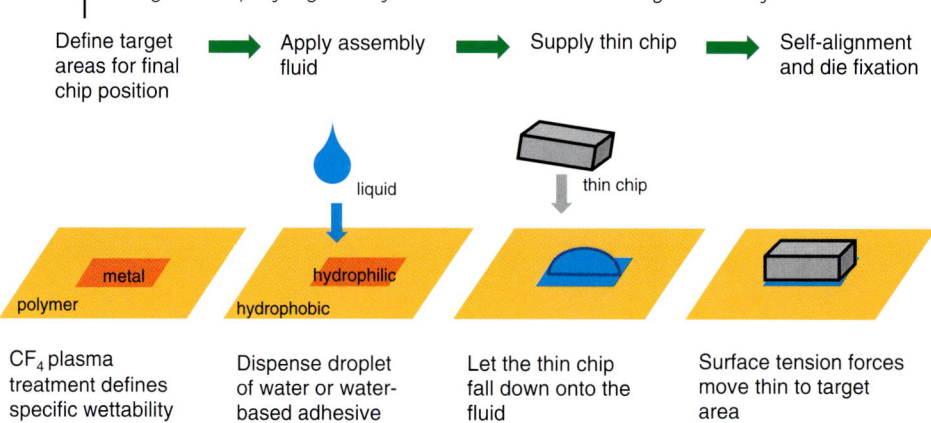

Figure 25.1 The principal process flow of the fluidic self-alignment process of thin silicon dies.

wafer is then placed in a plasma chamber with CF_4 gas as a reactive chemical. The fluorine plasma turns the polyimide area strongly hydrophobic, whereas the metal patterns become strongly hydrophilic. The next step is to apply a droplet of polar assembly liquid, such as water, on the metal target area. Then a thin die is supplied by simply letting it fall down onto the liquid film. The thin and lightweight chip swims on the fluid, which by itself cannot overcome the hydrophobic barrier at the interface of the metal pad and polymer. The system of chip, fluid, and metal pad will now turn into a configuration where the surface energy of the liquid will be minimized. As the contact areas of liquid and metal and liquid and chip are constant, the interface of liquid and air must be minimized. This state is reached when the chip is located directly above the metal pad.

25.3
Plasma Programming of the Surface

In earlier research work, it was found that fluorine plasma treatment at polymer and metal surfaces results in an opposite wetting behavior for droplets of water [6]. Figure 25.2 shows the result of measurements of the contact angle of water on polyimide and copper surfaces after plasma treatment in dependence of the gas composition. A mixture of CF_4: O_2 at a ratio of 80:20 results in a material selective wetting, in that the metal surface turns strongly hydrophilic whereas the polymer turns hydrophobic.

The hydrophilic effect at the metal areas lasts at least for two hours at ambient conditions and is dependent on the surrounding atmosphere [6]. Similar effects were found for the metals aluminum, copper, and gold and for the polymers polyimide, PET films, and others. Furthermore, it was found that in addition to pure water, aqueous mixtures of water and adhesives can also be used as an assembly liquid [7].

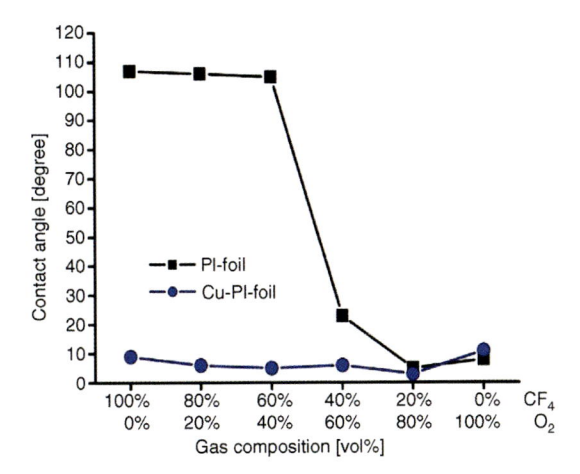

Figure 25.2 Selective wetting on polyimide and copper surfaces after fluorine plasma treatment.

X-ray photoelectron spectroscopy (XPS) measurements were conducted to analyze the molecular composition on metal and polymers after CF_4 plasma. It was found that fluorine and carbon are present on both the metal and polymer surfaces [7]. So, the explanation for the opposite wetting behavior might be related with the spatial configuration of C and F atoms. However, there is no final prove for this hypothesis yet.

25.4
Preparation of Materials for Self-Alignment Experiments

Preparation of target areas for self-assembly experiments was done by sputter deposition of aluminum on a few micrometers thick polyimide film on a silicon bottom wafer substrate (diameter 200 mm). Each "top chip" and each target area on the "bottom wafer" showed two alignment marks made of patterned thin film aluminum layers prepared by means of stepper lithography.

Thin dies (top chips) were prepared at Fraunhofer EMFT according to the concept "dicing-by-thinning" [8]. The size of the top chips was set to: $1 \times 1\,mm^2$, $2 \times 2\,mm^2$, $4 \times 4\,mm^2$, $7 \times 7\,mm^2$, $5 \times 9\,mm^2$, $6 \times 8\,mm^2$, and $10 \times 10\,mm^2$. Top chips were prepared with a standard surface passivation layer SiO_2/Si_3N_4. Separation lines are prepared at the front side of a standard silicon wafer by lithographic patterning and plasma dry etching of trenches. The trench width was set to 15 µm and the trench depth was set to 50 µm.

Wafer thinning was done by means of temporarily bonded carrier substrates using double-side adhesive tapes. We applied wafer grinding and wet chemical spin-etching at the backside of the plasma-trenched silicon wafers. Die separation took place during the final spin-etch process. After backside thinning, the complete set of thin and separated dies were transferred from the carrier substrate (bare silicon wafer) onto UV pickup tape mounted on a frame holder. Further details on material preparation are explained in Ref. [7].

25.5
Self-Alignment Experiments

Bottom wafers (diameter 200 mm) with Al target patterns were cut into quarters and were later on repeatedly introduced in the CF_4 plasma process for series of experiments. Self-alignment experiments generally were performed within a few hours after plasma activation. Pure water was used as the assembly liquid. Dispensing of liquid droplets and die handling were done by means of die bonding equipment (Panasonic flip-chip bonder FCB3). In a first set of tests, the useful amount of liquid was determined with respect to the different die sizes. In most cases, the volume of assembly liquid was set in the range of equal to the volume of the thin die to up to twice the volume of the die. The die bonding equipment used the x- and y-coordinates of the center positions of the target metal pads. At these positions, the liquid was first dispensed. Then a thin die was picked up from the UV tape frame, moved to the target position, and let fall onto the liquid. A specified offset between theoretical target center and the fall-down position was programmed. The offset value was varied between 0 mm and 2 mm. After the chips are released over the liquid, they self-aligned within a few seconds and the assembly liquid (DI water) dried off within minutes. The dies continued to stick at their individual positions. Figure 25.3 shows an example of self-aligned top chips on the metal target areas of the bottom wafer.

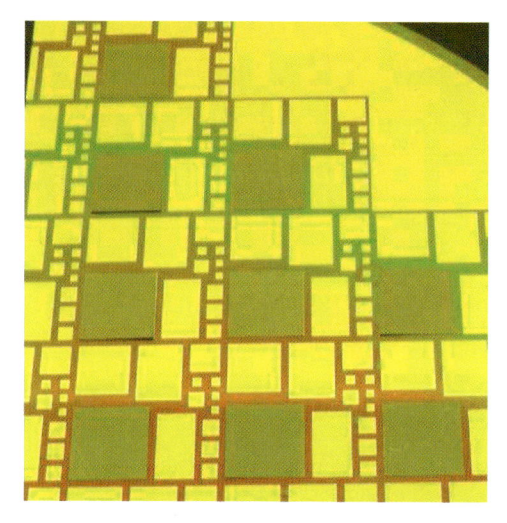

Figure 25.3 Example of self-aligned 50 μm thin top chips, placed face down on a bottom wafer; bright areas represent unoccupied aluminum metal target areas.

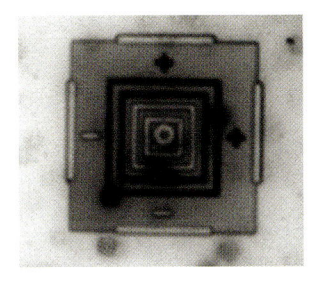

Figure 25.4 Infrared microscopy picture of a stacked die after self-alignment and design of related fiducial marks on top and bottom chips.

After a set of placements, the final positions of each die were determined by infrared (IR) microscopy. Figure 25.4 shows the IR picture with the resulting overlay of the alignment patterns from top and bottom chip. The distance of the inner circle to the inner square pattern of the alignment mark is 1 μm.

25.6
Results of Self-Alignment Experiments

Experiments were carried out for chips of a thickness of 50 μm. Figure 25.5 shows results of the self-alignment experiments for the chip sizes $7 \times 7\,\text{mm}^2$, $5 \times 9\,\text{mm}^2$, $4 \times 4\,\text{mm}^2$, and $2 \times 2\,\text{mm}^2$. Blue areas indicate an accuracy limit below 3 μm in the x and y directions.

Self-alignment accuracy most often was in the range of just 1 μm. Even in the case of a programmed offset for the fall-down position of 1–2 mm, the chips moved to a practically perfect aligned center position. However, there are also a few results that show larger misalignments in the range of 5–8 μm. This observation will be discussed in the following paragraph. It was observed that self-alignment accuracy depends on the volume of the water droplet. The volume of dispensed water V_l was set in relation to the volume V_c of the thin chip. The best results were obtained in the range of $V_c < V_l < 2V_c$. Figure 25.6 shows the corresponding measurements of self-alignment accuracy and volume of water droplet. According to this finding, the dispensed volume of assembly liquid needs to be adjusted to the individual size and thickness of thin dies.

As already mentioned, the hydrophilic behavior of the metal pads depends on the elapsed time after the CF_4 plasma treatment. To be able to estimate the minimum required difference in contact angles on metal and polymer, the self-alignment accuracy was analyzed in dependence of the actually existing contact angle on the metal target area. It was found that a contact angle below 30° generally resulted in precisely aligned chip components [9]. This information can be used to decide whether the plasma conditioning of the surfaces is still strong enough to continue the self-alignment process. Figure 25.7 shows this experimental result. Measured values of misalignment are given on a logarithmic scale.

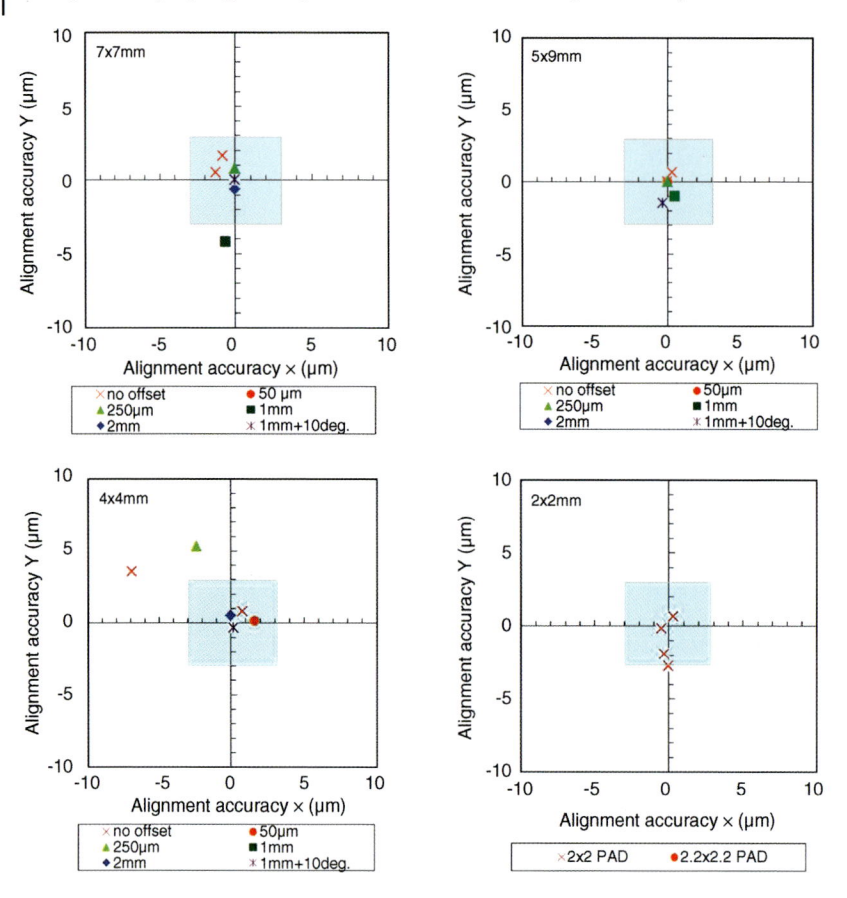

Figure 25.5 Measured placement accuracy after self-alignment of 50 μm thin chips of different size.

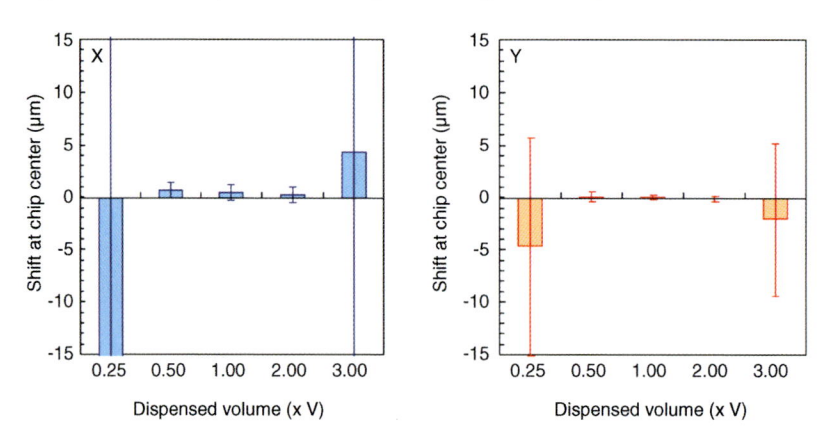

Figure 25.6 Influence of the volume of a dispensed water droplet on the alignment accuracy in the case of thin dies of a size of 6×8 mm^2.

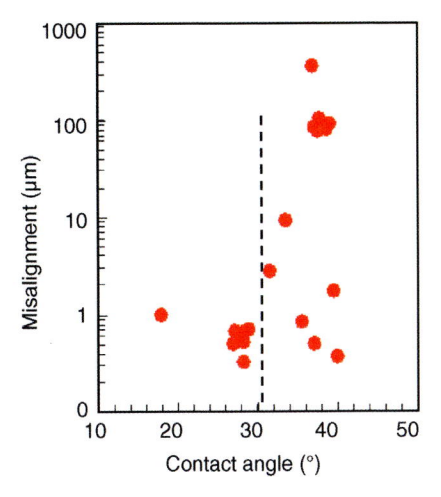

Figure 25.7 Influence of the hydrophilic contact angle of water on the metal target area on the alignment accuracy of thin dies of a size of 7 × 7 mm².

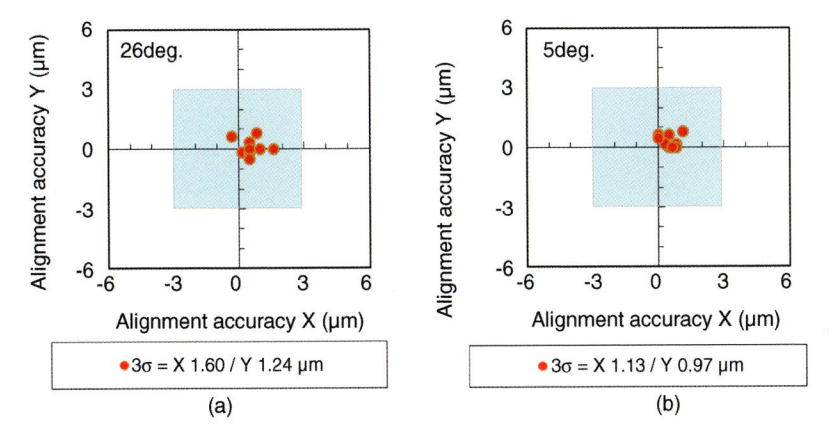

Figure 25.8 Self-alignment of thin dies (7 × 7 mm²) on metal target area with contact angles 26° (a) and 5° (b) shows similar good accuracy values.

Figure 25.8 shows the detailed results of measured self-alignment accuracy for dies that were self-assembled on metal target areas with contact angles of 26° (a) and 5° (b).

25.7 Discussion

With larger dies (>5 mm side length), we found many dies showing a self-alignment accuracy of less than 1 µm. Larger displacements seem to be linked with an increasing number of particulates and dust on the surfaces after repeated use of bottom substrates [9]. Also for small dies (1×1 mm^2, 2×2 mm^2), accurate self-alignment was observed. However, the percentage of dies shows larger displacements increased with decreasing chip size. A possible explanation for this observation is related to the imperfect chip sidewalls. As we used wet etching as the final step in wafer thinning (i.e., die separation), the chip edges undergo a certain amount of overetching. As a consequence, chip size can differ from the original design by a few micrometers. Furthermore, due to the selectivity of wet etching of Si and SiO$_2$, a very thin SiO$_2$ film at the surface may exceed the bulk Si material of the dies. The overhang of the thin oxide film is likely to break off and so the chip edge is no longer a perfect straight line. These imperfections in chip shape actually reduce self-alignment accuracy. This shape issue could be eliminated by some changes in the design and also by replacing wet etching with plasma dry etching as the last step in wafer backside thinning.

When using pure water as assembly liquid, the placed dies remain stuck to the target area due to surface adhesion. This is generally strong enough for further handling steps. If the dies should be securely kept on the bottom wafer for a longer time, then the use of low-viscosity water-based glues is possible as well. The good results of self-alignment were confirmed with adhesives diluted in water.

A further point of possible improvement is related to the control of the dispensed volume of the assembly liquid. The best choice has to be selected with respect to chip size and thickness.

25.8
Conclusions

Self-alignment of thin dies on selectively wetted surface patterns enables high accuracy chip bonding in the range of 1 µm. To reach high statistical confidence, chip edges need to be precisely defined and possible distortions by particulates or edge defects need to be eliminated. Plasma-diced chips offer optimum properties for further self-alignment. It is concluded that self-alignment processes offer a new technical solution for both high-accuracy chip bonding and low-cost placement techniques.

Possible applications for such processes could be self-alignment of a set of thin dies on a temporary carrier substrate (reconfigured wafer) for subsequent 3D integration [10]. A further application scenario could be the placing of thin dies on large area film substrates, for example, in roll-to-roll processing. In this case, the advantage of a parallel process for die attachment could be realized without the need for optical alignment and mechanical precision in die handling.

Acknowledgments

The authors acknowledge the wafer preparation work by Sabine Scherbaum, Robert Wieland, and the technological personnel at Fraunhofer EMFT in Munich and bonding machine operation work by Bernhard Oberhofer of Panasonic Factory Solutions Europe. We thank Panasonic Factory Solutions Japan for their collaboration and financial support.

References

1 Mastrangeli, M., Abbasi, S., Varel, C., Van Hoof, C., Celis, J.-P., and Böhringer, K.F. (2009) Self-assembly from milli- to nanoscales: methods and applications. *Journal of Micromechanics and Microengineering*, **19**, 083001.
2 Fukushima, T. *et al.* (2010) Evaluation of alignment accuracy on chip-to-wafer self-assembly and mechanism on the direct chip bonding at room temperature. 2010 IEEE International 3D Systems Integration Conference (3DIC), November 16–18, Munich, Germany, pp. 185–189.
3 Mastrangeli, M., Ruythooren, W., Van Hoof, C., and Cclis, J.-P. (2008) Characterization of Interconnects resulting from capillary die-to-substrate self-assembly. 2nd Electronic System Integration Technology Conference, 2008 (ESTC 2008), Greenwich, UK, pp. 135–140.
4 Noda, H., Usami, M., Sato, A., Terasaki, S., and Ishizaka, H. (2011) Self-aligned positioning technology to connect ultra-small RFID powder-chip to an antenna. 61st Electronics Components and Technology Conference, 2011 (ECTC 2011), pp. 1009–1014.
5 Landesberger, C., Yacoub-George, E., Hell, W., and Bock, K. (2010) Self-assembly and self-interconnection of thin RFID devices on plasma-programmed foil substrates. 3rd Electronic System Integration Technology Conference September 13–16, 2010, (ESTC 2010), Berlin, Germany.
6 Bock, K., Scherbaum, S., Yacoub-George, E., and Landesberger, C. (2008) Selective one-step plasma patterning process for fluidic self-assembly of silicon chips. 58th Electronics Components and Technology Conference, 2008 (ECTC 2008), Lake Buena Vista, FL, USA,.
7 Landesberger, C., Scherbaum, S., Bock, K., Hiroshima, M., and Oberhofer, B. (2012) Plasma dicing enables high accuracy self-alignment of thin silicon dies for 3D-device-integration. 4th Electronic System Integration Technology Conference, September 2012 (ESTC 2012), Amsterdam, Netherlands.
8 Landesberger, C., Scherbaum, S., and Bock, K. (2010) Ultra-thin wafer fabrication through dicing-by-thinning, in *Ultra-Thin Chip Technology and Application* (ed. J.N. Burghartz), Springer, Berlin, pp. 33–43. ISBN 978-1-4419-7275-0.
9 Hiroshima, M., Arita, K., Haji, H., Oberhofer, B., Landesberger, C., Scherbaum, S., Weber, J., and Bock, K. (2012) A robustness study on self-alignment of thin-Si dies using surface tension. Proceedings of the 22nd Micro Electronics Symposium (MES 2012), Osaka, Japan (in Japanese).
10 Klumpp, A. and Ramm, P. (2012) Temporary adhesive bonding with reconfiguration of known good dies for three-dimensional integrated systems, in *Handbook of Wafer Bonding* (eds P. Ramm, J.-Q. Lu, and M. Taklo), Wiley-VCH Verlag GmbH, Weinheim, pp. 347–354. ISBN 978-3-527-32646-4.

26
Challenges in 3D Fabrication
Douglas C.H. Yu

26.1
Introduction

The major distinctions between a high-volume manufacturing (HVM) 3D fabrication process and most of the reported literature on 3D are in two fundamental aspects, yield and cost. These two aspects separate production from R&D activities. This chapter describes the HVM 3D fabrication process challenges with emphases on the optimization of yield and cost.

HVM 3D production needs to meet stringent requirements on yield and reliability. Achieving a broad manufacturing window is required for each of the very many processing steps needed to complete the final product. Excellent yield (N9 yield) must be attained on each step so that, cumulatively, an acceptable final yield and cost can be achieved. 3D production is therefore best fitted for the semiconductor foundry environment, which is well disciplined on yield and cost control to manufacture the nanoscale devices. This is very different from performing the exploratory scientific studies. In addition to process development and optimization, a laundry list of infrastructure and logistic considerations must be carefully managed and put together from the selection of production tools and equipment, chemicals and materials, optimized daily operation routines, energy efficiency, green technologies and environmental impacts, down to the equipment uptime, waste and by-product handling, continuous quality monitoring and yield improvement protocols, and so on. Often over time, the most advanced technology may not necessarily yield the best results, particularly if the technology is new or using a prototyping tool. It is important, in the early technology development phase, to identify and understand the potential yield and cost issues of a new technology.

26.2
High-Volume Manufacturing for 3D Integration

Contrary to the general beliefs, the optimal technological option for the fabrication of a newly developed product is often using the simple process with existing materials and equipment and with the best results if it is based on a production-proven process flow, provided that the function and performance are capable of meeting the new product requirements. And 3D chip stacking fabrication is no exception to this rule.

3D integration using through-silicon via (TSV)-based chip stacking technology has been widely investigated on the following technical issues and challenges:

1) Front-side and backside wafer processes (including TSV, redistribution layer (RDL), etc.) and bumping (including creating microbumps and solder controlled collapse chip connection (C4) bumps on both sides of the same wafers, etc.) processes. This is done in wafer form.
2) Chip or wafer thinning and thin die and wafer handling and shipping. Thinning is done in wafer form mostly, but shipping and handling of thinned wafers and chips could be in wafer or chip forms or both.
3) Chip and wafer bonding, stacking, and protection (including underfill and molding compound). This could involve either chip or wafer form, or a combination of both.
4) Reliability tests and test methodology.
5) Final module or subsystem packaging and assembly. This mostly is in chip form, either onto the carrier substrate or organic printed circuit board (PCB).

26.3
Technology Challenges

There are multiple choices in selecting a process flow, each associated with different technical challenges and merits. The general discussion on 3D fabrication technical challenges in the following sections will focus on modular engineering issues and, when needed, with specific flow-dependent integration issues.

26.4
Front-Side and Backside Wafer Processes

Apart from the usual front-end-of-line (FEOL) and back-end-of-line (BEOL) silicon MOSFET processes, 3D chip stacking requires the TSV formation process to be inserted in the wafer process flow. The deep trenched via can be formed using either laser drill or lithography plus deep reactive ion etching (DRIE) processes. The well-known Bosch process [1] is capable of forming fine-diameter high aspect

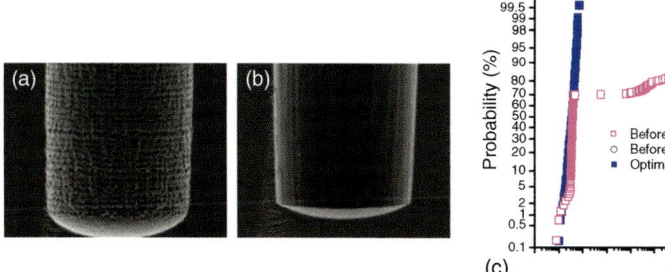

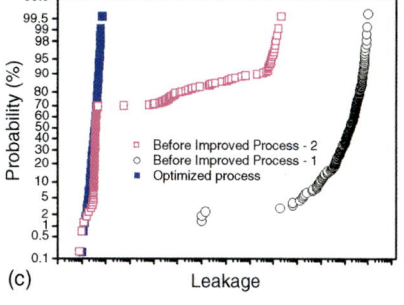

Figure 26.1 Cross-sectional SEM images of a TSV bottom (a) with a high scallop profile and (b) after improvement [2]. (c) Leakage current was measured between adjacent TSVs. The TSV etch profile and bottom cleaning have a significant impact on the integrity of liner deposition and TSV leakage. Reprinted with permission from IEEE. Courtesy of TSMC.

ratio trenches and becomes popular for TSV trench formation. Early process shows via sidewall surface roughness issues (Figure 26.1a), which may induce linear oxide local thickness variation and result in high leakage current (Figure 26.1c) [2]. This is no longer an issue for improved etching processes (Figure 26.1b). Electrical leakage current monitoring proves that the via-to-via leakage has been improved and minimized. An alternative option for via formation with larger dimension is using laser drill. Laser beam introduces highly focused local energy that smelts the materials (silicon or glass) and removes them to the desired depth [3]. Either high-speed pulsed laser or parallel broad-beam laser with mask can be used for via formation. Compared to lithographically fabricated trenches, laser vias, in general, may have lower cost of ownership (CoO) but have limitation in pitch, throughputs, and sidewall quality [4].

For integration into active devices, TSV formation could be inserted after FEOL transistor process but before BEOL Cu interconnect. This is generally called via-middle process. In the case of a silicon interposer where there is no active device, TSV is the first major process step and is called via-first process. Alternatively, TSV can also be formed from the backside of the wafer after the front-side process is finished. This is called via-last process. Most of the wafer fabs adopted front-side via processes while packaging houses adopted backside via-last process flow. Via-middle process flow is capable of creating smaller, denser TSV, poses fewer design and process constraints, and is more popular among the wafer fabs. Via-last formation from backside of the wafer often requires a large via-etch landing pad by using the front-side metallization layer. It creates nonuniform via size near the landing pads due to what is called "elephant foot" because of the nonuniformity in the etching process. For 3D ICs, enlarged via bottoms are difficult for seed layer deposition, and is therefore more prone to generating poor via contact. Additionally, Cu via extrusion issue to damage the BEOL layers on the front side cannot be easily eliminated when using via-last process from wafer backside.

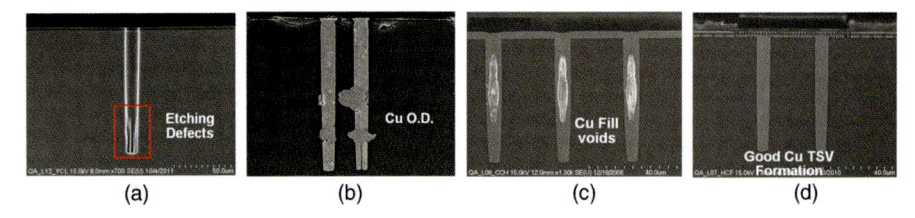

Figure 26.2 Trench formation defects observed during or after the processes: (a) etching defects; (b) Cu out-diffusion; (c) Cu fill voids; (d) good TSV formation. Courtesy of TSMC.

After via formation, liner dielectric deposition is followed by seed metal deposition and Cu filling the whole depth of the vias with electroplating. Several typical problems associated with Cu TSV formation are shown in Figure 26.2a–c. An optimized TSV process should take care of these issues and result in defect-free and highly reliable TSVs, as shown in Figure 26.2d.

Other than defects formed within the vias, Cu extrusion (or pumping) is a well-reported TSV issue [2]. It is generally believed that Cu electroplating with the subsequent high-temperature processing steps could induce microstructural and volumetric change, resulting in voids inside TSV (Figure 26.2c) or extrusion from TSV on the top surface, as shown in Figure 26.3a–c. Stress modeling also suggested that the extrusion is caused by Cu volume change due to microstructure evolution and volume expansion during processes (Figure 26.4) [2].

Integration of BEOL process into TSV wafers also introduced new process issues. Compared to BEOL Cu dimensions, TSV contains a huge volume of Cu inserted within silicon matrix (see, for example, Figure 26.3d). Such volumetric and

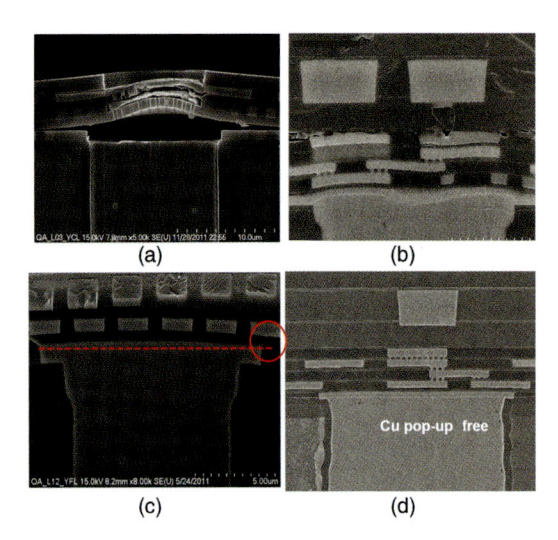

Figure 26.3 Cu extrusion impacts BEOL integrity after several thermal cycles at 400 °C/30 min: (a–c) before optimization and (d) after optimization [2]. Reprinted with permission from IEEE. Courtesy of TSMC.

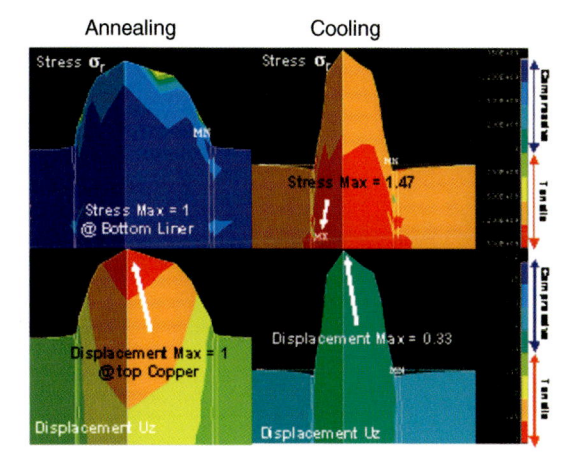

Figure 26.4 Modeling on a Cu extrusion mechanism. In both annealing and cooling processes, large extrusion at the TSV center is found [2]. Reprinted with permission from IEEE. Courtesy of TSMC.

mechanical discontinuity can cause nonuniform chemical–mechanical polishing (CMP) results (Figure 26.5) [2]. Nonuniform CMP topography, or dishing, has detrimental effects on the subsequent BEOL W-plug and Cu interconnect yield, as demonstrated in Figures 26.6 and 26.7. The issue can be resolved by improving CMP recipe and by locally optimizing the density of interconnect and layout design.

For the backside processes, the wafer needs to be thinned down from the wafer backside to the desired thickness and to reveal the Cu TSVs in order to form interconnection to the backside RDL. Wafer thinning and thin wafer protection during the backside process was a challenge. Detailed issues on wafer thinning and thin wafer handling are discussed in a later section. Backside RDL and bumping process are done using similar technologies for front-side processes, except for one additional limitation. Since the wafer front side is now protected by the carrier with temporary bonding layer, the backside processing using chemicals, temperatures,

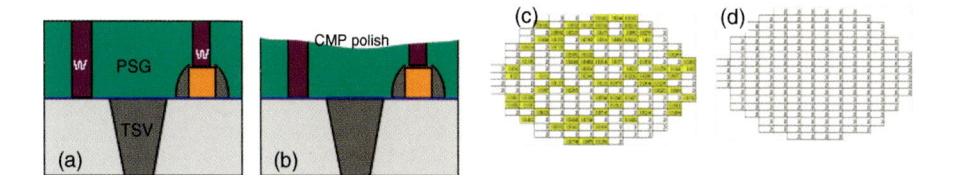

Figure 26.5 The PSG loss with schematic device structure (a) before CMP and (b) after CMP with Cu TSV underneath. (c) E-test mapping of the nonoptimized process, marked by yellow are failed dies and (d) E-test mapping of a wafer using optimized processes (no failed die) [2]. Reprinted with permission from IEEE. Courtesy of TSMC.

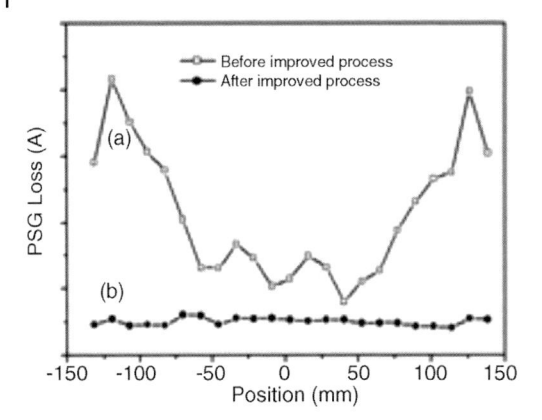

Figure 26.6 The PSG loss thickness (a) before improvement and (b) after improvement. Reprinted with permission from IEEE. Courtesy of TSMC.

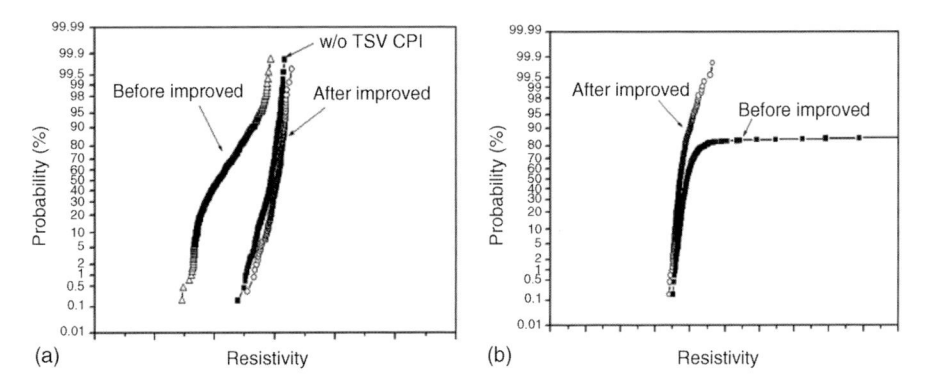

Figure 26.7 The impact of PSG loss on the (a) W-plug and (b) interconnect resistivity, before and after improvements. After improved process, the performance of W resistivity is same as without TSV process. Reprinted with permission from IEEE. Courtesy of TSMC.

and other wet or dry processes should not affect the carrier and the temporary bonding layer.

26.5
Bonding and Underfills

Once the front-side process is completed with solder microbumps, chip stacking on the bottom chip or wafer can be performed. The fact that both sides of the joints are silicon is with tremendous advantage in terms of bonding yield and reliability.

Figure 26.8 A demonstration of a large die size CoW homogeneous multichip integration using a solder microbump with Cu pillar assembly on the bottom wafer. Courtesy of TSMC.

The coefficient of thermal expansion (CTE) mismatch issue in traditional flip-chip bonding between silicon chip and organic substrate does not exist at the fine-pitch microbump joints. And the fact that in the near future, microbump size will soon shrink to sub-10 μm has put the chip-on-wafer (CoW)-first the most favorable technology in chip stacking process flow options [5]. Figure 26.8 demonstrates a production example of CoW stacking using very large rectangular top die (>200 mm^2) on an extremely large bottom die (>800 mm^2). Hundreds of thousands of full-array high-density microbumps on each chip can be joined with excellent yield throughout the whole wafer, owning to the production-proven pick-and-place and optimized solder reflow processes. Flux cleaning and capillary underfills applied after reflow are also being optimized. Traditional capillary underfills are employed in CoW solder microbump joints and have been optimized for high-yield volume production. Efforts have been made to improve the flux cleaning for the narrow-gapped microbump joints (Figure 26.9). Capillary underfill is in production for traditional C4 solder bump joints. High yield, high throughputs without the need for thermal compressive bonding, and proven reliability with mature production industry experiences are some of the obvious reasons why CoW is chosen for 3D integration. This is a good example of using

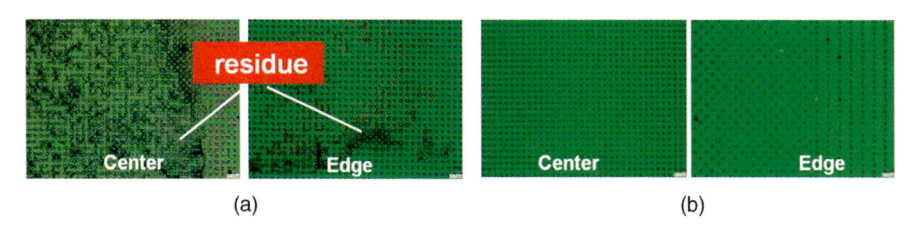

Figure 26.9 Large die (>200 mm^2) flux clean issue (a) before and (b) after improvement. Courtesy of TSMC.

conventional production-proven processes, with some modification and improvement, for new and novel production applications.

26.6
Multitier Stacking

Multitier, or multilevel, strata chip stacking is another completely new arena in 3D assembly. No similar process has been attempted by the industry. The challenge is twofold. First, a double-side-bumped chip needs to be stacked on top of the strata with sub-µm precision. Second, stacking, reflow, flux application, flux cleaning, and underfill application should be done for each and every tier stacking in the strata. The choices of underfill and flux and joining technology (reflow or thermal compressive bonding) determine if the multitier stacking processes are to be done in series or in one shot, or a mixture of both. Combining this with another technology question of whether one should choose face-to-face (F2F) or face-to-back (F2B) stacking, the issue becomes quite intricate. Figure 26.10 shows the definition of F2F and F2B stacking. The choice of F2F or F2B is more of a design and system requirement and perhaps specific to certain applications or products.

From a stacking process point of view, there is no fundamental difference between F2F and F2B stacking, although the process flow and sequence need to be adjusted accordingly. But from the production point of view, by considering cost, yield, and reliability, it is always advisable to have the stress-sensitive side, such as the front side with strain-sensitive transistors, low-k dielectrics, or the backside if a fragile, low-temperature polymer dielectric is used, of the chip facing away from the C4 bumps, which will be joined to the organic substrate, at the bottom of the strata (level 1) in order to minimize chip–package interaction (CPI) impact and packaging reliability concern.

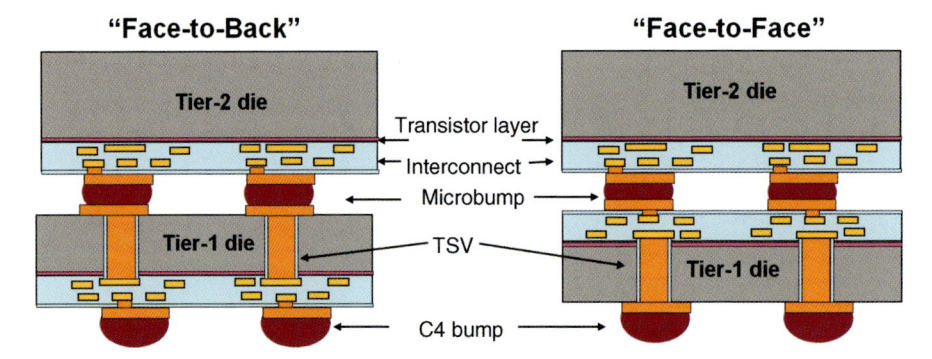

Figure 26.10 Face-to-face versus face-to-back. Chip front side is defined to be the side with the FEOL active device, or if for the interposer, the side with Cu BEOL interconnect, or passive components such as the capacitor, inductor, or resistor. Courtesy of TSMC.

26.7
Wafer Thinning and Thin Die and Wafer Handling

One of the most important technical challenges in 3D TSV chip stacking is wafer thinning and thin wafer handling. This is also the process that impacts the collaboration with supply chains. Wafers need to be thinned down to reveal the TSVs and subsequently processed with the backside interconnecting layers in order to complete the interconnects on both sides of the chips. One possible process flow arrangement suggests that wafer thinning and backside processing be done at the packaging house. Another possible flow is for the wafer thinning and backside processes to be done in wafer fab and the thinned wafer to be shipped to a different location for the next step, chip or wafer bonding and stacking. Both process flows are problematic with issues that need to be addressed. The process of thinned wafer handover protocols among different owners along the supply chain has proven to be tricky. Yield loss along the way is inevitable. Ownership of liability and responsibility is difficult to define. In order to reduce the yield loss and minimize the yield loss responsibility issue, it is highly desirable for the wafer, once it is thinned, to stay and complete the final assembly in the same facility.

Even staying within the same facility, the challenge of thin wafer handling remains daunting. Figure 26.11 shows the schematic cross section of a sharp Si edge after wafer thinning. Such sharp edges need to be polished or rounded before the wafer moves into the next process step. Figure 26.12 shows examples of edge chippings after the wafer is thinned down. Debonding of the temporary carrier (on the wafer front side) could easily cause mechanical damages at the wafer edge.

Another effect of the thinned wafer is wafer warpage without rigid carrier support. The thinner the wafer, the more warpage is observed (Figure 26.13). Wafer warpage causes subsequent process and alignment difficulties and results in yield loss. It is, thus, advisable not to thin down the wafer until it is necessary, and not allow the chance for handling the thinned wafer without rigid carrier protection. Long-distance shipping of the thinned wafer with the carrier is possible, but it has a substantial cost and logistic penalty. An obvious solution is to adopt a chip stacking process flow that minimizes the risk of thin wafer handling. This can be achieved

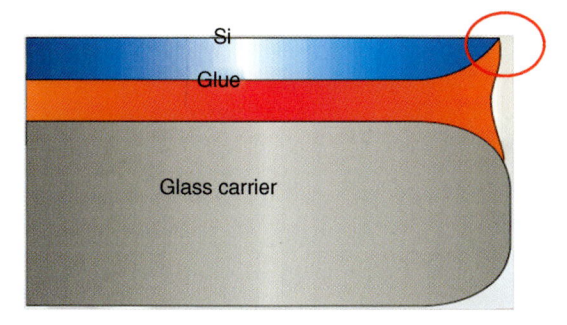

Figure 26.11 Schematic cross section showing that the thinned wafer sharp edge can be vulnerable, even with carrier support. Courtesy of TSMC.

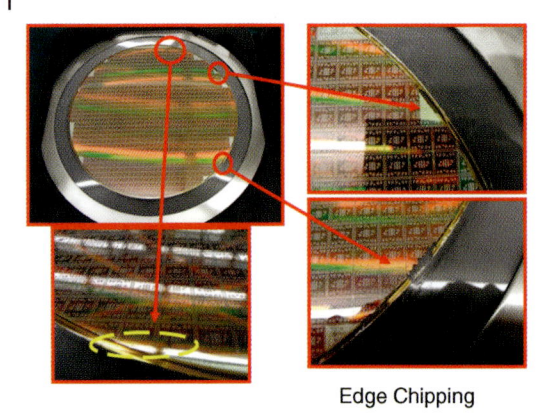

Figure 26.12 Wafer thinned down to less than 100 μm, showing several edge locations with chippings and cracks even before transportation. Courtesy of TSMC.

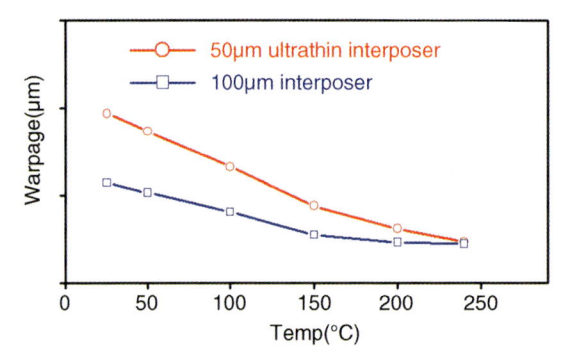

Figure 26.13 Thinner wafer shows larger wafer warpage, which makes microbump joining difficult with low yield [5]. Reprinted with permission from IEEE. Courtesy of TSMC.

by using CoW stacking first, dicing the CoW, and then joining the CoW on the substrate. The process, chip-on-wafer-on-substrate (CoWoS^{TM} [5]), is fully described in Chapter 20.

Wafer backside thinning requires a temporary supporting carrier to protect the wafer front side. A temporary bonded glue layer is needed between the wafer front side and the carrier. Bonding and debonding between the wafer and carrier is a challenging process. There are many solutions proposed by different suppliers in the industry. The challenge here is to protect the front-side topography during backside processes while retaining front-side silicon integrity during backside processes and debonding. An example of glue layer delamination and bump under-bump metallization (UBM) crack during backside process is shown in Figure 26.14. Glue layer delamination is often followed by stress concentration at solder bumps and subsequently causes bump cracks. Remember that, other than the interposer, a device wafer front side contains strain-sensitive active transistor

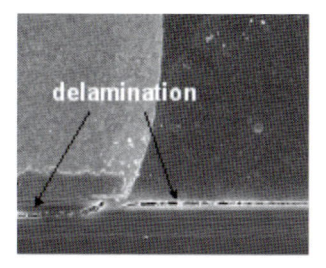

Figure 26.14 Typical thin wafer handling process shows severe delamination issue occurred during the backside or demount process [2]. Reprinted with permission from IEEE. Courtesy of TSMC.

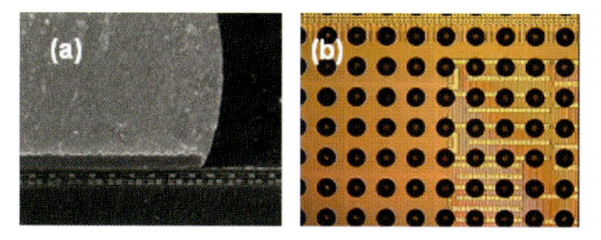

Figure 26.15 New thin wafer handling process shows (a) no delamination after mounting and backside thermal processes and (b) no post cleaning issue after the demount process on front-side bumps [2]. Reprinted with permission from IEEE. Courtesy of TSMC.

devices and mechanically fragile ultralow-k (ULK) or extremely low-k (ELK) BEOL interconnects. Protection of these fragile silicon assets to ensure high-yield bonding and debonding has become a crucial topic of yield management in thin wafer handling (Figure 26.15).

While it is absolutely essential to protect and maintain wafer front-side device integrity, it is not a guarantee that the device performance of the successfully debonded wafer remains the same after wafer thinning. An example is flicker noise benchmarking in analog devices before and after wafer thinning, shown in

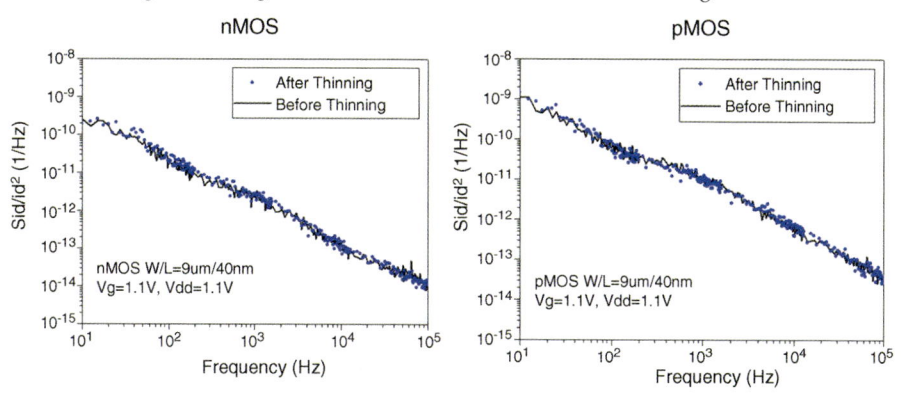

Figure 26.16 There is no degradation in the flicker noise on the devices in thinned wafers with thickness down to the 50 μm range [2]. Reprinted with permission from IEEE. Courtesy of TSMC.

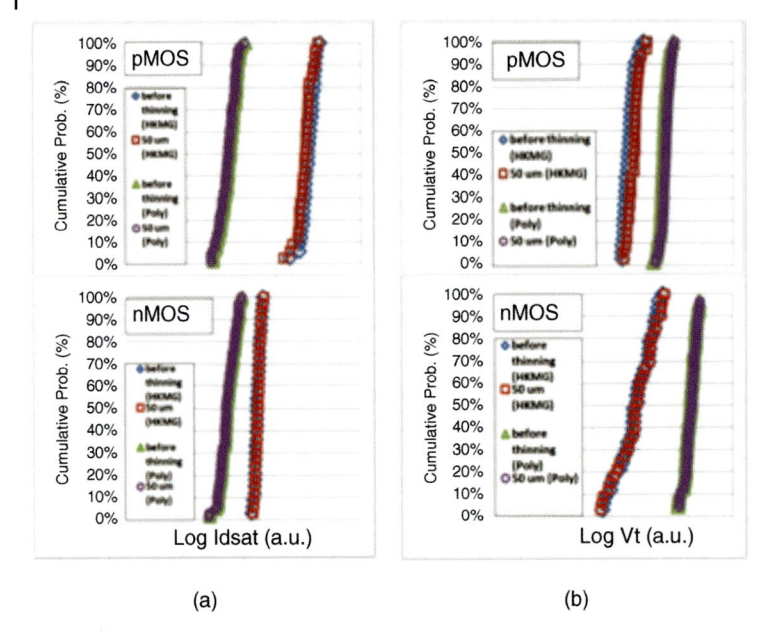

Figure 26.17 (a) Cumulative plots of I_{dsat} and (b) V_t of p- and nMOS before and after the thinning process for a poly and HKMG device with a mean wafer thickness of 50–100 μm. The thinning process has little or no degradation for both poly and HKMG devices [6]. Reprinted with permission from IEEE. Courtesy of TSMC.

Figure 26.16. In this case, post-thinning functional tests become necessary for certain types of 3D chips. Figure 26.17 shows a demonstration of a wafer thinned to 50–100 μm in thickness: Little or no impact is found on both poly and high-k/metal-gate (HKMG), p- and nMOS devices.

26.8
Strata Packaging and Assembly

The CoW strata stacks are diced and isolated and ready for flip-chip packaging onto an organic substrate. On the surface, the on-substrate (oS) C4 assembly process has no significant difference compared to the usual fcBGA or fcCSP packaging process. However, the same process flow could obtain a very different result if special attention is not taken in package design and process optimization. CoW strata on-substrate packaging and assembly needs to be considered as one integrated process (CoWoS) with design, materials and process selection, and assembly all considered and optimized simultaneously [5].

The important difference between the stacked-chip strata C4 assembly onto an organic substrate and the traditional flip-chip C4 assembly on substrate is that the CPI stress can propagate upward to the stacked chips. And the fact that contemporary 3D chip stacking strata may contain an ultralarge bottom chip only makes the issue more challenging than traditional fcBGA or fcCSP. Strata-stacked bottom chip thickness is very thin, most likely in the 50–200 μm range, due to the TSV aspect ratio requirement and depth limitation. Compared to most of the fcBGA and fcCSP, where silicon is thicker, CPI stress due to CTE mismatch between the organic substrate and 3D silicon strata inevitably propagates upward to the upper layers of the strata. The subtle but critical difference in a stacked strata is that all silicon layers (except for the top level chip) contain TSV and all layers are joined together with low height profile microbumps. Silicon with different TSV densities has different CTE compared to bulk silicon. And ultralow stand-off microbump joints make the strata layers very sensitive to warpage bending force induced by CTE mismatch between the organic substrate and silicon stacks. These facts have made CPI a far more complex challenge in 3D stacking than conventional flip-chip packaging and require a more coherent and integrated package design and process consideration. Figure 26.18 demonstrates a successful pilot production integration of 3D CoWoS stacking.

The successful stacking has been proven with various homogeneous (same top dies) and heterogeneous (combining different top dies) integrations (also described in Chapter 20). An adaptive and robust CoWoS integration flow, which is capable of coping with different chip dimensions and design, is necessary for 3D stacking production.

A final CoWoS assembly is shown in Figure 26.19, quite similar to a conventional fcBGA. This is the final form of the product that is ready to be shipped to the customer.

The most important priorities for successful volume production are yield and cost optimization. As discussed in the previous sections, various technology

Figure 26.18 3D X-ray tomographic view on a stacked 3D strata. The top and bottom dies are joined with solder microbumps and the bottom die with TSV is joined to the organic substrate using C4 bumps. Noticed how close the distance is between the organic substrate and the thin bottom die with TSV and the distance between the C4 bumps and microbumps. Also notice the large dimension difference between the microbumps and C4 bumps [7]. Reprinted with permission from IEEE. Courtesy of TSMC.

Figure 26.19 A demonstration of a finished CoWoS package. Also shown for comparison is an isolated bottom chip strata with a face-up C4 array. Courtesy of TSMC.

challenges have been overcome using proper chip stacking sequences and well-planned process flow. From a system point of view, traditional chip and package design are now extended into 3D stacking strata design. For package design, a risk-free CPI can only be achieved by considering the whole 3D strata simultaneously. Different chips brought together under the same strata need to be optimized under a single CPI model. New elements needed for high-density 3D interconnection, such as TSV, microbump, and interposer, are also brought together. The issue is made easier if the chip and package design as well as the wafer and package processes are integrated. The traditional design–foundry–package–test supply chain can no longer meet the challenges under stringent yield and cost control requirement for 3D integration. An integrated solution is more attractive than a distributed supply chain for the reasons already discussed. As technology advances from generation to generation with shorter and shorter product life cycles, integrated solutions are even more important to meet the schedule and a large variety of application needs.

The issue is complicated as the processes are divided into the front side and backside on the chips. As highlighted, the wafer needs to be thinned down to the desired thickness before backside processes, such as RDL and bumping, can occur. If a thick TSV wafer is to be shipped, the packaging house will need to be able to perform the wafer thinning and backside processes. In this case, at least partial wafer process capability needs to be duplicated in the packaging house. If the wafer is to be thinned and backside process finished in wafer fab before shipping to the packaging house, thin wafer shipping and handling becomes an inevitable part of the process. The thin wafer handling dilemma is an issue if the traditional supply chain arrangement is to be followed. Figure 26.20 summarizes the above-mentioned thin wafer handling dilemma. The most cost-effective solution with

	Pros	Cons
Thick TSV Wafer	1. Avoid thinned wafer handling issues	1. Unclear liability for TSV 2. Complicated hand-shaking 3. High investment for OSAT
Thinned TSV Wafer	1. "Conventional" supply chain model 2. Less "third party die" handling barrier	1. Thin wafer handling issues 2. Microbump interface hand-shaking issues
Full Turnkey	1. Simplified integration that favors yield and time to market 2. Avoid thinned wafer handling issues	1. "Third party die" integration issue

Figure 26.20 A wafer handling dilemma for thick and thin TSV wafers. An integrated turnkey is the best supply chain solution. Courtesy of TSMC.

clear-cut liability and responsibility is the one with high yield and the most viable ecosystem solution.

26.9
Yield Management

Without doubt, yield management is of the highest priority for 3D integration. The stake is high and the cost associated with yield loss is even higher. The elements of the key 3D constituents such as TSV, microbump joint, and C4 bump joint are more than hundreds of thousands or even millions in each strata stack. Yield needs to be almost near perfection for each of the very many process steps in order to cumulatively obtain reasonable high yield for the final 3D assembly (Figure 26.21).

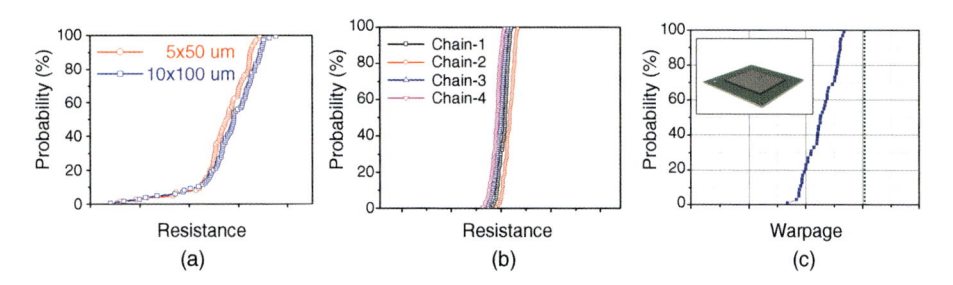

Figure 26.21 High yield for individual components in 3D processes is a fundamental requirement. (a) 100% TSV yield at different TSV dimensions; (b) 100% microbump joining yield at different daisy chain length; (c) 100% C4 joining yield at final CoWoS stack. Thanks to CoW-first assembly with excellent warpage control, both microbump joining and C4 bump joining can achieve 100% yield [5]. Reprinted with permission from IEEE. Courtesy of TSMC.

In traditional flip-chip C4 bump packaging, die and substrate warpage are controlled parameters for high-yield assembly, especially for large thin die packages. In 3D assembly, large die size and ultrathin die thickness is commonplace. High assembly yield with warpage concern seems to be a mission impossible for chip-on-chip (CoC) or chip-on-substrate (CoS). However, if chip stacking on a wafer (CoW) is done with a silicon-to-silicon joint first, the CTE mismatch is removed from the yield equation during microbump assembly. High CoW yield is a natural result caused by silicon-on-silicon stacking. Once the stacked CoW strata are sawed and isolated for oS assembly, the stack is protected and the oS assembly is similar to traditional flip-chip C4 assembly. Yield is optimized in this CoWoS process flow [5].

26.10
Reliability

Semiconductor products exhibit good correlation between yield and reliability. High yield represents tighter distribution in product variation, which in turn represents fewer early failures according to the bathtub curve and excellent long-term product reliability. In terms of reliability, 3D systems are different from traditional microelectronics systems in two aspects. First, there are new building blocks in 3D chip stacking packages, such as TSV, RDL, interposer, integrated passive device (if any), and microbump joints. And second, the new stacking scheme introduces new mechanical, thermal, electrical, and environmental stresses caused by different geometric and structural designs of the new package. Design for reliability (DfR) and design for yield (DfY) are most needed for 3D system integration. To optimize the overall system performance, yield, and reliability, it is important to have every aspect of the system integration considered and potential solutions implemented from the beginning of the system design phase. The designer may need to know about the materials and processes to be used and have overall system–package interaction (SPI) knowledge to optimize the system partition and configuration. Traditional distributed product supply chain could become a nightmare in practicing DfR and DfY. Integrated modeling design approach, such as Figure 26.22, could largely improve the design cycle time for fast realization and product manufacturing.

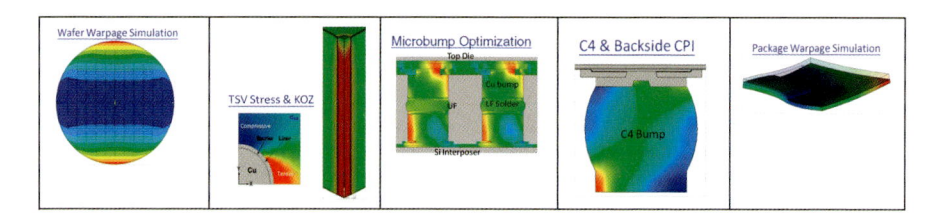

Figure 26.22 All-in-one materials and process database to ensure model accuracy, including wafer-level mechanical properties, process- and material-related stress, microbump and C4 bump CPI modeling, and overall silicon strata and package warpage simulation are integrated under one 3D package simulation platform. Courtesy of TSMC.

26.10 Reliability

A good example of 3D system new reliability concern is with the diminishing solder microbump dimension used in chip-to-chip interconnection. While RDL and TSV are "all Cu" interconnection, solder microbumps with Cu pillars are certainly susceptible to electromigration for their small dimension. However, studies showed that with optimized UBM, chip stacking assembly process tuning, and accelerated current stressing conditions, solder microbump exhibits immortal electromigration performance [8].

Such astonishing electromigration performance that is far superior to C4 solder bumps using the same UBM metallurgy is due to one fundamental difference in microbumps, that is, the dimension. Traditionally, when a conductor dimension is shrunk, electromigration worsens because of smaller cross-sectional area or higher current density effects. However, for the solder microbump, the effect is the opposite. Much smaller solder microbumps show far better electromigration results than larger C4 bumps. Detailed studies at the microstructure evolution during current stressing show that all the solder in the microbumps can transform into intermetallic compound and sustain much higher current and temperature stressing without further microstructure change or evolve into voids or failures (Figure 26.23) [8]. This is a good lesson learned, in that dimensional shrinkage can result in much better performance or reliability improvements, similar to what was observed in transistor performance, TSV keep-out-zone (KOZ) effects, and other examples in FEOL.

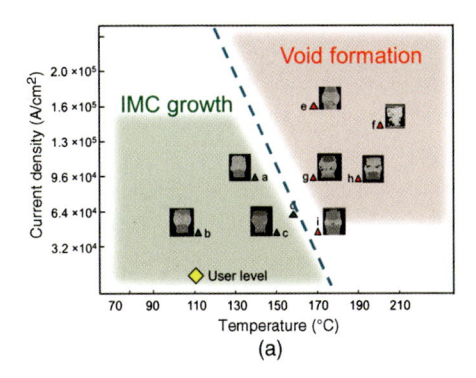

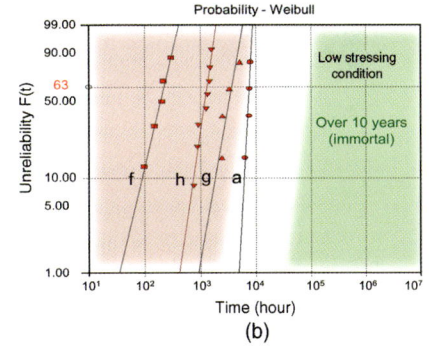

Figure 26.23 (a) Correlation of the degradation mechanisms with the stressing conditions: modes versus stressing conditions. Intermetallic compound (IMC) formation dominates the green area when the stressing conditions, a, b, and c, are low. The longest stressing time has been over 17 months. Accelerated conditions, a, b, and c, are still far more aggressive than any possible user levels. (b) Weibull distribution of EM lifetime. For excessively high stressing conditions, e–i, void formation leads to continued resistance increase and eventual open failure with a much shorter projected lifetime [8]. Reprinted with permission from IEEE. Courtesy of TSMC.

26.11
Cost Management

Cost management is a continuous and ubiquitous effort in 3D development. From basic materials and chemicals, to the equipment and process optimization, to the integration, stacking, package, and assembly, and to the final product testing, yield improvement, and reliability, 3D integration cost management links all these factors into an overall production management task. One of the most important aspects in cost management is to improve the overall production yield. One percent of product yield improvement often translates into several times of cost percentage reduction. It is, thus, imperative to make every effort to improve the yield. Examples already given in this discussion are using CoWoS instead of CoCoS, in which the warpage-dominant assembly yield can be optimized by using CoW-first. A seemingly minor difference of CoW-first process approach can significantly improve the process yield – not to mention that all the accompanied benefits one can get, from enabling large die assembly, microbump pitch shrinkage for the future generation, to good reliability due to less residue stress.

CoWoS together with integrated design and manufacturing solutions for 3D assembly can remove all the yield-harmful and cost-adding factors in 3D integration, and achieve the lowest cost solution available. 3D system integration has added cost to the product because of the introduction of various sophisticated process steps and flows such as TSV, microbumping, ultrathin wafers, and so on. It is highly desirable to efficiently improve and manage the yield from each and every step to justify the value-added 3D integration scheme. Removing the intermittent product shipping and testing is one important aspect (particularly when these semi-products are thin and fragile). Using integrated design and production flow is another. And clear ownership on the production yield and cost control is the final.

26.12
Future Perspectives

Like FEOL transistors and BEOL Cu/low-k, 3D integration needs to evolve along with technology node shrinkage. As Moore's law is only concerned with horizontal dimensional shrinkage, 3D integration also needs to involve the third and vertical dimensional shrinkage. Figure 26.24 demonstrates that the third dimension (z dimension) shrinks, which is coupled with horizontal (x and y dimensions) shrinkage in order to keep the geometry and aspect ratio within process control for the key interconnection components (TSV, microbumps, and RDL).

As Moore's law demonstrates, dimensional shrinkage causes performance gains, changes in form factor, and most importantly, cost benefits, which is imperative to motivating the whole industry forward. Likewise in 3D miniaturization, the same force will drive the system into smaller form factors in all three dimensions (x, y, and z) to gain performance and cost benefits. Dimensional shrinkage of TSV and microbumps has proven to be beneficial to the devices. Figure 26.25 shows a

26.12 Future Perspectives

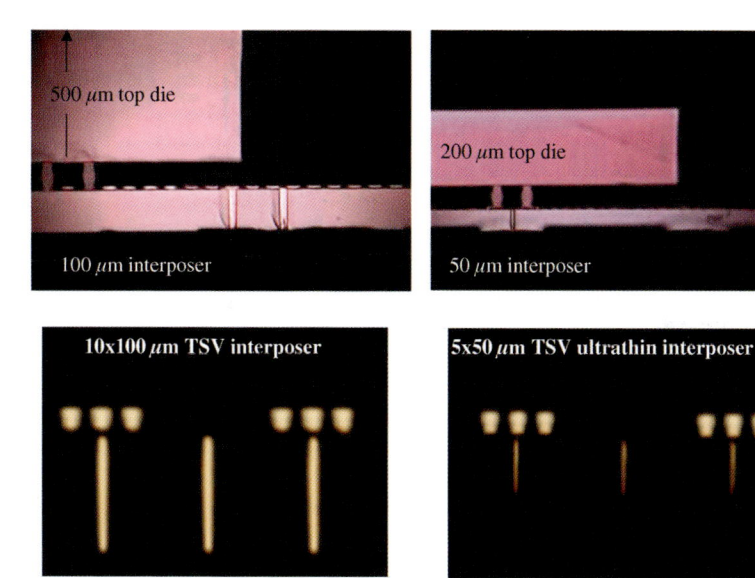

Figure 26.24 TSV and interposer technology evolution are demonstrated, from the current (a) 500 μm thick top die and 100 μm thick interposer with 10 μm diameter TSV, down to (b) a 200 μm thick top die and 50 μm thick interposer with 5 μm diameter TSV [5]. Reprinted with permission from IEEE. Courtesy of TSMC.

comparison of TSV-induced strain effects on both short-channel p- and nMOS devices around a 6 μm diameter TSV and a 2 μm diameter TSV. Although nMOS is slightly less sensitive than pMOS, smaller TSVs induce much less stress affecting the surrounding transistors and a smaller KOZ for both n- and pMOS. This is possibly due to less Cu volume inside the TSV, which results in less tensile strain field surrounding the TSV. Similarly, smaller microbumps also show less stress on the transistors. Other beneficial effects from miniaturization, such as solder microbump electromigration immortality, were discussed in previous sections.

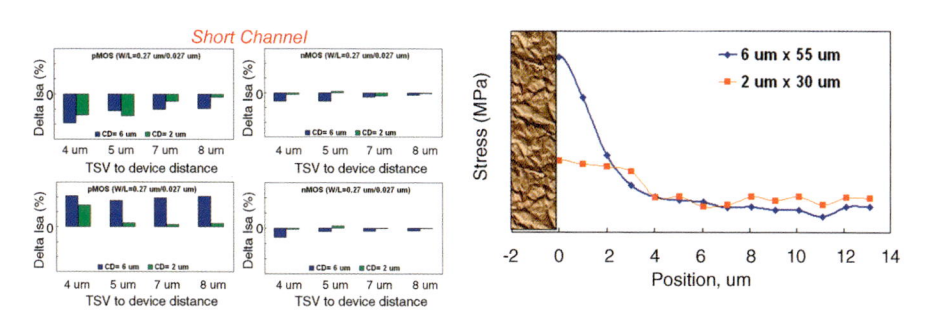

Figure 26.25 A small 2 μm diameter TSV dramatically reduces stress on the surrounding devices and the impact on the short-channel device performance [9]. Reprinted with permission from IEEE. Courtesy of TSMC.

3D chip stack packaging and integration is in production. Despite all the daunting challenges, all of them have been solved one by one. Different solutions have been proposed for different applications. The most cost-competitive solution with clear technological, performance, and cost benefits will be ultimately successful in the highly competitive and exciting semiconductor industry. With the increasingly fast-growing mobile device market, the opportunity for 3D integration is tremendous.

References

1 Puech, M., Thevenoud, J.M., Gruffat, J.M., Launay, N., Arnal, N., and Godinat, P. (2008) *DTIP of MEMS and MOEMS*, EDA Publishing/DTIP.

2 Lin, J.C., Chiou, W.C., Yang, K.F., Chang, H. B., Lin, Y.C., Liao, E.B., Hung, J.P., Lin, Y.L., Tsai, P.H., Shih, Y.C., Wu, T.J., Wu, W.J., Tsai, F.W., Huang, Y.H., Wang, T.Y., Yu, C.L., Chang, C.H., Chen, M.F., Hou, S.Y., Tung, C.H., Jeng, S.P., and Yu, D.C.H. (2010) IEEE International Electron Devices Meeting (IEDM), pp. 2.1.1–2.1.4.

3 Bullema, J.E., Bressers, P.M.M.C., Oosterhui's, G., Mueller, M., Huisin't Veld, A.J., and Roozeboom, F. (2011) 18th European Microelectronics and Packaging Conference (EMPC), pp. 1–5.

4 Tang, C.-W., Li, K.-M., and Young, H.-T. (2012) Micro & nano letters. *IET*, 7 (7), 693–696.

5 Chiou, W.C., Yang, K.F., Yeh, J.L., Wang, S.H., Liou, Y.H., Wu, T.J., Lin, J.C., Huang, C.L., Lu, S.W., Hsieh, C.C., Teng, H.A., Chiu, C.C., Chang, H.B., Wei, T.S., Lin, Y.C., Chen, Y.H., Tu, H.J., Ko, H.D., Yu, T.H., Hung, J.P., Tsai, P.H., Yeh, D.C., Wu, W.C., Su, A.J., Chiu, S.L., Hou, S.Y., Shih, D.Y., Chen, K.H., Jeng, S.P., and Yu, C.H. (2012) Symposium on VLSI Technology (VLSIT), pp. 107–108.

6 Lo, T., Chen, M.F., Jan, S.B., Tsai, W.C., Tseng, Y.C., Lin, C.S., Chiu, T.J., Lu, W.S., Teng, H.A., Chen, S.M., Hou, S.Y., Jeng, S. P., and Yu, C.H. (2012) IEEE International Electron Devices Meeting (IEDM), 2012, pp. 793–795.

7 Yang, K.F., Wu, T.J., Chiou, W.C., Chen, M. F., Lin, Y.C., Tsai, F.W., Hsieh, C.C., Chang, C.H., Wu, W.J., Chen, Y.H., Chen, T.Y., Wang, H.R., Lin, I.C., Jan, S.B., Wang, R.D., Lu, Y.J., Shih, Y.C., Teng, H.A., Tsai, C.S., Chang, M.N., Chen, K., Hou, S.Y., Jeng, S.P., and Yu, C.H. (2011) Symposium on VLSI Technology (VLSIT), pp. 140–141.

8 Chen, H.Y., Shih, D.Y., Wei, C.C., Tung, C. H., Hsiao, Y.L., Yu, D.C.H., Liang, Y.C., and Chen, C. (2013) Proceedings of the 63rd Electronic Components and Technology Conference (ECTC), 2013, pp. 49–57.

9 Chang, H.B., Chen, H.Y., Kuo, P.C., Chien, C.H., Liao, E.B., Lin, T.C., Wei, T.S., Lin, Y.C., Chen, Y.H., Yang, K.F., Teng, H.A., Tsai, W. C., Tseng, Y.C., Chen, S.Y., Hsieh, C.C., Chen, M.F., Liu, Y.H., Wu, T.J., Hou, S.Y., Chiou, W.C., Jeng, S.P., and Yu, C.H. (2012) Symposium on VLSI Technology (VLSIT), pp. 173–174.

27
Cu TSV Stress: Avoiding Cu Protrusion and Impact on Devices

Eric Beyne, Joke De Messemaeker, and Wei Guo

27.1
Introduction

The large difference in the coefficient of thermal expansion (CTE), between Cu and Si will result in significant thermomechanical stress in the Cu through-silicon via (TSV) and the surrounding Si. This may be a concern for product quality and reliability and may also influence the performance of devices embedded in the proximity of the TSVs. As a result, it may be required to limiting back-end-of-line (BEOL) routing above a TSV or define areas around the TSVs where no active devices may be placed, the so-called keep out zone (KOZ) [1,2,3].

To mitigate the negative impact of the Cu–Si CTE difference, it is important to understand the nature and evolution of stresses in the Cu TSV during processing. This allows definition of the appropriate annealing procedures for the TSV to avoid problems with Cu protrusion during further device processing and 3D stacking. Similarly, understanding the TSV-induced stress in the Si allows us to predict the impact of TSVs on active devices and define the appropriate KOZ [4,5].

27.2
Cu Stress in TSV

The evolution of stress with temperature of the Cu in a TSV differs significantly from that of a Cu layer on a silicon surface. A Cu film on a silicon wafer surface is constrained in plane by the silicon, but is free to deform in the thickness direction. The temperature-stress behavior of a typical electroplated Cu film on Si is shown in Figure 27.1 [6]. This measurement was derived from wafer curvature measurements during thermal cycling of the test sample. The stress values of the Cu film are obtained using the Stoney equation. The as-plated Cu film exhibits a low initial tensile stress state. When heating up the sample, stress decreases and become compressive. The slope of

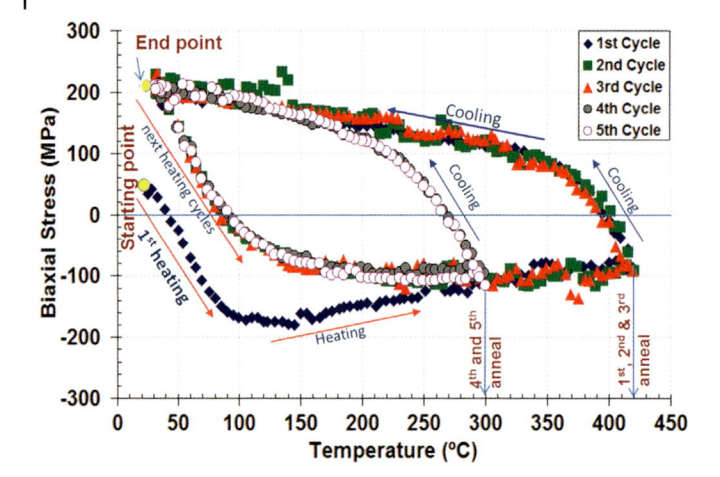

Figure 27.1 Temperature–stress evolution of a 5 μm thick electroplated Cu film on a Si wafer. The stress values are derived from wafer bowing measurement during temperature cycling using the Stoney equation. The sample was subjected to three thermal cycles up to 420 °C and two cycles up to 300 °C [6].

the measured stress change is approximately proportional to the elastic modulus of Cu multiplied by the CTE difference with the Si substrate. When reaching a compressive stress of about 200 MPa, the Cu yield stress is reached and further heating does not increase the compressive stress as the Cu film will plastically deform. Actually, as the yield stress decreases with temperature, the compressive stress will reduce with increasing temperature. When cooling down from the maximum temperature reached, the Cu stress will become tensile and reach high values. The slope of the stress change is the same as the slope during initial ramp-up. This continues until, once again, the yield stress limit of Cu is reached. Below this temperature, the Cu will deform. As it is cooled down to room temperature, the stress increases slightly as the yield stress increases with decreasing temperature. The result of this first thermal cycle to a high temperature is that a high tensile stress is observed in the Cu film. Subsequent thermal cycles will be similar, reaching the same compressive and tensile Cu plastic yield lines. The stress after each cycle remains, however, the same. The area of the curve can be considered as a parameter of plastic deformation (damage) during each thermal cycle and can be used as a parameter describing thermal-cycling fatigue.

The Cu in a TSV structure resides in a rather different stress condition [6,7,8]. In Figure 27.2 the proposed temperature–stress cycle for Cu TSV is given schematically. Similarly, as in the Cu film case, the initial stress of the Cu in the TSVs is rather low, and stress will become compressive when heating the wafer. Cu at the top of the wafer and the Cu "overburden" will

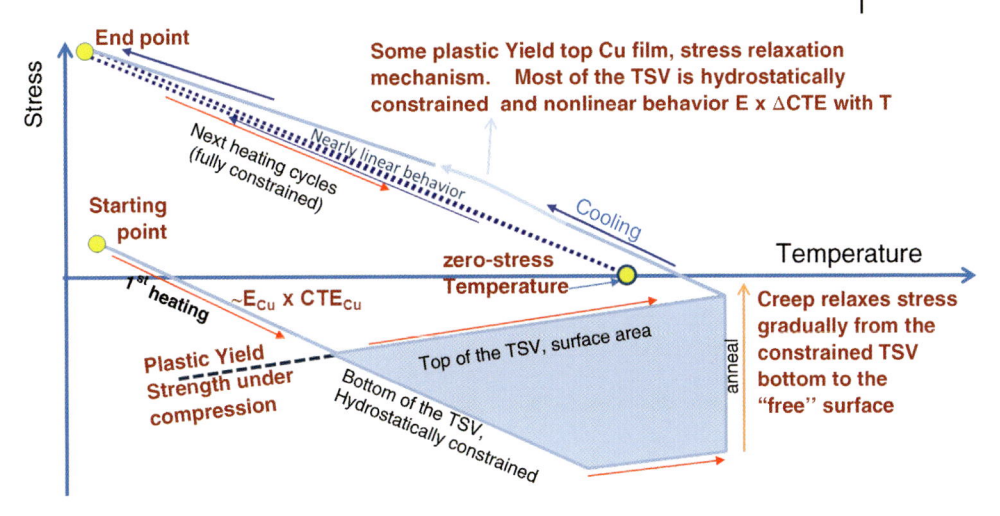

Figure 27.2 Schematic representation of the stress–temperature evolution for Cu-TSVs during repeated thermal cycling.

follow the same behavior as described above for the Cu film. Cu at the bottom of the TSV will however behave in a very different way as it is confined by the Si surrounding the TSV or by the thick, high aspect ratio Cu above. The Cu will, therefore, reach very high values of hydrostatic compressive stress. The enclosure is, however, not perfect and the Cu will be allowed to relax stress by deforming in the axial direction. This results in a creep-like behavior of the Cu in the TSV. When keeping the TSV for a longer period of time at an elevated temperature, the stress will relax and a relatively low compressive stress will be obtained. When cooling down to room temperature from this annealed condition, the stress will rapidly increase to high tensile values. Again, as the Cu in the TSV is hydrostatically confined, the stress cannot relax. Furthermore, as the temperatures are now much lower and below the copper recrystallization temperature, creep stress relaxation becomes very slow and the stress increases approximately linearly to high values at room temperature. Subsequent thermal cycles will become nearly elastic thermal cycling, causing minimal damage to the Cu in the TSV.

This model was experimentally verified by measuring the temperature-bowing of Si wafers with integrated TSVs. In Figures 27.3 and 27.4 the results are shown for a wafer with TSVs and Cu overburden and a wafer after overburden chemical–mechanical polishing (CMP). The results are in agreement with the proposed Cu stress model. Experiments focused on the mitigation of Cu pumping (see Section 27.3) offer further support for this hypothesis.

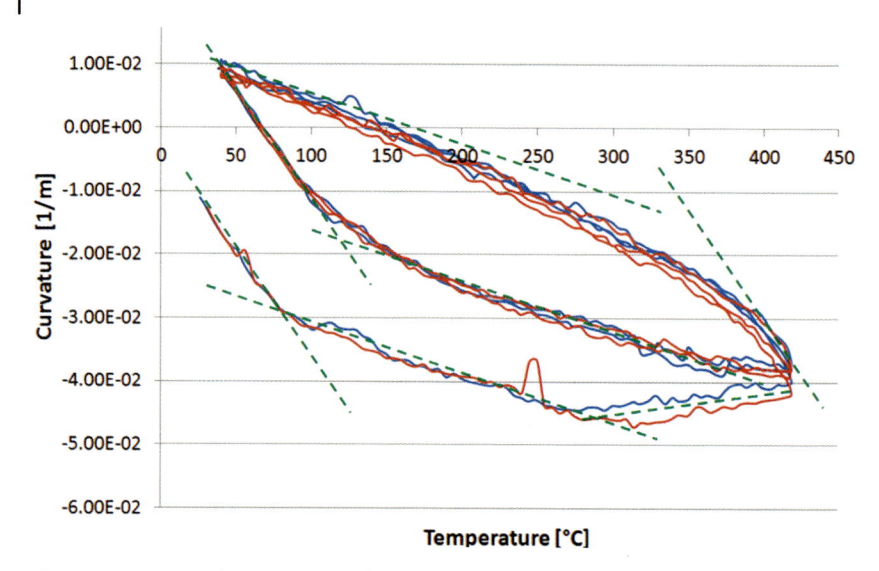

Figure 27.3 Measured temperature–wafer curvature behavior (∼ stress) of a Si wafer with Cu TSVs after Cu plating, before CMP processing. Three thermal cycles are performed up to 420 °C.

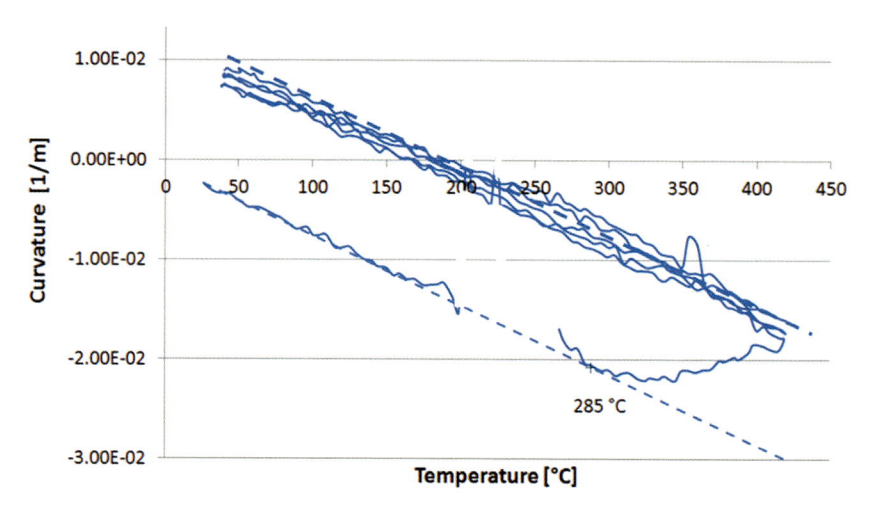

Figure 27.4 Measured temperature–wafer curvature behavior (∼ stress) of a Si wafer with Cu TSVs after CMP processing. Three thermal cycles are performed up to 420 °C.

27.3
Mitigation of Cu Pumping

Without the anneal of the Cu after TSV processing, the Cu in the TSV may protrude during subsequent device processing, as illustrated in Figure 27.5. This is generally referred to in literature as "Cu pumping." The driving mechanism for Cu pumping is the Cu-stress evolution during temperature cycling, as described in Section 27.2. Two

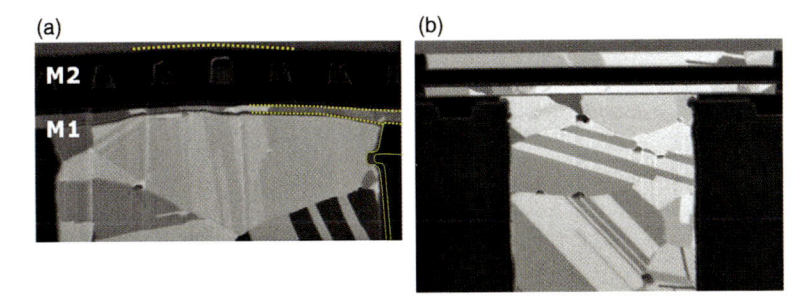

Figure 27.5 Illustration of the Cu pumping effect. (a) Integrated TSV without proper anneal post-TSV processing. Protrusion of the Cu TSV in the BEOL layers and "bulging" of the BEOL layers above the TSV is observed after a postprocess sinter anneal step. (b) With a proper post-TSV process anneal, no Cu pumping is observed after post-wafer integration sinter anneal.

different types of Cu pumping are reported. The first one is illustrated in Figure 27.5 and is characterized by a bulging of the Cu TSV surface without changes at the edge of the TSV. The second type occurs if adhesion of the Cu on the liner-side wall is weak and the Cu is pushed out of the TSV as a cylinder. This requires an improvement of the Cu–liner adhesion mechanism. In this section, we only discus the first type.

To study this effect, we define a novel Cu pumping test methodology, as illustrated in Figure 27.6 [9,10,11]. Cu TSVs are realized in a blanket Si wafer using the standard method. After TSV Cu fill plating the wafer is annealed at a specific temperature for a specific time (these are the first experimental variable parameters). Next, CMP is used to remove the Cu overburden from the wafer, using the same method as used for actual device wafers. In a next step, a thin oxide layer (180 nm SiO_2) is deposited over the wafer to realize a thin membrane over the Cu TSVs (mimics the deposition of a first BEOL dielectric layer). This also avoids Cu pumping of individual grains from the Cu TSV CMP surface. A thin metallic metal layer (e.g., 25 nm Ta) is also deposited to provide the necessary reference layer for a precise height metrology using WLIR. The surface profile of the TSVs is then measured for a first time. This profile is mainly a signature of the TSV CMP (e.g., Cu dishing) and the topography of the deposited thin capping layers. Next, the wafer is annealed for a second time at a specific "sinter" temperature for a specific "sinter" time. (This is the second experimental parameter.) A second measurement of the

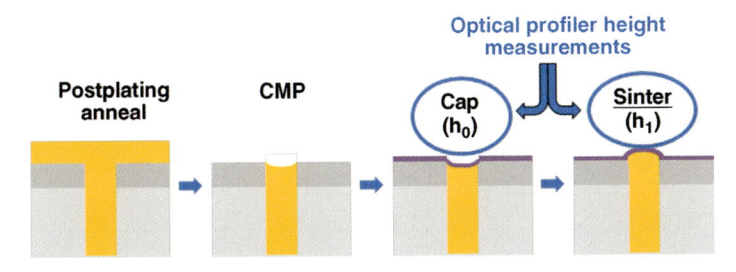

Figure 27.6 Schematic representation of the experimental method used to study Cu pumping.

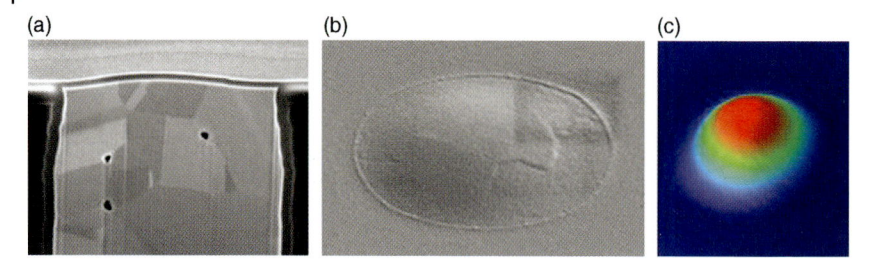

Figure 27.7 Example of a WLIR topography measurement (a) on 5 μm diameter TSV after sinter anneal, exhibiting Cu pumping (b). A Fib cross section of the TSV is also shown (c).

TSV profile is performed and the difference with the first recorded profile is calculated. The difference is the result of Cu pumping and can be used as response parameter in an anneal and sinter study of the Cu TSVs. An example of a Cu showing a "Cu-pumping" behavior and a corresponding WLIR measurement is shown in Figure 27.7. Typical results are shown in Figure 27.8 for 5 μm diameter, 50 μm deep (5 × 50) Cu TSVs after sintering at 420 °C for 20 min and for different postplating anneal conditions. Cu pumping is significantly reduced when using a higher postplating anneal temperature. The amount of Cu pumping has a bi-normal distribution with a broad spread, related to the microstructure of the Cu in the TSV (ECTC) [9,10,11].

Figure 27.9 shows the 99.9th percentile of the residual Cu pumping distributions for the $5 \times 50\,\mu m^2$ and $10 \times 100\,\mu m^2$ TSVs as a function of postplating anneal

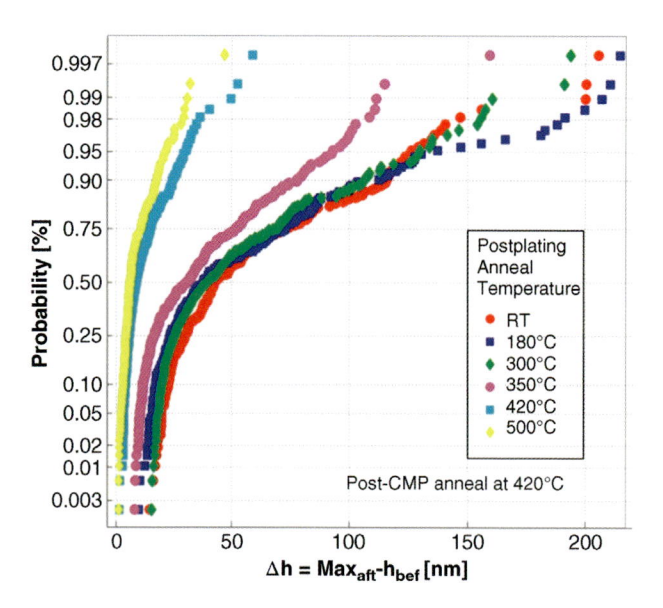

Figure 27.8 Measured Cu pumping statistics as a function of the postplating anneal temperature and for a postprocess sinter anneal temperature of 420 °C. For each condition, 344 TSVs were measured. The statistics of the Cu pumping values fits a bi-normal distribution.

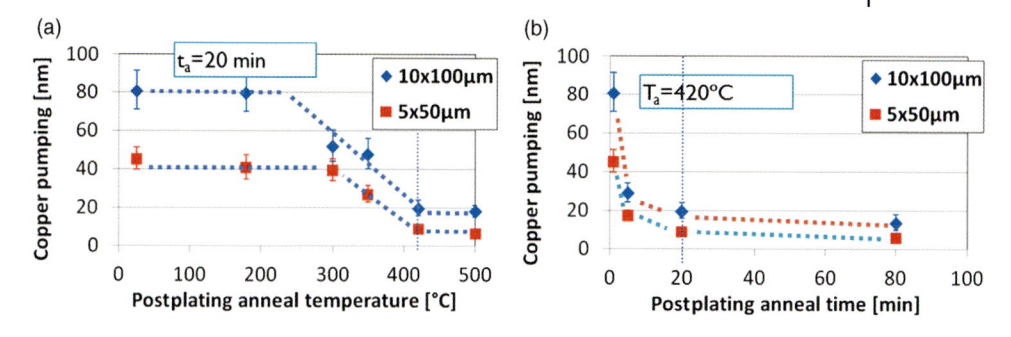

Figure 27.9 Measured maximum Cu pumping (99.9% statistics) for 5 × 50 and 10 × 100 μm² TSVs as a function of the postplating anneal temperature (a) and postprocess anneal time (at 420 °C postplating anneal temperature) (b) and for a postprocess sinter anneal temperature of 420 °C for 20 min. For each condition, 344 TSVs were measured and the 99.9% value was derived from the bi-normal statistic distribution.

temperature (for constant anneal time of 20 min) and as a function of postplating anneal time (for a constant annealing temperature of 420 °C). This 99.9th percentile of the distributions shows the same dependencies on postplating anneal temperature and time as p50 (not shown), but the values are about eight times higher. This large difference is a result of the bi-normal nature of the distribution.

For postplating anneals below approximately 300 °C for 5 × 50 TSVs, limited impact on the Cu-pumping amplitude is observed. In the 300–400 °C temperature range, significant changes occur and Cu pumping is significantly reduced when annealing at a temperature equal to the postsinter temperature (420 °C). Higher temperature postplating anneals (up to 500 °C) only result in small improvements in Cu pumping. We can, therefore, conclude that an increase of the postplating anneal temperature above the sinter temperature of 420 °C would only increase the thermal budget with little gain in terms of Cu pumping.

The impact of the postplating anneal time at 420 °C is also shown in Figure 27.9. The curves exhibit an exponential behavior with time, in agreement with the creep-deformation hypothesis for Cu stress relaxation at high temperatures during annealing. For rather short times, the residual pumping height sharply decreases, but the additional reduction between 20 and 80 min is limited and insufficient to justify this fourfold increase in time investment. Together with the data for the postplating anneal temperature, these results confirm that the optimum postplating anneal conditions are those equal to the sinter conditions of 20 min at 420 °C, as was first proposed in Ref. [3].

27.4
Impact of TSVs on FEOL Devices

For further minimizing the TSV stress-induced KOZ for semiconductor devices, both the TSV and the device stress-sensitive properties need to be considered. As was previously established, the in-plane stress induced at the device locations by the thermal mismatch of the TSV Cu and the silicon substrate can be approximated by the Lamé equation (Equation 27.1), neglecting the smaller out-off plane stress components. This formula relates the radial tensile TSV Cu stress, σ_{TSV}, to the axial, σ_r, and circumferential, σ_θ, stress components of the cylindrical stress field in the Si substrate:

$$\sigma_r \approx \sigma_\theta \approx \sigma_{TSV} \left(\frac{\emptyset}{2r}\right)^2, \qquad (27.1)$$

where r is the radial distance from the TSV center to the device center and $\emptyset$ is the diameter of the TSV Cu.

Both stress components are approximately equal in size but opposite in sign. The stress levels in the silicon will decay quadratically with increasing distance from the TSV center. The cylindrical stress field (see Equation 27.1) can be rewritten in orthogonal stress components for the related stresses in the active devices, in line and parallel to the device channel:

$$\sigma_{xx} \approx \sigma_{TSV} (\cos^2\theta - \sin^2\theta) \cdot \left(\frac{\emptyset}{2r}\right)^2,$$

$$\sigma_{yy} \approx \sigma_{TSV} (\sin^2\theta - \cos^2\theta) \cdot \left(\frac{\emptyset}{2r}\right)^2. \qquad (27.2)$$

The device drain current variation δI_d can be expressed as a function of the mechanical stresses in line and parallel to the device channel:

$$\delta I_d = \frac{\Delta I_d}{I_d} \approx (\pi'_{11} \cdot \sigma_{xx} + \pi'_{22} \cdot \sigma_{yy}), \qquad (27.3)$$

where π'_{11} and π'_{22} are the effective piezo-coefficients, describing the sensitivity of the device currents to stresses in line and parallel to the device channel σ_{xx} and σ_{yy}.

By substituting Equation 27.2 in Equation 27.3, we obtain a relation between δI_d and the geometry parameters r and θ:

$$\delta I_d = \frac{\Delta I_d}{I_d} = K(2\cos^2\theta - 1) \cdot \left(\frac{1}{r}\right)^2, \qquad (27.4)$$

where we define the proportionality constant K as a device and TSV-dependent parameter:

$$K = (\pi'_{11} - \pi'_{12})\sigma_{TSV} \left(\frac{\emptyset}{2}\right)^2. \qquad (27.5)$$

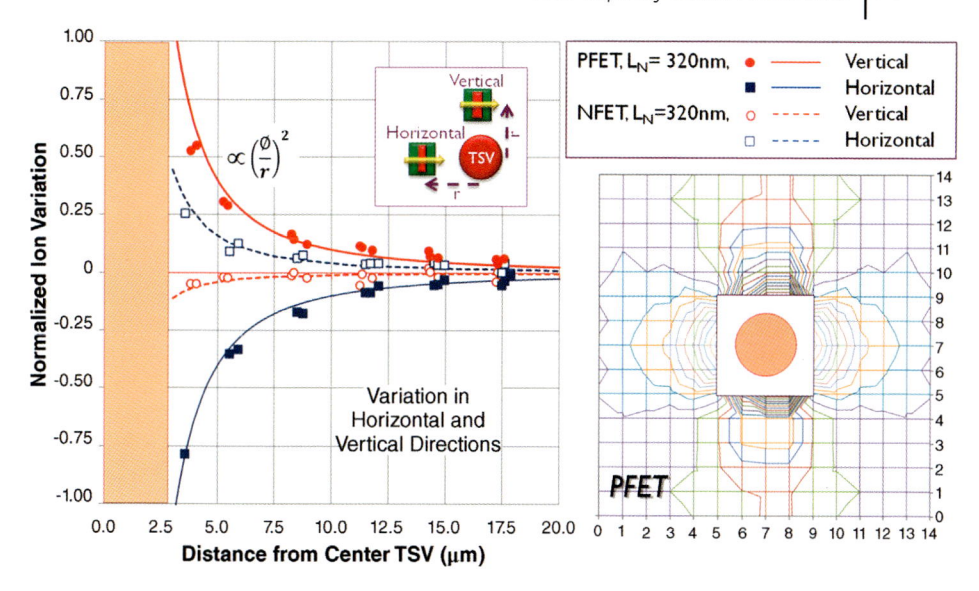

Figure 27.10 Measured drain current of planar high-k/metal gate FET devices as a function of the distance to the TSV center. TSV FET array test structure. The TSV measures 5 μm in diameter and is 50 μm deep in the silicon [4].

Equation 27.4 relates to the typical flower-shape geometry of the $\delta I_d(r, \theta)$ or $\delta I_d(x,y)$ relation. The technology constant K, as defined in Equation 27.5, quantifies the respective contributions of the TSV and the device properties on the drain current variations. This K constant contains two parts: The first part $(\pi'_{11}-\pi'_{22})$ relates to device technology and the second part $\sigma_{TSV}(\emptyset/2)^2$ depends on the TSV technology.

We derived this K-factor directly from measurements using FET array test structures, where a TSV is placed in the center of the structure and properties of individual FETs can be addressed by addressing them individually through row and column selectors [4]. An example of such measurement for high-k/metal gate FET devices is shown in Figure 27.10. As can be seen in Figure 27.10, the drive current and distance to TSV dependency follows a K/r^2 law.

It is important to minimize the TSV-induced KOZ. TSV diameter downscaling is an effective solution to reduce the KOZ. As can be seen from Equation 27.4 the K-factor scales with $\emptyset^2$, therefore a strong impact is to be expected. The high-k/metal gate test vehicle described in Figure 27.10 was also realized with 3 μm diameter TSVs [3]. The electrical measurements, summarized in Figure 27.11, confirm the reduction of the KOZ area (K) by a factor of 3, as predicted by the theory set forward above.

For a known TSV technology, with a given diameter $\emptyset$ and stress σ_{TSV}, the K-value can also be estimated in a different way, using the effective piezo-coefficients π'_{11} and π'_{22}. These can be derived from by four-point bending

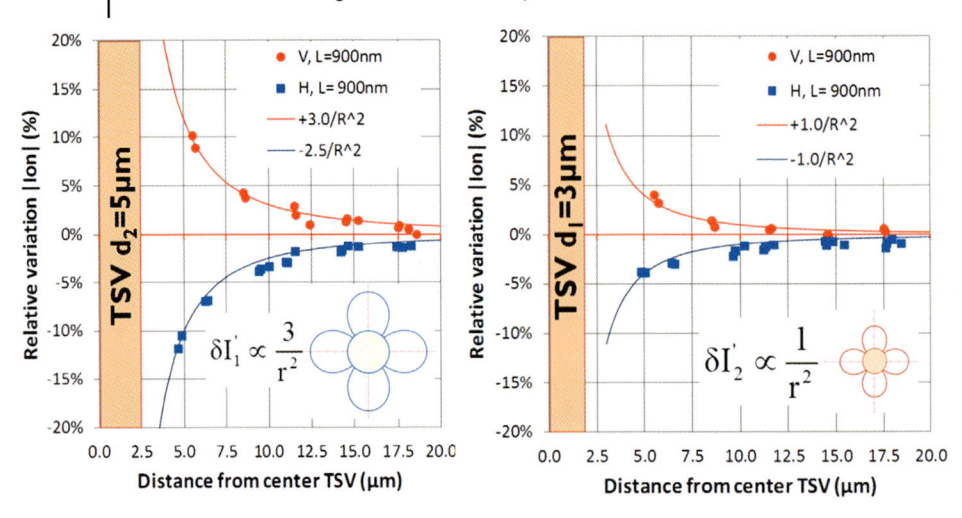

Figure 27.11 Impact TSV diameter scaling on the KOZ size and parameter K, as derived from FET array test structures. On the left is a 5 × 50 μm TSV and on the right is a 3 × 40 μm TSV integrated in a planar high-k/metal gate FET test vehicle [3].

experiments on wafers of any specific device technology (without TSV), by applying uniaxial stress and measuring the drain current variation of the devices.

In Figure 27.12, a comparison between the calculated and measured K factors is given, an excellent agreement obtained. This figure also shows that the K constant decreases with decreasing device channel length, as predicted by the piezo-coefficients measurements.

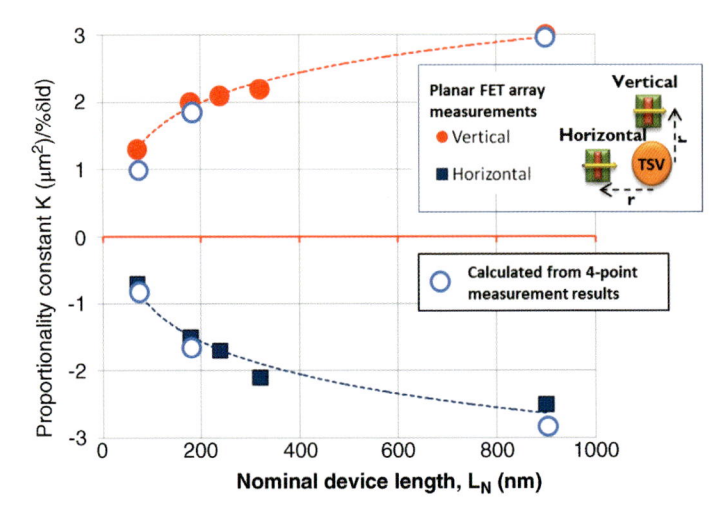

Figure 27.12 Correlation between the estimated and calculated K-factor values for planar FET devices with different channel lengths. The TSV measures 5 μm diameter and is 50 μm deep in the silicon [4].

27.4 Impact of TSVs on FEOL Devices

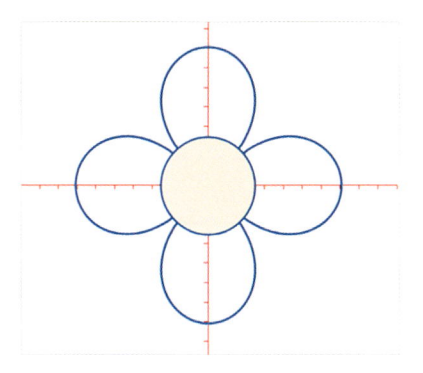

Figure 27.13 Shape of the KOZ, r_{KOZ}, around a TSV for a specific (r, θ).

Based on Equation 27.4, a model for the TSV KOZ can be proposed where the factor K, as defined in Equation 27.5, and the maximum allowed device drain current variation δI_d are the only variables. The KOZ is then defined by the distance r_{KOZ}, which is dependent on the angle θ, and varies according to (r, θ), as shown in Figure 27.13.

$$r < r_{KOZ} = \sqrt{\frac{K}{\delta I_d}(2\cos^2\theta - 1)}. \tag{27.6}$$

This model was extensively validated by Si data on advanced planar and FinFET device technologies. The model allows prediction of the KOZ induced by single or multiple TSVs. For a single TSV, the KOZ shown in Figure 27.13 can be approximated by a piecewise linear approximation, as shown in Figure 27.14, resulting in a cross-shaped KOZ area.

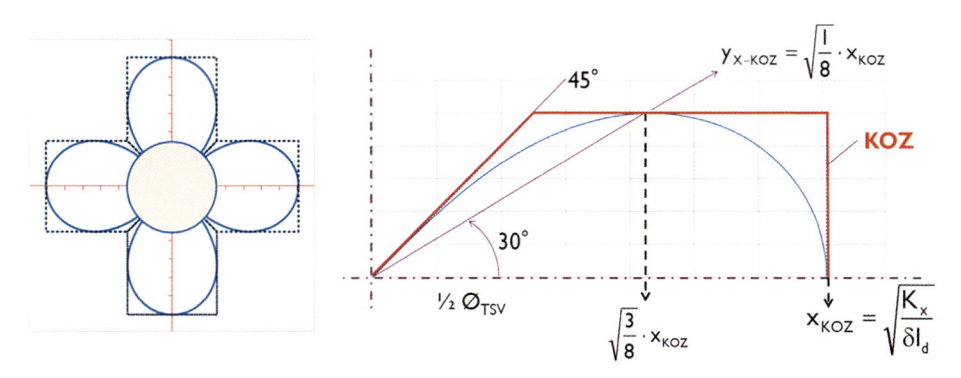

Figure 27.14 Piecewise linear approximation of the KOZ approximation around an isolated TSV.

The Equations 27.1–27.6 assumed perfect cylinder symmetry. In practice there may be small differences in K-value for the x- and y-directions due to the specific geometry of the devices and the stress gradient across the devices, which may differ for devices with current flow radial to the TSV (horizontal, x) from devices with a perpendicular current flow (vertical, y). This can be observed in Figure 27.10. Two K-values may be derived, K_x and K_y, resulting in a KOZ cross region with different sizes in x and y. $x_{KOZ} = r_{KOZ}$ (0°) and $y_{KOZ} = r_{KOZ}$ (90°). Figure 27.14 shows details of one-eighth of the KOZ area around an isolated TSV and defines the x_{KOZ} and $y_{x\text{-}KOZ}$ KOZ geometry parameters. A fixed relation is found between these parameters. The KOZ area is uniquely defined by the K_x and K_y parameters for any maximum allowed device drain current variation δI_d. The KOZ area A_{KOZ} is given by

$$A_{KOZ} = \left(\sqrt{2} - \frac{1}{4}\right)\left[\frac{K_x + K_y}{\delta I_d}\right]. \tag{27.7}$$

This equation illustrates the physical meaning of the K-factors: They are proportional to the KOZ area and are typically expressed in μm^2 % δI_d. Defining the KOZ as a cross shape, rather than a rectangle, reduces the KOZ area about 40%.

When considering multiple TSVs, the stress fields of the different TSVs will overlap. First we can assume a linear behavior and add the effects of different TSVs. Then we consider the nearest-neighbor interaction. In Figure 27.15, the device drain current variation δI_d is plotted for two closely spaced TSVs with pitch P and equal K factor, using Equation 27.4.

In Figure 27.15 one can distinguish three different cases: isolated TSV ($r_{KOZ} \ll P$: KOZ_1), weakly coupled TSV ($r_{KOZ} < 0.35\ P$: KOZ_2), and strongly coupled TSV ($r_{KOZ} > 0.35\ P$: KOZ_3). For each of these cases, analytical or semianalytical formulas can be derived for the x_{KOZ}, $y_{x\text{-}KOZ}$, y_{KOZ}, and $x_{y\text{-}KOZ}$ parameters defining the KOZ cross-geometry. These formulas are listed in as a function of K_x, K_y, and δI_d is for both the x- and y-directions (Table 27.1).

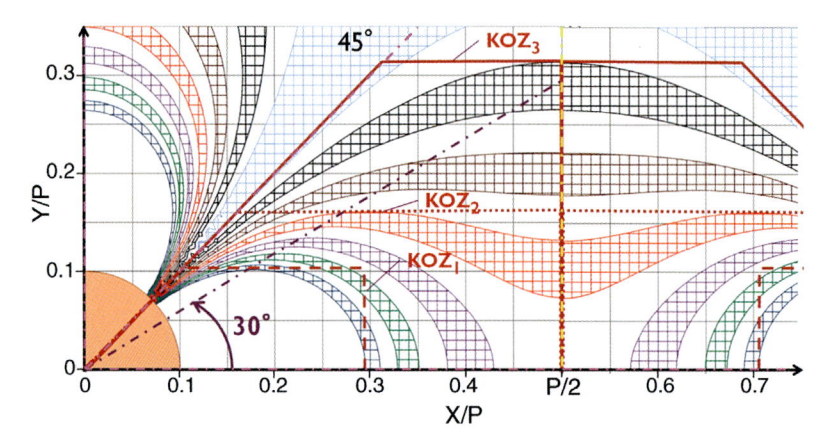

Figure 27.15 Impact of TSV proximity on FET drive current. The contour lines correspond to equal driver current locations between two TSVs with spacing P.

Table 27.1 Overview formulas for calculating the KOZ parameters x_{KOZ}, $y_{x\text{-}KOZ}$, y_{KOZ}, and $x_{y\text{-}KOZ}$ as a function of K_x, K_y, and δI_d, for both x- and y-directions for isolated, weakly coupled, and strongly coupled TSVs.

	Width KOZ in x-direction x^*_{KOZ}	Height KOZ in y direction $y^*_{x\text{-}KOZ}$
Isolated TSV	$x_{KOZ} = \sqrt{\dfrac{K_x}{\delta I_d}}$	$y_{x\text{-}KOZ} = \sqrt{\dfrac{1}{8}} \cdot x_{KOZ}$
Weakly coupled TSV $P_x > \sqrt{8} \cdot x_{KOZ}$	$x'_{KOZ} = \dfrac{P_x}{2} - x_{KOZ}\sqrt{1 + \left(\dfrac{P_x^2}{4x_{KOZ}^2}\right) - \sqrt{1 + \dfrac{P_x^2}{x_{KOZ}^2}}}$	$y'_{x\text{-}KOZ} \approx y_{30°\text{-}KOZ}$ $y_{30°\text{-}KOZ} \approx \sqrt{\dfrac{1}{8}} \cdot x_{KOZ} + \dfrac{x_{KOZ}^4}{P_x^3}$
Strongly coupled TSV $P_x < \sqrt{8} \cdot x_{KOZ}$	$x''_{KOZ} = \dfrac{P_x}{2}$	$y''_{x\text{-}KOZ} = \max\{y_{30°\text{-}KOZ}; y_{P/2\text{-}KOZ}\}$ $y_{P/2\text{-}KOZ} = x_{KOZ}\sqrt{-1 - \left(\dfrac{P_x^2}{4x_{KOZ}^2}\right) + \sqrt{1 + \dfrac{P_x^2}{x_{KOZ}^2}}}$

	Width KOZ in y direction y^*_{KOZ}	Height KOZ in x-direction $x^*_{y\text{-}KOZ}$
Isolated TSV	$y_{KOZ} = \sqrt{\dfrac{K_y}{\delta I_d}}$	$x_{y\text{-}KOZ} = \sqrt{\dfrac{1}{8}} \cdot y_{KOZ}$
Weakly coupled TSV $P_y > \sqrt{8} \cdot y_{KOZ}$	$y'_{KOZ} = \dfrac{P_y}{2} - y_{KOZ}\sqrt{1 + \left(\dfrac{P_y^2}{4y_{KOZ}^2}\right) - \sqrt{1 + \dfrac{P_y^2}{y_{KOZ}^2}}}$	$x'_{y\text{-}KOZ} \approx y_{30°\text{-}KOZ}$ $x_{30°\text{-}KOZ} \approx \sqrt{\dfrac{1}{8}} \cdot y_{KOZ} + \dfrac{y_{KOZ}^4}{P_y^3}$
Strongly coupled TSV $P_y < \sqrt{8} \cdot y_{KOZ}$	$y''_{KOZ} = \dfrac{P_y}{2}$	$x''_{y\text{-}KOZ} = \max\{x_{30°\text{-}KOZ}; x_{P/2\text{-}KOZ}\}$ $x_{P/2\text{-}KOZ} = y_{KOZ}\sqrt{-1 - \left(\dfrac{P_y^2}{4y_{KOZ}^2}\right) + \sqrt{1 + \dfrac{P_y^2}{y_{KOZ}^2}}}$

References

1 Mercha, A. *et al.* (2010) Comprehensive analysis of the impact of single and arrays of through silicon vias induced stress on high-k/metal gate CMOS performance. IEEE International Electron Devices Meeting (IEEE 2010), pp. 2.2.1–2.2.4.

2 Van Olmen, J. *et al.* (2008) 3D stacked IC demonstration using a through silicon via first approach. IEEE International Electron Devices Meeting (IEEE 2008), pp. 1–4.

3 Beyne, E. (2011) Electrical, thermal and mechanical impact of 3D TSV and 3D stacking technology on advanced CMOS devices: technology directions. IEEE International 3D Systems Integration Conference (3DIC), IEEE-CPMT, 31 January–2 February 2012, Osaka, Japan.

4 Guo, W., Van Der Plas, G., Ivankovic, A., Eneman, G., Cherman, V., De Wachter, B., Mercha, A., Gonzalez, M., Civale, Y., Redolfi, A., Buisson, T., Jourdan, A., Vandevelde, B., Rebibis, K.J., De Wolf, I., La Manna, A., Beyer, G., Beyne, E., and Swinnen, B. (2012) 3D chip package interaction thermo-mechanical challenges: proximity effects of through silicon vias and μ-bumps. IEEE International Conference on IC Design & Technology (ICICDT 2012), May 30–June 1, 2012, Austin, TX, USA.

5 Guo, W., Van der Plas, G., Ivankovic, A., Cherman, V., Eneman, G., De Wachter, B., Togo, M., Redolfi, A., Kubicek, S., Civale, Y., Chiarella, T., Vandevelde, B., Croes, K., De Wolf, I., Debusschere, I., Mercha, A., Thean, A., Beyer, G., Swinnen, B., and Beyne, E. (2012) Impact of through silicon via induced mechanical stress on fully depleted bulk FinFET technology. International Electron Devices Meeting (IEDM 2012), IEEE-EDS, December 10–12, 2012, San Francisco, CA, USA pp. 18.4.

6 Okoro, C. *et al.* (2010) Elimination of the axial deformation problem of Cu-TSV in 3D integration. 11th International Workshop on Stress-Induced Phenomena in Metallization, April 12–14, 2010, Dresden, Germany, pp. 214–220D.

7 Wilson, C.J., De Wolf, I., Vandevelde, B., De Messemaeker, J., Ablett, J.M., Redolfi, A., Simons, V., Beyne, E., and Croes, K. (2012) Comparison of X-ray diffraction, wafer curvature and Raman spectroscopy to evaluate the stress evolution in copper TSVs. IEEE-EDS International Interconnect Technology Conference (IITC), June 4–6, 2012, San Jose, CA, USA.

8 De Wolf, I., Simons, V., Cherman, V., Labie, R., Vandevelde, B., and Beyne, E. (2012) In-depth Raman spectroscopy analysis of various parameters affecting the mechanical stress near the surface and bulk of Cu-TSVs. IEEE-CPMT 62nd Electronic Components and Technology Conference (ECTC), May 29–June 1, 2012, San Diego, CA, USA, pp. 331–337.

9 De Wolf, I. *et al.* (2011) Cu pumping in TSVs: effect of pre-CMP thermal budget. *Microelectronics Reliability*, 51 (9–11), 1856–1859.

10 De Messemaeker, J., Croes, K., Vandevelde, B., Velenis, D., Redolfi, A., Jourdain, A., Beyer, G., Swinnen, B., Beyne, E., and De Wolf, I. (2012) Thermal mismatch induced reliability issues for Cu filled through-silicon vias. IEEE-CPMT 4th Electronics System Integration Technologies Conference (ESTC), September 17–20, 2012, Amsterdam, The Netherlands.

11 De Messemaeker, J., Valera Pedira, O., Vandevelde, B., Philipsen, H., De Wolf, I., Beyne, E., and Croes, K. (2013) Impact of post-plating anneal and through-Si via dimensions on Cu pumping. IEEE-CPMT Electronic Components & Technology Conference, 2013 (ECTC 2013), Las Vegas, NV, USA, pp. 586–591.

28
Implications of Stress/Strain and Metal Contamination on Thinned Die

Kangwook Lee and Mariappan Murugesan

28.1
Introduction

There are many challenges in 3D LSIs with through-silicon vias (TSVs) and metal bumps to be solved before the volume production starts. The most serious concerns are implications of stress and strain and metal contamination on device reliabilities in 3D stacked chips. Influences of mechanical stress and strain are introduced in the wafer-thinning process for 3D integration. Cu TSVs and metal bumps introduce significant mechanical stress and strain into thinned Si wafers. Active regions in the 3D LSI with a thinned Si wafer might be more easily affected by metal impurity contamination. An extrinsic gettering region for gettering metallic contaminants during LSI process is eliminated by the wafer-thinning process and Cu atoms diffuse from Cu TSV when the blocking property of the barrier layer in the TSV to Cu is not sufficient. These Cu atoms may diffuse into both dielectric and active regions of Si substrate during the back-end process and cause the performance degradation and early breakdown of devices. In this section, the influences of mechanical stress and strain and Cu impurity contamination on device reliabilities in thinned 3D LSIs are discussed.

28.2
Impacts of Cu Contamination on Device Reliabilities in Thinned 3DLSI

To fabricate 3D LSIs, each functional LSI wafer should be thinned to 10–50 μm thickness by mechanical grinding and stress-relief polishing methods. However, this may cause severe degradation in device reliability since an extrinsic gettering region for gettering metallic contaminants might be removed by the wafer-thinning process, and consequently, active regions in the 3D LSI might be more easily contaminated by metallic impurities such as Cu. Cu atoms stuck on the background surface are not completely eliminated even after the cleaning process.

380 | *28 Implications of Stress/Strain and Metal Contamination on Thinned Die*

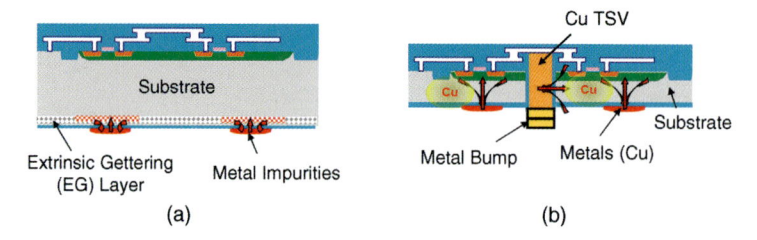

Figure 28.1 Conceptual structures of LSI wafers before the wafer thinning (a) and after the wafer thinning for 3D LSIs with Cu TSV and metal bumps (b).

Cu atoms also diffuse from Cu TSVs when the blocking property of the barrier layer to Cu is not sufficient as shown in Figure 28.1. These Cu atoms may diffuse into both dielectric and active regions of the Si substrate even at low temperature postprocessing and cause the performance degradation and early breakdown of devices [1,2].

The impact of Cu contamination in 3D integration has attracted attention in recent years. Hozawa *et al.* reported that Cu diffusion effect is significantly enhanced in the thinned wafer thickness [3]. Secondary ion mass spectrometry (SIMS) is a popular method to evaluate metal diffusion behavior. Figure 28.2 shows Cu concentration profiles measured from the back surface and front surface of the thinned Si wafer of 50 μm thickness by SIMS measurement after the annealing at 300 °C for various times [4].

Almost all Cu atoms are diffused into the region within 400 nm depth from the back surface after annealing at 300 °C for 60 min. Cu concentrations measured from more than 400 nm depth from the back surface and the front surface show below 1×10^{17} atoms cm^{-3}, which is the resolution limitation of SIMS measurement. SIMS method may not accurately detect Cu impurities owing to the

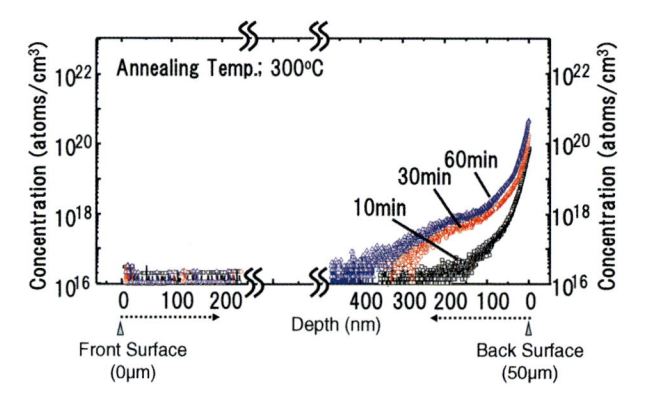

Figure 28.2 Cu concentration profiles measured from the back surface and front surface of the thinned Si wafer of 50 μm thickness by SIMS measurement after the annealing at 300 °C for various times.

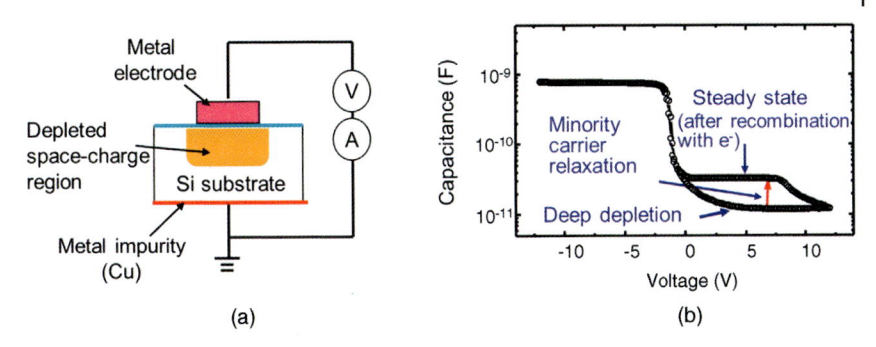

Figure 28.3 Configuration of C–t analysis using a MOS capacitor (a) and C–V plot obtained for the minority carrier relaxation.

resolution limitation. Total reflection X-ray fluorescence (TRXF) analysis has attracted attention due to the high measuring sensitivity for Cu diffusion behavior [3,4]. However, TRXF could not directly characterize Cu diffusion effect on device reliability in fabricated device wafers.

To measure sensitively Cu contamination effects on device reliability in 3D LSIs, a transient capacitance measurement, which is called a capacitance–time (C–t) analysis, has been suggested as an electrical evaluation method [4]. This method can quantitatively define the generation lifetime of minority carriers in the depletion region, where Cu impurities generate deep levels at midgap in the energy band and increase the generation–recombination probability between electrons and holes [5,6]. In the C-t analysis, the capacitance change of the MOS capacitor is measured after applying a step voltage to the gate electrode, as shown in Figure 28.3. Just after applying the step voltage, the MOS capacitor is in the deep depletion condition; hence, it represents a smaller capacitance. The capacitance increases with time to the final value of C_f, as minority carriers are generated in the depletion region. When C reaches C_f, the transient time t_f, which is the time required for the capacitance to reach the inversion state from the initial deep depletion state, is important in the C–t analysis.

The transient time t_f decreases as the generation lifetime of the minority carrier becomes shorter. Hence, a shorter t_f implies more metallic contamination. Therefore, it can sensitively and electrically characterize the lifetime degradation of minority carriers caused by Cu contamination in the fabricated device wafer. In this section, we review the Cu diffusion characteristics and the impacts of Cu contamination on device reliabilities in 3D LSIs by the C–t analysis.

Figure 28.4 shows the measured C–t curves of the capacitor formed in the thinned wafer of 50 µm thickness comprised with a defect-free denuded zone (DZ) layer (Figure 28.4a) and generation lifetime (τ_g) of minority carrier measured C–t analysis versus surface concentration of Cu atoms after the intentional diffuse of Cu atoms from the backside surface at 300 °C (Figure 28.4b). The C–t curves show a severe degradation with a shorter t_f even after the initial annealing for 5 min. It means that Cu atoms easily diffuse into the active region from the back surface of

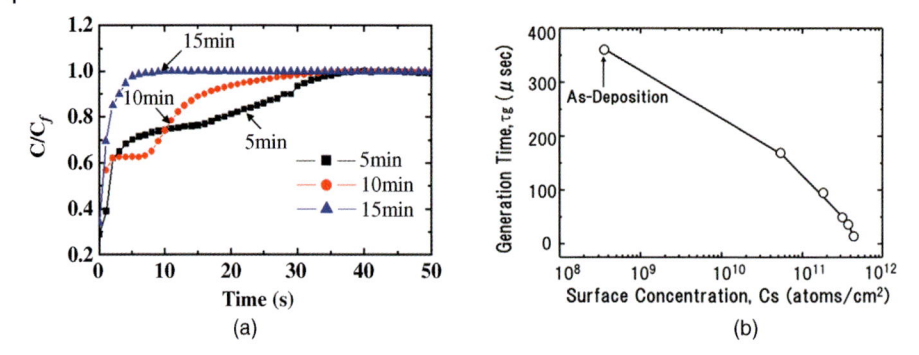

Figure 28.4 Measured C–t curves (a) and generation lifetime (τ_g) of the minority carrier versus the surface concentration of Cu atoms (b) after the intentional Cu diffusion from the backside surface.

the polished wafer and consequently the generation lifetime of the minority carrier is significantly reduced. The quantitative relationship between the generation lifetime of the minority carrier and surface concentration of Cu atoms is shown in Figure 28.4b. The generation lifetime is significantly reduced from the as-deposition condition with the surface concentration of Cu atoms after the annealing [4].

To suppress Cu diffusion from the backside of the thinned wafer, intrinsic gettering (IG) [4,7] and extrinsic gettering (EG) layers [8] have been reported. To electrically characterize the Cu blocking property of the IG layer, the MOS capacitor was fabricated on an annealed wafer of 50 μm thickness comprising the DZ and IG layers, where the IG layer region is 20 μm deep from the back surface. The C–t curves of the MOS capacitor exhibit only a little change from the as-deposition condition even after annealing up to 350 min, as shown in Figure 28.5. It indicates that Cu atoms hardly diffuse into the active region owing to Cu retardation by the IG region and the generation lifetime of the minor carrier is only minutely reduced even after the long annealing time. The IG layer can effectively prevent Cu diffusion since it is a defected zone with sufficiently high density of oxygen precipitates [4,7].

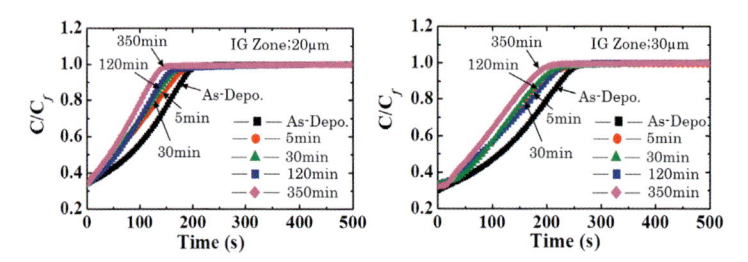

Figure 28.5 Measured C–t curves of the MOS capacitor formed on the Si substrate of 50 μm thickness with an IG layer after the intentional Cu diffusion from the backside surface at 300 °C.

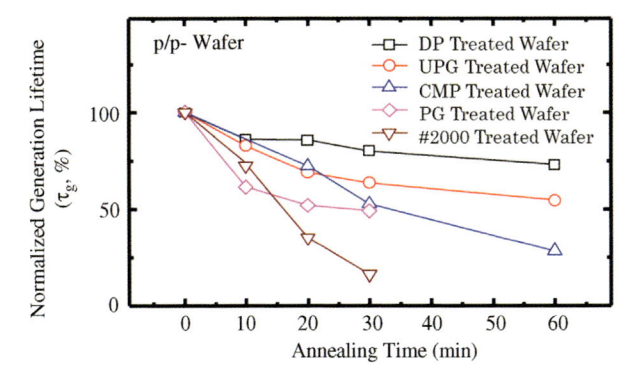

Figure 28.6 Normalized generation lifetime, τ_g of the MOS capacitors formed on the thinned wafers treated with DP, UPG, CMP, PG, and #2000, after the intentional Cu diffusion at 300 °C.

Generally, the stress-relief polishing process is required to release a residual stress in the thinned wafer after a mechanical grinding process. Depending on the polishing condition, the defect band zone in the range of 0.1–1 µm thickness remains near the grinded surface of the thinned wafer. This defect zone could be used as a gettering layer to Cu diffusion, because it has high-density point defects and dislocations. The behaviors of Cu diffusion at the backside surface of the thinned wafers with various extrinsic gettering layers were electrically evaluated. The normalized generation lifetime, τ_g of the minority carrier is plotted versus annealing time after the intentional Cu diffusion from the backside surface at 300 °C, as shown in Figure 28.6 [8]. In the chemical–mechanical polishing (CMP), ultrapoligrind (UPG), poligrind (PG), and #2000 treated wafers, τ_g is significantly decreased compared to the as-deposition condition according to the annealing time. Meanwhile, the dry polish (DP)-treated wafer shows the relatively strong immunity to Cu diffusion. The poor gettering ability of the CMP treated wafer is induced by the shallow damaged zone within 50 nm thickness, which is not enough to block Cu atoms. The poor gettering abilities of the UPG, PG, #2000 treated wafers are induced by deep microcracks, severe point defects, and dislocations in the damaged zone of more than 300 nm thickness near the grinded surface, where Cu atoms may diffuse easily into the depletion region passed through these severe defects during the annealing. Meanwhile, the DP-treated wafer shows most good gettering ability to Cu diffusion. It indicates that the crystal defects damaged zone with around 100 nm thickness could be acted as a good EG layer.

However, even the DP treated wafer, the damaged zone is not completely uniform, as shown in Figure 28.7, and it induces relatively unstable retardation to Cu diffusion for longer annealing time.

In the case of line A with a 50 nm thick damaged zone, Cu atoms diffused into the Si substrate passed through a relatively shallow damaged zone. In the case of line B with a 100 nm thick damaged zone, meanwhile, Cu atoms are blocked within the damaged zone and not diffused into Si substrate (see Figure 28.7). It indicates that the crystal defects damaged zone of around 100 nm thickness is required to

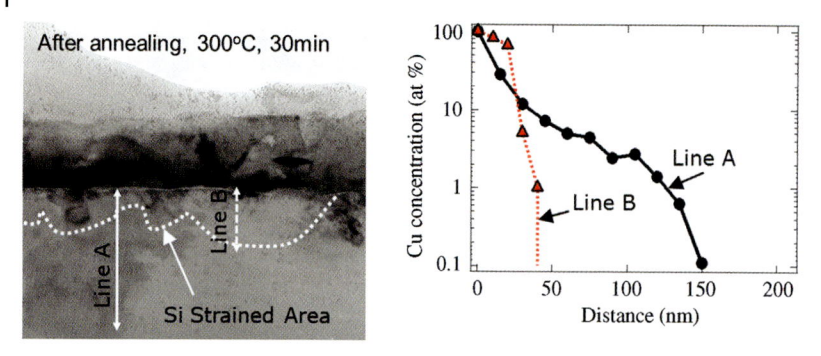

Figure 28.7 TEM cross-sectional image of the back surface in the PG-treated wafer after Cu diffusion at 300 °C for 30 min (a) and Cu concentration profiles measured by SIMS (b).

acts as a good EG layer to Cu diffusion. However, it is another challenge to form a 100 nm damaged zone uniformly by the conventional polishing method.

Cu TSV formation is a key technology for 3D LSI fabrication, because TSV is an important factor to determine 3D LSI performance. However, Cu atoms from TSV can easily diffuse and contaminate the nearby devices when the blocking property of the Cu barrier layer is not sufficient. The TSV is formed commonly by the BOSCH process using a deep reactive ion etcher for high aspect ratio via etching. However, via etching results in a sidewall surface roughness called scalloping, caused by the cyclic etching and passivation in the BOSCH process. If the scalloping roughness is high, it would be challenging for the conformal deposition of the dielectric liner and barrier layer. Especially, poor coverage of the barrier layer may induce the diffusion of Cu atoms from the Cu TSV during a postannealing process. To avoid Cu diffusion from the Cu TSV, minimizing the sidewall scalloping, followed by the conformal deposition of sputtered barrier layer is required. The influence of Cu diffusion from the Cu TSV was electrically characterized by the C–t analysis using a trench MOS capacitor composed of a Cu/Ta gate electrode and Cu TSVs. Two types of the sidewall scalloping with average roughness of 30 and 200 nm were prepared, as shown in Figure 28.8. After the formation of a 100 nm thick oxide liner into via holes, two types of Ta barrier layers with thickness 10 nm and 100 nm at the surface were formed. The minimal Ta barrier layer thickness at the trench sidewall was approximately 3 and 20 nm, respectively.

The generation lifetime of the minority carrier is plotted versus the annealing time as in Figure 28.9. In the trench capacitor with 10 nm thick Ta, the generation lifetime is significantly reduced from the as-deposition condition after initial annealing for 5 min at 300 °C even in the small scalloping roughness of 30 nm. Meanwhile, in the trench capacitor with 100 nm thick Ta, the generation lifetime is not reduced after annealing up to 60 min even in the large scalloping roughness of 200 nm (Figure 28.9a). However, the generation lifetime is reduced to 50% level from the as-deposition condition after annealing for 30 min at 400 °C even with a thick Ta layer and the small scallop roughness (Figure 28.9b).

28.2 Impacts of Cu Contamination on Device Reliabilities in Thinned 3DLSI

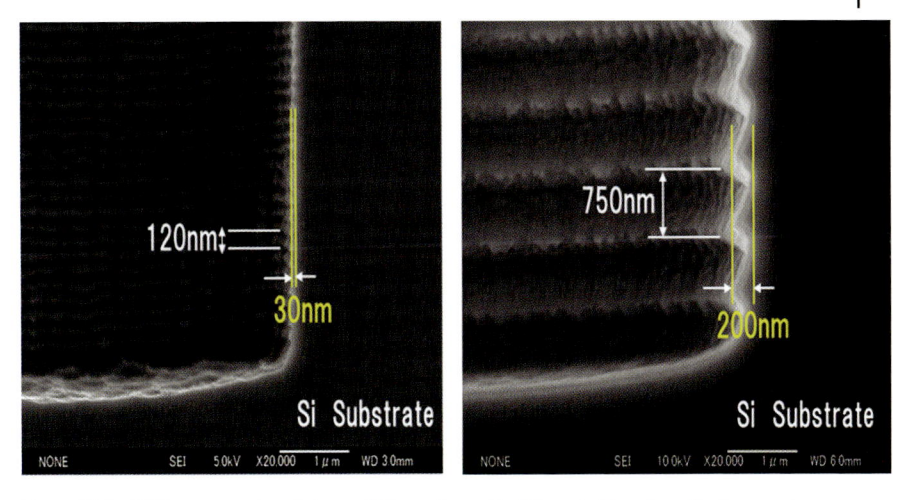

Figure 28.8 SEM cross-sectional views of Si trenches with different sidewall scalloping roughness.

DRAM stores electronic charges as the information data. The control of the retention time (refresh) for the stored charge is a key issue for realizing reliable 3D DRAM. This requirement derives from needs to keep the refresh interval constant even if the thin memory chips stack vertically to achieve 3D DRAM. Therefore, one of the most critical reliability issues for high reliable 3D DRAM is the data retention characteristics attributed to the electron leakage in the storage capacitor. The electron leakage is induced by several mechanisms. One of the critical origins for the electron leakage is metallic impurities such as Cu introduced in the 3D integration process. Especially, Cu is extremely mobile in silicon and silicon dioxide [9,10]. It exhibits the highest diffusivity in silicon among all metal elements. When present in active device regions, Cu impurities could cause functional failures through a variety of mechanisms – increasing leakage currents or carrier generation–recombination rates – and can also cause loss of functionality [11]. In

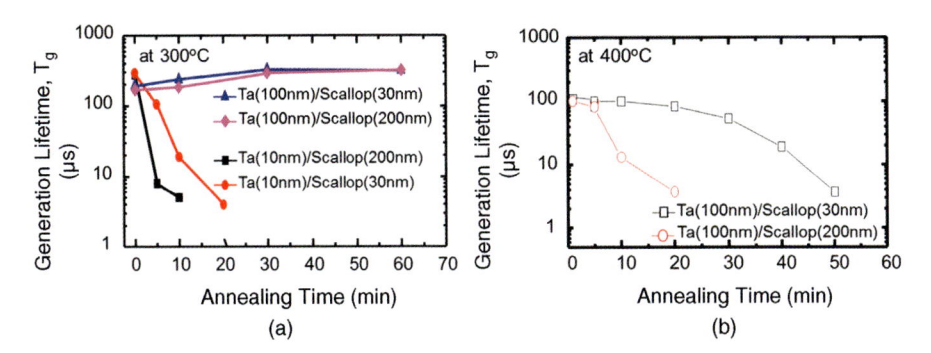

Figure 28.9 Generation lifetime of minority carrier obtained from the C–t analysis versus annealing times at 300 °C (a) and 400 °C (b) with various times.

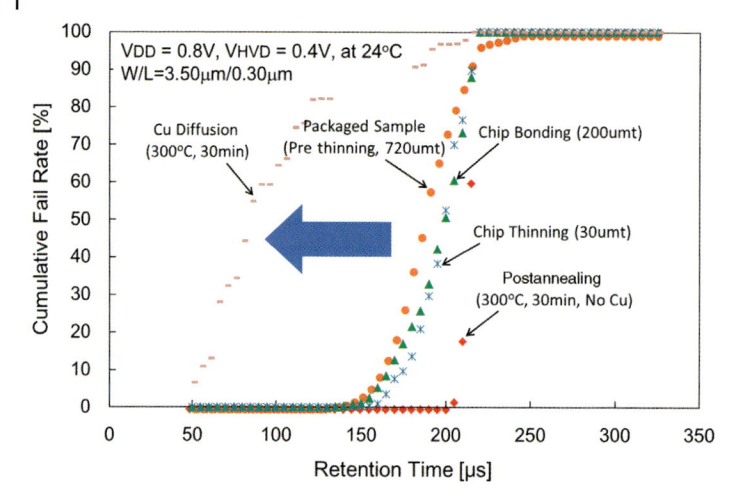

Figure 28.10 The failure rate of DRAM cell array (W/L = 3.50/0.30 μm) as a function of retention time at 24 °C measured after chip bonding, chip thinning, and Cu diffusion for 30 min at 300 °C, respectively.

extremely sensitive devices, such as DRAM cells, even small contamination levels can shorten retention time.

The influence of Cu diffusion from the backside surface on the DRAM cell reliability was electrically characterized, as shown in Figure 28.10 [12]. The DRAM cell array shows 0% failure up to 130 μs and 50% failure at 200 μs after chip bonding and chip thinning conditions. The retention time of the DRAM cell exhibits no change from the as-bonding condition even after chip thinning down to approximately 30 μm. However, the DRAM cell array shows 50% failure at 70 μs after Cu diffusion for 30 min at 300 °C and consequently shows up as a severe degradation of the retention characteristics. It indicates that Cu atoms diffused from the back surface reach to the Si–SiO$_2$ interface of the front surface in active areas and cause increasing carrier generation–recombination rates, consequently shortening retention time.

To realize a highly reliable 3D DRAM with good retention characteristics, the reduction of Cu contamination introduced during the 3D integration process is crucial.

28.3
Impacts of Local Stress and Strain on Device Reliabilities in Thinned 3DLSI

The most justified advantage of 3D is the reduction in the interconnect distance between chip functions, which leads to substantial decreases in RC delay and power consumption [13]. Three-dimensional integration involves a vertical stacking of a die or wafer onto another die or wafer along with (i) formation of permanent electrical connection between the input and output pins of the devices using TSVs and microbumps, and (ii) mechanical stability by injecting the organic underfill (OUF) at the interchip region. Since the coefficients of thermal expansion (CTE) of Si ($2.6 \times 10^{-6}\,\text{K}^{-1}$), Cu ($17 \times 10^{-6}\,\text{K}^{-1}$), and OUF ($35 \sim 50 \times 10^{-6}\,\text{K}^{-1}$) vary enormously, it is bound to introduce thermomechanical stress (TMS) in the thinned IC chip. Although the implications of TMS induced by Cu TSVs on the vicinal active Si are extensively studied by several authors [14–18], the reliability issues related to TMS induced by microbumps are hardly investigated.

28.3.1
Microbump-Induced Stresses in Stacked LSIs

Microbumps in the 3D stack may cause two different stresses, namely, TMS and locally induced mechanical stress (LMS). TMS is due to the difference in the CTE values between Si and the bump metal. Conversely, LMS is the result of local deformation in the thinned die around the microbump region, and this is due to the CTE mismatch between the bump metal/Si and OUF. In 3D integration, the die and wafer must be thinned down to less than 50 µm for various reasons. Such ultrathin dies are easily affected by TMS induced by microbumps (Figure 28.11a). It is worth mentioning that when such ultrathin dies are bonded through high-density microbumping, the microbump-induced TMS not only reaches the active region of the die but also overlaps with the TMS produced by the adjacent microbump [15,19]. As can be seen in Figure 28.11b, the CTE mismatch between OUF and the microbump and Si causes the local deformation of the stacked die, especially when the die thickness is reduced below 50 µm. The magnitude of TMS induced by TSVs and microbumps as well as LMS produced in the stacked die around the microbump region can be calculated via finite element analysis [20]. Therefore, one can predict the keep out zone for 3D stacked devices well in advance. The quantitative measurement of TMS and LMS in the stacked LSI Si is generally carried out nondestructively either by using piezo-resistive stress sensor [21] or by employing micro-Raman spectroscopic (µ-RS) technique [22].

In general, in the typical Raman spectrum of crystalline Si, mainly one single degenerated longitudinal optical (LO) peak is observed, whose frequency is located at around $521\,\text{cm}^{-1}$. The µ-RS results presented in this chapter are obtained using the excitation laser with the wavelength of 488 or 785 nm. It is well known that a

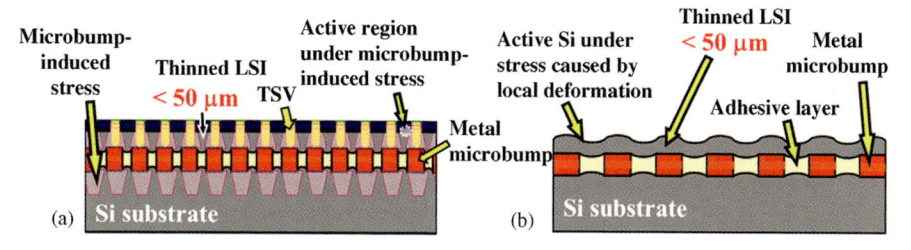

Figure 28.11 Schematic view of (a) 3D LSI with high-density microbumps and (b) local deformation of the stacked die.

tensile strain will shift the Si Raman peak toward lower frequency ($\Delta\omega < 0$), while a compressive stress will result in a high frequency shift ($\Delta\omega > 0$) for an Si Raman peak. Since the frequency of the lattice vibrations of a material will change when the material is subjected to compressive stress and/or tensile strain, the shift in the Si Raman peak frequency can be directly related to the kind of stress present in the material. Also, from the magnitude of the peak–frequency shift, one can quantify the amount of stress. It should be noted that the relation between stress and strain in Si and the Raman frequency is quite complex, since all the nonzero strain tensor components influence the position of the Raman peak. However, in most of the cases, it is assumed to be linear for stress determination. Here, we have used σ (MPa) $= -434 \times \Delta\omega$ (cm^{-1}) and $\sigma xx + \sigma yy$ (MPa) $= -434 \times \Delta\omega$ (cm^{-1}) respectively for uniaxial and biaxial stress in the (100) plane of Si. In the following sections, the reliability issues arising due to the microbumps in 3D LSIs are reviewed.

28.3.2
Microbump-Induced TMS in LSI

CuSn microbumping is inalienable interconnect material for a low-temperature back-end-of-line (BEOL) integration process. Although interconnect metal-induced TMS in 3D LSIs has been known for decades, very few reports discussed TMS caused by metal microbumps; nevertheless, such microbumps are also copiously used in die and wafer stacking for face-to-face bonding [15,19].

In the following, 2D stress distribution for cross-sectional 3D LSI samples containing an array (100 × 10 microbumps) of CuSn microbumps with different sizes such as 5 × 5, 10 × 10, and 20 × 20 μm^2 will be discussed. The top and bottom chip sizes are respectively 5 × 5 and 7 × 7 mm^2, and the die thickness is around 280 μm. The cross-sectional scanning electron microscope (SEM) image of bonded CuSn microbumps, where the under bump metal Cu is formed by electroplating followed by the evaporation of Sn, clearly reveals the formation of Cu$_6$Sn$_5$ and Cu$_3$Sn intermetallic compounds at the interface (Figure 28.12a), [20]. Figure 28.12b shows the cross-sectional 2D TMS distribution image for 3D LSI with 20 × 20 μm^2-sized CuSn bumps. Similar to Cu TSVs, there exists severe compressive stress in the Si that resides in the

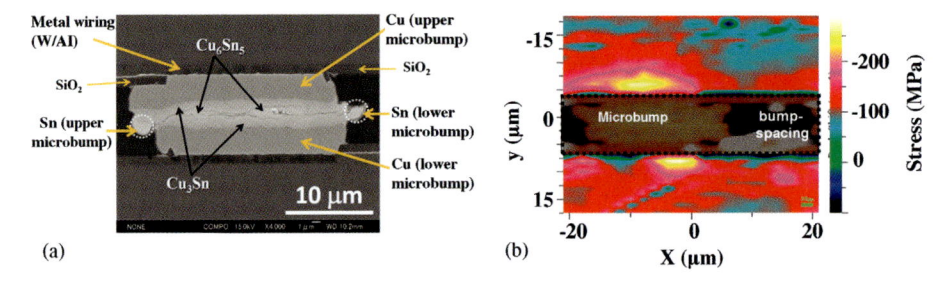

Figure 28.12 Cross-sectional view of EEB-formed CuSn microbump SEM image (a) and 2D stress distribution (b).

immediate vicinity of the microbump. At the bump-space region it induces either less compressive stress or tensile stress. Although the magnitude of the induced stress increased after annealing, the stress distribution pattern remained the same even after the postheat treatment [19]. The maximum stress values of 125, ~250, and >350 MPa compressive stress for before bonding, after bonding, and after postheat treatment, respectively, have been observed. Both the magnitude and the in-depth distribution of the compressive stress induced by the microbump increased with the increase of the bump size. In the case of finer size (5 μm × 5 μm) and high-density CuSn bumps, the compressive stress produced by the two adjacent microbumps along the plane parallel to Cu–Si interface overlapped each other at the bump space region.

28.3.3
Microbump-Induced LMS

As compared to TMS induced by metal microbumps and TSVs in the active Si, the magnitude of LMS arising due to the local deformation of thinned LSI after underfill injection and curing is tremendously large. This causes serious reliability problems before realizing the high-density 3D LSIs. The degree of local deformation depends on various factors such as CTE and modulus of underfill as well as the microbump density, pitch, and height, and to some extent the surface morphology of the stress-relieved thinned die.

For example, the 10 μm thick LSI die experienced a maximum ~225 nm bending for the plasma-etched relief process, while it was nearly half for the CMP process (Figure 28.13). It was proved that the grinding grooves left behind on die surface after stress relief is the main cause for this kind of maximum bending [15]. Such local bending induces a large amount of LMS at the active Si of the stacked LSI as shown in Figure 28.14, where +1.8 GPa of tensile stress around the microbump region and <−0.5 GPa of compressive stress at the bump space are noticed.

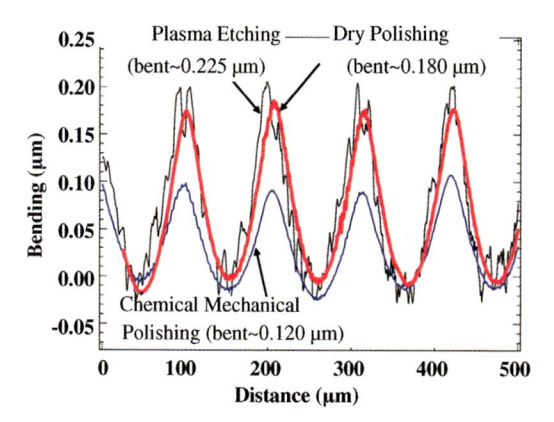

Figure 28.13 Line profile revealing the local deformation formed in 3D LSIs, with variously stress-relieved LSI chips.

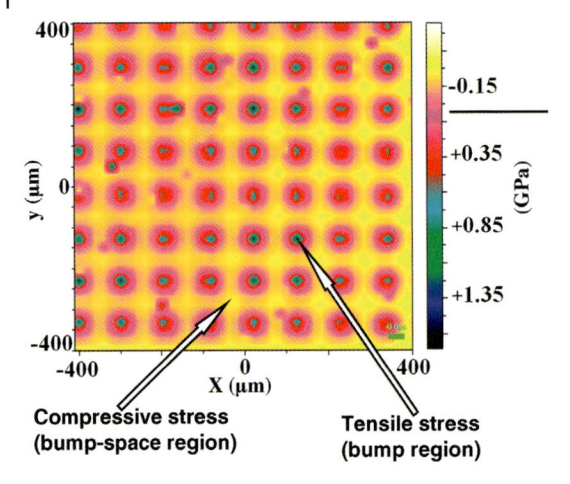

Figure 28.14 Two-dimensional stress distribution image obtained for the top die integrated over the bottom die having an array of microbumps.

A 500 MPa of tensile stress in the 30 μm thick LSI die stacked over an array of microbumps (bump size: 5 μm × 5 μm) caused ∼3–4% increase in the drain current I_d of NMOS transistor (Figure 28.15), which is equivalent to the 14% increase in the electron mobility) [14] and a 10% decrease in the drain current I_d of the PMOS transistor [15]. It is expected that the very large effective mass of the hole as compared to the electron is responsible for the pronounced stress effect observed for the PMOS transistor. The FEM parametric study also showed that the magnitude of n-type carrier mobility change correlates with the increase in the underfill CTE and modulus as well as microbump pitch and height, but decreases with the microbump radius [20]. Recently, over 40% shift in the I_d of MOSFET is reported in the 3D LSIs owing to the stress induced by backside metal microbumps [24].

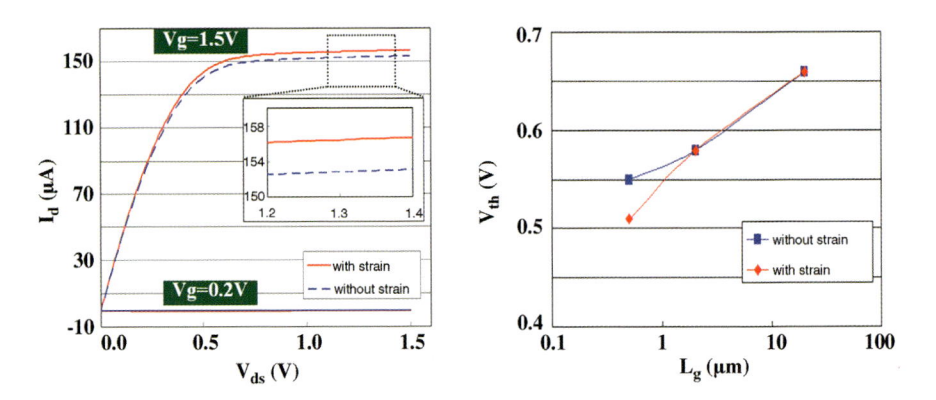

Figure 28.15 Change in drain current I_d and the role of characteristics for NMOS Tr. due to microbump-induced stress.

References

1. Hozawa, K. *et al.* (2002) True influence of wafer backside copper contamination during the back-end process on device characteristics. Technical Digest: International Electron Devices Meeting (IEDM), pp. 737–740.
2. Istratova, A.A. and Weberb, E.R. (2002) Physics of copper in silicon. *Journal of the Electrochemical Society*, **149** (1), G21–G30.
3. Hozawa, K. *et al.* (2009) Impact of backside Cu contamination in the 3D integration process. Symposium on VLSI Technology. pp. 172.
4. Bea, J. *et al.* (2011) Evaluation of Cu contamination at backside surface of thinned wafer in 3-D integration by transient-capacitance measurement. *IEEE Electron Device Letters*, **32** (1), 66–68.
5. Heiman, F.P. (1967) On the determination of minor carrier lifetime from the transient response of an MOS capacitor. *IEEE Transactions on Electron Devices*, **14**, 781–784.
6. Lee, S.-Y. *et al.* (1999) Measurement time reduction for generation lifetime. *IEEE Transactions on Electron Devices*, **46**, 1016.
7. Lee, K.-W. *et al.* (2011) Evaluation of Cu diffusion characteristics at backside surface of thinned wafer for reliable three-dimensional circuits. *Semiconductor Science and Technology*, **26**, 025007.
8. Lee, K.W. *et al.* (2011) Cu retardation performance of extrinsic gettering layers in thinned wafers evaluated by transient capacitance measurement. *Journal of the Electrochemical Society*, **158** (8), H795–H799.
9. Weber, E.R. (1983) Transition metals in silicon. *Applied Physics A*, **30**, 1–22.
10. Helneder, H. *et al.* (2001) Comparison of copper damascene and aluminum RIE metallization in BICMOS technology. *Microelectronic Engineering*, **55**, 257–268.
11. Ramappa, D.A. and Henley, W.B. (1999) Effects of copper contamination in silicon on thin oxide breakdown. *Journal of the Electrochemical Society*, **146** (6), 2258–2260.
12. Lee, K.-W. *et al.* (2012) Impact of Cu contamination on memory retention characteristics in thinned DRAM chip for 3-D integration. *IEEE Electron Device Letters*, **33**, 1297–1299.
13. Banerjee, K. *et al.* (2001) 3-D ICs: a novel chip design for improving deep-submicrometer interconnect performance and systems-on-chip integration. *Proceeding of IEEE* **89**, 602.
14. Murugesan, M. *et al.* (2009) Impact of remnant stress/strain and metal contamination in 3D-LSIs with through Si vias fabricated by wafer thinning and bonding. IEEE International Electron Devices Meeting (IEDM), p. 361.
15. Murugesan, M. *et al.* (2010) Wafer thinning, bonding, and interconnects induced local strain/stress in 3D-LSIs with fine-pitch high-density μ-bumps and through-Si vias. IEEE International Electron Devices Meeting (IEDM), p. 30.
16. Murugesan, M. *et al.* (2011) High density 3D LSI technology using W/Cu hybrid TSVs. IEEE International Electron Devices Meeting (IEDM), p. 139.
17. De Wolf, I. *et al.* (2012) In-depth Raman spectroscopy analysis of various parameters affecting the mechanical stress near the surface and bulk of Cu-TSVs. IEEE Electronic Components and Technology Conference (ECTC), p. 331.
18. Gambino J. *et al.* (2011) Stress from W-TSVs measured by Raman spectroscopy on cross-sectional samples. SSDM, 2011, p. 793.
19. Murugesan, M. *et al.* (2010) Impact of μ-bump induced stress in thinned 3D-LSIs after wafer bonding, IEEE 3D-IC. doi: 10.1109/3DIC.2010.5751432.
20. Aditya, P. *et al.* (2011) Microbump impact on reliability and performance in TSV stacks. MRS Proceedings, 1335.
21. Kumar, A. *et al.* (2011) Residual stress analysis in thin device wafer using piezoresistive stress sensor. *IEEE Transactions on Components, Packaging, and Manufacturing Technology*, **1** (6), 841.
22. De Wolf, I. (1996) Micro-Raman spectroscopy to study local mechanical stress in silicon integrated circuits. *Semiconductor Science and Technology*, **11**, 139.
23. Murugesan, M. *et al.* (2012) Low resistance CuSn μ-bump for 3D chip stacking formed by electroplated-evaporation bumping technology. *Journal of Electronic Materials*, **41**, 720.

29
Metrology Needs for 2.5D/3D Interconnects

Victor H. Vartanian, Richard A. Allen, Larry Smith, Klaus Hummler, Steve Olson, and Brian Sapp

29.1
Introduction: 2.5D and 3D Reference Flows

This chapter will focus on the metrology operations needed to support 2.5D and 3D reference flows employing via-middle copper TSV (through-silicon via) fabrication, wafer thinning, TSV reveal processing using a handle wafer, and chip-level assembly and packaging. This chapter will not address reference flows that utilize via formation after thinning or wafer-to-wafer 3D integration. A fundamental decision on inspection/metrology strategy is whether 100% in-line coverage or a sampling strategy is to be employed. The choice depends primarily on the criticality of a process step, the yield impact of a defect, the number of measurement targets, the number of wafers involved, and the throughput of the metrology/inspection tool. For general descriptions of metrology techniques used for IC processing, we recommend reference books by A. Diebold [1], D. Schroder [2], I. Herman [3], and W. Runyan and T. Shaffner [4], as well as the document repository at the Frontiers of Characterization and Metrology for Nanoelectronics web site [5].

For the purposes of this chapter, "via-middle" processing consists of etching, metalizing, and planarizing the copper TSVs after the contact level and before the upper metal layers [6,7]. Some practitioners perform these steps immediately before M1, whereas others perform them at a higher metal level. For logic chips, typical TSV dimensions are 5–10 μm diameter and ∼50 μm depth; typical TSVs for interposers have ∼10 μm diameter and ∼100 μm depth.

Etch depth metrology is critical because this is a blind etch step,[1] and the reference flow requires that the TSV reveal height after wafer thinning be tightly controlled across the wafer. The etch profile needs to be monitored and well controlled in high aspect ratio (HAR) TSVs to ensure defect-free coverage by the liner, barrier, and seed (L/B/S) layers. L/B/S metrology presents new challenges due to the need to measure thicknesses of tens of nanometers in vias 50–100 μm deep. Void-free copper filling of HAR TSVs requires tight process control over the

---

1) A blind etch step is one that does not etch to a stop layer, but etches for a given time.

copper plating process, as well as a proper etch profile and seed layer. Inspection of TSVs for voids is extremely important and represents another metrology challenge. After annealing and CMP steps, the top of the TSVs should be inspected for defects such as surface protrusions that could impact the yield or reliability of the subsequent metal layers.

Residual stress in the TSVs and/or the adjoining silicon needs to be characterized because of its impact on TSV reliability and the electrical performance of nearby devices [8].

After completing the TSV formation and the remaining front-side processing steps including bumping, the wafers are bonded to a handle wafer using a temporary adhesive. This bond interface must be verified to be free of bond voids, and the bond strength should be sufficient, otherwise the wafers will be rendered inutile, resulting in wafers being scrapped after thinning or during backside processing. The thickness, bow, and warp of the bonded wafer pair must be characterized before and after wafer thinning. The thinning process requires metrology and inspection to monitor wafer thickness as well as the edge quality of the wafer perimeter after edge trim. The TSV reveal process requires careful control of the final thinning process to ensure that the TSV reveal heights fall within the required process limits across the entire wafer surface. After completion of backside processing, the TSV wafers are debonded from the handle wafers and placed on a dicing tape.

The process flow may involve different companies for the front-end and via-middle processing [9], the so-called MEOL processes of wafer bumping, thinning, and backside processing, and the back-end stacking and assembly steps. Outgoing and incoming quality assurance inspection will be needed for bumped and microbumped thinned wafers mounted on dicing frames to verify that the wafers meet all product requirements, including acceptable limits on bump dimensions and surface quality.

Package-level assembly of, for example, memory cubes requires handling 50–100 μm thick die and interposers patterned with microbumps with diameters and pitches as small as 20 and 40 μm, respectively. These aggressive die thicknesses and bump diameters, as well as the use of multitier stacks place new demands on inspection metrology to identify defective bond joints and underfill defects.

29.2
TSV Formation

Device yield, performance, and reliability depend on an optimal TSV etch profile, liner uniformity and integrity, void-free TSV fill, and minimal residual stress. Metrology and inspection related to the TSV etch process include measurement of critical dimension (CD), etch depth and profile, and detection of etch artifacts and defects such as scalloping and residue. Having an optimal TSV profile ensures that the TSV Cu fill will be free of voids or other defects and extend to the desired depth.

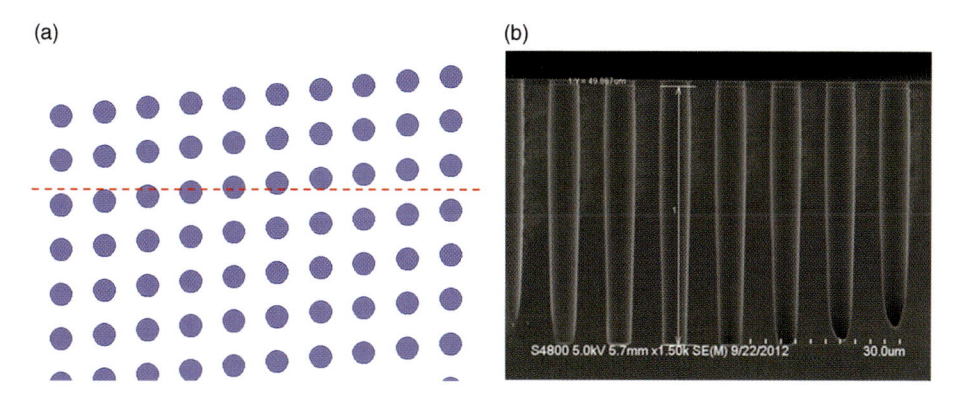

Figure 29.1 A SEM bar, an array of dense staggered-pitch TSVs is shown in (a), allowing multiple TSV cross sections to be generated with one random cleave. The SEM (b) shows 50 μm deep TSVs (5 μm diameter).

Metrology and inspection needs related to liner deposition and via fill integrity include L/B/S thickness, coverage, and profile, identification and characterization of voids or crystalline defects in the copper fill, and process- and temperature-induced stress in the region of the TSV.

29.2.1
TSV Etch Metrology

The challenges of dimensional metrology for TSVs are particularly related to the high aspect ratio of the TSV. Top CD can be determined using methods commonly employed for top CD of conventional vias, including microscopy, various optical imaging methods, or optical profilometry. TSV depth can be determined using a variety of techniques, including spectral reflectometry from the top side, infrared (IR) interferometry from the backside [10], and mass metrology.

Destructive techniques include cross-sectional scanning electron microscopy (SEM) (Figure 29.1). Here, an offset array of TSVs can be used to obtain a cleave line through the center of a TSV for SEM metrology (left). These features can be built into the die and located by naked eye for cleaving. Multiple TSV cross-sections can thus be generated with a single cleave.

Note that although white light interferometry (WLI) can be successfully applied to larger diameter (>5 μm) TSVs, its use may be limited for smaller TSVs due to the requirement of obtaining a measurable collimated beam reflected from the bottom of the TSV.

Because monochromatic interferometry has excellent vertical resolution, it has been applied to TSV depth measurement [11]. Backside IR interferometry (Figure 29.2) has become an important technique to measure TSV depth as it does not require light to penetrate to the bottom of the TSV and reflect back out to a detector. Rather, IR light reflects off the front surface of the wafer from the backside and from the bottom of the TSV. The difference in distance is the TSV

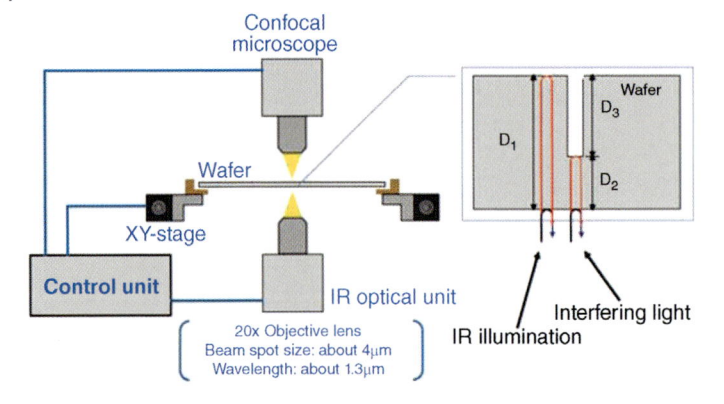

Figure 29.2 A backside IR interferometry system for TSV metrology. Courtesy of Lasertec.

depth. Rapid acquisition times allow real-time wafer mapping of TSV depth (Figure 29.3) to help optimize wafer grinding. However, IR interferometry provides no profile or CD information, and the signal-to-noise (S/N) ratio can be a limitation as is spatial resolution due to the longer wavelengths used.

White light confocal systems are excellent for surface topography, bump metrology, and roughness measurements owing to high lateral resolution (0.01 μm), stemming from shorter wavelengths and lower noise levels compared with IR illumination. However, white light cannot penetrate through Si, has slower vertical scan rates, and does not provide good vertical resolution in HAR applications [12].

White light interferometric optical profilers have fast vertical scan speeds, but are limited in feature width and aspect ratio to about 4 : 1 for a 5 μm diameter via with 1σ variation reported to be as low as 0.05%. The optical elements must also be well

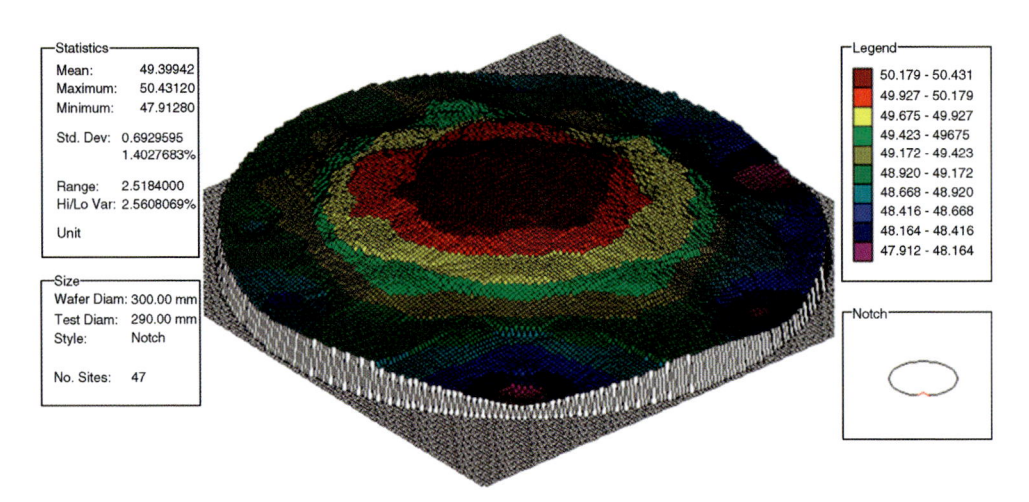

Figure 29.3 A 5 μm diameter TSV depth map using IR interferometry.

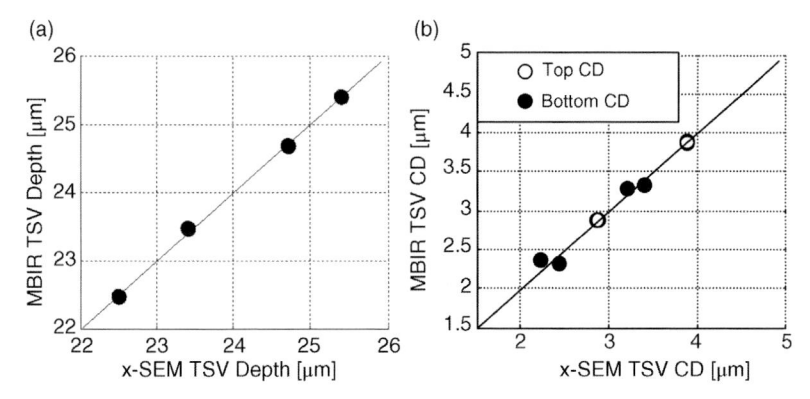

Figure 29.4 Correlation between TSV depth (a) and top and bottom CD (b) achieved with MBIR. Courtesy of Semilab.

aligned to the TSV to maximize the light that reaches the bottom of the feature. Collimated white light interferometers have successfully reported depth measurements for up to 15 : 1 aspect ratio TSVs.

New techniques are being applied to TSV profiles and bottom CDs, including dark-field reflectometry, which requires model-based interpretation based on a comparison of measured spectra to a spectral library. The dark-field spectrum has been shown to be sensitive to variation in sidewall angle (SWA), bottom CD (BCD), and bottom curvature [13].

In addition, model-based IR (MBIR) measurements on TSV arrays have demonstrated that it can match the performance of cross-sectional SEM in determining TSV depth up to 25 μm (Figure 29.4a) and TSV top and bottom diameters from 2.5 to 4 μm (Figure 29.4b) [14]. However, MBIR provides average dimensions in the illumination area and not information on discrete devices.

Mass metrology is a method by which the effect of a process step on the mass of the wafer is accurately measured. This technique can be used in the determination of CD, depth, and/or profile in TSV formation [15,16]. Figure 29.5 is a plot of mass loss for 5 μm diameter vias at various depths. Once a baseline is established for a particular process, mass metrology can be used to monitor whether the change on a particular wafer follows the expected historical value. Note that this technique provides average information about all of the structures on a wafer and that additional dimensional information is needed to separate out the different dimensions.

29.2.2
Liner, Barrier, and Seed Metrology

Measurement of liner, barrier, and seed thickness and uniformity presents the challenge of determining whether the extremely thin films deposited on the sidewalls and bottom of a HAR TSV opening are sufficiently thick and uniform to

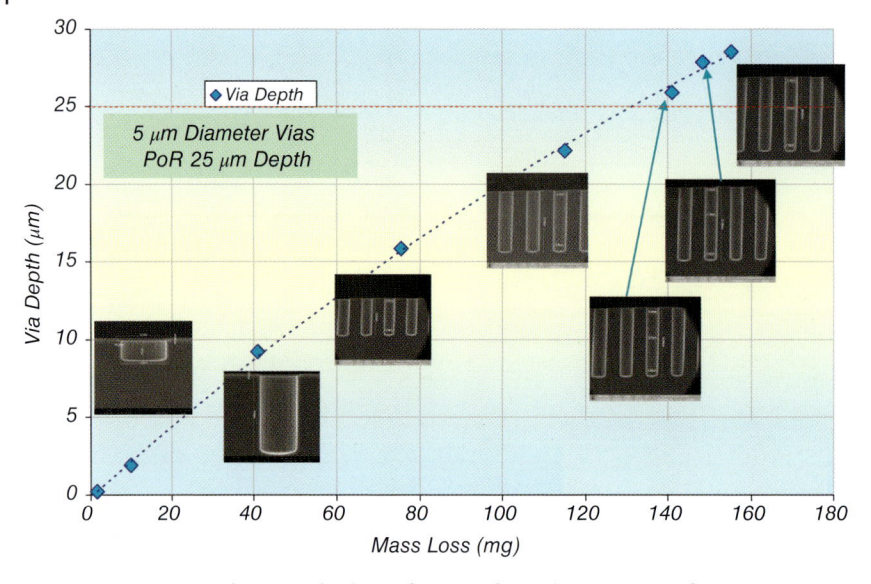

Figure 29.5 Mass metrology – via depth as a function of mass loss. Courtesy of Metryx.

provide coverage over the entire surface of the TSV opening. The only metrology technique mature enough at the time of this writing is cross-sectional transmission electron microscopy (TEM), as shown in Figure 29.6. In this figure, the region of the TEOS liner, diffusion barrier, and seed layer can be seen between the surrounding silicon and the copper fill. Although image contrast is insufficient to differentiate the diffusion barrier and seed layers, energy-dispersive X-ray

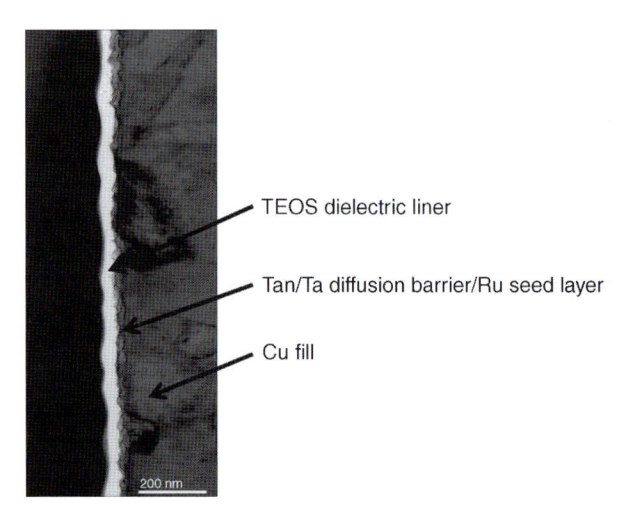

Figure 29.6 TEM of a TSV sidewall showing the TEOS dielectric liner, TaN/Ta diffusion barrier/Ru seed layer, and Cu fill.

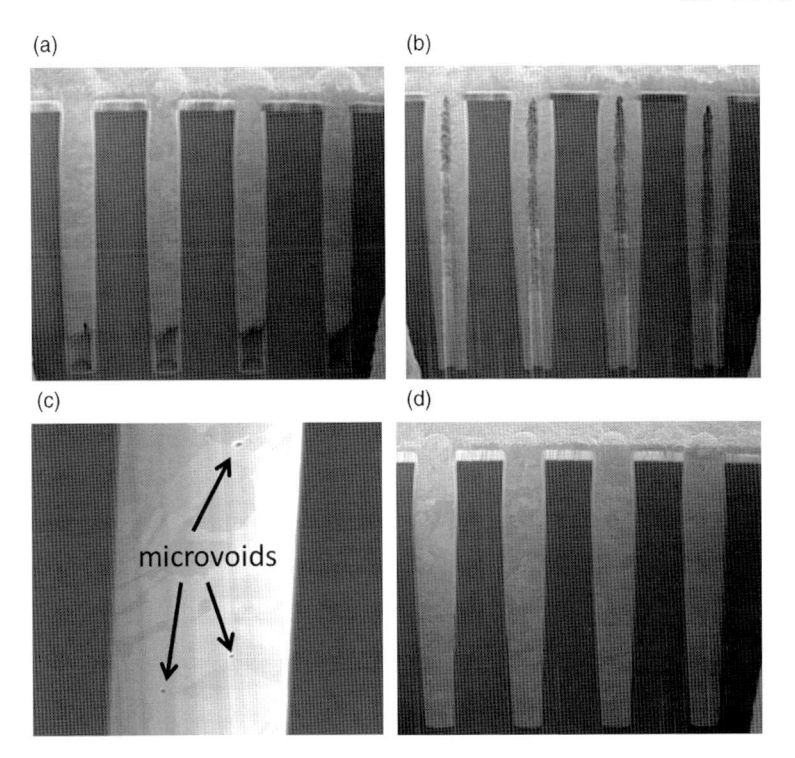

Figure 29.7 Dual-beam FIB cross-sectional SEMs of TSVs showing bottom voids (a), seam-line voids with loss of bottom-up fill (b), microvoids (c) and void-free, bottom-up fill (d).

spectroscopy (EDS) is typically employed. In addition to cross-sectional TEM, in-line, nondestructive optical techniques such as dark-field reflectometry are under development by metrology suppliers to determine liner thickness and uniformity, sidewall roughness, and defects.

29.2.3
Copper Fill Metrology (TSV Voids)

Metrology to ensure void-free deposition of copper to the full depth of the TSV opening is critical to avoid yield and reliability problems. Three types of TSV copper defects–bottom voids (a), seamline voids with oss of bottom-up fill (b), microvoids (c) and void-free, bottom-up fill (d) are shown in the dual-beam focused ion beam (FIB) cross-sectional SEMs shown in Figure 29.7 [17].

Cross-sectional SEM and X-ray computed tomography (CT) can be used to image the interiors of small numbers of TSVs to visually identify and characterize voids; details of these techniques are provided below. Mass metrology, as described earlier, is also applicable to TSV deposition giving a measure of total material deposited in the TSV holes. If the total volume available for TSV fill is known, by determining the total copper deposited by mass metrology, an estimate of the total void volume over the wafer can be made.

29.2.4
Cross-Sectional SEM (Focused Ion Beam Milling Sample Preparation)

As sample preparation is a destructive technique, cross-sectional SEM is not considered useful for in-line high-volume manufacturing (HVM) metrology but for offline metrology to support process development and, in periodic inspections, to verify that copper plating bath chemistries are not contributing to TSV void formation. An array of staggered TSVs can be built into the die and located by naked eye for cleaving, resulting in some TSVs that can be used for void detection.

Subsurface SEM imaging is possible on whole 300 mm wafers when an in-line dual-beam FIB SEM is employed to create a local cross section by gallium beam ion mill (liquid metal ion source (LMIS)). The FIB removes material (wells) from the wafer surface, exposing the cross section of the copper-filled TSV. Although the wafer does not need to be cleaved, the wells probably prevent further processing of the wafers in the fab.

Recent developments in plasma FIB (PFIB) milling using XeF_2 to replace gallium in an inductively coupled plasma (ICP) source have resulted in 20 times faster milling rates compared with conventional liquid metal ion sources, improved turnaround times for void detection in copper-plated TSVs, and decreased surface contamination. Figure 29.8 shows TSVs in a milling volume of 55 μm width, 5 μm length, and 60 μm depth that were milled in 3 min [17].

29.2.5
X-Ray Microscopy and CT Inspection

Various methods use X-rays to allow imaging of TSVs without cross-sectioning. These include X-ray microscopy, computed tomography, and computed laminography (CL). X-ray microscopy relies on the detection of X-rays that have been transmitted through

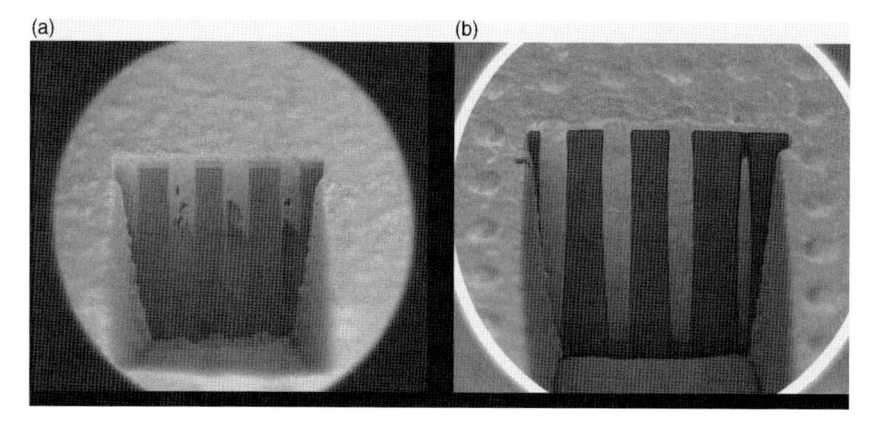

Figure 29.8 PFIB milling to create a well for subsurface imaging and Cu plating void detection – 5 μm × 50 μm deep TSVs with (a) and without (b) voids.

the sample, giving a two-dimensional image similar to optical microscopy. The size of the sample that can be imaged using this technique is determined by the penetration depth, which is the maximum distance X-rays will travel through a material before being fully absorbed.

Most commercially available laboratory-based X-ray microscopes are based on the principle of point projection, where the incident X-ray source is transmitted through the sample and onto a detector. The magnification is a simple function of distance from the source to the sample and the sample to the detector [18]. Flat panel detectors, which have greatly improved the sensitivity of X-ray detectors, are typically either 20.3 or 40.6 cm in size, with up to 100 frames per second (fps) (20.3 cm), for fast image acquisition for S/N improvement energy range from 20 kV to 15 MV, 14- or 16-bit contrast resolution with 200 or 400 μm pixels, and an image size up to 1024 × 1024 pixels (20.3 cm) or 2048 × 2048 pixels (40.6 cm). However, at higher magnifications, the resolution is limited by blurring of the image at the detector plane. A simple X-ray microscope can achieve a spatial resolution of about 100 nm.

Improved spatial resolution of up to 30 nm can be obtained using zone plates [19] so that constructive interference, like a diffraction grating rather than refractive optics, is used to focus X-rays at the desired focus. Alternatively, a lens-based X-ray microscope, implemented using diffraction instead of refraction to focus the X-rays, can provide spatial resolution of approximately 50 nm.

CT uses a series of X-ray images, taken around a single axis of rotation, to reconstruct a full three-dimensional view of the sample. As CT scanning requires the sample to be rotated during measurement, the working distance increases as the sample size increases. Thus, some destructive physical sample preparation must be done to properly size the sample to fit the X-ray microscope, limiting its use to offline metrology to support process development and periodic inspections.

X-ray laminography can be used to provide the 3D position of objects. In classical laminography, the source and detector are moved simultaneously in synchronous motion either rotationally or translationally in opposite directions to create the focal slice that contains the data of interest. However, rotational laminography is more complex mechanically than translational laminography; the angular region swept by each scan is larger. The advantage to laminography is that it does not require a full 360° angular rotation [20]. Computed laminography has been shown to provide improved resolution over CT for imaging voids in TSVs [21].

X-ray imaging allows for visual inspection of individual TSVs without requiring them to be physically cross-sectioned. At 500 μm probing depth, CT scanning allows for visualization of the total volume of voids within a single TSV or an array of TSVs [22], as shown in Figure 29.9.

Since CT scanning does not require physical cross-sectioning of the TSVs, the same structure can be imaged before and after specific process steps. Figure 29.10 shows X-ray images before and after annealing, showing effects such as delamination, void growth, and TSV extrusion [23].

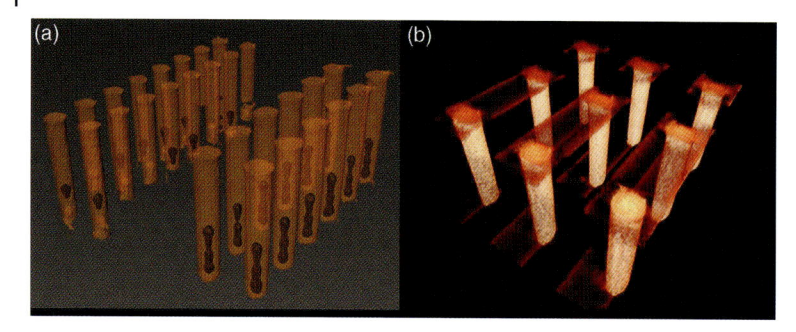

Figure 29.9 Examples of use of CT scanning to measure (a) TSV void volume and (b) overlay between TSVs and underlying metal 1 layer. Courtesy of Xradia.

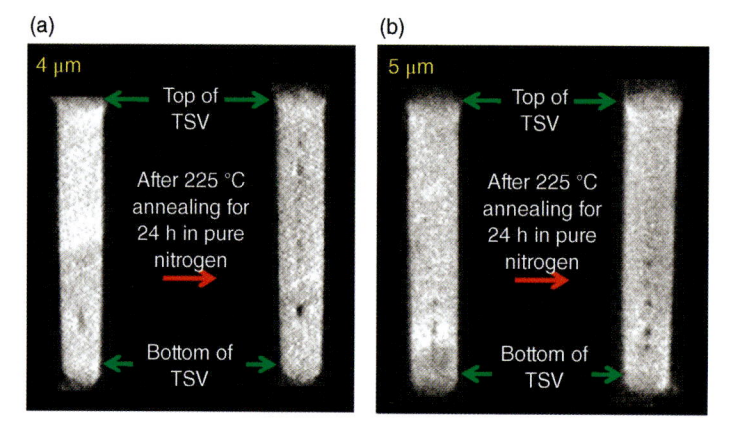

Figure 29.10 Examples of use of X-ray microscope to image the same feature before (left TSV) and after annealing (right TSV) showing (a) TSV void growth and delamination and (b) TSV extrusion.

29.2.6
Stress Metrology in Cu and Si

Process- and temperature-induced stresses in the TSV copper fill and the surrounding silicon are important factors in technology performance, reliability, and cost. Stress in the silicon surrounding a TSV can impact performance because of shifts in carrier mobility, necessitating a so-called keep-out zone (KOZ) surrounding each TSV where no CMOS devices are located. Additionally, this stress can cause yield and reliability issues such as cracking of Si and TSV dielectrics, delamination of the copper fill from the TSV sidewall, stress voiding within the TSV, and copper protrusion during and after thermal treatments associated with downstream process steps (Figure 29.11).

Stress associated with TSV formation is not expected to be a typical in-line measurement, but is important for process development and process monitoring. Methods for measuring the overall film stress, before removal of the overburden by CMP, include X-ray diffraction (XRD) and wafer bending. Both methods average over relatively large areas and are useful for process monitoring and both can be performed before and after the copper anneal.

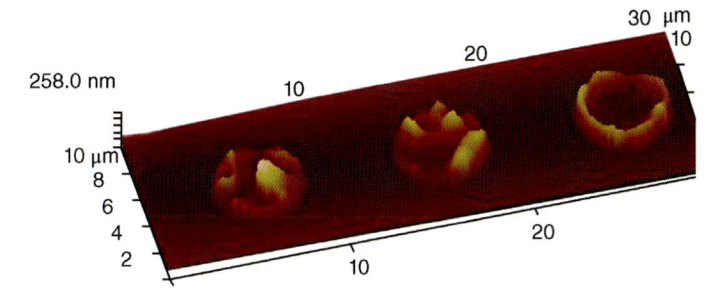

Figure 29.11 A 3D AFM profile of Cu protrusion.

For a detailed analysis of stress localized to a small number of TSVs, several methods are available, including micro-Raman spectroscopy (µRS), which is the only practical method for in-line or laboratory, nondestructive silicon stress measurements at the micrometer scale [24–26]. Micro-Raman spectroscopy can measure stress close to the surface of the silicon (depending on the incident wavelength), where stress effects are most relevant to CMOS device shifts, using spot sizes of 1 µm or less. A strong Raman signal is derived from the surrounding silicon with µRS but the stress state of the copper in the TSV cannot be obtained directly; it is inferred from the measured silicon stress by modeling. Shown in Figure 29.12 are Si stress contour maps before and after a 10 min anneal at 400 °C obtained using 457.9 nm Raman excitation, showing the dependence of compressive stress magnitude on the distance to the TSV.

Other methods for localized stress analysis are cross-sectional TEM and synchrotron X-ray microdiffraction (µSXRD). Cross-sectional TEM can provide stress analysis at an extremely high resolution, but the needed sample preparation is both destructive and likely to alter the stress state of the sample. However, µSXRD, with spot sizes below 1 µm, can resolve individual grains in single TSVs [27,28], measuring stress directly in both copper and silicon. µSXRD can be applied to TSVs in cross section without the risk of stress relaxation, because the beam can penetrate

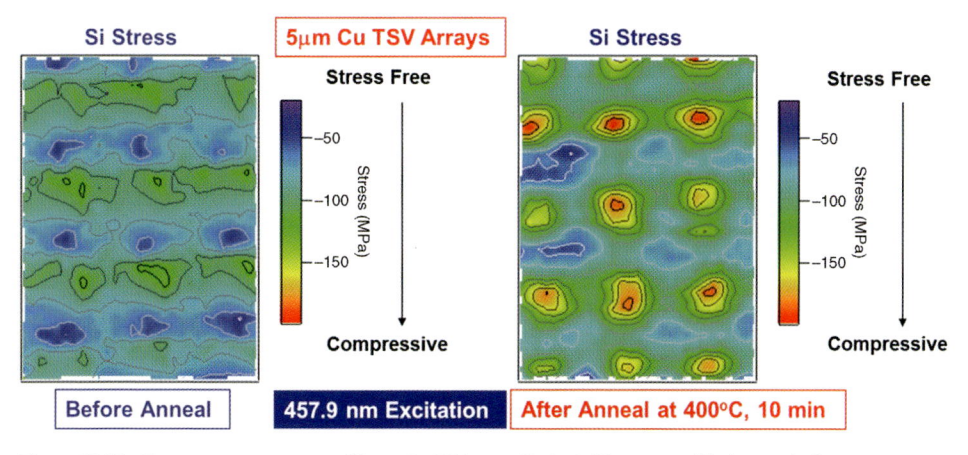

Figure 29.12 Si stress measurement of 5 µm Cu TSV array (1 : 1 pitch) measured before and after anneal showing increased compressive stress around TSVs. Courtesy of WaferMasters.

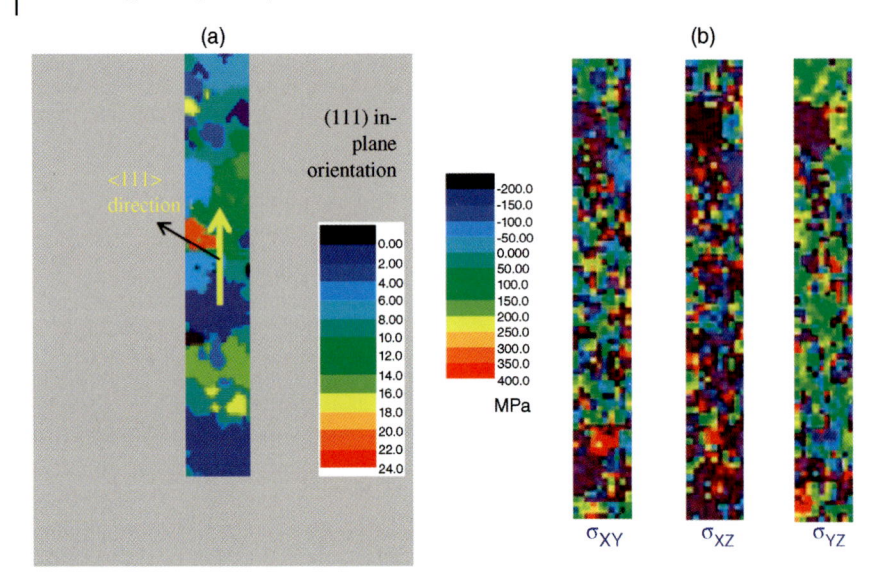

Figure 29.13 Crystal grain orientation mapping (a) and Cu stress states in TSVs by synchrotron X-ray microdiffraction (b).

several micrometers of silicon and still generate sufficient signal from an individual copper grain inside an individual TSV. By combining data obtained with monochromatic and polychromatic X-ray light, μSXRD can map the complete strain tensor and determine grain sizes and orientations in individual TSVs (Figure 29.13). However, even though it is by far the most powerful, it is also the most expensive method of stress analysis, requiring access to a synchrotron radiation source.

Note that none of these aforementioned techniques measure stress in Si or Cu directly. Raman spectroscopy provides phonon frequency shifts; material parameters are used to convert these into stress values. The diffraction-based techniques measure elements of the strain tensor; these are converted to elements of the stress tensor mathematically, in most cases using the linear elastic properties of macroscopic single crystalline samples. In some cases, microscopic measurements such as micro- or nanoindentation can provide additional support.

29.3
MEOL Metrology

Successful completion of the MEOL process steps requires a number of metrology and inspection steps, including the following:

- Edge trim inspection for profile and defects.
- Bond void detection and characterization.

- Stack or wafer thickness, total thickness variation (TTV), bow, and warp. This may be required at one or more points in the process: after completion of temporary bonding, thinning, TSV reveal, RDL and related process steps, and debonding processes.
- TSV reveal height and condition after CMP.

In addition, the wafer must be inspected for TSV reveal damage following debonding and cleaning.

29.3.1
Edge Trim Inspection

The physical condition of the edge of a thinned product wafer is important for successful handling and further processing, as chipping at the edge of the wafer can create stress concentration regions that cause the wafer to break. Full thickness silicon wafers have beveled edges to mitigate these defects. Depending on the integration flow, a vertical edge can be formed on the thinned device wafer (Figure 29.14) or a truncated version of the original bevel is left. Various metrology methods are used to identify defects at the edge of thin wafers.

Inspecting the edge of a wafer for defects requires inspecting a large amount of surface area (front, back, and vertical face). If edge inspection is limited to a region 1 mm from the wafer edge on the front and back sides, approximately 940 mm^2 of wafer surface will be inspected. This area is roughly equivalent to a 30 mm square die. The large area necessitates the use of optical methods for defect inspection. Optical or scanning electron microscopy can be employed for defect review.

Thin wafers are mounted to a carrier wafer or a dicing tape during the inspection. This prevents at least one of the three surfaces from being inspected. Multiple inspections at different stages of fabrication may be required to fully monitor the edge of the thin wafer.

Figure 29.14 An SEM image of edge grind of a device wafer.

29.3.2
Bond Voids and Bond Strength Metrology

Voids between the bonded wafers arise from two primary sources: material inhomogeneity and trapped gases (which may be generated during a post-bond anneal) in the bond layer. Material inhomogeneity can be further divided into two types: nonplanarities that result from the process (e.g., surface topography or revealed copper pillars) and particulate contamination. During bonding, the high force applied to the wafer stack renders the void surrounding the particle (or bubble of trapped gas) negligible or even absent. However, after the bond force is removed, the particle or trapped gas acts as a counterforce applied to the bond interface, causing the wafers to delaminate in a nearly circular pattern. The dimension of this delamination depends on both the size of the particle (volume of gas) and the bond strength. A number of metrology techniques have been developed for bond void metrology.[2] The most common methods are scanning acoustic microscopy and IR imaging. Each technique involves a trade-off between speed of measurement and resolution. Time to measure a single wafer is directly related to the sensitivity. For example, an acoustic microscope can scan a single wafer in about 1 h with a pixel size of about 50 μm, which allows for the detection of voids on the order of 25 μm. If the pixel size is halved to 25 μm, the scan time increases by a factor of 4.

Voids in bonded wafer stacks (BWS) can have serious consequences during subsequent thinning operations. The thinned silicon wafer on the top of the BWS can delaminate, manifesting itself as a bubble of silicon that is not attached to the bottom wafer. Potentially the thinned silicon wafer can shred, leaving an open defect of cracked silicon approximately the size and shape of the original void. Examples of different bond defects, here imaged using scanning acoustic microscopy, are shown in Figure 29.15.

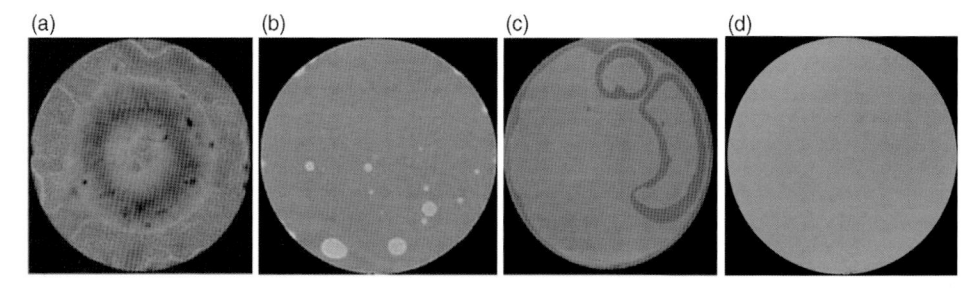

Figure 29.15 Scanning acoustic microscope-identified defects in bonded wafer stacks: dendritic structures from (a) improperly cured adhesive, (b) particle-induced voids, (c) adhesive thickness nonuniformity, and (d) defect-free. Figure 29.15 (d) courtesy of Shin-Etsu.

---

2) Several of the metrologies to identify and characterize bond voids were first applied to wafer bonding for MEMS packaging applications, focusing on >50 μm voids that can cause failure of the hermetic seal.

29.3.2.1 Acoustic Microscopy: Operation

Acoustic microscopy uses ultrasound to survey the interface of a bonded wafer pair [29]. The frequency is typically in the range of 5–500 MHz. A particular frequency is chosen considering the specific metrology requirements, as low frequencies have deep penetration depths and long focal lengths, but large spot sizes and lower resolution, whereas high frequencies have shallow penetration depths and short focal lengths, but small spot sizes and sub-10 µm resolution.

In most instances, BWS will undergo acoustic microscopy metrology immediately after bonding; consequently, a transducer frequency and focal length that provide optimal imaging of the BWS interface, which is beneath 775 µm of silicon, should be chosen (e.g., a 110 MHz transducer meets these conflicting requirements of resolution and penetration depth). As voids (air gaps) between the bonded wafer pairs do not transmit ultrasound, ~100% of the energy in the ultrasonic signal is reflected to the transducer and can be mapped on the wafer pair as a void. Using a 230 MHz transducer, 6 µm features were resolved in a bonded wafer with various void sizes created by lithography and etch [30].

Ultrasound is reflected, transmitted, or refracted through bonded wafer pair materials. Differences in acoustic impedance of materials allow a wafer pair interface to be probed using a transducer that both generates the ultrasound and measures its reflected signal.

Because air does not transmit ultrasound, a couplant fluid must be used between the transducer and the bonded wafer pair, typically deionized water. Temperature control for the acoustic microscopy transducer couplant liquid has been shown to improve measurement stability; a temperature control module for the transducer couplant liquid may be provided by the tool supplier. Care must be taken to ensure that residual contamination does not remain from the couplant fluid after the measurement. The couplant fluid should be maintained by performing frequent water flushes or by using filtration to maintain purity, so that the water leaves only minimal contamination on the bonded wafer stacks once the fluid has dried. Couplant fluid encroachment through the wafer edge is also a concern.

A BWS with etched voids of known sizes was used to test the resolution of a scanning acoustic microscope (Figure 29.16). Voids range from 0.5 to 300 µm in diameter and have semi-dense (5 : 1), isolated, and dense (1 : 1) pitches.

29.3.2.2 Acoustic Microscopy for Defect Inspection and Review

Scanning acoustic microscope software can identify bond voids and locate them on a Cartesian coordinate system based on the center of the defect relative to the bonded wafer pair center, providing the results in KLARF (KLA results file) or some other file format.

Since the defects that cause voids are often too small to be resolved by the acoustic microscope, this defect report will facilitate subsequent defect review using a higher resolution metrology tool (e.g., IR microscopy).

Defect cluster analysis tools allow binning to analyze and quantify defects. Cluster software allows custom filters to be applied to the defect map to increase or decrease sensitivity for what is considered a defect. Cluster analysis allows different

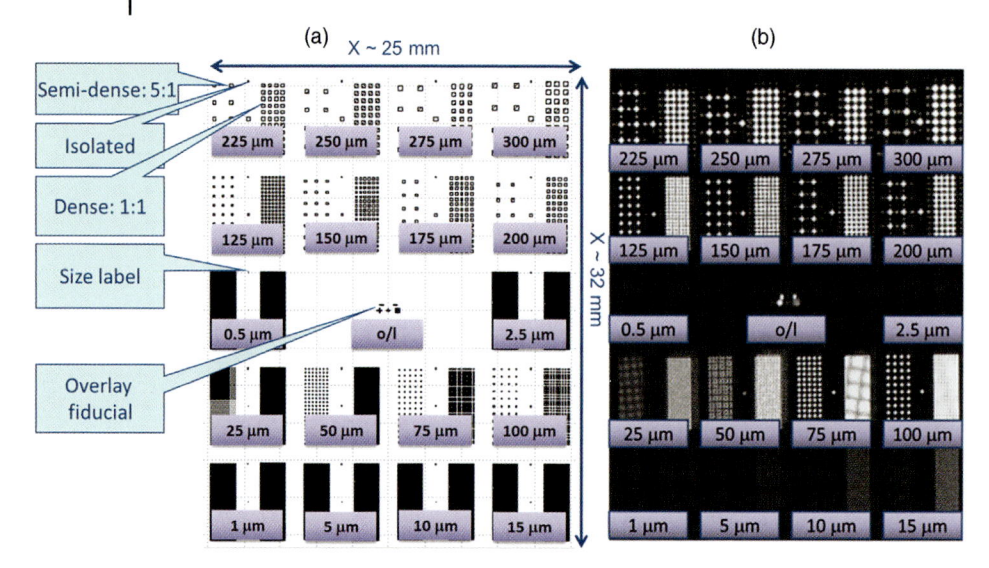

Figure 29.16 A schematic (a) and a scanning acoustic microscopy image (b) of a bonded wafer with various void sizes created by lithography and etch used for resolution testing. Defects (voids) range from 0.5 to 300 μm in diameter with semi-dense (5 : 1), isolated, and dense (1 : 1) pitches.

process tool conditions to be compared, assisting process development and subsequent in-line metrology monitoring. Different process tool contributions to bonded wafer stack defectivity can be assessed and corrective actions implemented when defect maps show that the bonded wafer process is out of control. The total defect area is summed, allowing wafer-to-wafer comparisons.

Once the acoustic microscopy software identifies and locates defects for a BWS map, defect review is enabled by driving to the individual areas in the locator map. Offline analysis of the defect map is also supported.

One option for improving the speed of acoustic microscopy is using multiple transducer heads, which reduces the acquisition time by the number of transducers.

29.3.2.3 Other Bond Void Detection Techniques

IR microscopy, described earlier, can also be applied to resolve voids [31] between bonded silicon wafers; optical microscopy can be used if the handle wafer is made of optically transparent glass. Full-wafer IR illumination techniques are applicable to full-wafer scanning. Confocal microscopy and coherence interferometry can be used for high-resolution evaluation of localized areas where voids have been identified using full-wafer techniques.

X-ray tomography, also described earlier, can be applied to the evaluation of small voids with resolution of approximately 1 μm or less [18].[3]

---

3) Several commercial CT suppliers quote a voxel resolution of <1 μm, including General Electric, Nikon, Nordson Dage, North Star Imaging, and Xradia (now Zeiss).

Residual stress metrologies use photoelastic stress measurement (e.g., a gray-field polariscope) to identify regions where the wafer is under stress. Since voids are typically caused by trapped particles or gas, these stressed regions can be mapped to the presence of voids.

Photoacoustic microscopy uses a picosecond laser pulse to create an acoustic wave in the wafer stack. The wave propagates through the stack and is reflected when it intersects with voids. A second beam detects the acoustic signal when the reflection returns to the surface. Photoacoustic microscopy is a noncontact and nondestructive technique capable of sub-10 µm resolution.

29.3.3
Bond Strength Metrology

Bond strength is the measure of the resistance of the bond interface against crack propagation as measured in units of energy per unit area. There are two classes of metrology methods – those that can be performed on an unpatterned wafer and those that require a patterned test structure for measurement. All of the techniques that can be performed on unpatterned wafers are destructive. Some of the test structure techniques are nondestructive. All of these tests assume that the fracture strength of the bond is less than the fracture strength of the weaker wafer in the pair. If the wafer is weaker than the bond, the wafer itself will crack before the bond fails, giving only a lower limit to the bond strength.

Of the destructive test methods, many of them make use of a precrack. None of the methods described herein is strictly quantitative; that is, interpretation of the data acquired using each method requires the output of finite element analysis (FEA). This FEA must take into account the composition of the wafers as well as the dimensions of the test structure. Accurate FEA depends on an accurate description of the precrack.

Unpatterned (field)-based tests include the razor blade test and the four-point bend. The razor blade test [32] is the simplest method of measuring wafer bond strength. After bonding, a razor blade is inserted between the wafers at the bond plane to induce local delamination. Comparing the shape and depth of the region of delamination with the results of FEA gives the bond strength.

The four-point bend test [33] uses a test coupon cut from the bonded wafer pair. The dimensions of the test coupon are arbitrary, but each configuration must be modeled separately. A notch is sawn partially through one of the wafers to serve as a precrack and force is applied as shown in Figure 29.17. The applied force causes the crack to propagate vertically until it reaches the bond plane and then horizontally as the wafers delaminate. The force as a function of time is recorded.

Test structures include microchevrons [34] (SEMI Standard MS5-0813 2013), shown in Figure 29.18 [35]. Microchevrons use a notch as the precrack. A wedge is etched into one of the wafers before bonding. The remaining surface, which is now shaped like a chevron, is bonded to an unpatterned wafer. The point of the chevron serves as the precrack to focus the load on a known location. The load is applied at the top and the bottom over the opening with the force effectively applied to the

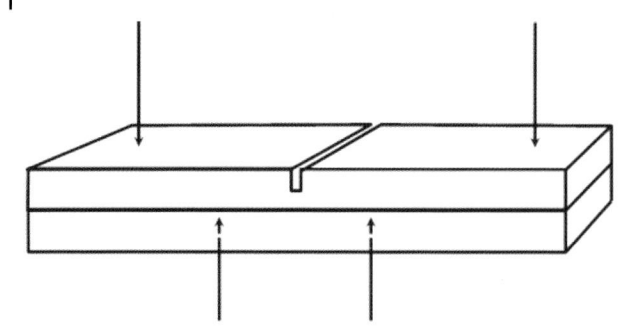

Figure 29.17 A four-point bend test structure.

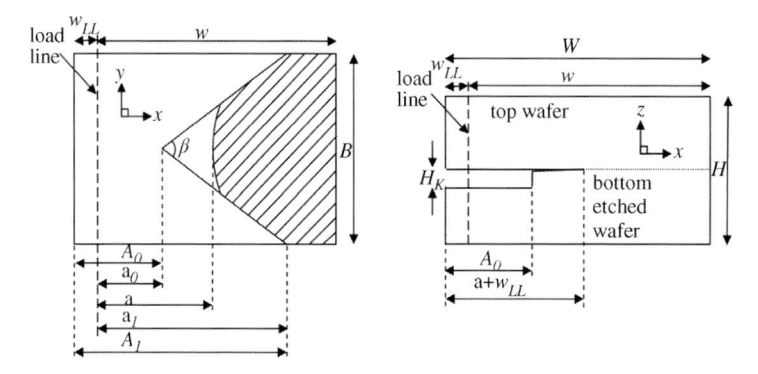

Figure 29.18 Microchevron test structures described in SEMI MS5.

load line as shown in Figure 29.18. The dimensions of the microchevron test structures described in SEMI MS5 are not specified; by convention, the dimensions of W and B are 10 mm each and the angle of the notch (β) is 90°.

29.3.4
Bonded Wafer Thickness, Bow, and Warp

Metrology for control of BWS parameters, such as BWS thickness,[4] TTV,[5] bow,[6] warp[7]/sori,[8] and flatness, is essential to the successful implementation of a wafer

---

4) Thickness is the distance through the wafer between corresponding points on the front and back surfaces (SEMI MF59).
5) TTV is the difference between the maximum and minimum values of the thickness of a wafer within the fixed quality area (SEMI MF59).
6) Bow is the deviation of the center point of the median surface of a free, unclamped wafer from a median surface reference plane established by three points equally spaced on a circle with diameter a specified amount less than the nominal diameter of the wafer (SEMI MF23).
7) Warp is the difference between the most positive and most negative distances of the median surface of a free, unclamped wafer from a reference plane (SEMI MF1390).
8) Sori refers to the algebraic difference between the most positive and most negative deviations of the front surface of a wafer that is not chucked from a reference plane that is a least squares fit to the front surface within the fixed quality area (SEMI MF1451).

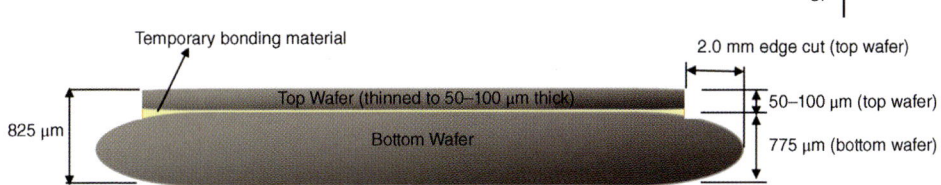

Figure 29.19 A bonded wafer pair – temporary-bonded, edge-trimmed, and thinned.

bonding process. These parameters provide meaningful information about the quality of the wafer thinning process (if used), the uniformity of the bonding, and the amount of deformation induced on the wafer stack by the bonding process. Total thickness variation is also critical in certain bonded wafer manufacturing steps, since nonplanarity can cause problems in subsequent processing steps, including lithography and quality of electrical contact between metal layers on the bonded wafers.

The wafer stacks considered here include carrier and product wafers and bonding layers, including instances of more than two wafers in a stack. Bonded wafers can be classified as either temporarily bonded (i.e., a device to a carrier wafer) or permanently bonded. Temporary bonding uses a temporary adhesive; permanent bonding could be done with adhesive, oxide, metal–metal (e.g., Cu–Cu), or hybrid bonding (e.g., combination of metal–metal and adhesive bonding). A representative two-wafer stack is depicted in Figure 29.19 showing a product wafer bonded to a 775 μm carrier (or handle) wafer using a temporary adhesive and thinned to 50–100 μm. Figure 29.20 shows a 60 μm thick product wafer after debonding from the carrier wafer and transferred to a dicing tape.

29.3.4.1 Chromatic White Light

A chromatic white light sensor, or optical stylus probe metrology, is based on the principle of wavelength-dependent focal length to determine distance (Figure 29.21).

Figure 29.20 A SEMATECH wafer thinned to 60 μm and transferred to a dicing tape.

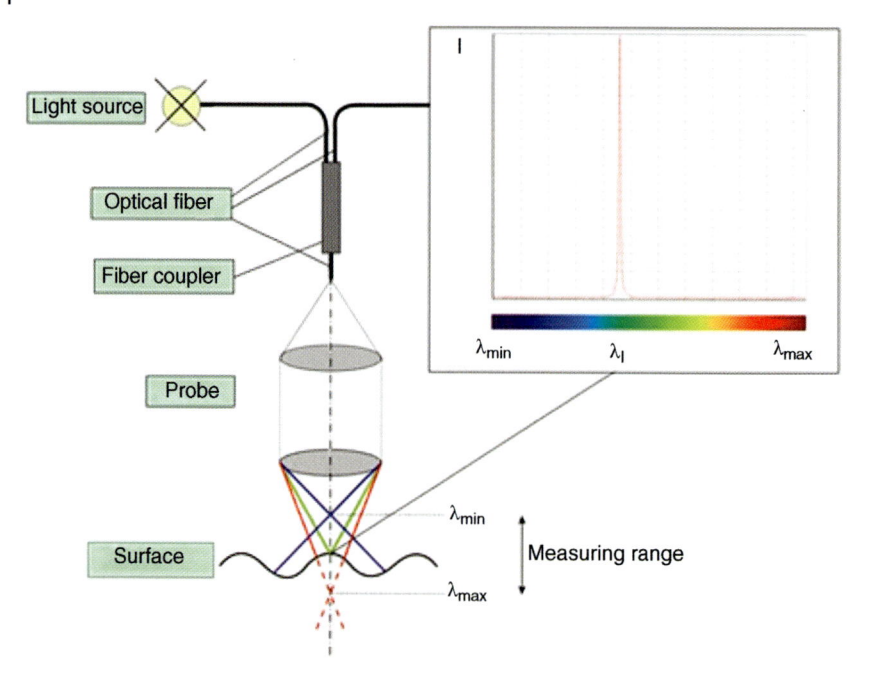

Figure 29.21 A chromatic white light sensor used for bow, warp, and topographical measurements. Focal lengths are wavelength dependent, providing distance measurements. Courtesy of Precitec.

The spectrum of light reflected on a surface generates a peak that is used to determine distance to the sample surface. The peak occurs at the optimal focal point for each wavelength. Resolution varies with objective (from micrometers to nanometers). This technique is useful in determining 2D profile, 3D topography, planarity, roughness, and wafer contour (bow and warp). A resulting 3D thickness contour map for an 825 μm thick bonded wafer is shown in Figure 29.22, using a chromatic white light topography sensor.

29.3.4.2 Infrared Interferometry

Interferometry is a well-known technique for measuring distances by measuring the interference between two coherent light waves that follow different paths [36]. When two light waves recombine, the resulting pattern is determined by the phase difference between the two waves. The phase difference creates the interference pattern between the initially identical waves. Light waves that are in phase undergo constructive interference, whereas waves that are out of phase undergo destructive interference. If a single beam has been split along two paths, then the phase difference is diagnostic of anything that changes the phase along the paths. This could be a physical change in the path length itself or a change in the refractive index along the path. By analyzing the pattern of constructive versus destructive interference, the difference in the distances traveled by the two beams can be determined. For a thin transparent medium, the reflection from the top surface

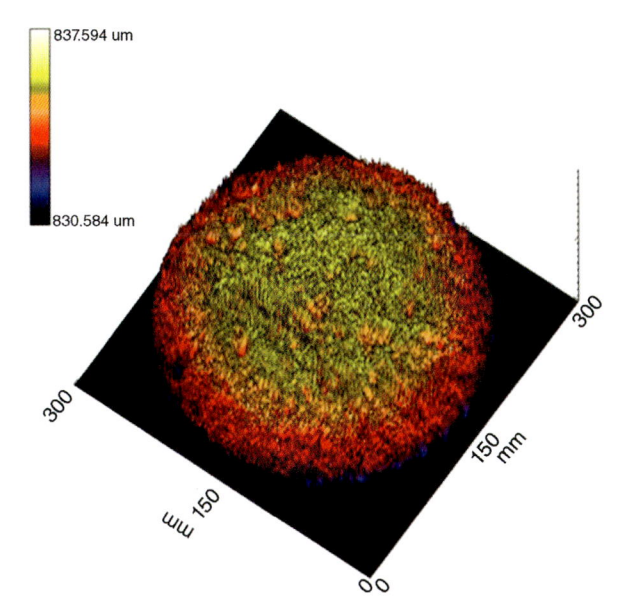

Figure 29.22 A total stack thickness contour map for an 825 μm thin bonded wafer pair. Courtesy of FRT.

serves as the reference and interferes with the reflection from the bottom surface according to a Fabry–Perot etalon [37].

The light source with a wavelength of 1.3 μm is placed facing the back of the wafer. The sensor detects the interference pattern and returns a thickness measurement (Figure 29.23). The fast (millisecond) and nondestructive measurements based on

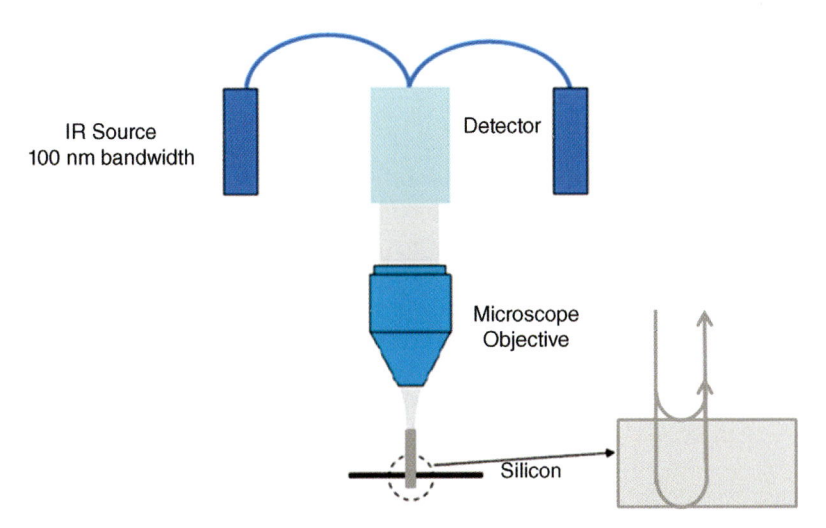

Figure 29.23 An infrared interferometric sensor used for distance measurements for wafer thickness, bow, warp, and TSV depth. Courtesy of Tamar.

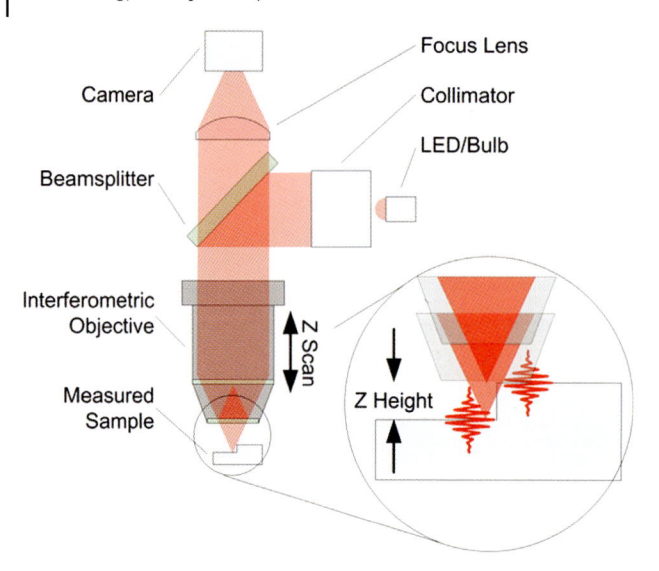

Figure 29.24 White light interferometry for wafer thickness metrology. Courtesy of R.A. Smythe, LLC.

backside IR illumination allow immediate feedback, sufficient for use in a HVM process. IR interferometry can be used to measure wafer thickness (both thick and thin), thicknesses of individual wafers in a bonded wafer stack, wafer bow, warp, TTV, and TSV etch depth.

29.3.4.3 White Light Interferometry (or Coherence Scanning Interferometry)

White light interferometry or coherence scanning interferometry (CSI), depicted in Figure 29.24, is a noncontact optical profiling system for measuring step heights and surface roughness in precision engineering applications [38–40]. The technology uses a white light beam that passes through a filter and then a microscope objective lens to the surface of the wafer. The light reflecting back from the surface is combined with the reference beam and captured for software analysis. As the objective scans in the z-direction, interference is detected by the imaging camera as the measured sample comes into focus. At focus, each pixel detects the interference peak. Analysis software produces a map of surface height variation from the pixel-by-pixel interference peak detection. WLI can measure polished and rough surfaces. WLI can measure and position three-dimensional features to submicrometer repeatability in X and Y, combined with subnanometer repeatability in Z [41].

After obtaining data for each point, the system can generate a 3D image (topography) of the surface. The vertical (i.e. height) resolution of this technique is extremely good, better than 0.3 nm (3 Å), which makes it a potentially practical tool for assessing semiconductor surfaces. However, the lateral resolution can be limited to the spot size, in the range of 0.35–0.50 μm [42]. With its broad capability, it is possible to measure local step height, CD, overlay, multilayer film thickness and optical properties, combined topography and film thickness, and wafer bow. The technology has particular strength in new advanced packaging applications for process control of TSVs, microbumps, redistribution layers, copper studs, and pillars. Microbump/pillar measurements have been demonstrated in production with 0.5 s focus/acquire/measure/move (FAMM) times, while maintaining nanometer Z performance. The presence of films and film stacks is both an opportunity and technology limitation of CSI. Present CSI algorithms are limited to reliably measure in the presence of films to thicknesses greater than 1.5 μm optical thickness (thickness times index of refraction). Some manufacturers claim this capability down to 1 μm optical thickness. Reported measurement results are film thickness (if the index of refraction is known) and surface profile of the top and also the bottom of the film. Advanced work is in progress to measure features in thin film structures, improved CD, and other edge of resolution features. As these technologies emerge, new applications will be possible. When films have less than the measurable optical thickness, unpredictable errors occur in the measured data, and the data are unreliable in Z height.

Measuring nonhomogeneous materials limits WLI accuracy. White light interferometry measures Z height by measuring interferometric phase. The phase depends on the material measured. Insulators experience a constant 180° phase shift, whereas conductors and semiconductors induce varying phase shifts. These phase shifts appear as surface topography errors up to tens of nanometers. It is possible to correct for these by identifying material types by regions and applying a correction factor region by region. Some manufacturers have been successful in correcting these algorithmically, but this has seen little commercial utilization. For process control, phase change on reflection is a constant offset and therefore can be ignored when process variation is being tracked when absolute values are not critical.

29.3.4.4 Laser Profiling

The principle of laser profiling [43] is shown in Figure 29.25 in which two focused laser beams are used, one focused on the top and the other on the bottom of the wafer stack. The measurement is made by moving the focus position from the surface of the top wafer to the top surface of the bottom wafer. The dual optical measurement system provides accurate wafer thickness measurements independent of material properties, which is especially useful for patterned wafers, bumped wafers, GaAs, and other wafer substrates, including wafers after backgrinding and dicing [43].

29.3.4.5 Capacitance Probes

Noncontact, capacitance sensors are commonly used in wafer fabrication for wafer thickness, TTV, warp, and bow measurements [2]. A typical measurement utilizes of two capacitance probes, one on either side of the target being measured (Figure 29.26). The difference between the outputs of the two sensors is directly related to the thickness of the material being measured. By taking a differential measurement, the effect of the position of the material within the probe gap is canceled. If the target cannot be grounded, thickness can still be measured by synchronizing the sensor's amplifiers 180° out of phase, thus balancing the target charges and creating a net electrical potential value of 0 V at the target.

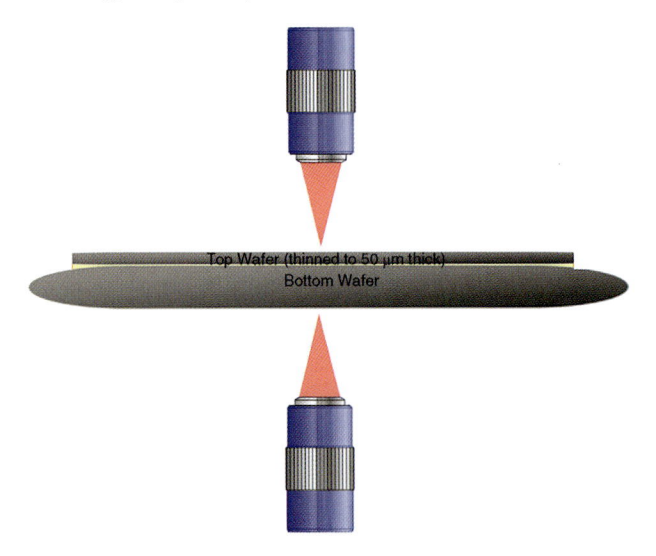

Figure 29.25 A dual-beam configuration for wafer thickness metrology. Courtesy of Chapman Instruments.

Single-sided thickness measurements can also be made using capacitive sensors if the backside of the material being measured can be referenced to some fixed plane. The product thickness is directly proportional to the gap between the probe and the surface of the material. Figure 29.27 illustrates a typical 825 μm bonded pair wafer bow map made with a single probe.

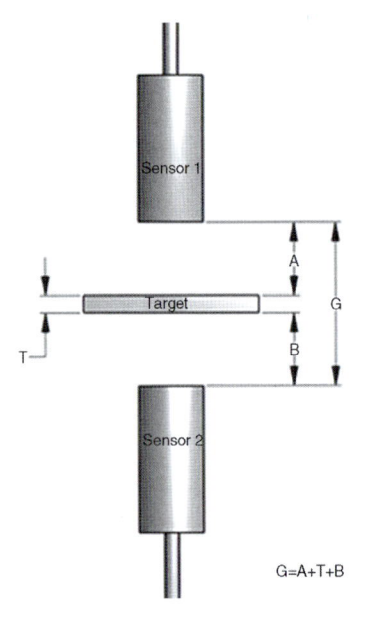

$$G = A + T + B$$

Figure 29.26 Typical thickness measurement using dual capacitance probes. Courtesy of MTI Instruments.

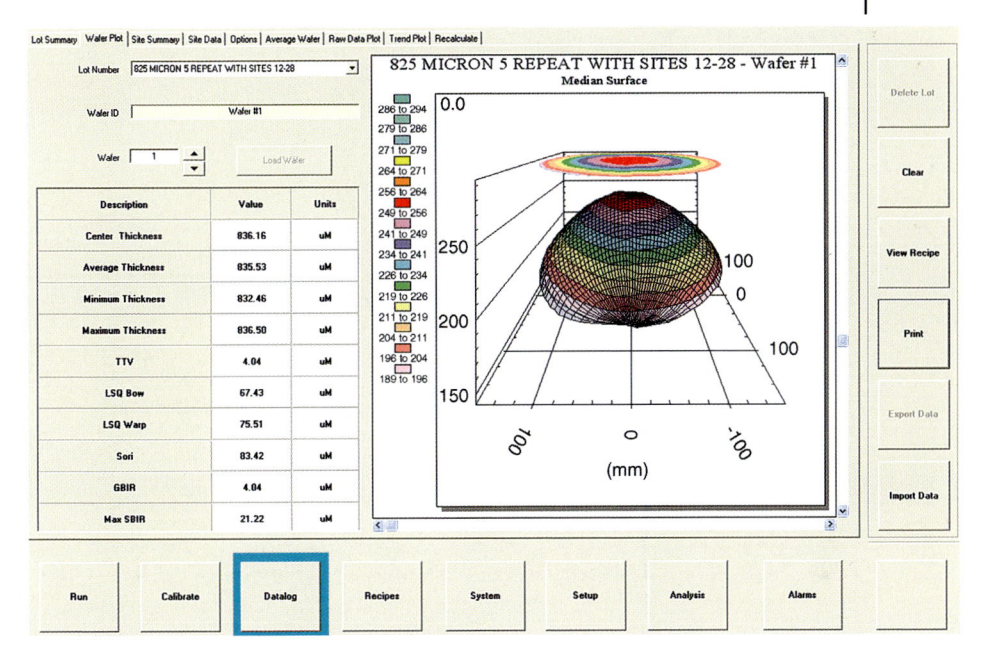

Figure 29.27 An 825 μm bonded pair wafer bow map.

29.3.4.6 Differential Backpressure Metrology

The dual backpressure sensor can be used to measure thickness bow and warp independently of any surface condition (smooth or rough) or material property (conductive or nonconductive) requiring only a low supply pressure (10 psi or less) and only controlled dry air (CDA) or nitrogen [44]. In this technique, backpressure is converted to voltage, a calibration curve is generated, and the operating point is determined on the curve. A reference backpressure measurement is made on a gauge block before every wafer measurement to compensate for ambient temperature variation effects.

29.3.4.7 Acoustic Microscopy for Measuring Bonded Wafer Thickness

With acoustic microscopy, the intrinsic data used to image the bond quality of bonded wafers also contain time/distance information. To image at a particular interface of interest, the time/distance to that interface can be displayed in an A-scan waveform format. The A-scan at each X–Y location can be processed to determine the thickness of each layer at that point and the surface contours of the front, internal interface, or back of the wafer stack. In addition, thickness of a layer within a wafer stack can be measured, as shown in Figure 29.28. In this image, the C-scan image is shown on the left, while example A-scan waveforms showing the full bonded wafer thickness at location #1, the delamination region at location #2, and defective regions within the stack at location #3 are shown on the right.

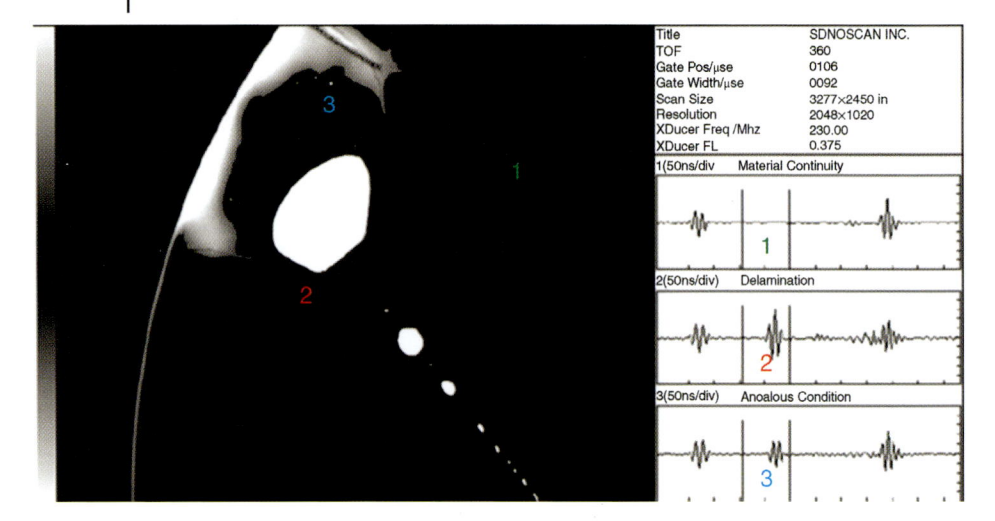

Figure 29.28 A scanning acoustic microscopy C-scan image of a stacked wafer showing three A-scan waveforms. Waveform #1 shows the complete thickness of the stacked wafer. Waveforms #2 and #3 show the distance from the top surface to the delamination and anomalous regions within the wafer stack. Courtesy of Sonoscan.

29.3.5
TSV Reveal Metrology

TSV reveal requires removal of most of a full thickness silicon wafer. A mechanical grind process, followed by a reveal etch, is used to remove this silicon revealing the copper pillars. The thickness and uniformity of the wafer are monitored to adequately control this process to ensure uniform TSV reveal height, which can be measured using methods such as white light interferometry [45,46], laser profiling [43], confocal chromatic imaging [47], laser triangulation [48], and profilometry [49]. In the case of the data shown in Figure 29.29, TSV wafer thickness was measured by IR interferometry, including backside measurements of TSV depth. In this plot, the initial wafer thickness and center-to-edge profile are shown after grind and CMP and prior to the TSV reveal etch process. The difference between the wafer thickness after the TSV reveal etch process and the TSV depth is the TSV reveal height [50]. This information is used to control the final wafer thickness and profile after thinning, so that the TSV reveal pillar height is optimized across the wafer.

If the TSVs are not revealed to adequate height, they may not make electrical contact to the redistribution layer (RDL) during bonding; if the TSVs are revealed to excess height, they may be susceptible to shearing during CMP.

Etch defects (surface roughness as well as angular, smooth, and circle defects) can be monitored using a combination of optical, SEM, and profilometry methods (Figure 29.30). SEM is useful in providing more detailed information than optical inspection, and profilometry provides height/depth information. Profilometry is fast, can scan large areas, and can accommodate z-axis deflections from 10 nm to 1 mm.

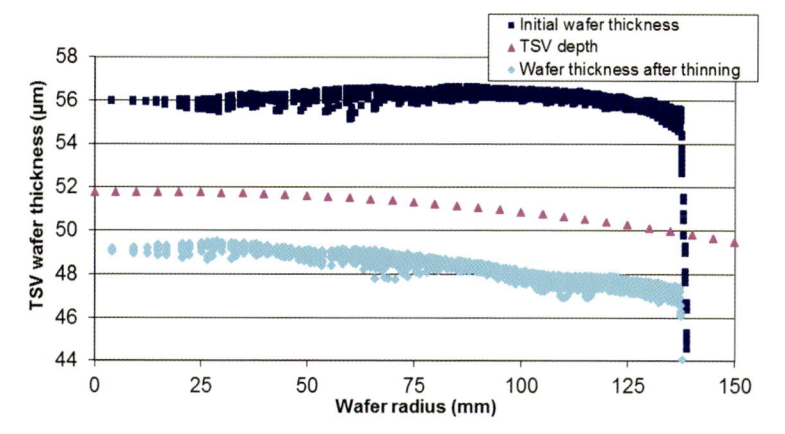

Figure 29.29 TSV wafer thickness and center-to-edge profile are monitored during the wafer thinning process. The plot shows initial wafer thickness before TSV reveal etch, TSV depth, and wafer thickness after reveal etch.

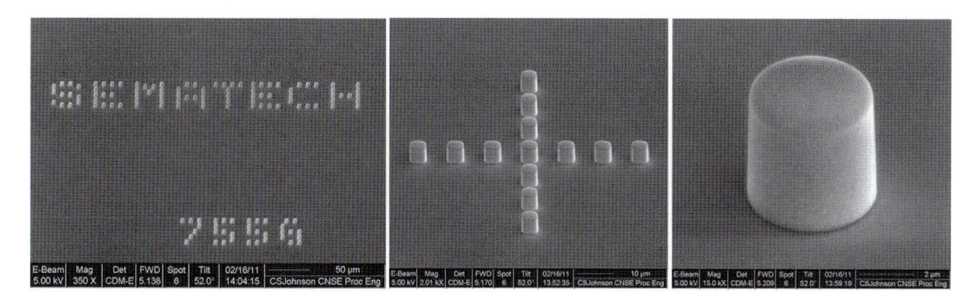

Figure 29.30 Tilt SEM images showing that the silicon surface is smooth, with no defects around the via. As TSV reveal is performed on the backside of the wafer, the image on the left has been reversed for readability.

Wafer inspection tools relevant to 3D interconnect metrology include dual-channel tools with scattered and reflected light imaging. With a large depth of focus, substrate depths up to 2 mm can be inspected; with multiple wavelengths possible in dark field, penetration depth is tunable. Applications include postgrind topography, edge chipping, and inspection of residual Si thickness and TSV reveal features, as well as voids, delamination, and embedded particles. Figure 29.31 shows surface contamination on an 850 μm thick bonded and thinned wafer pair obtained with dark-field illumination with Cu reveal pillars exposed at the wafer center (b) compared with the edge (c).

After reveal etch, a TSV reveal height map can be generated using optical microscopy. Figure 29.32 shows the across-wafer TSV heights for 31 sites using confocal chromatic microscopy. TSV reveal height ranges from 2.1 to 4.0 μm.

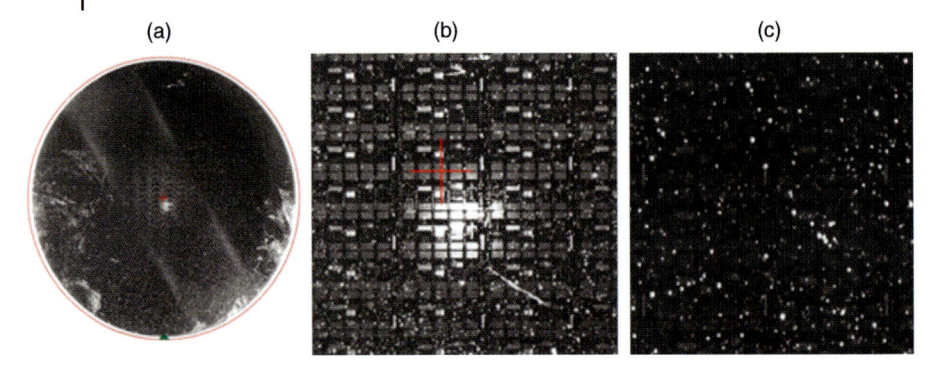

Figure 29.31 Surface contamination shown in dark field image (a) with exposed copper reveal pillars at center of wafer (b) and compared with edge (c). Courtesy of Nanometrics.

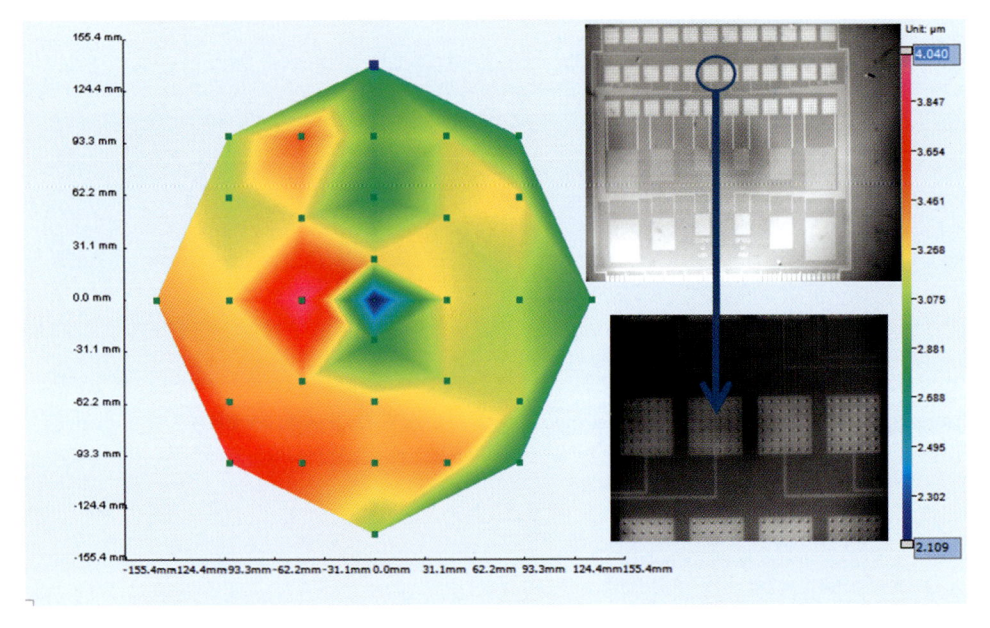

Figure 29.32 A map of TSV reveal height by confocal chromatic microscopy. Courtesy of Fogale nanotech.

29.4
Assembly and Packaging Metrology

This section includes the inspection steps that need to be performed at the back-end assembly facility to characterize the assembly processes and identify defective bond joints and underfill defects. Also included in this section is outgoing and incoming quality assurance inspection of the bumped and microbumped wafers, which needs to be performed at both the MEOL and back-end facilities.

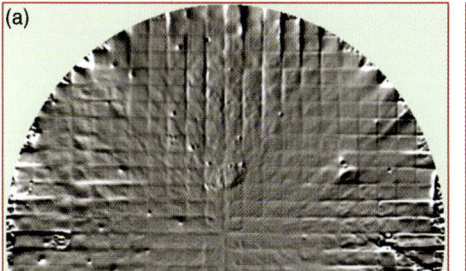

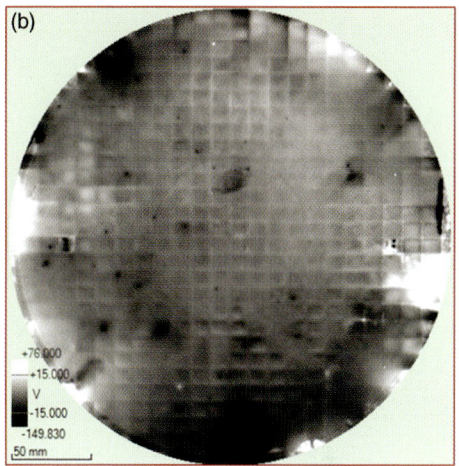

Figure 29.33 A C4-bump wafer exhibiting several nonvisual defect signatures. Courtesy of Qcept Technologies.

29.4.1
Wafer-Level C4 Bump and Microbump Metrology and Inspection

As 3D devices require many thousands of I/O connections per die, controlled collapse chip connection (C4) and microbump metrology and inspection are challenging due to the size, pitch, and large number of features. Microbump diameters can be as large as 20 μm and as small as 5 μm; future scaling will reduce these dimensions and reduce minimum pitch, allowing for higher feature density. Process control will require both metrology and inspection. An example of nonvisual defects (NVD) is shown in Figure 29.33, in which derivative work function images (a) depict surface nonuniformities that include both positive (white) and negative (black) work function signatures. These nonuniformities exhibit different shapes and sizes and may represent surface residues, process-induced charging of dielectric films, or some combination of the two.

Optical metrology and inspection tools can provide information on bump dimensions (diameter, height, and coplanarity) and defects such as missing bumps (Figure 29.34), bridging between bumps, double bumps, nodules, damage to bump(s), contamination including residue, foreign material, holes in passivation layer, particles, and oxidation. Figure 29.35 shows a bump defect classification map on a wafer containing more than 1.4 million microbumps using a 2D inspection system. In addition, 3D metrology can be performed using chromatic white light triangulation to determine bump CD, height, and coplanarity, and a wafer map can be generated as shown in Figure 29.36.

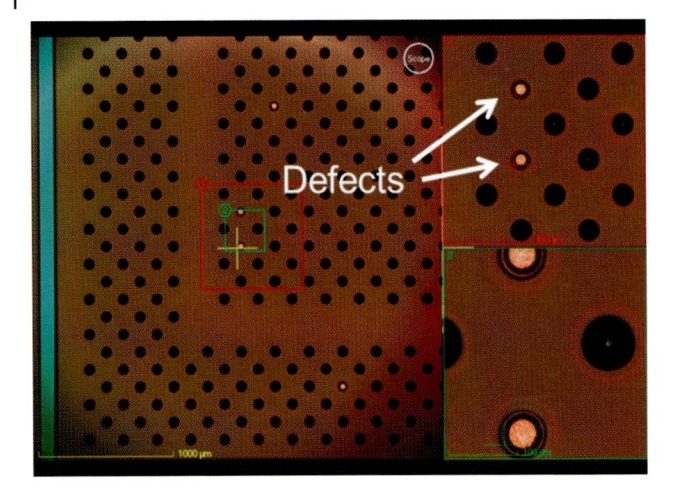

Figure 29.34 An optical microscope image of a C4-bump wafer showing defective or missing bumps.

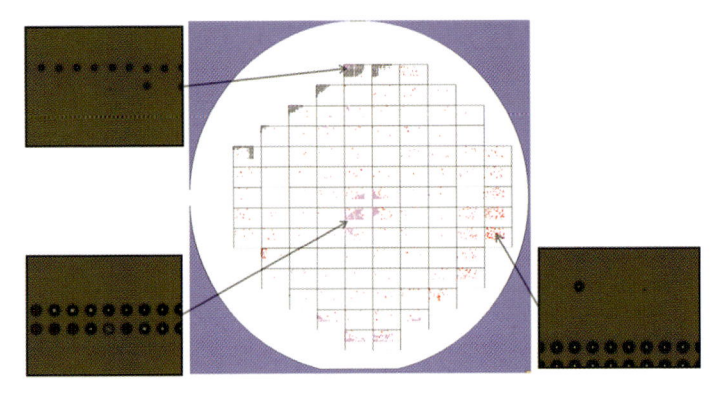

Figure 29.35 A microbump wafer defect map showing various bump defects. Courtesy of Camtek.

29.4.2
Package-Level Inspection: Scanning Acoustic Microscopy

An inspection step following underfill encapsulation is critical to ensure reliability of chip-to-chip bonding [51].[9]

A variety of defects can be detected using SAM including delamination of the package from the semiconductor device [52], and bond joint defects such as delamination of the underfill layer, particulates or voids in the underfill, and microcracks and striations of the underfill. Figure 29.37 shows scanning acoustic microscopy images obtained using a 200 MHz transducer with an 8 mm focal length, in a 20 mm × 20 mm scanning area, and 10 µm pixel resolution. The black

---

9) The data shown in this section on SAM and in the next section on X-ray inspection are obtained from a packaged die assembly, not a 3D stacked die assembly.

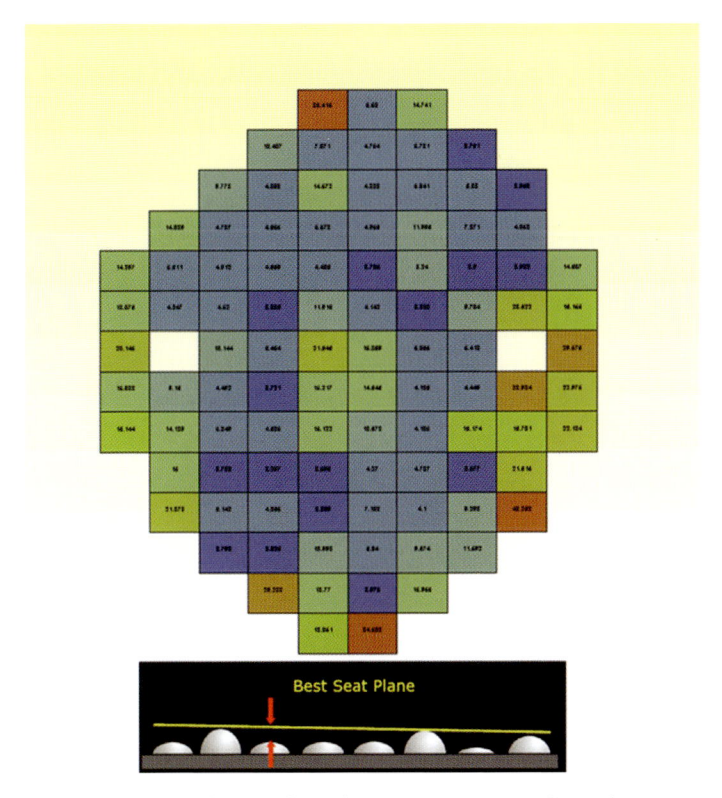

Figure 29.36 A microbump wafer coplanarity map. Courtesy of Camtek.

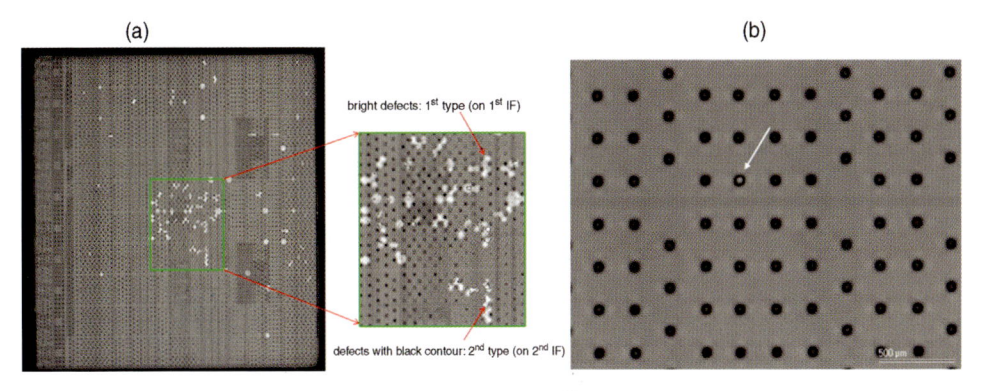

Figure 29.37 (a) A C-scan image showing underfill defects (white spots). Dark spots are defects at the first interface. (b) A C-scan image showing a defective bump (arrow). Courtesy of Endicott Interconnect Technologies (now i3 Electronics) and PVA Tepla.

spots correspond to defects at interface 1 (C4 to die interface), whereas the white spots are defects that penetrate into the underfill volume (Figure 29.37a).

Bump defects can also be detected by acoustic microscopy (Figure 29.37b). Broken bumps can be detected by comparison of the reflected acoustic signals from a defective and a defect-free bump.

These include voids in the underfill caused by the interplay between underfill viscosity and substrate wettability, as well as the number and density of bumps and their surface energy. Voids reflect ultrasound, generating a bright spot, so that no information below the void can be obtained. The sample must be inverted to obtain information from the other side; porosity in the underfill easily generates an ultrasound signature due to scattering. Underfill cracks scatter the ultrasound away from the detector and appear as dark areas.

29.4.3
Package-Level Inspection: X-Ray

X-ray microscopy, CT, and X-ray laminography can also be used for nondestructive package-level inspection and metrology of 3D stacks or to determine a location to perform cross-section SEM for failure analysis. They can also provide bump dimension and coplanarity measurements, identify open contacts in bumps (dry joints) due to insufficient solder or UBM, bridging/shorts due to excess solder, and detect voids due to gas bubbles in the solder. Bump alignment can also be verified by 2D X-ray metrology. Large arrays of bumps can be measured and statistically analyzed for size distribution and crack or void volume, as shown by the green areas in Figure 29.38. In addition, phase contrast is useful in X-ray imaging applied to underfill or in compositional differences in bump material. The main advantage of X-ray techniques is that silicon is transparent to X-rays, whereas copper and other metals are either translucent or opaque.

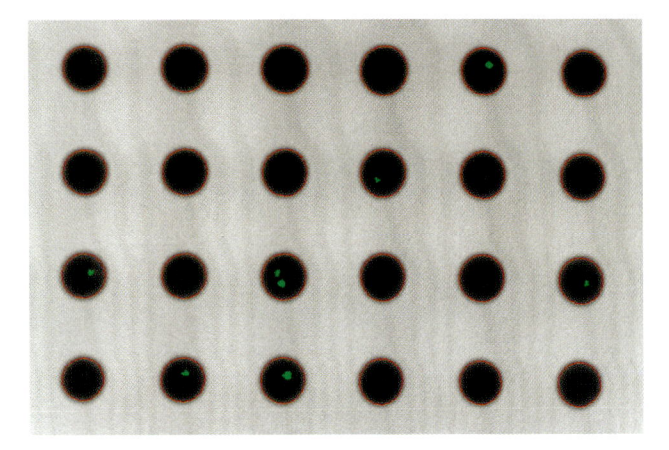

Figure 29.38 A 2D X-ray image of a C4-bump array, highlighting voids. Courtesy of Endicott Interconnect Technologies (now i3 Electronics).

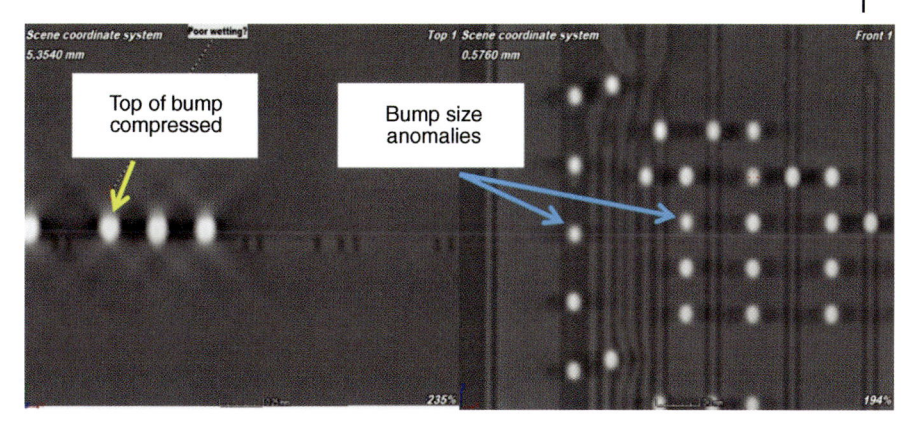

Figure 29.39 A 2D X-ray image of a C4-bump array, highlighting bump anomalies. Courtesy of Endicott Interconnect Technologies (now i3 Electronics) and Hadland Technologies.

Soldering imperfections fall into the following categories:

- Dry joints due to insufficient solder.
- Bridging/shorts due to excess solder.
- Voiding due to gas bubbles within the solder.
- Misplacement/misalignment due to inaccurate placement of components.

Joint inspection requires high resolution, sample tilt, and rotation of the device or the imaging system, as well as image processing software that can extract information about solder bumps such as dry joints due to insufficient solder (Figure 29.39), bridging/shorts due to excess solder, voids due to gas bubbles in the solder, and misalignment [53]. Bumps can then be selected for subsequent destructive analysis.

A full 3D rotation CT image may require up to 3000 2D images; thus, capturing such an image of a flip chip array can take many hours at the highest resolution (Figure 29.40). Recent advances in flat panel detector technology, source design,

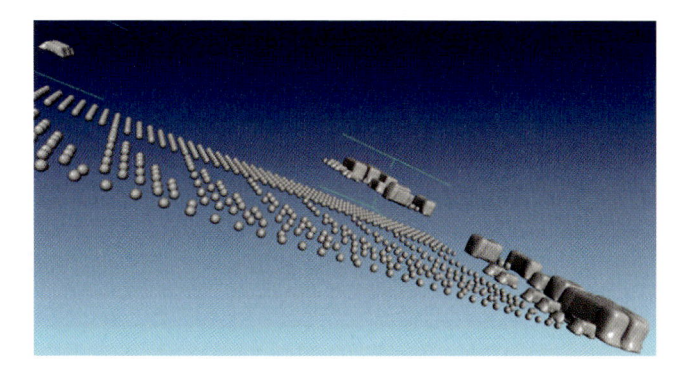

Figure 29.40 A 3D CT scan of a packaged die assembly. Courtesy of Endicott Interconnect Technologies (now i3 Electronics) and North Star Imaging.

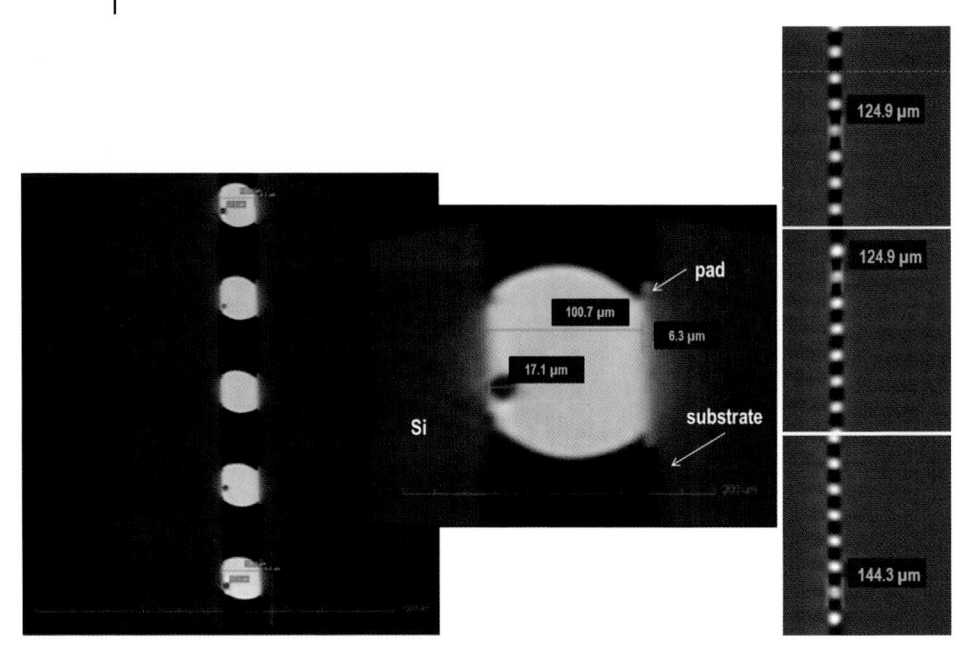

Figure 29.41 A virtual cross section showing C4 bumps (with voids) and pads in a packaged assembly. The enhanced contrast between the pad and the substrate is visible. Courtesy of Endicott Interconnnect Technologies (now i3 Electronics) and Xradia (now Zeiss).

image processing algorithms, and computer hardware have greatly improved the speed of CT acquisition so that today such images can be acquired in 1–2 h.

CT images shown in Figure 29.41 indicate that by using the appropriate X-ray energy, enhanced contrast between various similar, low-attenuation layer materials may be achieved. Void formation can also be observed in some of the bumps. No gaps between bumps and pads can be observed. The field of view (FOV) of the image was 1.2 mm × 1.2 mm, the voxel size was 1.2 μm, and the scan time was 4.5 h.

29.5
Summary

The evolution of 2.5D and 3D packaging for improvement in device performance (bandwidth and interconnect delay), power consumption, cost, and footprint requires new applications of metrology and inspection to characterize features and interfaces that are inaccessible to conventional metrology tools. These features are hidden in high aspect ratio TSVs, or in the bond interface layers between wafers. Although techniques for investigating bonds have been developed for MEMS technologies [54,55] and techniques for characterizing bumps have been developed for packaging and circuit board technologies [56], 2.5D and 3D integration deals with significantly smaller features and higher tolerance requirements than either of these applications. Metrology and inspection requirements for TSV fabrication, wafer thinning, TSV reveal processing, and chip-level assembly and packaging are formidable. Aggressive TSV diameter and pitch scaling, die thicknesses, bump diameters, and gap sizes, as well as the use of multitier stacks, place new demands on HVM metrology and inspection to identify TSV defects, defective bond joints and underfill defects.

Acknowledgments

The authors would like to acknowledge the following individuals who contributed to this chapter: Sitaram Arkalgud, Andy Rudack, Laurie Modrey, Sue Gnat (SEMATECH), Lay Wai Kong (GLOBALFOUNDRIES assignee to SEMATECH), Pete Moschack, Harry Lazier, and Elizabeth Lorenzini (CNSE assignees to SEMATECH), David Read (NIST, retired), and Yaw Obeng (NIST). In addition, the authors would like to acknowledge Paul Hart from Endicott Interconnect Technologies (now i3 Electronics) for supplying packaged die assemblies.

References

1 Diebold, A. (2001) *Handbook of Silicon Semiconductor Metrology*, Marcel Dekker, New York.
2 Schroder, D. (1998) *Semiconductor Material and Device Characterization*, 2nd edn, John Wiley & Sons, Inc., Hoboken, NJ.
3 Herman, I. (1996) *Optical Diagnostics for Thin Film Processing*, Academic Press, San Diego, CA.
4 Runyan, W. and Shaffner, T. (1998) *Semiconductor Measurements and Instrumentation*, 2nd edn, McGraw-Hill, New York.
5 Frontiers of Characterization and Metrology for Nanoelectronics, NIST, conference archives for 1995-2013: http://www.nist.gov/pml/div683/conference/index.cfm (accessed February 2014).
6 Arkalgud, S. (2010) Via mid through silicon vias – the manufacturability outlook. Proceedings of the 2010 International Symposium on VLSI Technology Systems and Applications, April 26–28, 2010, pp. 106–107.
7 Redolfi, A., Velenis, D., Thangaraju, S., Nolmans, P., Jaenen, P., Kostermans, M., Baier, U., Van Besien, E., Dekkers, H., Witters, T., Jourdan, N., Van Ammel, A., Vandersmissen, K., Rodet, S., Philipsen, H., Radisic, A., Heylen, N., Travaly, Y., Swinnen, B., and Beyne, E. (2011) Implementation of an industry compliant, $5 \times 50\,\mu m$, via-middle TSV technology on 300mm wafers. Proceedings of the Electronic Components and Technology Conference, pp. 1384–1388.
8 Zschech, E., Radojcic, R., Sukharev, V., and Smith, L. (eds) (2011) *Stress Management for 3D ICs Using Through Silicon Vias*, American Institute of Physics.
9 Yoon, S., Na, D., Choi, W., Kang, K., Yong, C., Kim, Y., and Marimuthu, P. (2011) 2.5D/3D TSV processes development and assembly/packaging technology. Proceedings of the 13th IEEE Electronics Packaging Technology Conference, pp. 336–340.
10 Teh, W.H., Marx, D., Grant, D., and Dudley, R. (2010) Backside infrared interferometric patterned wafer thickness sensing for through silicon via (TSV) etch metrology. *IEEE Transactions on Semiconductor Manufacturing*, **23** (3), 419–422.
11 Teh, W.H., Caramto, R., Qureshi, J., Arkalgud, S., O'Brien, M., Gilday, T., Maekawa, K., Saito, T., Maruyama, K.,

Chidambaram, T., Wang, W., Marx, D., Grant, D., and Dudley, R. (2009) A route toward production-worthy 5 μm × 25 μm and 1 μm × 20 μm non-Bosch through-silicon-via (TSV) etch, TSV metrology, and TSV integration. Proceedings of the IEEE International 3D Systems Integration Conference (3DIC), September 2009, pp. 1–5.

12 Novak, E. and Schmit, J. (2010) TSV metrology with white light interferometry microscopy. SEMATECH Workshop on 3D Interconnect Metrology, July 14, 2010. Available at http://www.sematech.org/meetings/archives/3d/index.htm (accessed February 2014).

13 Graves-Abe, T., Collins, C., Moore, V., Farooq, M., Hannon, R., Barak, G., Dotan, E., Aloni, A., Belleli, A., Rybski, M., and Petrucci, J. (2012) Novel method for TSV profile metrology using spectral reflectometry. SEMATECH Workshop on 3D Interconnect Metrology, July 11, 2012. Available at http://www.sematech.org/meetings/archives/3d/index.htm (accessed February 2014).

14 Höglund, J. (2012) 3D metrology by MBIR. SEMATECH Workshop on 3D Interconnect Metrology, July 11, 2012. Available at http://www.sematech.org/meetings/archives/3d/index.htm (accessed February 2014).

15 Cunnane, L., Kiermasz, A., and Ditmer, G. (2009) Characterisation of through silicon via (TSV) processes utilizing mass metrology. Proceedings of the 2009 ICEP.

16 Berry, M., Halder, S., Leunissen, P., Miller, A., Maenhoudt, M., Beyne, E., Kiermasz, A., and Ditmer, G. (2012) Using mass metrology for process monitoring and control during 3D stacking of IC's. SEMATECH Workshop on 3D Interconnect Metrology, July 11, 2012. Available at http://www.sematech.org/meetings/archives/3d/index.htm (accessed February 2014).

17 Rudack, A., Nadeau, J., Routh, R., and Young, R. (2012) Through silicon via (TSV) plating void metrology using focused ion beam mill. *Proceedings of SPIE*, **8324**, 832413.

18 Kalender, W.A. (2011) *Computed Tomography*, 3rd revised edn, Publicis Publishing, Erlangen.

19 Chao, W.L., Harteneck, B.D., Liddle, J.A., Anderson, E.H., and Attwood, D.T. (2005) Soft X-ray microscopy at a spatial resolution better than 15 nm. *Nature*, **435**, 1210–1213.

20 Gondrom, S., Schröpfer, S., and Saarbrücken, D. (1999) Digital computed laminography and tomosynthesis – functional principles and industrial applications. International Symposium on Computerized Tomography for Industrial Applications and Image Processing in Radiology, March 15–17, 1999, Berlin, Germany. Available at http://www.ndt.net/article/v04n07/bb67_11/bb67_11.htm (accessed February 2014).

21 Zschech, E. and Diebold, A. (2011) Metrology and failure analysis for 3D IC integration. Frontiers of Characterization and Metrology for Nanoelectronics. Available at http://www.nist.gov/pml/div683/conference/upload/zschech_2011.pdf (accessed February 2014).

22 Kong, L., Krueger, P., Zschech, E., Rudack, A., Arkalgud, S., and Diebold, A. (2010) Sub-imaging techniques for 3D-interconnects on bonded wafer pairs, in *Stress-Induced Phenomena in Metallization: 11th International Workshop* (eds E. Zschech, P.S. Ho, and S. Ogawa), American Institute of Physics, pp. 221–228.

23 Kong, L., Lloyd, J., Rudack, A., Yeap, Y., Zschech, E., Liehr, M., and Diebold, A. (2011) Visualizing and identifying the mechanism of stress-assisted void growth in through silicon vias (TSV) by X-ray microscopy and finite element modeling. SEMATECH Workshop on 3D Interconnect Metrology, July 13, 2011. Available at http://www.sematech.org/meetings/archives/3d/10125/pres/11%20-%203D%20Metrology%20SemiconWest_LWK.pdf (accessed February 2014).

24 McDonough, C., Capulong, J., Backes, B., Singh, P., Smith, L., Wang, W., and Geer, R. (2010) Profiling of process-induced stress in Cu through-silicon vias (TSVs) for wafer-scale, 3D integration. SEMATECH Workshop on 3D Interconnect Metrology, July 13, 2010. Available at http://www.sematech.org/meetings/archives/3d/8996/pres/Geer.pdf (accessed February 2014).

25 Trigg, A., Yu, L., Cheng, C., Kumar, R., Kwong, D., Ueda, T., Ishigaki, T., Kang, K., and Yoo, W. (2010) Three dimensional stress mapping of silicon surrounded by copper filled through silicon vias using polychromator-based multi-wavelength micro Raman spectroscopy. *Applied Physics Express*, **3**, 086601.

26 Kwon, W., Alastair, D., Teo, K., Gao, S., Ueda, T., Ishihaki, T., Kang, K., and Yoo, W. (2011) Stress evolution in surrounding silicon of Cu-filled through silicon via (TSV) undergoing thermal annealing by multiwavelength micro-Raman spectroscopy. *Applied Physics Letters*, **98**, 232106.

27 Nakatsuka, O., Kitada, H., Kim, Y.S., Mizushima, Y., Nakamura, T., Ohba, T., and Zaima, S. (2010) Characterization of local strain around through silicon via (TSV) interconnects by using X-ray microdiffraction. Advanced Metallization Conference, pp. 125–126.

28 Budiman, a., Shin, H.-A.-S., Kim, B.-J., Hwang, S.-H., Son, H.-Y., Suh, M.-S., Chung, Q.-H., Byun, K.-Y., Tamura, N., Kunz, M., and Joo, Y.-C. (2012) Measurement of stresses in Cu and Si around through-silicon via by synchrotron X-ray microdiffraction for 3-dimensional integrated circuits. *Microelectronics Reliability*, **52**, 530–553

29 Sood, S., Adams, T., and Thomas, R. (2008) Acoustic characterization of bonded wafers. *ECS Transactions*, **16** (8), 425–431.

30 Ramanathan, S., Semmens, J., and Kessler, L. (2006) High-frequency acoustic microscopy studies of buried interfaces in silico. Proceedings of the Electronic Components and Technology Conference, pp. 1865–1868.

31 Allen, R., Rudack, A., Read, D., and Baylies, W. (2010) Intercomparison of methods for detecting and characterizing voids in bonded wafer pairs. *ECS Transactions*, **33** (4), 581–589.

32 Horning, R., Burns, D., and Akinwande, A. (1992) A test structure for strength measurement and bond strength measurement and process diagnostics. Proceedings of the Electrochemical Society, pp. 386–393.

33 Charalambides, P., Lumd, J., Evans, A., and McMeeking, R. (1989) A test specimen for determining the fracture resistance of bimaterial interfaces. *Journal of Applied Mechanics*, **56**, 77–82.

34 Knechtel, R., Knaup, M., and Bagdahn, J. (2006) A test structure for characterization of the interface energy of the anodically bonded silicon-glass wafers. *Microsystems Technology*, **12**, 462–467.

35 MS5-0813 (2013), Test Method for Wafer Bond Strength Measurements Using Micro-Chevron Test Structures, SEMI, San Jose, CA.

36 Wyant, J. (1995) Computerized interferometric measurement of surface microstructure. *Proceedings of SPIE*, **2576**, 122–130.

37 Born, M.E. and Wolf, E. (1993) *Principles of Optics*, 6th edn, Pergamon Press, Oxford, pp. 36–51.

38 Hariharan, P. (2003) *Optical Interferometry*, 2nd edn, Academic Press, London, pp. 151–157.

39 Wyant, J. (2002) White light interferometry. *Proceedings of SPIE*, **4737**, 98–107.

40 de Groot, P. and Deck, L. (1993) Three-dimensional imaging by sub-Nyquist sampling of white-light interferograms. *Optics Letters*, **18** (17), 1462–1464.

41 R.A. Smythe, LLC. Available at www.RASmythe.com (accessed February 2014).

42 Schmit, J., Creath, K., and Wyant, J.C. (2007) in Surface profilers, multiple wavelength, and white light interferometry, in *Optical Shop Testing* (ed. D. Malacara), John Wiley & Sons, Inc., Hoboken, NJ, Table 15.1.

43 Chapman Instruments. Wafer Thickness: Available at http://www.chapinst.com/ApplicationNotes/MPT1000%20Thickness%20Measurement.pdf (accessed February 2014).

44 SigmaTech. Available at http://www.sigma9600.com/Products-services/Sigmatech-9600M.htm (accessed February 2014).

45 Montgomery, P., Montaner, D., Manzardo, O., Flury, M., and Herzig, H. (2004) The metrology of a miniature FT spectrometer

MOEMS device using white light scanning interference microscopy. *Thin Solid Films*, **450** (1), 79–83.

46 Darwin, M. (2009) Optical metrology for TSV process control. SEMATECH 2009 Workshop on 3D Interconnect Metrology, July 15, 2009. Available at http://www.sematech.org/meetings/archives/3d/index.htm (accessed February 2014).

47 Tiziani, H. and Uhde, H.-M. (1994) Three-dimensional image sensing by chromatic confocal microscopy. *Applied Optics*, **33** (10), 1838–1843.

48 Asgari, R. (2008) 3-D laser metrology: supporting micro-bump technology. Advanced Packaging, August/September.

49 Bennett, J. (2007) Light scattering and nanoscale surface roughness, in *Nanostructure Science and Technology* (ed. A. Maradudin), Springer, New York, pp. 1–33.

50 Olson, O. and Hummler, K. (2012) TSV reveal etch for 3D integration. Proceedings of the IEEE International 3D Systems Integration Conference (3DIC), pp. 1–4.

51 Iniewski, K. (ed.) (2012) Chapter 7, in *Nano-Semiconductors: Devices and Technology*, CRC Press, Boca Raton, FL.

52 Kim, S. (1991) The role of plastic package adhesion in IC performance. Proceedings of the 41st IEEE Electronic Components and Technology Conference, pp. 750–758.

53 Nikon, X-ray inspection of BGA, wirebonds, MEMS, loaded PCB: http://www.nikonmetrology.com/en_US/Applications/Electronics/X-ray-inspection-of-BGA-wirebonds-MEMS-loaded-PCB (accessed February 2014).

54 Suni, T., Henttinen, K., Lipsanena, A., Dekker, J., Luoto, H., and Kulawski, M. (2006) Wafer scale packaging of MEMS by using plasma-activated wafer bonding. *Journal of the Electrochemical Society*, **153** (1), G78–G82.

55 Cakmak, E., Dragoi, V., Capsuto, E., McEwen, C., and Pabo, E. (2010) Adhesive wafer bonding with photosensitive polymers for MEMS fabrication. *Microsystem Technologies*, **16** (5), 799–808.

56 Asgari, R. (2002) Semiconductor backend flip chip processing, inspection requirements and challenges. Proceedings of the 27th Annual IEEE/SEMI International Electronics Manufacturing Technology Symposium, pp. 18–22.

Index

a

acoustic microscopy 406, 407
- bonded wafer thickness, measuring 417
- measuring bonded wafer thickness
-- contamination issues 419, 420
-- defect inspection 407–408
-- high-volume manufacturing 400
-- KLARF format 407
-- locates defects, BWS map 408
acrylate 148
AD/DA converters 104
adhesion control agents 182, 183
- bulk layer WSS adhesive 182
- elimination 182, 183
- thin primer layer 182
-- wafer surface 182
adhesives 148, 149, 151, 152, 154–156, 178, 179, 182
- bonding 185, 262
- carrier separation, removal 171
- chemical resistance, for different chemicals 173
- damage chemicals 172
- experimental 182
- solubility test, at high temperature 171
- thermal characteristics 171
- warming 179
Advanced Semiconductor Engineering (ASE) 4
aligned bonding 299–302
- die-to-wafer bonding, pick-and-place equipment 299, 300
-- bonded chips on a 300mm wafer 300
-- chip-to-wafer (C2W) direct bonding 299
-- chip-to-wafer misalignment 300
-- microenvironment realized, illustration of 300
-- SET FC300 equipment 299
- die-to-wafer, self-assembly technique 300–302

-- alignment accuracy 302
-- based on 300
-- dummy silicon oxidized chip, process yield 300
-- hydrophobic 301
-- infrared observation, daisy chain after bonding 302
-- self-aligned C2W structures, observation of 301
-- SEM observation, bonding interface 301, 302
-- steps 300, 301
- wafer-to-wafer bonding 299
-- alignment marks, patterned wafers bonding 299
-- bonded pair misalignment 299
Amkor 9, 20
analog ICs 27
anneal temperature 263
Aptina Imaging Corporation 24
Asahi Glass Corporation (AGC) 51
ASET, research-focused consortium
- determine "parallel with offset" M1/M2 placing 108, 109
- developing TSV/3D process technologies 101
- developing W2W stacking process 103
- examine positions for M1 and M2 108
- organization for project 100
- research-focused consortium 99
- via-last TSV for, D2D and W2W Processes 103–105, 112
- working groups, functions 101, 102
assembly 9
ASTRI (Hong Kong) 20
atomic-scale flatness 263
Au–In bonding 271
automotive sensors 24
Avago 26

Handbook of 3D Integration: 3D Process Technology, First Edition.
Edited by Philip Garrou, Mitsumasa Koyanagi, and Peter Ramm.
© 2014 Wiley-VCH Verlag GmbH & Co. KGaA. Published 2014 by Wiley-VCH Verlag GmbH & Co. KGaA.

b

back-end-of-line (BEOL) processes 23, 263, 296
– low-temperature 388
backpressure sensor, configuration 417
backside imagers (BSIs) 23
backside processing, back-via process 232–234
– copper (Cu) contamination 233
–– possibilities of 234
–– removing steps 233
– fabricated Cu-TSV
–– first-level metallization (M1) 233
–– SEM micrograph 233, 234
– processing temperature 233
– three-dimensional integration process flow 232, 233
–– backside microbump formation 232, 233
–– copper (Cu) chemical-mechaincal polishing 232, 233
–– copper (Cu) electroplating 232, 233
–– deep reactive ion etching (DRIE) 232, 233
–– multilayer stacking 233
–– nitride/oxide film deposition 232, 233
–– oxide reactive ion etching (RIE), trench bottom 232, 233
–– Si substrate thinning 232, 233
–– support material 232, 233
–– temporary bonding, support material 232, 233
ball grid arrays (BGAs) 17
– package 41
bandwidth 32
baseband chips 10
BCB–BCB bonding 263, 264, 265
benzocyclobutene 262
BEOL-based wafer processes 51
BEOL damascene processing 89
BGAs. *see* ball grid arrays (BGAs)
bidirectional insulated gate bipolar transistor (BD-IGBT) 267
blade dicing 250–253
– blade 251, 252
–– electroplating, manufactured by 251
–– fine blade, finer diamond grit 252
–– hub blade 252
–– rough blade, rough diamond grit 251
– dicer 252
–– alignment process 252, 253
–– fully automated 252
– dual dicing applications 252, 253
–– dual cut 252
–– step cut 253
– method 250
–– bonding 250
–– definition 250
–– device layer 250
–– test element group 250
– point 250, 251
–– basic items, dicing process 251
–– cut position 251
–– ultrathin diamond blade 250
–– working 250
– process control, optimization of 252
–– balance productivity, importance of 252
blanket metal direct bonding principle 302–304
– bonding energies
–– double cantilever technique 302
–– metal bonding for different temperatures 303
– bonding tests 302
– chemical-mechanical polishing (CMP) 302
– Cu–Cu bonding 302
– different metal bonding after a 400 °C annealing 298
–– scanning acoustic microscopy 302, 303
– getter effect 303
–– titanium 303
– metal bonding interface, TEM observation 304
– metal surfaces including think oxide layer
–– confirmed by 302
–– measured by 302
– oxide dissolution 303
– SiO_2–SiO_2 bonding 302
– thickness, metal oxide layer before bonding 302
– Ti–Ti bonding 302
– W-W bonding 302
blanket Si wafer, Cu TSVs 369
bond 176, 350–352
– LTHC coating, release layer 176
– metrology 147
–– package-level 424
– processes 152–155
– program 140
– quality characterization 267–269
– reliability 269
–– tests 268
– strength 268, 409
–– characterization 267
–– delamination 401, 402
–– management 147
–– resolution 412
– surfaces, impacts of 264

- UV-curable adhesive 176
- voids 406
-- bonded wafers 406
-- bonded wafer stacks 406
-- resolution 412
bonded CuSn microbumps, SEM image 388
bonded wafer
- Bow 410
- classified, temporarily bonded 411
- pairs 407
-- bow map 417
-- edge-trimmed 411
-- temporary bonded 411
-- thinned 411
-- total stack thickness contour map 413
- stack
-- acoustic microscope-identified defects 406
-- defects (voids) 408
-- process tool contributions 408
- thickness 410
- warp 410
bonded wafer thickness, A-scan waveforms 417, 418
bonder tool processing in situ 183
bonding equipment 140
- contamination 141
Bosch process 346–347
brittle bonding materials 268
BSI CMOS image sensors 23, 25
BSI sensors 24
bump 18
- damage 144
- D2D/W2W structure 105
- topography 169
- under-bump metallization (UBM) 354
bumped wafers
- thickness 179
bump interconnect 313
- ductile solder 313
- early development 313, 314
- face-to-face electrical connection 313
- IBM 313
- primary bump structures 314
- variants for 314
-- 2.5D integration 314
-- 3D integration 314

c

capacitance 80
capacitance probes 415, 416
- bow measurements 415
- measurements, single-sided thickness 416
- using, typical thickness measurement 416

- wafer thickness 416
capacitance–time (C–t) analysis 381
- configuration of 381
- obtain, minority carrier
-- generation lifetime 385
- use, MOS capacitor 381
carrier wafer 147–149, 152, 153
- material selection 148
cartesian coordinate system 407
C4 bump 421
- array 2D X-ray image 424
- wafer exhibiting 421
- wafers, showing defective 422
CEA-Leti (France) 19
chemical debonding 136
chemical-mechanical grinding (CMG) 214
- grinding wheels 214
-- ceria particles (CeO_2) 214
-- chemically active abrasives 214
chemical–mechanical planarization 80
chemical–mechanical polishing (CMP) 170, 171, 178, 296–298, 383
- adhesives 170, 171
- based preparation 297
- blanket Cu–SiO_2 direct bonding principle 296–298
- bonded pair ground down 298
-- after 200 °C postbonding anneal 298
- bonded patterned wafers, acoustic image of 298
-- after 400 °C anneal 298
-- at room temperatures 298
- CEA-LETI 297
- copperpads vertical displacement at 200° C, simulation of 297, 298
- damascene copper 296
- dishing controlled on copper pads 296, 297
- EVG SmartViewTM alignment tool 297
- optimizations 297
- TEM image, bonded copper pad 297
chemical vapor deposition (CVD) 80
Chinese Academy of Science (China) 19
chip-level approaches 16
chip-on-chip (CoC) 280
- challenges 280
chip-on-chip-on-substrate (CoCoS) 283, 284
- alternative assembly 283
- C4 bumps 283
- CoC-first 283
- CoS-first 283
-- three-dimensional assembly 283, 284
- vs. chip-on-wafer-on-substrate (CoWoS) 283, 284

chip-on-wafer (CoW) 351
– chip stacking 360
– – silicon-to-silicon joint first 360
– demonstration, large die size CoW 351
– homogeneous multichip integration 351
– stacking 351
chip-on-wafer-on-substrate (CoWoS™) 280
– key enablers 280
– process flow 280
– – steps 280
– Si interposers, extremely large 285
– stacking 283
– – application 285
– substrate stacking, completion of final 281
– traditional flip-chip packaging processes 283
– *vs.* chip-on-chip (CoC) 280
– *vs.* chip-on-substrate (CoS) 280
chip-on-wafer (CoW) process 280
– stacking 280
chip-on-wafer stacking 281–283
– face-to-back (F2B) 281
– face-to-face (F2F) 281
– nonflow underfill (NUF) 281
– post-CoW test result 282
– process flow, C4 bumps 282
– under-bump-metallization (UBM) 281
– wafer-level underfill (WLUF) 281
chipping 135, 142
chip-scale packaging (CSP) 17
chipsets 37
chip-to-chip (C2C) bonding 79
chip-to-chip interconnection
– diminishing solder microbump dimension 361
chip-to-wafer (C2W) bonding 79
chip-to-wafer 3D integration 325
– known good dies (KGDs) 325
– traditiona pick-and-place chip assembly
– – *vs.* surface-tension-driven multichip self-assembly 326
cleanliness, higher level 248, 249
– chemical-mechanical polishing (CMP) 248
– – particle adhesion to ground wafers 248
– silicon particles, to remove 249
– – brush cleaning 249
– – chemical cleaning 249
– wafer, wet conditions 248
clock distribution layers 104
cluster analysis 407
cluster software 407
CMOS image sensors (CISs) 22, 29, 67
CMOS scaling 14
CMOS technologies 13, 14

CMOS transistors 17
CMP polishing 211, 212
– conventional CMP process 211
– grinding cluster tools 212
– removal rate 211
– results in 212
– soft rotating brushes (scrubber) 212
– used as 211
CoC bonding 266, 270
coefficients of thermal expansion (CTE) 143, 280, 351, 365, 387
– microbump joints 351
– mismatch 143
– – traditional flip-chip bonding 351
compound annual growth rate (CAGR) 17
computed tomography (CT) 400–402
conductivity 51
contaminants 139
conventional IC package 42
CoO. *see* cost, of ownership (CoO)
co-optimization of unit processes
– in backside processing and via-reveal flow 89–91
– in via formation sequence 88, 89
copper 51
– protrusion 89
copper (Cu) bumps 319, 320
– bonding methodologies, practical application 319
– bonding yield 319
– electroplating process, formed in 319
– Fujitsu precision Cu bump-cutting process 319, 320
– high bonding pressures 319
– lower temperature bonding 319
– surface preparation 319
– thermocompression bonding 319
– – at 300 °C 319, 320
copper fill 399
– TSV voids 399
copper–oxide surface direct bonding 295
– advantages 295
copper (Cu) pillar bumps 316–319
– advantages 317
– 2.5D applications 317, 318
– – 45μm pitch Cu pillar bumps 317
– – developed by 316
– forming process similar to 316
– – electroplated solder bumps 316
– pitch and assembly methodology 317
– primary variations 317
– – flip-chip (FC) Cu pillar bumps 317
– – TC Cu pillar bumps 317

– replacements for 316
– solder layer
– – Cu–Sn intermetallic 318
– – intermetallic phases 318
– – thickness 318
– solid–liquid interdiffusion (SLID) bonding 318
– – using Sn 318, 319
– standoff height 318
– – assembly yield, impact on 318
– structure illustration 316
– thermocompression bonding 318
– underfill 318
copper–tin SLID system 266
corrosive chemicals 141
cost 17
– of future scaling 34–37
– management 362
– – effort 3D development 362
– – improve, production yield 362
– of ownership (CoO) 122, 347
– reduction 273
– remain impediment to 2.5D and 3D product introduction 37, 38
CoW bonding 266, 271
– advantages 271
CoW 3D integration platform 270, 272
CoW-first process approach 362
CoWoS^{TM} (chip-on-wafer-onsubstrate) technology 6
– assembly 357
– conventional fcBGA 357
– integrated design 362
– package 358
– – finished, demonstration 358
crack opening method 267
crystal grain orientation mapping 404
C4 solder bumps 315, 316
– fabrication of 315
– implementation limitations 315
– Si interposer (2.5D) applications 315, 316
– vertical stacking applications 315
CTE. see coefficients of thermal expansion (CTE)
Cu atoms 379, 380
– barrier layer 384
– diffuse 381, 386
– Si substrate 383
Cu–BCB hybrid bonding 265
Cu concentration profiles 380
Cu contamination 379–386
– impacts of 379
– – 3D integration 380

– reduction 386
– thickness 379
– – mechanical grinding 379
– – stress-relief polishing methods 379
Cu–Cu bonding 262, 263, 265, 272
– TEM cross section 268
Cu diffusion 380
– backside surface 382
– behaviors 383
– effect 380
– IG layer 382
– layer 383
– PG-treated wafer, TEM cross-sectional image 384
– thickness 382
Cu extrusion
– impacts, BEOL integrity 348
– modeling 349
Cu–liner adhesion mechanism, improvement 369
Cu–PECVD oxide hybrid bonding 265
Cu pillars 17
Cu protrusion 365, 402
– avoiding 365
– 3D AFM profile 403
Cu pumping
– bulging, Cu TSV surface 369
– Cu adhesion
– – liner-side wall 369
– Cu TSV CMP surface 369
– experimental method, representation 369
– measured maximum 371
– measured, statistics function 370
– mitigation of 368–371
– observe, amplitude 371
– postplating anneal temperature 370, 371
– postplating anneal time, impact of 371
– postsinter temperature 371
– test methodology 369
Cu–Si CTE difference 365
Cu–Sn IMC bonding 264
Cu/Sn microbumps 189
– SEM image 388
Cu–Sn system 266, 271
Cu stress 365–368, 371
– creep deformation hypothesis 371
– CTE difference 366
– high tensile stress 366
– hydrostatic compressive stress 366, 367
– illustration of effect 368
– Si substrate 366
– slope 366
– temperature 366

–– cycling 368
– tensile 366
Cu TSV
– 3D LSI fabrication 384
– formation 384
– in Si interposer layer 66
– technology 384
CVD. *see* chemical vapor deposition (CVD)
C–V plot, minority carrier relaxation 381

d

3D assembly, manufacturing solutions 362
data processing system 100
data traffic growth predictions 99
3D bonding approaches 262
3D chip penetration, of IC semiconductor market 30
3D chip stacking 191, 360
– integrated passive device 360
– interposer 360
– microbump joints 360
– RDL 360
– through-silicon via (TSV) connections, creation of 191
– TSV 360
3D CoWoS stacking, integration 357
3D DRAM 385
– data retention 385
– stack, memory chips 385
2.5D/3D stacking, infrastructure movements 100
D2D stacking technology 103, 168
debonding 135, 136, 147, 150, 153, 176
– cleaning 155, 156, 185, 188, 189
– cluster 155
– debonded thinned device wafer on a tape frame 155
– force as displacement function 154
– force *versus* roller displacement 154
– integrated thin wafer cleaning inside a debonding cluster 155
– parameters, in slide-off debonding 144
– processes 152–155, 178
–– flow for 148
–– raster, standard IR marking laser 178
–– remove, WSS adhesive 178
– using 3M wafer de-taping tape 176
– using release layers 137
– wave line 154
– ZoneBOND room-temperature debonding 163–165
–– EZR step 164, 165
decontamination 139

defect cluster analysis 407
delamination, high-temperature processing 180
design–foundry–package–test 358
device-level approaches 16
device reliabilities 379–386
– Cu diffusion effect 381
– 3D LSIs 381
– local stress, impacts 386
3D fabrication 345–364
– bonding, underfills 350–352
– constituents 359
–– C4 bump joint 359
–– microbump joint 359
–– TSV 359
– cost management 362
–– microbump pitch shrinkage 362
– future perspectives 362–364
–– 3D chip stack packaging 363
–– FEOL transistors 362
–– technology node shrinkage 362
– multitier stacking 352
– reliability 360, 361
– strata packaging 356–359
– yield management 359
3D IC application roadmaps 10–11
dicing 250–260
– blade dicing (*see* blade dicing)
– 3D integrated chips, through silicon via (TSV) device 257–260
–– blade dicing 258 (*see also* blade dicing)
–– cutting major issues 258
–– cutting wafer into 257
–– stealth dicing (*see* stealth dicing)
– laser dicing (*see* laser dicing)
– low-k dicing 254
–– dielectric constant 254
–– low-*k* oxide, mechanical strength 254
– thin wafer dicing 253, 254
–– blade loading, cutting 254
–– die attach film (DAF) 254
–– die strength 254
–– process condition, examples 254
–– test element group (TEG) 254
3D IC IP activities 20
– academic institutions 20
– 3D IC patents 21
–– academic assignees 22
–– patent assignees 21
3D IC nomenclature 1, 2
3D IC packaging 9
3D IC technology, geographical distribution of patent filing for 20

die-to-wafer integration scheme 296
– oxide bonding 296
diffusion 263
digital and analog 3D integration technology 101
3D integrated chips, through silicon via (TSV) device 257–260
– blade dicing 258 (*See also* blade dicing)
– cutting major issues 258
– cutting wafer into 257
– – die 257
– – encapsulated packages dicing 257
– – stacked die 257
– stealth dicing (*See* stealth dicing)
3D integrated passive devices (IPDs) 26
3D integrated products, application development 31
3D integration 18, 172, 346, 362, 385, 393. *See also* 3D through-silicon via (TSV) technology
– challenges 346
– cost-adding factors 362
– CoW 351
– high-volume manufacturing 346
– – chip 346
– – collapse chip connection (C4) bumps 346
– – wafer bonding 346
– – wafer processes 346
– links, cost management 362
– reliability (DfR), design 360
– stacking technology 346
– technologies 4, 14, 16, 30, 101, 346
– – platforms 262
– through-silicon via (TSV)-based chip 346
– vertical stacking 386
– wafer-to-wafer 393
– yield (DfY), design 360
– yield-harmful 362
2.5D interposers 29
– major application drivers for 43
direct bonding 262, 264
direct copper experiments 296
direct wafer bonding 295
direct-write deposition, of redistribution layers 126
– characteristics 127, 128
– laser-induced forward transfer 128–130
– LIFT results 130, 131
– redistribution layers, overview 126
2.5D packaging architectures
3D performance advantages 33
3D power/RF/analog passive interposers 28
drain current variation
– device 374

– measure 374
DRAM cell
– array 386
– Cu diffusion 386
– failure rate of 386
DRAM chip 7
DRAM memory 79
dry polish (DP) 213
– development, basic idea of 213
– rotation geometry, schematic 214
– treated wafer, immunity 383
2D SoC *vs.* 3D IC comparison 15
3D stacking 16, 79
– production 357
– strata design 358
3D through-silicon via (TSV) technology 287
– mobile platforms, solution development 289–293
– – chemical–mechanical polishing (CMP) 291
– – film-assisted molding (FAM) technology 292
– – major challenges 289
– – outsourced semiconductor assembly and test (OSAT) 289
– – plasma-enhanced chemical vapor deposition (PECVD) 291
– – plated microbumping process 290
– – process flow and capabilities, STATS ChipPAC 289
– – stealth dicing process 292
– – temporary bonding/debonding processing 291
– – thinning illustration 291
– – via reveal, illustration 291
– package test vehicle 290
– process flow 290
– STATS ChipPAC 289, 290
– thermocompression bonding process 292
– – fine-pitch interconnection 292
2.5D TSV integration 292, 293
– alternative approaches 294
– – eWLB 294
– – fan-out wafer level packaging (FOWLP) 294
– process flow 293
– – large thin through-Si interposer (large TSI) 293
– STATS ChipPAC 292
3D TSV integration 22, 23, 30
– applications using 22
3D TSV market value 30
3D WLCSP platforms 29
D-W techniques 127

e

ECD. *see* electrochemical deposition (ECD)
economics
- for interposer use in mobile products 38
- of scaling 33, 34

edge bead 140
edge cleaning 142, 143
- exposed temporary bonding material after backgrinding 142
- temporary bonding material
-- carrier wafer after high-temperature treatment 143

edge trimming 142, 247, 248
- chipping 247
- grooves 247
- pretrimming before coating and bonding 142
- process diagram 247
- profiles, without 247
- straight wafer edge 247

Edge Zone Debond (EZD®) step 164
effective piezo-coefficients 372
electrical characterization 304–310
- die-to-wafer (D2W) copper-bonding 304–307
-- bonded die, deported electrical characterization probes 305
-- daisy chains 304, 305, 307
-- NIST (stand-alone and bonded) 304, 305, 306
-- SiN encapsulation 305
-- tested dice full integration 304
-- *vs.* wafer-to-wafer bonding 305, 307
- package-level electromigration test 309, 310
-- achieved on 309
-- Cu–TiN interface 309
-- daisy chain 309, 310
-- stress and matter concentration 310
-- time to failure (TTF) 309
- reliability 307
- stress voiding (SIV) test 308, 309
- thermal cycling 307, 308
- wafer-to-wafer copper-bonding 304–307
-- NIST (stand-alone and bonded) 304–306
-- tested dice full integration, schematic diagram 304

electrochemical deposition (ECD) 80
electromigration 320–322
- cross-sectional analysis 321, 322
- damage in SnPb solder bump 320, 321
- Darveaux data 321
- improved performance 321
-- Cu pillar bumps 321
-- lead-free solders 321
- performance comparison
-- solder bump alloys *vs.* Cu pillar 321

electromigration performance 361
electromigration (EM) test 269
electroplating 142
elephant foot 347
- nonuniformity, etching process 347
Elpida Memory 19
end-of-line (FEOL) processes 263
energy-dispersive x-ray spectroscopy (EDS) 398, 399
equipment
- development 155
- and process integration 155
- SUSS MicroTec's XBS300 temporary bonding platform 156
etching 82, 83
ETRI (Korea) 19
European e-BRAINS consortium 31

f

face-to-face bonding, wafer stacking 388
FBAR filters 26
FEOL devices 371–377
FET array test
- impact TSV diameter scaling 374
- K-factor 373
- KOZ size 374
- measurements 373
- parameter K 374
- structures 373
FIB milling 400
field programmable gate arrays (FPGAs) 269
- devices 28, 79, 101
- products 287
finite element analysis (FEA) 409
- bond strength 409
- depends precrack, description of 409
- wafers, composition 409
flip-chip C4
- assembly 356
- bump packaging 360
- high-yield assembly 360
flower-shape geometry 373
fluidic self-alignment process 335, 336
- chip position, final 335
-- target area for 335
- metal patterns, preparation of 335
- polar assembly liquid 336
-- droplets of 336
- principal process flow 335, 336
- silicon chips, thin 335

flux clean issue 351
– large die 351
focused ion beam (FIB) 399
foundries 6
four-point bend test 267
– structure 409, 410
– use, test coupon 409
– wafers 373, 374
fracture tests 214, 215
– bending geometry, FEM modeling 215
– breaking force values, measurement of 215
– four-point bending tests 214
– fracture strength, thinned silicon 214
–– experimental determination 215
– mathematical formualtion 215
– ring-on-ring tests 214, 215
– thin silicon wafers 214
– three-point bending tests 214, 215
Fraunhofer-Gesellschaft München (Germany) 19
front-end-of-line (FEOL) process 67
front-side topography 354
front-side TSV-last approach 69, 70

g
gaming applications 23
generic process flow 192
– wafer backside processing 192
– wafer thinning processing 192
glass interposers
– challenges 49
– cost-reduction potential 48
– glass interposer fabrication with TPV and RDL 53
– lower cost interposers 48, 49
– metallization of glass TPV 51, 52
– reliability of copper TPVs in 52, 53
– small-pitch through-package via hole formation 49–51
– thermal dissipation of glass 53
– with through-glass via for hermetic MEMS 3D integration 60
– ultrathin glass handling 49–51
GlobalFoundries (GF) 7
glue layer delamination 354
– stress concentration 354
grinding 140, 142, 241–250
– 3D integrated chips, through silicon via (TSV) devices 246–250
– fine 242, 243
– grinder polisher 243
–– chuck table 243
–– grinding point 243, 244

–– spindles 243
–– work flow 244
– method 241, 242
–– back grinding (BG) 241
–– creep feed method 241, 242
–– in-feed method 241, 242
– rough 242, 243
–– surface protection tape 242
– thinning 243–246
–– edge chipping 245
–– grinding damage 245
–– low mechanical strength 244
–– wafer breakage, causes (See wafer breakage)
–– wafer warpage 243–245
– wheel, illustration 243
grinding capability 172

h
handle wafer edge grind, SEM image 405
HB-LED submount integration on silicon interposer with conformal TSVs 60
heterogeneous 3D integration technology 14, 16, 31, 101
heterogeneous 3D stacking system 101
heterogeneous integration 31
high-bandwidth interconnections 41
high-density silicon interposers with TSVs 60
high-end BSI sensors 23
high-speed I/O devices 101
high-temperature material solutions 178, 179
high-volume manufacturing (HVM) 29, 345, 346
hybrid bonding 262–264, 263, 265, 273, 274
hybrid memory cube (HMC) 8, 9, 31, 59, 99
– consortium 31
– electrical performance 9
hydrophobic bonding 265
Hynix 9

i
IC fabrication 65, 167
image processing software 425
image sensors using TSVs 269
IMEC (Belgium) 19
impurity gettering, backside processing
– chemical-mechanical polishing (CMP) 235, 236
– copper (Cu) contamination 234
–– silicon nitride (Si_3N_4) film deposition 234
– copper (Cu) diffusion 234
–– prevention 235
– 3D LSI fabrication, importance 235

- epiwafers p-on-p$^+$ structure 235
- extrinsic gettering (EG) 235
-- back-via process, formation of 236
-- crystal defects 235
-- 3D LSIs implementation, challenge 235
-- dry polish (DP) 235
-- mechanical grinding 235
-- Poligrind (PG) 235
-- UltraPoliGrind (UPG) 235
-- via-middle process, formation of 236
- intrinsic gettering (IG) 235
- LSI wafer/chip backside thinning 234
- plasma etching 235, 236
-- thinning 230, 231, 236
- stress-relieved wafers
-- defect depth 235
-- remnant stress 235
-- stress field region, HRTEM image 235
- Young modulus change
-- Si wafer thickness, reducing 236, 237
indirect bonding 262, 264
infrared (IR) interferometry 414
- sensor 416
- using TSV depth map 397
Innovative Micro Technology (IMT) 26
inorganic dielectrics 179
- LC-5200-F11 179
- LC-5200-F1035 179
integrated circuit 1, 3
integrated MEMS systems 32
integrated modeling design approach 360
integrated thin wafer cleaning
- inside a debonding cluster 155
integration 88, 89, 279
- and co-optimization of unit processes in via-last flow 91, 92
- heterogeneous 279
- homogeneous 279
- with packaging 92
- process performance 172
Intel 19
interferometry 414
- TSV reveal height, map 418–420
intermetallic compound (IMC) formation 361
interposers 4–6, 9, 10, 41, 42
- applications 58
- configurations 6
- 3D glass interposer for MEMS integration with ASIC 59
- EDA tools to design with interposers 58
- fabrication 81
-- via-middle TSVs 81
-- via-reveal flow 82

- forecast by application 29
- glass interposers for RF applications 59
- IBM high-performance 2.5D silicon interposer 58, 59
- integration 57
- manufacturing supply chain readiness 57, 58
- materials, comparison of 44, 45
- NEC high-bandwidth SMAFTI 3D interposer 59
- silicon interposer for high-brightness LED submounts 59
- structures 42
- technical and manufacturing challenges 56, 57, 79
- testability, and test cost 57
- thermal management 57
- in various regimes 61
I/O interconnect gap 17
ISIS Sentronics 140
ITRI 19
- 2009 ITRS roadmap 3–4

k

KAIST (Korea) 19
keep-out zone (KOZ) 371, 373, 375–377, 402
- approximation, piecewise linear approximation 375
- area 373, 375, 376
- cross-geometry, defined 375, 376
- parameters formulas, overview 377
- reduction of 373
KGDs. *see* known-good-dies (KGDs)
KGS. *see* known good stacks (KGS)
known-good-dies (KGDs) 270, 284, 285
known good stacks (KGS) 284, 285
- design-for-testing (DfT) 285
- 3D TSV integration 284
-- *vs.* traditional packaging technology-based 3D integration 284
- testing 284, 285
KOZ. *see* keep-out zone (KOZ)

l

lame equation, for silicon substrate 372
lamination 138
laser dicing 254–257
- ablation 254, 255
-- process 254, 255
- laser full cut application 255
-- beam shaping system (BSS) 255
-- laser ablation, processing example 255

– – vs. blade dicing die strength 256
– stealth dicing (SD) 256
– – kerf width 256
– – perfect dry process 256
– – process 256, 257
– – vs. blade dicing die strength 257
– used for 254
laser drilling, of TSVs 121
– cost of ownership comparison 121–123
– drilling strategy 124
– – experimental drilling results 126
– – mechanical 124, 125
– – optical 125
– requirements for an industrial TSV laser driller 123
laser energy 137
laser-free release layer 183
– glass carrier, removal 183
laser-induced forward transfer (LIFT) 121, 131, 132
– challenges 130
– conductive line produced by 131
– confocal image of copper structures deposited on 131
laser profiling 415
– dual optical measurement system 415, 416
– laser beams 415
leakage current 96
legal status of 3D IC patents 21
LIFT. see laser-induced forward transfer (LIFT)
light-to-heat conversion (LTHC) 175, 178, 181, 182
– coating 175
– free process 183
– glass delamination 181
– laser-free release layer 183
liquid metal ion source (LMIS) 400
lithography 185
– fabricated trenches 347
– scaling 79
LMS. see locally induced mechanical stress (LMS)
locally induced mechanical stress (LMS) 387
– microbump-induced 389
– microbump region 387
– stacked LSI Si 387
logic–memory 79
low-cost silicon interposer 56
lower cost interposers 48
– glass interposers 48, 49
low-TCE organic interposers 53, 54

3D-LSI. see Three-dimensional LSI fabrication process
LTHC. see light-to-heat conversion (LTHC)

m

material choice, from vendors 186
materials inhomogeneity 406
– nonplanarities 406
mechanical damage, to interconnects 144
mechanical scanning techniques 124, 125
mechanical shock test 269
mechanical stability 143, 263
mechanical strength 138
mechanical via reveal 202
– Cu/Si chemical–mechanical polish (CMP), final step 202
– damaged Si layer, to remove 202
– used when 202
memory cube 8, 9
memory on logic 28, 31
MEMS applications 23
MEMS inertial sensors 26
MEMS technologies 430
metal bonding 262
metal etches 185
metallization 51
metrology 393
– acoustic microscopy 407
– advantages, of x-ray 424
– assembly and packaging 420–426
– barrier 397, 398
– bond strength 409, 410
– bumped 394
– – diameters 427
– classes 409
– 2.5D, 3D Interconnect 393
– differential backpressure 417
– 3D packaging 426
– 2D x-ray 424
– edge inspection 405
– laser profiling 415
– liner 393, 397, 398
– mass 395
– – vs. etch depth 394
– MEOL process 404–405
– – edge trim inspection 405
– – bond voids and bond strength 406–409
– microbump 421
– overlay 402
– seed 397, 398
– stress in Cu and Si 402–404
– through-silicon via (TSV) 393
– TSV etch 394–397

- underfill defects, to identify 427
- wafer thickness 416
- wafer thinning 411
microbump-induced stress 390
- drain current change 390
microbumps 17, 346, 350, 357, 361, 362
- array of 387
- bonding 271
- high-density
-- 3DLSI 387
- technology 271
microchevrons 409
- bond strengths, range 409
- test structures 410
- use, notch 409
micron hybrid memory cube (HMC) 9
micron technology 31
microsolder technology 271
microstereolithography 128
microstructure evolution 361
miniaturization 16
mobile technologies 23, 27, 79
model-based IR (MBIR) 397
moisture 139
Moore's law 13, 16, 79
- causes, performance gains 362
- horizontal dimensional shrinkage 362
More than Moore (MtM) 14
MOS capacitor 381
- C–t curves of 382
- form, thinned wafers 383
- measured C–t curves of 382
MtM. *see* More than Moore (MtM)
multichip modules (MCMs) 41
multitier stacking 352
- choose face-to-face (F2F) 352
- double-side-bumped chip 352
- face-to-back (F2B) 352
- flux application 352
- flux cleaning 352
- sub-µm precision 352
3M wafer support system (WSS) 175–183
- adhesives 178
-- designed 178
- applications, range of 182
- bonding technologies 175
- debonding processes 177
- DRAM wafers, creation 178
- future directions 177, 181–183
- general advantages 177
- high-temperature material solutions 178–180
- laser process 178

-- debonding 183
- layer material, release 183
- LED, creation 178
- light-to-heat conversion (LTHC) coating 175
- liquid UV-curable adhesive 175
- LTHC-free process 183
- LTHC glass delamination 181
-- LTHC, carrier void 181
- primer coat 182
- process considerations 180, 181
-- adhesive delamination 180, 181
- process steps 176, 177
- reliable solution
-- temporary wafer bonding 183
- second-generation WSS process 182
- simple bonding 177
- system description 175–177
- thin power device, creation 178
- tool builders 178
- use fot production processes 178
- wafer de-taping tape 176
- wafer types, application 179, 180

n

nonuniform chemical–mechanical polishing 349
- recipe 349
- topography 349

o

off-chip interconnection 43
Omnivision 23
on-chip interlayer dielectric (ILD) 43
on-substrate (oS) C4 assembly process 356
optical techniques 399
- chromatic white light 411
- wavelength-dependent focal length, principle 411
optoelectronic 3D devices 23
outsourced semiconductor assembly and tests (OSATs) 7, 104, 289, 314
- future perspective 293
oxide matrix 263

p

package level
- bump 422, 424
- coplanarity measurements 424
- underfill inspection 424
package-on-packages (PoPs) 17
packaging 14
- technologies 16
Panasonic 23

PCB. *see* printed circuit board (PCB)
peak–frequency shift 388
– magnitude of 388
PECVD. *see* plasma-enhanced chemical vapor deposition (PECVD)
permanent bonds 273
photoacoustic microscopy 409
– noncontact 409
– nondestructive technique 409
– use, picosecond laser pulse 409
photolithography 147
physical vapor deposition (PVD) 52, 80, 84, 86, 96, 104, 110, 111, 227, 232
piezo coefficients, measurements 374
planar FinFET devices
– calculated K-factor, values 374
– correlation 374
– Si data 375
planarization 140, 249, 250
– process point 250
– surface bumps 249
– uniformity improvement 249
plasma dry etching 212, 213
– advantages 212
– AFM pictures, fine-ground silicon wafer surface 213
– etching rate, Fraunhofer EMFT 213
– ground silicon stress relief
– – breaking force 217
– limitation 212
– plasma chamber 212
– – gaseous SiF_4 molecules 212
– roughness value 213
– sulfur hexafluoride SF_6 212
plasma-enhanced chemical vapor deposition (PECVD) 71, 84, 91, 167, 186, 188, 262, 265, 291
– depositions 186, 188
– oxide bonding 265
plasma FIB (PFIB) milling 400
– create, subsurface imaging 400
– developments 400
plasma programming, surface 336, 337
– fluorine plasma treatment 336
– – selective wetting on polyimide/copper surfaces 336, 337
– hydrophilic effect 336
– x-ray photoelectron spectroscopy (XPS) measurements 336
polishing 135
polycrystalline silicon interposer technology 55–57
polyimide 148

PoPs. *see* package-on-packages (PoPs)
post-bonded TTV 169
power
– delivery 104
– devices 29
– distribution network (PDN) design 101
– integrity (PI) 101
Powertech 9
pretrimming 142
printed circuit board (PCB) 17
printed wiring board (PWB) 41
process flow, for wafer thinning and dicing
– dicing by thinning (DbyT) 220–222, 221
– – before 221
– – after 220
– – backside chipping 221
– – drawbacks 221
– – before thinning 220
– laser dicing 222
– plasma dicing 221, 222
– wafer thinning 220–222
– – for thick wafers 220
– – for thin wafers 220
processors 13, 31, 38
process standardization 2
PSG loss 349
– impact 350
– interconnect resistivity 350
– schematic device structure 349
– thickness 350
– W-plug 350
pulsed laser deposition (PLD) 128
PVD. *see* physical vapor deposition (PVD)
PWB. *see* printed wiring board (PWB)

r
Raman spectrum 387, 388
razor blade test 267
RDLs. *see* redistribution layers (RDLs)
redistribution layers (RDLs) 4, 17, 41, 121, 237–238
– back-via process 238
– CMOS technology 237
– copper (Cu)-RDL, formation of 237
– cross-sectional structure and photomicrograph 237, 238
– Cu/Sn microbumps 237
– 3D image sensor 237, 238
– importance 237
– layout 127
– processing 80
reliability 360, 361
– bathtub curve 360

– building blocks, 3D chip stacking packages 360
– 3D system 361
– long-term product 360

S

Samsung, technology advencement 23
– 3D TSV chip-stacking technology 8
– 4 Gb wide I/O mobile DRAM 7
– Micron hybrid memory cube consortium 9
scallop roughness 384, 385
scanning electron microscopy (SEM) 395
– in line high-volume manufacturing (HVM) metrology 400
– sample preparation 400
secondary ion mass spectrometry (SIMS) 380
– measurement 380
– thickness 380
second-generation WSS process 182
self-alignment experiments 337–341
– bottom wafers, Al target patterns 338
– die bonding equipment 338
–– Panasonic flip-chip bonder FCB3, 338
– liquid amount determination, resepect to 338
–– different die sizes 338
– materials preparation 337, 338
– plasma activation 338
– results 339–341
–– for different chip sizes 339
–– hydrophilic behavior, metal pads 339
–– self-alignment accuracy 339–341
– self-aligned 50 μm thin top chips 338
– stacked die, final positions of 339
–– infrared microscopy 339
self-assembled chips to chips/wafers, interconnection 328–332
– flip-chip-to-wafer 3D integration 329, 330
–– array of self-assembled chips 330
–– facedown self-assembly process, representation 329, 330
–– fine-pitch flip-chip self-assembly 330
–– known good dies (KGDs) 329
–– MEMS chips 330
–– microbumps, cross-sectional views 330
–– Ti/Au pads to wafer 330
– reconfigured-wafer-to-wafer 3D integration 331, 332
–– faceup chip bonding 331
–– known good dies (KGDs), self-assembled 331
–– LSI wafers 331
–– microbumps-to-microbumps bonding 331

–– multichip-to-wafer 3D integration methodology 331
–– process flow 331, 332
–– reconfigured wafers 331
–– SAM technology 332
self-assembly on alignment accuracies, parameters 327, 328
– chip size accuracies 328
–– standard saw dicing, ceramic blades 328
– driving force F_x, known good dies (KGDs) 327
– higher alignment accuracies 328
–– forced wetting 328
– impact on 329
–– hydrophilic/hydrophobic areas on chips/wafers design 329
–– liquid volume 329
– liquid surface tension 327
–– water vs. molten solders 327
– liquid volume 328
– wafer tilt 328
– wettability contrast 327, 328
self-assembly process 325–327
– hosting fluids 327
–– diluted fluoric acid 327
–– pure water 327
–– water-soluble adhesives 327
–– water-soluble flux 327
– requirements 325
– self-assembled monolayer (SAM)
–– hydrophobic regions 325
– short movie of chip self-assembly 326
– steps 325
semiconductor IC wafers 17
semiconductor industry 16
Semiconductor Manufacturing International Corporation (SMIC) 26
semiconductor packaging industry 16
semiconductor technology 65
sensor network 99, 100
sheer test 269
sidewall angle (SWA) 397
signal-to-noise (S/N) ratio 396
silicon 148, 365, 372–374
– interposers 17, 65, 79
–– pricing 38, 39
–– with TSV 45–48
– layers microbumps 357
– wafers 262
siliconware 9
Si–OH bonds 265
Si–O–Si bonds 265
SiO_2–SiO_2 bonding 262, 263

Si–Si hydrophilic bonding 262
Si–Si hydrophobic bonding 262
Si–SiO$_2$ bonding 262
SK Hynix 19
SLID bonding 271
slide debonding 136
slide-off approach 137
SLID technology
– advantages 266
smaller through-silicon vias 147
SoC-type integration 14
soft bonding surfaces 263
soft via reveal 202, 203
– backside knock-off CMP TSV reveal process 203, 204
– backside soft TSV reveal process 203
– – self-aligned Cu contact exposure 203, 204
– chemical vapor deposition (CVD) 203
– mechanical grinding stopped 202
– – before Cu TSVs 202
– residual silicon thickness (RST) 202
– TSV nails 202, 203
– wet/dry etching techniques 202
solder-bumped wafer 172
solid–liquid interdiffusion (SLID) bonding 266
spectral coherence technology 140
spin coating 136, 138
spray coating 138
squeeze out 140, 141
stacked 3D strata, x-ray tomographic 357
stacked heterogeneous chips 16
stacked LSIs 387
– microbump-induced stresses 387
stacked memories 28, 30, 31
stacked wafer, C-scan image 418
standard 3D IC process flow 5
standard IR marking laser 178
– glass carrier 178
STATS ChipPAC 19, 288–290, 292
stealth dicing 259, 260
– bonding process, de facto standard 259
– bumps 259
– dicing tape 259
– – removal method 259
– improvement 259
– process flow, example 259
– singulation 259
– – chip-on-wafer (CoW) 259
– – filling material 259, 260
– – low-k layers 260
– – major issue 259, 260
– – wafer-on-wafer (WoW) 259, 260

– thermal removal 259
– $vs.$ blade dicing 260
stealth dicing process 292
STMicroelectronics accelerometer 27
strata packaging 356–359
– CoW strata stacks 356
– CPI stress 356, 357
– flip-chip packaging 356
– on-substrate (oS) C4 assembly process 356
– silicon layers 357
– – contain, TSV 357
– thickness, bottom chip 357
stresses 387, 402
– copper 402
– thermomechanical stress 387
stress–temperature evolution, representation of Cu-TSVs 367
structural integrity 80
surface mount devices (SMDs) 16
synchrotron x-ray microdiffraction (μSXRD)
– stress analysis 403
system-in-package (SiP) 287
– architectures 13
system-on-chip (SoC) 13, 287
– die 279

t
tablets 23
tapes 136
TechSearch roadmap 123
temperature excursions, in plasma processes 143
temperature–stress cycle 366
temperature–stress evolution 366
– thick electroplated Cu film Si wafer 366
temporary bond 405
– total thickness variation (TTV) 405
temporary bonded wafers 150
temporary bonding 135, 139, 167–174, 185
– adhesive 139
– – categories 151
– – selection of 151, 152
– application 139
– carrier selection 148–151
– – device wafer, undergo edge trimming 149
– – silicon $vs.$ glass, physical properties 149
– – SPARK 150
– – thickness variation, overcome 151
– – TTV mapping 150
– debonding 270
– on edge of carrier wafer 143
– process flow for 148
– strength 267

– TGA plot 139
temporary debonding 147
temporary wafer bonding 266
– systems 175, 186, 187
temporary wafer debonding methods 187, 188
test houses 9
testing
– known good stacks (KGS) 284, 285
thermal budget, for TSV process steps 65
thermal coefficient of expansion 43
thermal conductivity 43, 48, 53, 59, 143, 148, 149, 152, 195
thermal desorption spectroscopy (TDS) 151
thermal gravimetric analysis (TGA) 151
thermal–mechanical stress 273
thermal oxide bonding 265
thermal shock, and cycling test 269
thermal stability 45, 138, 151, 171, 181, 194
thermocompression bonding technique 292
– fine-pitch interconnection 292
– nonconductive film (NCF) 292
– nonconductive paste (NCP) 292
thermogravimetric analysis (TGA) 138, 139
– data 178
– weight loss 179, 182
thermomechanical stress (TMS) 387
– CTE value 387
– distribution image, 3D LSI 387
– implications 387
– microbump-induced 387
– quantitative measurement 387
– thinned IC chip 387
thermoplastic adhesive 138
– rheology curve 139
thermoplastic bonding material 136
– key requirements 138, 139
thermoset polymers 172
thin semiconductor devices 207, 208
– interposer substrates, preparation of 208
– thin wafers 207
– – benefit for different products types 208
– – future predictions 207
– through-silicon via (TSV), preparation of 208
– ultrathin silicon devices 207
thin wafer backside processing 202–205
– via-last 203–205
– – backside TSV processes 205
– – key challenge 203
– – notching, bottom oxide/Si interface 204
– via-reveal process, via-middle 202
thin wafer processing 159–161
thin wafer support systems 191, 194–198

– approaches, based on debonding method 195
– glass carrier support system, laser debonding approach 196
– – light-to-heat conversion (LTHC) coating 196
– – liquid bonding layers, use of 196
– – wafer thinning 196
– glue-less methods 195
– – electrostatic carrier solutions 195
– – membrane thinning 195
– requirements, complex set 194, 195
– room-temperature, peel-debondable 197, 198
– – based on 197
– – non-zone-bond systems 198
– – zone-bond systems 197, 198
– silicon advantage 195
– temporary wafer bonding 194
– thermal slide debondable system, thermoplastic glue 196, 197
– – debonding wafer stack, steps 197
– – process flow 197
– – WaferBOND® HT10.10 debonding material 196, 197
– wafers, sub-100 μm thickness 194
thin wafer technology 101
thin wafer total thickness variation 161, 162
three-dimensional integrated circuit (3DIC) technology 261, 279
three-dimensional LSI fabrication process 380, 388, 389
– microbump-induced TMS 388, 389
– wafers, conceptual structures 380
through-package vias (TPVs) 41
through-silicon vias (TSVS) technology 1, 17, 41, 65, 79, 99, 135, 175, 191, 295, 346, 347, 362, 363, 373, 384, 393, 394
– across-wafer, heights 419
– application of 287
– – for partitioning/reintegration 287
– backside reveal 418, 419
– backside TSV-last approach 68
– based RDIMM 9
– BEOL Cu interconnect 347
– copper-filled 399
– Cu atoms 384
– Cu extrusion 348
– Cu reveal pillars 419
– Cu stress states 404
– demonstrate, evolution 363
– depth metrology, radial etch-rate control 393
– design 65

– determine 3D LSI performance 384
– diameters 80, 373, 394
– – center 372
– – Cu behavior of 367
– – Cu film silicon wafer surface 365
– – Cu layer silicon surface 365
– – Cu stress 365–368
– – device 372
– diffuse 384
– dimensional shrinkage 362
– in 2.5D Si interposer application 66
– 3D stacking 365
– dual-beam FIB cross-sectional SEMs 399, 400
– electrical characterization 92–96
– electrical performance 394
– etching 88
– fabrication steps (*see* TSV fabrication steps)
– flow 80
– formation 347
– – at ASET 99
– – BOSCH process 384
– – redistribution layers (RDLs) 405, 418
– geometry 89
– Holy Grail 289
– impact 371–377
– ImpactDevices 373
– implementation in IC wafers 67
– induced KOZ 373, 375
– – minimize 373
– – prediction of 375
– induced strain effects 363
– insert, FEOL transistor process 347
– interconnect technologies 17
– interferometry measure 414, 415
– isolation step 21
– in MEMSIC 27
– micro-Raman compressive stress map 403
– middle approach 68
– motivation 26
– offset array 395
– processes (*see* TSV process for D2D; TSV process for W2W; TSV unit processes)
– profile 403
– reduce stress
– – devices 363
– reliability and manufacturability 66
– revealed 418
– scanning electron microscope (SEM) micrographs 126
– SEM imaging 395
– – focused ion beam (FIB) 400
– sidewall, SEM 398

– STATS ChipPAC' view, for different variants 288
– stress, evolution 365
– structure 366
– target characteristics 105
– technology 28
– thermal budget, for process steps 65
– thermal mismatch 372
– via-first process 347
time-dependent dielectric breakdown (TDDB) 81
TOK 167, 168, 172, 185, 186, 187
– bonding 167–174
– thermoplastic polymer 172
– total solution 168
total reflection X-ray fluorescence (TRXF) 381
total thickness variation (TTV) 136, 140, 147, 178, 185, 405
– diameter bumped wafers 178
– effect of edge bead on thin wafer uniformity 141
Touch Microsystems Technology 26
trapped gases 406
trench formation
– Cu TSV formation 348, 394
– defects 348
trend of patent filing, for 3D IC 17
TSMC demonstrator vehicle 6, 19
TSMC reference flows 6
TSMC Si interposer 66
TSMC Xilinx program 10
TSV bottom 347
– cross-sectional SEM images 347
TSV depth 395
– correlation 397
– cross-sectional SEM 397
TSV/3D stacking technologies 99
TSV fabrication steps 70
– anneal 73
– etching 70, 71
– images of Cu protrusion effect in TSVs 74
– insulation 71
– metallization 71, 72
– overburden removal by CMP 72, 73
– reliability concerns 75, 76
– temporary carrier wafer bonding and debonding 74
– wafer thinning and TSV reveal 74, 75
– yield considerations 75
TSV keep-out-zone (KOZ) effects 361
TSV process for D2D 105, 106
– attach WSS and thinning 106, 107
– backside bump 111

– barrier metal and seed layer deposition by PVD 110
– CMP 110, 111
– Cu electroplating 110
– deep Si etching from backside 107
– detach WSS 111
– dicing 112–113
– front-side bump forming 106
– liner deposition 107
– removal of SiO$_2$ at bottom of via 107–109
TSV process for W2W 113
– barrier metal and seed layer deposition 114
– barrier metal and seed layer deposition and Cu plating 117
– CMP 115, 117
– Cu plating 114, 115
– deep Si etching 116
– next W2W stacking 118
– polymer layer coat and development 114
– TSV liner deposition and SiO$_2$ etching of via bottom 117
– wafer thinning 116
– W2W-F2F stacking 116
TSVS technology. *see* through-silicon vias (TSVS) technology
TSV unit processes 82
– backside RDL 87
– CMP of copper 85
– defect review 87, 88
– etching 82, 83
– inspection 87
– insulator deposition with CVD 83, 84
– metal liner/barrier deposition with PVD 84
– metrology 87
– temporary bonding between carrier and device wafer 86
– via filling by ECD of copper 84, 85
– wafer backside thinning 86, 87
TSV/WLP reality in high-end 24, 25
TTV. *see* total thickness variation (TTV)
tungsten TSV (W-TSV) 31, 66
two-dimensional stress
– distribution image 390
typical post thinning processes 178
– chemical vapor deposition (CVD) 178
– physical vapor deposition (PVD) 178
– silicon (Si) etch 178

u

ultraviolet (UV) light 137
ultrawide bus SiP 3D integration technology 101
UMC 7, 9, 26, 36

under-bump metallization (UBM) 127, 428
– C4 solder bumps 361
– metallurgy 361
underfill defects, A-scan image 428
underfills 350–352
– CoW solder microbump joints 351
– traditional capillary 351
– traditional C4 solder bump joints 351
unpatterned (field)-based tests 409
– razor blade test 409
UV-curable adhesives
– 3M 183
– temporary bonding 183

v

vacuum atmosphere 168
– thermal bond 168
vacuum depositions 185
via-middle process 227–232
– copper (Cu) contamination, backside processing 228, 229
– – capacitance-time *(C–t)* curves, degradation of 229
– – deep level at mig gap of Si energy band, generation of 228
– – Si thinning 228
– copper (Cu) diffusion
– – oxide liner, backside 232
– three-dimensional (3D) integration process flow 227, 228
– – back-end-of-line (BEOL) process/metal microbumps formation 228
– – backside microbump formation 228
– – barrier metal deposition 227, 228
– – copper (Cu) chemical-mechanical polishing (CMP) 227, 228
– – copper (Cu) electroplating 227, 228
– – dicing 228
– – logic/memory multilayer stacking, ball grid array (BGA) substrate 228
– – oxide linear deposition 227, 228
– – Si substrate thinning 228
– – Si trench deep reactive ion etching (DRIE) 227, 228
– – support material detaching 228
– – temporary bonding, support material 228
– via reveal process to avoid Cu contamination, CMP method 230
– – backside microbump formation 230
– – exposing copper (Cu) 230
– – nitride/oxide film deposition, PE-CVD 230
– – plasma etching 230

–– Si substrate thinning, chemical-mechanical polishing (CMP) 230
–– temporary bonding to support material 230
– via reveal process to avoid Cu contamination, resist lift-off method 230, 231
–– backside microbump formation 230, 231
–– barrier metal/Cu, exposing of 230, 231
–– Cu-TSV base, micrographs of 231
–– nitride/oxide film deposition, PE-CVD 230, 231
–– photoresist coating 230, 231
–– photoresist etch-back 230, 231
–– plasma etching, Si substrate thinning 230, 231
via reveal 249
– copper contamination 249
– copper vias 249
– countermeasure 249
– passivation layer 249
viscosity 138
voids/defects, affecting bonding quality 267
VTI and ST microelectronics 26

w

wafer alignment 163
wafer backside thinning, stress-relief techniques 216–219
– CMP polishing, silicon wafers 217
–– possible alternatives 217
– dicing technology 219
–– important point 219
– dry-polish parameters 217
– fine-ground silicon wafers 216, 218
–– strength properties comparison 218
– grinding-induced damage 216
– manufacture technology comparison 218, 219
– plasma dry etching 217, 218
– resulting breaking force 218
–– comparison 218, 219
– rough grinding 216
–– postprocessing techniques summary 216
– surface roughness R_a 218
– ultrafine grinding 216
– Weibull plot 218, 219
– wet-chemical spin etching 216
wafer-bending methods 402, 403
wafer bonder 168–170
– equipment lineups 170
– glass carrier 172
wafer bonding, schemes for WoW 3D integration with TSVs 273

wafer breakage 245, 246
– all-in-one system 244, 246
– dicing before grinding (DBG) 246
–– die singulation 246
–– die attach film (DAF) 246
–– stacking multiple die 246
– stress relief 245, 246
–– chemical-mechanical polishing (CMP) 246
–– dry polishing 245
wafer chipping 172
wafer curvature behavior 368
wafer debonder 170
– equipment lineups 170
– removes, glass
–– thinned wafer 170
wafer edge 353, 405
– defects 405
– trimming 192–194
–– after wafer bonding 193
–– 3D integrations 192
–– edge-trimmed wafer edge 193
–– razor sharp thin wafer edge 193
–– rectangular edge profile, thinned wafer 193
–– Si wafers 192, 193
–– wafer dicer 193
–– wafer thinning impact, bonded wafer pair edges 193
wafer flatness 248
– influenced by 248
– noncontact gauge 248
– through silicon via (TSV) devices, for 248
– total thickness variation (TTV), silicon 248
–– first vs. second ground thickness 249
– wafer thickening control 248
– wafer thickness measurement, coventional 248
wafer grinding 209, 210
– current trends 210
– geometry
–– abrasive wheel 209
–– rotating wafer 209
– ginders 209
–– diamonds embedded grinding wheels 209
–– multimaterial blocks, grinding wheel 209
– grinding wheel 209
–– mesh value 209
– rough-ground silicon wafers 209
– semiconductor grinders 209
–– spindle (Z2), fine grinding 209
–– spindle (Z1), rough grinding 209
– TAIKO grinding process 210
–– interesting aspect 210

– wafer backside thinning 209
wafer handling 353–356
– challenge 353
– delamination 354
– dilemma 358
– new thin 354
– supply chain arrangement 358
wafer-level approaches 16
wafer-level chip-scale packaging (WLCSP) 17, 126
wafer-level packaging technologies
– usingTSV vertical interconnects 18
– wafer-level packaging, in semiconductor IC processing industry 17
wafer-on-wafer (WoW) 261, 262
wafer processes
– bumped
– – inspection 421
– – optical metrology 421
– deep reactive ion etching (DRIE) 346
– desired thickness 349
– electrical leakage current 347
– flow 346
– front-end-of-line (FEOL) 346
– front-side, backside 346–350
– full thickness, edges 418
– improve, etching processes 347
– inspection tools, 3D interconnect metrology 421
– processes, back-end-of-line (BEOL) silicon MOSFET 346
– stacking, 3D chip 346
– thinned 411
– thinning 349
– TSV trench formation 347
wafer Si interposer 56
wafer support systems (WSSs) 185, 246, 247
– bonded wafer, wafer with substrate 246, 247
– method for 246
– wafer breakage risk, to reduce 246
wafer surface analysis 170
wafer thinning 168, 198–201, 353–356, 393, 394
– amorphous Si (a-Si), formation of 199
– cumulative plots 356
– degradation
– – flicker noise 355
– 3D TSV chip stacking 353
– grinding mechanisms 199
– – brittle grinding 199, 200
– – ductile grinding 199, 200
– mechanical grinding 198

– – composed of 198
– – polygrind 198
– – process-induced damages 198, 199
– micro-Raman spectroscopy 199
– multilayer damaged surface structure 200, 201
– packaging house 358
– show, larger wafer warpage 354
– show, schematic cross section 353
– show, severaledge locations
– – chippings 354
– silicon surface
– – brittle fracture 199, 200
– – sequential phase transformation 199
– surface SEM pictures 199
– – polygrind Si grinding 199
– – rough Si grinding 199
– temporary carrier
– – debonding 353
wafer thinning techniques 208–214
wafer-to-wafer (W2W) bonding 79
wafer warpage 353
– CTE mismatch 143, 144
warpage improvement, using low-CTE organic cores 55
warping 135
waxes 136
Weibull distribution, EM lifetime 361
wet-chemical spin etching 210, 211
– etchant consumption 211
– Fraunhofer EMFT 211
– material removal 210
– mixture used 210
– principal mechanism 210
– schematic of 211
– silicon wafers, industrially applied stress-relief process 210
– uniform etching behavior 210
– vs. etching bath 210
white light interferometry 414
– noncontact optical profiling system 414
– white light beam, technology 414
wireline communications 79
wiring density interposer 41
WLCSP. see wafer-level chip-scale packaging (WLCSP)
WLIR topography 370
WoW 3D platform 272
WSSs. see wafer support systems (WSSs)
W2W stacking
– cross section of stack 104
– with via-last process 103

x

x-ray diffraction (XRD) 402, 403
x-ray microscopy 400, 424
x-ray tomography 408

y

yield management 359, 360
– chip-onchip (CoC) 360
– chip-on-substrate (CoS) 360
– 3D integration, priority 359
– equation microbump assembly 360
– high CoW, causing silicon-on-silicon stacking 360
– individual components 3D processes 359
– remove, CTE mismatch 360
– substrate warpage 360
Yole development prediction 32
Young's modulus 49

z

zero newton 167, 168, 171
– bonding TTV 169
– debonder 168
– debonding technology 167
– equipment 171
– thinned wafer handling 168
– total integration 169
– – carrier 169
– – chemicals 169
– – equipment 169
– wafer bonder 168
ZoneBOND® process 137
ZoneBOND temporary bonding–debonding 163–165
– process flow for thin wafer handling 270